SCIENCE

SMITHSONIAN

SCIENCE

THE DEFINITIVE VISUAL GUIDE

EDITOR-IN-CHIEF
ADAM HART-DAVIS

DK | Penguin Random House

FIRST EDITION

Senior Art Editors
Stephen Knowlden, Vicky Short

Senior Editors
Janet Mohun, Kathryn Wilkinson

Section Designers
Vivienne Brar, Paul Drislane,
Mandy Earey, Clare Joyce,
Mark Lloyd, Heather McCarry, Matthew
Robbins, Alison Shackleton

Section Editors
Ann Baggaley, Kim Dennis-Bryan,
Jemima Dunne, Martha Evatt,
Nicola Hodgson, Neil Lockley,
Martyn Page

Designers
Keith Davis, Supriya Sahai, Silke Spingies

Editors
Katie John, Patrick Newman, Anna
Osborn, Frank Ritter, Nikki Sims, Sarah
Tomley, Angela Wilkes

Illustration Visualizer
Peter Laws

US Editor
Jane Perlmutter

Jacket Designer
Duncan Turner

Production Editors
Phil Sergeant, Marc Staples

Picture Researchers
Ria Jones, Sarah Smithies, Louise Thomas

Production Controller
Sophie Argyris

Managing Art Editor
Louise Dick

Associate Publisher
Liz Wheeler

Art Director
Bryn Walls

Managing Editor
Julie Oughton

Publisher
Jonathan Metcalf

COBALT ID
www.cobaltid.co.uk

Art Editors
Rebecca Johns
Darren Bland

Editors
Simon Clarke
John Murray

Managing Art Editor
Paul Reid

Managing Editor
Marek Walisiewicz

Illustrators
Oliver Burston and Jurgen Ziewe
at www.debutart.com
Tom Coulson and Martin Darlison
at Encompass Graphics Ltd

THIRD EDITION
DK UK

Jacket Designer
Surabhi Wadhwa

Editor
Annie Moss

Jacket Design Development Manager
Sophia MTT

Senior Editor
Peter Frances

Production Editor
Gillian Reid

US Editor
Jennette ElNaggar

Senior Production Controller
Meskerem Berhane

Managing Editor
Angeles Gavira Guerrero

Managing Art Editor
Michael Duffy

Associate Publishing Director
Liz Wheeler

Art Director
Karen Self

Publishing Director
Jonathan Metcalf

Design Director
Phil Ormerod

DK DELHI

Senior Art Editor
Pooja Pipil

Editor
Ishita Jha

Managing Art Editor
Sudakshina Basu

Senior Editor
Dharini Ganesh

DTP Designers
Vikram Singh, Bimlesh Tiwary

Senior Managing Editor
Rohan Sinha

Pre-production Manager
Balwant Singh

Senior Picture Researcher
Surya Sankash Sarangi

Production Manager
Pankaj Sharma

Picture Research Manager
Taiyaba Khatoon

Design Head
Malavika Talukder

Editorial Head
Glenda R. Fernandes

This American edition, 2021
First American Edition, 2009
Published in the United States by DK Publishing
1450 Broadway, Suite 801, New York, NY 10018

Copyright © 2009, 2016, 2021 Dorling Kindersley Limited
DK, a division of Penguin Random House LLC

21 22 23 24 25 10 9 8 7 6 5 4 3 2 1
001–319133–Nov/2021

A catalog record for this book
is available from the Library of Congress.
ISBN 978-0-7440-4561-1

DK books are available at special discounts when purchased
in bulk for sales promotions, premiums, fund-raising, or educational
use. For details, contact: DK Publishing Special Markets,
1450 Broadway, Suite 801, New York, NY 10018
SpecialSales@dk.com

Printed and bound in UAE

For the curious
www.dk.com

MIX
Paper from
responsible sources
FSC™ C018179

This book was made with Forest Stewardship
Council ™ certified paper—one small step
in DK's commitment to a sustainable future.
For more information go to
www.dk.com/our-green-pledge

Editor-in-Chief Adam Hart-Davis

Adam Hart-Davis is a writer, broadcaster, and photographer, and one of the world's most popular and respected "explainers" of science. His TV work includes *What the Romans, Victorians, Tudors and Stuarts, and Ancients Did For Us*; *Tomorrow's World*; *Science Shack*; *The Cosmos: A Beginner's Guide*; and *Just Another Day*. He is the author of more than 30 books on science, invention, and history.

Main Consultants

John Gribbin

Physics

Popular science writer, astrophysicist, and Visiting Fellow in Astronomy at the University of Sussex, UK.

Jeremy Cherfas

Biology

Writer and broadcaster in biological subjects, with a PhD in animal behavior.

Marty Jopson

Biology

Science communicator and TV presenter, with a PhD in plant cell biology.

David Bradley

Chemistry

Science writer and editorial consultant, chartered chemist, and member of the Royal Society of Chemistry, UK.

Douglas Palmer

Earth Sciences

Science writer and lecturer for the University of Cambridge Institute of Continuing Education, UK, specializing in earth science and paleontology.

Iain Nicolson

Astronomy and Space Technology

Formerly Principal Lecturer in Astronomy at the University of Hertfordshire, UK; writer, lecturer, and occasional broadcaster on astronomy and space science.

Barry Lewis

Math

Formerly Director of Math Year 2000 and now President of the Mathematical Association, UK.

For second edition

Andrew K. Johnston Geographer, Center for Earth and Planetary Studies, National Air and Space Museum, Smithsonian
John Gribbin Physics
Derek Harvey Biology and Medicine
Carole Stott Astronomy
David Bradley Chemistry
Douglas Palmer Earth Sciences
Edouard Marchais Mathematics

Contributors

David Burnie Biology and Medicine
Jack Challoner Physics
Robert Dinwiddie Earth Sciences and Physics
Derek Harvey Biology and Chemistry
David Hughes Astronomy
Giles Sparrow Physics and Space Technology
Carole Stott Astronomy
Marcus Weeks Maths and Technology

Other contributions Hayley Birch, John Farndon, Ian Graham, Andrew Impey, Jane McIntosh, Georgia Mills, Douglas Palmer, Steve Parker, Sally Regan, Frank Ritter, Mark Steer, Amber Tokeley, Martin Toseland, James Urquhart, Diana Vowles.

For third edition

Roger Bridgeman Technology and engineering
Andy Extance Chemistry
Derek Harvey Biology and medicine
Hilary Lamb Physics
Jacqueline Mitton Astronomy
Douglas Palmer Earth sciences and human evolution
Giles Sparrow Astronomy
Karl Warsi Mathematics and computing

✳ Smithsonian

The Smithsonian is the world's largest museum and research complex, dedicated to public education, national service, and scholarship in the arts, sciences, and history.

Contents

1
THE DAWN OF SCIENCE
PREHISTORY TO 1500

2
RENAISSANCE & ENLIGHTENMENT
1500–1700

3

THE INDUSTRIAL REVOLUTION
1700–1890
124

4

THE ATOMIC AGE
1890–1970

5

THE INFORMATION AGE
1970 ONWARD

REFERENCE

Foreword

have always been fascinated by science, and for the last 25 years, by the history of science; so when I was asked to take part in the preparation of this book I jumped at the chance.

And here it all is; this superbly illustrated book paints a broad picture of the whole of science and its history. Arranged in chronological order, according to when a scientific principle was first laid out or when a process became technically possible, it begins with the ancient Babylonians, Chinese, and Greeks, with the idea of the four elements, and goes all the way through to string theory and space telescopes.

Science is not just a collection of answers, but an ongoing search for the truth about how the Universe works; it is not merely about the facts, but also about the struggle to discover them. One scientific idea often leads to another, and then another. This was especially true of the vacuum: theories and inventions followed one another rapidly in the mid-17th century; steam engines were a natural consequence in the 18th, cathode-ray tubes in the 19th, and today we have many more pieces of vacuum technology. The book is divided into five chapters, from the dawn of science through to the present day. Each chapter has its own timeline to help you find the various threads that make up that particular period.

Scientific ideas often occur to more than one person at a time, which has led to some disputes—over the invention of calculus, for example, or the discovery of oxygen. All these events are included. Alongside the ideas and theories in this book are the people who dreamed them up, from Pythagoras and Aristotle to Einstein and Marie Curie. There are double-page biographies of 18 major characters, and around 100 features on other great pioneers, from Eratosthenes to Richter. At the end of the book is a 54-page reference section, including brief biographies of all the major scientists, past and present, plus a plethora of scientific facts.

Because of its sheer size and complexity, this is the toughest book I have ever worked on, and it would never have been completed without a small army of writers, editors, designers, artists, and picture researchers. I thank them all, but particularly Janet Mohun and her team in the DK office.

Adam Hart-Davis

1

THE DAWN OF SCIENCE

PREHISTORY TO 1500

The ancient world saw the first breakthrough moments in science,
as the coming of age of great civilizations from Egypt to Babylon
proved a decisive spur to invention, people learned to write,
and scholars from Aristotle to Zeno had time and space to
think deeply about the world around them.

THE DAWN OF SCIENCE

PREHISTORY TO 1500

18,000 BCE	3000 BCE	2000 BCE	1000 BCE

c. 18,000 BCE
Early pottery created in China and Japan.

c.3000 BCE
Acupuncture needles first used in China. Babylonians create a number system based on 60; it survives in the way hours and minutes are measured today.

Early acupuncture chart ▶

c.2000 BCE
First wheels with spokes made. Baked bricks used for important public buildings in Mesopotamia.

Ziggurat of Ur, Iraq ▶

◀ Ancient Egyptian astronomical calendar

c.600 BCE
Oldest known world map drawn by Babylonians.

▼ Map by Herodotus, c.450 BCE

▼ Egyptian figurine

Pythagoras of Samos ▶

c.5500 BCE
Copper smelting begins in the Balkans and West Asia.
c.3500–3200 BCE
In Mesopotamia solid wheels used for transportation.

▲ Jomon bowl

c.2700 BCE
Abacus first appears in Mesopotamia and becomes widely used for calculations.

▲ 16th-century abacus

1500 BCE
Early astronomical calendar created in Egypt.

500 BCE
Babylonian astronomers note an eclipse cycle known as the Saros.
495 BCE
Pythagoras introduces the concept of mathematical proof.

▼ Mesopotamian chariot on the Standard of Ur

c.1300 BCE
Vertical obelisks used as clocks and compasses in Egypt.

c.400 BCE
Empedocles states that all matter is composed of four elements: fire, earth, water, and air. Hippocrates establishes medical school on the Greek island of Cos.

c.350 BCE
Aristotle provides scientific evidence that Earth is a sphere.

c.3000–2500 BCE
Ring of standing stones erected and possibly used as an astronomical calendar at Callanish, Scotland.

▲ Early plank wheel

▼ Callanish stones

▼ Earth's curved shadow cast on the Moon

c.3000 BCE
Chariots with solid wheels used in battle in Mesopotamia.

▼ Chinese bronze vessel

c.1500 BCE
Bronze cast in furnaces in China.

As empires rose and fell, from Egypt to China, the practical demands of the first great cities and armies stimulated a wave of inventions—bronze for making tools and weapons, wheels for moving loads and milling grain, gears for making machines, and the water and windmills to power those machines. Meanwhile, the administrative needs of the new rulers for everything from accurate calendars to tax and land inheritance calculations prompted the first great stirrings of science. At the same time early stargazers such as Hipparchus were mapping the night sky with astonishing precision, and brilliant scholars such as Euclid and Al-Khwarizmi were laying the foundations of mathematics.

300 BCE | 100 CE | 700 CE | 1000 CE »

c.300 BCE
Early use of coordinates in cartography by Dicaearchus of Messana. Euclid establishes the basic principles of geometry in *The Elements*.

▼ Woodcut depicting Archimedes' discovery

c.240 BCE
Archimedes discovers how to measure the volume of irregular-shaped objects. Eratosthenes formulates a system for finding prime numbers, known as the sieve of Eratosthenes.

230 BCE
Early geared mechanisms in use in China.

▼ Chinese chariot with early gearing mechanism

c.200 BCE
Hindu–Arabic numerals and decimal system evolve into a recognizable ancestor of our modern numerals.

c.160 BCE
Hipparchus maps the positions of about 1,000 stars and calculates how far Earth is from the Sun.

c.100
Sculpture of Atlas shows the oldest known representation of the celestial globe.

▲ 2nd-century sculpture of Atlas holding the globe

132
Zhang Heng invents the first seismometer, which could not only detect a distant earthquake but also indicate in which direction it occurred.

▼ World map by Ptolemy

150
World map drawn by Ptolemy. He also divides the stars into 48 constellations.

c.300
Mayans develop a number system based on the numbers 5 and 20, with symbols that include zero.

705
Map of the constellations drawn in China.

▼ Chinese map of the constellations

▼ Early Chinese rocket

c.800
Gunpowder invented in China. Hindu symbol for zero adopted in the Middle East. Persian mathematician Al-Khwarizmi invents algebra and the rules for solving equations.

c.900
Al-Zahrawi develops the ligature technique in surgery.

NERVVS OPTICVS.

▲ Diagram of eye from the *Book of Optics*

c.1011–21
Alhazen writes his *Book of Optics*.

900-1200
Windmills with vertical sails used to drive machinery.

▼ Wooden post mills

1279
Pierre de Maricourt records the basic facts of magnetism.

1154
Earliest known mechanical clock built in Damascus, Syria.

1202
Fibonacci publishes *Liber Abaci*, proposing calculations using zero to nine and place value.

Fibonacci ▶

c.1315
Mondino de Liuzzi performs the world's first public dissection.

c.1440
Johannes Gutenberg creates the first Western European printing press.

◀ Gutenberg Bible

Fire is a chemical reaction between a fuel, such as wood, coal, or oil, and oxygen in the air; it usually gives off energy in the form of heat and light. This can be harnessed and used to positive effect. The reaction starts when the fuel is ignited, either naturally or by human effort by another fire, a spark, or heat. Particles of burning materials and ash that glow and discolor are visible as flames and smoke. A fire generally sustains itself until it has exhausted the supply of fuel or oxygen.

Early humans and fire

Charcoal and burned bones from several African sites suggest people may have controlled and used fire as early as 1.5 million years ago and certainly by 400,000 years ago. For the last 100,000 years fire has been in common use. Control over fire brought many benefits, including the ability to live in colder regions and protection against predators. People's diet was improved by eating cooked food, roasted over a fire or baked in the embers. Fire also provided a social focus. Some groups used torches or lit fires to stampede animals into places where they could be trapped and killed.

From the earliest times when open fires were lit, people began to develop hearths, often ringed by stones. After 40,000 BCE more efficient hearths with clay surrounds and air-intake channels were invented to control and increase heat. Simple lamps were made from stones with a natural hollow: in this animal fat was burned, using a plant-fiber wick. Some groups began controlled burning to clear vegetation, facilitating hunting and the growth of useful plants.

Making fire
Heat is needed to start a fire. One common method of creating heat is by friction: here, a stick is being rotated fast on a piece of wood until a glowing ember is created.

Fired clay

Using clay in domestic hearths led people to discover the transformative power of fire on clay. This knowledge was used to create figurines of baked clay. Later, the technology was developed further, to produce pottery. Prepared clay was mixed with a temper, such

Dolni Vestonice Venus
The Palaeolithic inhabitants of the Czech site of Dolni Vestonice produced many human and animal figurines c.20,000 years ago, including this example, modeled from a mixture of clay and carbonized bone before being baked.

BEFORE

In certain circumstances fire occurs naturally. Humans in prehistory saw its potential and learned to control its power.

TECHNOLOGY

By 2.5 million years ago, when they began making stone tools, early humans had understood that they could **alter the natural world** to their own advantage. By **employing tools** they could extend their abilities.

FIRE'S POTENTIAL

When early humans encountered naturally occurring bush fires, they came to appreciate not only fire's **power to destroy** the landscape and everything in it, but also its potential as a **source of heat, light, and defense**, and as a **tool for shaping the world**.

Fire Power

Fire is a terrifying and potentially devastating natural phenomenon. Early mastery of fire offered many benefits to humankind including protection against wild animals, heat that enabled them to spread into cooler regions, and the ability to cook food.

IN PRACTICE

FIRED BRICKS

Clay, as daub or sun-dried bricks, was an important ancient building material. For more durable and impressive structures, the technology of ceramic production was used to make baked bricks. Hand- or mold-formed, the bricks were allowed to dry, then fired in large stacks encased in fuel. Fired-brick buildings and walls defended people from enemies and the elements. In the 2nd millennium BCE, the Harappans in the Indus Valley used baked bricks to protect against flooding, and for wells and bathroom floors, while the Mesopotamians used them for public buildings such as the ziggurat (stepped temple-mound) of Ur, Iraq (right).

Jomon pot
Some of the world's first pots were created in Japan around 18,000 BCE and later developed into a well-made product, Jomon ware, of which this pedestaled bowl is a late example.

FAIENCE AND GLASS

Before 4000 BCE people in western Eurasia began producing faience objects, such as this Egyptian figurine (c.1300 BCE). This precursor to glass was made with silica, such as quartz sand, mixed with a stabilizer (lime) and an alkaline flux (often soda). The paste, colored with copper oxide, was shaped, then heated to about 1500–1800°F (850–1000°C), causing the grains inside the paste to sinter (adhere) and the object's surface to fuse as a thin glaze.

From c.1600 BCE, furnaces reaching higher temperatures enabled a similar mixture to be melted into a viscous liquid, true glass. During the 1st millennium BCE in the Levant, glassblowing was invented, revolutionizing the production of glass.

as sand, to toughen it. After shaping, pots were left to dry and then were fired in a clamp (bonfire kiln). The pots were stacked on and covered with a layer of fuel, then sealed by a layer of clay. Holes in the top and around the sides allowed air to circulate so the fuel would burn. More permanent kilns appeared from c.6000 BCE in western Asia. A fired vessel's color depended on the type of clay and the firing conditions: circulating air produced reddish hues, while the absence of air turned the pottery black.

The pyrotechnical skills needed to control kiln temperature and air flow and to achieve high temperatures, coupled with a growing knowledge of other natural materials, were the prerequisites for the development of other transformational industries, including metallurgy and glassmaking.

Cuisine

Fire made food easier to chew, more digestible and often more palatable, and improved human health by killing bacteria and parasites. Heat enabled foodstuffs to be preserved for later use by drying or smoking, and was used to remove or neutralize poisons present in foodstuffs. By 40,000 BCE some food

> "When the earth was young… **human beings**… did not know yet how to enlist the aid of fire."
>
> LUCRETIUS (TITUS LUCRETIUS CARUS), ROMAN PHILOSOPHER, 100–55 BCE

was cooked in an animal skin or skin-lined pit containing water by adding heated stones. The invention of pottery made boiling easier: fire could be applied directly to the vessel of food and water. Unleavened bread was cooked on hot stones or pottery dishes over the fire. Clay ovens appeared from the 6th millennium BCE, heated by lighting a fire inside.

Through time, people experimented with different ways of cooking. Bread and cakes were leavened with yeast, which created carbon dioxide gas, causing the dough to rise; cooking hardened the risen form.

 AFTER

Mastery of fire has enabled civilizations to transform the world, through technology and controlled destruction.

DIET
Over the last 2,000 years people have developed ever more **elaborate ways of preparing food** and combining ingredients, using heat in increasingly sophisticated ways.

MEDIEVAL KITCHEN

TRANSFORMING MATERIALS
Heat has been used to transform many materials, including metal ores 18–19 ». Advances in ceramics include **metallic glazes** and **kilns** capable of achieving higher and better controlled temperatures than before.

HARNESSING ENERGY
A broadening **range of fuels**, including coal and natural gas, have been used to create heat and power. Applications include sophisticated **ways of heating**, such as Roman hypocausts (underfloor heating). Steam power brought about Europe's **Industrial Revolution**.

SEE ALSO »
pp.18–19 EARLY METALWORKERS
pp.132–33 STEAM POWER TO STEAM ENGINE

◁ **Controlling fire**
Harnessing the power of fire was an important part of the development of civilization and has influenced every part of human life.

Egyptian figure
This figurine (c.2500 BCE) shows a servant girl grinding grain into flour to be baked in bread.

Early Metalworkers

Metallurgy started at different times across the globe, but by the 1st millennium BCE it was extensively practiced. The use of metals proved revolutionary: unlike stone, metal could be worked into any shape, and broken objects could easily be mended or recycled to make new things.

North American copper arrowheads
Around 3000–2500 BCE many American communities made tools and ornaments by cold-hammering pieces of the native copper abundantly available around Lake Superior.

BEFORE

At first metals were valued for their attractive appearance, but their other useful properties were quickly appreciated.

TOOL MANUFACTURE
During early times and, in some areas, until recently, **tools were shaped from stone, wood, and other nonmetallic materials,** by cutting or percussion (striking).

METALS AS EXOTIC STONES
When metals were first encountered, in their native (natural pure) form, they were **treated as attractive stones** that, instead of fracturing, changed shape when struck.

PYROTECHNOLOGY
The development of **pottery kilns** created the pyrotechnical skills (achieving high temperatures; controlling firing conditions) required for **smelting and melting metals.**

Native (naturally occurring pure) copper, gold, silver, lead, and tin were exploited first, to make small objects, including jewelry and daggers; copper was also used for tools. Copper could be worked by cold hammering but was easier to shape if heated until soft. Hammering made the metal brittle; its toughness could be restored by annealing (heating then cooling slowly), though this reduced its hardness.

Smelting and casting
Copper was also extracted from ore using easily smelted copper oxide and carbonate ores (rock containing metal compounds mixed with other minerals). Smelting involved heating crushed ore and charcoal in a clay furnace. Air pumped in through a clay pipe using bellows helped raise the furnace temperature. Carbon displaced the metal from its oxide or carbonate and escaped as carbon dioxide (CO_2), leaving mineral impurities (as slag) and pure copper, which was denser and collected in the crucible base. When these ores were unobtainable, sulfide ores were used: before smelting, these required roasting in an open bonfire, to drive off the sulfur as an oxide gas.

For casting, smelted copper was melted (at 1982°F/1083°C) and poured into stone or fired clay molds or impressions in wet sand. At first, open molds were used, producing simple shapes, flat on one face. Later, two-piece molds enabled objects to be cast in the round.

5500–5000 BCE
The time when copper smelting began in the Balkans and western Asia. Iron was not smelted until the 2nd millennium BCE.

More elaborate objects could be made by lost-wax casting: a wax model was coated in clay, which was then fired, allowing the wax to run out, creating a mold. Metal was then poured into the mold, which was smashed to remove the finished object.

Using alloys
Copper is relatively soft, so is unsuitable for heavy tools such as axes. Alloyed with certain other metals, however, it becomes harder. Early metallurgists often

Tapping hole, plugged during heating

Gas and smoke outlet

Casting channel

Crucible

Fire chamber with charcoal fuel

Stoke-hole and fueling pit

Bronze-casting furnace, China
Bronze was melted in a crucible, from which it was poured into a mold. In sophisticated furnaces, such as this Chinese example (c.1000–800 BCE), the metal was tapped directly from the crucible along a channel into the mold.

Ancient Greek blacksmith
On this 6th-century BCE Greek vase, a blacksmith is removing the bloom from a shaft furnace in which iron ore has been heated with charcoal.

stimulus to trade. While bronze was widely employed for tools and weapons, other alloys were also used. For example, adding lead lowered copper's melting point and increased its fluidity, producing an alloy suitable for casting complex shapes where strength was not required (such as ornaments).

preferentially exploited copper ores naturally containing arsenic, to produce a strengthened copper alloy. Generally alloyed in the proportion 5–10 percent, tin produced a superior metal, bronze, but was rare: once tin–bronze was developed (around 3000 BCE in western Asia) demand for tin became a major

Iron
Ironworking developed late, since it presented technological challenges. Pure iron melts at a high temperature, around 2,912°F (1,535°C). In antiquity only the Chinese achieved the temperature needed to smelt iron so only they could produce cast iron: they

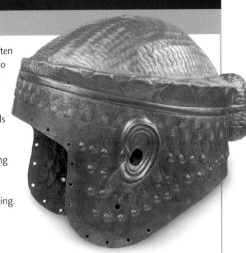

GOLDWORKING

Gold generally occurs as a native metal, often as electrum (gold-silver alloy, sometimes also containing copper). In South America and Turkey, golden objects were made from electrum, using gold's low reactivity: one method was depletion gilding, where acids removed baser metals from the surface. Gold's malleability makes it easy to work, using techniques such as punching (making holes for decoration), filigree (twisted threads of gold) and repoussé (creating a raised pattern), wiremaking, and hammering.

GOLD HELMET, MESOPOTAMIA, C.2500 BCE

built blast furnaces of good refractory clay capable of withstanding high temperatures, and created the necessary draft using water-driven piston bellows.

Elsewhere, only wrought iron could be produced. Iron ore and charcoal were heated in a furnace to around 2,000–2,100°F (1,100–1,150°C). At this temperature some of the impurities were given off, but others remained mixed with the iron in a spongy mass known as the bloom. Iron was laboriously extracted from the bloom by repeated hammering and heating to high temperatures. This process was also used to shape the resulting metal and to weld (join together) pieces of iron. Heated in a charcoal fire,

the iron was converted into steel (carbon-iron alloy), which is harder and tougher than wrought iron. Hardness was increased by quenching (plunging the white-hot object into water to cool it rapidly), although this also made it brittle. This was countered by tempering (reheating and cooling slowly), reducing both the brittleness and the hardness.

Successful working required a balance between tempering and quenching.

Once these technologies were mastered, iron, which is far more common than copper and tin, rapidly became the main material for tools and weapons.

Molten metal
Metal was circulated in antiquity as ingots which could be melted. The molten metal was poured from the crucible into a mold.

AFTER

Chinese bronze vessel
A Western Zhou (1100–771 BCE) bronze ritual vessel. The Chinese created elaborate bronzes in molds made of many interlocking pieces, which were disassembled after casting, for repeated use.

For millennia metals have been a dominant material for manufacturing strong, durable tools and weapons, and attractive ornaments.

OTHER METALS
Since antiquity, a wider range of metals, such as **zinc, aluminum, and tungsten**, have come into use and new alloys, such as **brass and pewter**, have been created.

TECHNOLOGICAL ADVANCES
From the 16th century onward the West also developed the technology to make **cast iron**. A more efficient blast furnace, using **coke as fuel**, was devised by Abraham Darby in 1709. Other technological advances included **stainless steel** and **tinned steel cans** for preserving food.

BLAST FURNACE IN IRONBRIDGE, ENGLAND

NEW USES FOR METALS
In recent centuries metals have been **put to new uses**, for example, to construct buildings, ships, airlanes, and rockets; also, in microcircuitry.

SEE ALSO »
pp.178–79 THE NATURE OF HEAT

Development of the wheel
From their origins in ancient Mesopotamia, wheels have come to be used all around the world, and are still being developed and refined today.

3000 BCE Metal strips or nails are added to wheels to make hard rims; this protects the rims so that wheels last longer, even if it does nothing to improve the ride.

c.1600 BCE Spoked wheels first appear on Egyptian chariots. These wheels also seem to have developed independently in Europe some 200 years later.

800–600 BCE The Celts invent pivoting front axles, as found in burial mounds on archaeological sites. Such axles give vehicles much greater maneuverability than traditional fixed axles.

3500 BCE	3000 BCE	2500 BCE	2000 BCE	1500 BCE	1000 BCE	1400 CE

3500–3200 BCE An unknown Mesopotamian innovator takes a solid potter's wheel and turns it through 90 degrees. Joining two solid wheels with an axle results in the first transportation wheel.

2600 BCE Thinner and lighter plank wheels (right) begin to become more popular than their solid counterparts.

1800 BCE The crossbar wheel (right) appears, although it is unclear if this is a step toward the first spoked wheel or a parallel innovation. The oldest known one was found in Italy.

c.1400–1500 CE Solid iron bands are used to reinforce wheel rims on carts and wagons. These bands are the first "tires."

‹‹ BEFORE

Even before the wheel had been invented, humans had devised various ways to move heavy objects around.

BEASTS OF BURDEN
Humans first domesticated animals between 9000 and 7000 BCE, and people began using **oxen** to draw **plows** around 4000 BCE. Camels, elephants, horses, llamas, yaks, and goats have all been used as beasts of burden.

ALL-TERRAIN SLEIGHS
Known from at least 7000 BCE, **sleighs** were used by **hunting** and **fishing** communities in **northern Europe**. Whether pulled by people, dogs, or deer, their long, thin runners spread the weight of a heavy load over almost any terrain, including snow or ice, sliding freely like skis.

Evolution of the Wheel

Probably the most important mechanical invention of all time, the wheel has a long and varied history. For 5,500 years we have been thinking up ever more inventive ways of using wheels to improve our lives. From water mills to jet planes, a whole range of things that we use routinely rely in some way on the wheel.

Ancient Mesopotamian potters were the first people, as far as we know, to make use of wheels—for spinning clay to fashion their wares—as long ago as 3500 BCE. Evidence from diagrams found on clay tablets suggests wheels did not catch on as a handy aid to transport for another 300 years, when the Mesopotamians started to build chariots in 3200 BCE. Even with this intellectual barrier breached, the development of our most useful asset progressed at a very slow pace. It

would be another 1,600 years before the ancient Egyptians invented chariot wheels with spokes.

While the concept might seem simple to us today, there must, however, be something about the wheel that is conceptually difficult. Neither the Mayan, the Aztec, nor the Incan civilizations, all of which were highly developed, used the wheel. In fact, there is no evidence that the Americas ever saw a wheel before it trundled over with the Europeans.

In Europe, wheels evolved steadily through the ages until the early 19th century and the beginning of the Industrial Revolution, which saw rapid advances in their development.

▽ **Sumerian battle chariot**
The artefact known as the Standard of Ur, dating from c.2500 BCE, shows one of the oldest existing images of a chariot. This war chariot, from Sumer in Mesopotamia (now part of Iraq), is drawn by onagers (wild asses), and has solid wheels made from flat pieces of wood held together by pegs. The much lighter spoked wheels did not appear on chariots until around 1600 BCE.

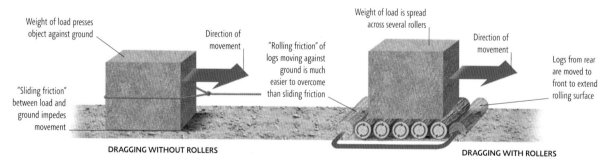

Weight of load presses object against ground

Direction of movement

"Rolling friction" of logs moving against ground is much easier to overcome than sliding friction

Weight of load is spread across several rollers

Direction of movement

Logs from rear are moved to front to extend rolling surface

"Sliding friction" between load and ground impedes movement

DRAGGING WITHOUT ROLLERS

DRAGGING WITH ROLLERS

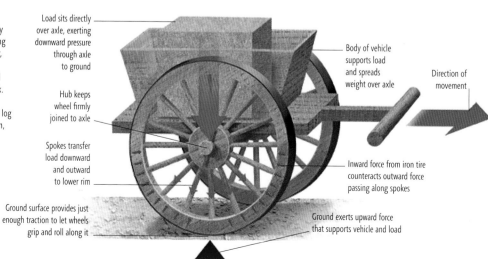

Wheeled transportation
Before wheels were used, one way to transport loads was by dragging them along the ground. However, the weight of the load, and the "sliding friction" from the ground surface, made this very hard work. One advance, possibly predating wheels, was to pull the load over log rollers, which lessened the friction, but wheeled vehicles are much more efficient.

Load sits directly over axle, exerting downward pressure through axle to ground

Hub keeps wheel firmly joined to axle

Spokes transfer load downward and outward to lower rim

Ground surface provides just enough traction to let wheels grip and roll along it

Body of vehicle supports load and spreads weight over axle

Direction of movement

Inward force from iron tire counteracts outward force passing along spokes

Ground exerts upward force that supports vehicle and load

1820s John Loudon McAdam and Thomas Telford build roads using compacted broken stone aggregate—the first macadamized roads.

1846 The pneumatic tire, a hollow tube of inflated India rubber, is patented by Robert William Thomson. It is reinvented in 1888 by John Boyd Dunlop.

1901 Edgar Purnell Hooley patents tarmac, a tar and aggregate mix spread out and rolled to make a much tougher road surface than just aggregate.

1967 The first alloy wheels are made, for racing cars. Lighter than steel wheels, they improve steering and speed. Better heat conductors, they allow heat to dissipate from the brakes.

| 1800 | 1850 | 1900 | 1950 |

1820s The artillery wheel (right), which has a metal hub, is invented. First used to move heavy steam vehicles without the spokes snapping, it is soon adopted for artillery pieces.

1870 Improvements in metalworking enable the creation of fine metal wheels with wire spokes (right), invented by George Cayley in the 1850s. This allows for lightweight, fast bicycles.

2005 The Tweel, an experimental polyurethane tire-wheel hybrid, is invented by Michelin. Flexible spokes connect the hub to a thin, flexible rim and take on the shock-absorbing role of a traditional tire's sidewall.

How wheels work

There are three reasons why wheels make moving a load over the ground considerably easier than just dragging or pushing it. Most importantly, they greatly reduce friction. While the small part of a wheel that is in contact with the ground will not roll, the rest of the wheel can roll on over the top of it, carrying the whole structure forward or back. Second, wheels make it much easier to change the direction of the pushing or pulling of the load. Third, wheels raise the load, so reducing the angle at which the force required to move the load must be applied. All three factors are evident when pushing a loaded wheelbarrow, for example.

Importance of the wheel

Wheels have had a huge influence on the evolution of human society. The potter's wheel—the first known type of wheel—the water mill's turbine, and the spinning wheel have all played a significant part in shaping history. Think also of the wheels on the chariots that armies used to subjugate enemies, or the wheeled plows and farm vehicles that transformed agricultural practice. One of the wheel's most important forms is the cogwheel. With a history dating back more than 3,000 years, to the first rudimentary gears made from wooden wheels with pegs driven into the rims, cogs have been central to the development of transportation and timepieces.

Modern uses

Once you start to look closely at the modern world, you see that a host of things that we take for granted depend on wheels to function. Gas, diesel, and jet engines, disk drives, even the electric toothbrush—none of these everyday things could work without an internal wheel to keep them moving.

AFTER »

Given the wheel's central place in everyday life, there is never likely to be a post-wheel period in human history.

IMPROVING ROADS

Wheels work much better when they can roll across a smooth surface. Rudimentary roads began to appear soon after the first chariots. Even today, a great deal of money is still spent on **developing even better road surfaces**.

HARNESSING ENERGY

Whether it is wind power, water power, or steam power that turns the blades, **wheels are at the heart of turbines**, which play a vital part in producing most of the electricity that we use around the world, in fact.

SPREADING THE LOAD

Narrow wheels sink into soft ground under the burden of a heavy load. To spread heavy loads, tractors have very wide wheels, while tanks and other extremely heavy armored vehicles have **caterpillar tracks over wide wheels**.

SEE ALSO »
pp.40–41 SIMPLE MACHINES
pp.42–43 HOW GEARS WORK
pp.50–51 WATER AND WIND POWER

◁ **Early spinning wheel**
The first spinning wheels may have come from China, like this example from the Song dynasty (960–1279 CE). They were large, rimless, hand-cranked wheels, which were driven by a belt and turned a horizontally mounted spindle. Each revolution of the wheel produced several turns of the much smaller spindle.

« **BEFORE**

In ancient times even mundane happenings were seen to be part of huge cosmic events, far beyond human understanding.

To our ancient ancestors, the world was **dominated by powerful gods**, and Earth, the stars, and the sky were full of mystery. They wondered at what was going on around them, and attributed the events of the natural world to supernatural forces. The **ancient Greeks** brought organization, experimentation, and the search for **simple universal laws**. They began to realize that to understand the world one needed to know its nature (*physis*, hence the modern word "physics"), and that **natural phenomena** had logical explanations. This was a giant step from the assumptions of the old world that the supernatural determined almost everything.

The fundamental question of what the Universe is made of has occupied the minds of the greatest thinkers for thousands of years. The ancient Greeks, tireless and ingenious speculators about the Universe, turned their attention to the structure of the materials composing it. The Greek philosopher Thales of Miletus, who lived from about 625 to 546 BCE, is generally credited with being the first to speculate on the nature of the elements. He suggested that the basic principle of the Universe was water, and that all known natural substances were modifications of it. Water surrounded the land, and trickled through the soil—clearly, life

Anaximander
A student of Thales, Anaximander thought that an invisible substance, *apeiron*, was the source of all things. He is known to have conducted the first recorded scientific experiment.

be universal and free from any specific characteristics. Anaximander identified an invisible, all-purpose plasma, which he called *apeiron*, from the Greek for "infinite." This mysterious substance could morph into all the materials on Earth.

Anaximander's own student, Anaximenes, rejected *apeiron*. He believed that everything had to be

Elements of Life

The Universe is made from a relatively small number of naturally occurring elements. Our understanding of elements stemmed from what the ancient Greeks believed, and for over 2,000 years it was thought that everything was made from just four elements: water, fire, air, and earth.

Thales of Miletus
The earliest of the natural philosophers of ancient Greece, Thales was a mathematician and astronomer. His assertion that the world started from water was the first to consider the elements of the Universe.

could not exist without it. He described Earth itself as a flat slab, floating on top of an infinite mass of water.

Fundamental substances
Thales's view of a water-based universe attracted fierce opposition. Anaximander, a pupil of Thales, rejected this theory and scoffed at the idea that Earth was like a floating log resting on a vast sea. He believed that Earth was curved, and that it dangled unsupported in space. Water could not be the main element because it was not versatile enough. According to Anaximander, if a particular element in nature, such as water, was the origin, a substance with an opposite nature, such as fire, could not emerge or coexist. The origin must

composed of something, and proposed that the missing "something" was air. Air could be experienced, and turned into smoke and fire; when condensed, it became mist and water. The debate spread to Ephesus, which was home to Heraclites, who had a different theory. For him, fire was the key element, since it was dynamic and caused change in other materials. Xenophanes, from neighboring Colophon, favored earth. Although Earth changes slowly over time and its shape may shift, its fundamental essence stays the same.

The Milesian School
Thales, Anaximander, and Anaximenes had much in common. They were all from Miletus and, despite their fierce disagreements, shared an intellectual approach. This later became known as the Milesian School. All three sought

to understand the true nature of reality, and to explain phenomena that most humans deemed terrifying, such as thunder, lightning, and earthquakes. The traditional view attributed such events to whims of the gods, but the Milesian School proposed natural explanations: lightning and thunder resulted from wind; rainbows were the result of the Sun's rays falling on clouds; and earthquakes were caused by the cracking of the ground when it dried out after being moistened by rain.

Not one element, but four
The Greek philosopher Empedocles adhered to the teachings of Pythagoras (see pp.32–33). He, too, struggled with the problem of a fundamental substance, but concluded that the Universe was made of several elements: the water of Thales, the fire of Heraclites, the air of Anaximenes, and the earth of Xenophanes. These essential four accounted for all matter on Earth. Empedocles also proposed that two active principles united and divided the elements: love and strife. Love was the uniting principle that kept the elements together in different substances, and strife was the principle that divided them. No one believed that an empty space could exist, so love and strife were also counted as elements, filling the void between the other four. Empedocles called these substances *rhizomata*, the "roots" of matter. It was only after Aristotle, in the 4th century BCE, that they became known as elements.

SCIENTIST AND PHILOSOPHER (490–430 BCE)

EMPEDOCLES

Empedocles was a philosopher, physiologist, and religious teacher. A citizen of Acragas in Sicily, he was given to wearing a gold girdle, a laurel wreath, and bronze sandals. He was a famous healer and brilliant orator; Aristotle described him as the inventor of rhetoric. Empedocles is best known for his belief that all matter was composed of four elements—

fire, air, water, and earth—that were mixed and parted by the personified cosmic forces of love and strife. He believed in reincarnation, and insisted that he was a divine being. According to legend, he perished when he threw himself into the flames of Mount Etna, in an attempt to prove that he would return as a god.

"For from these [elements] come all things that are, or have been, or shall be..."

▽ The four elements
This illustration from a 1472 edition of Lucretius's *De Rerum Natura* ("On the Nature of Things") shows the four elements of Empedocles: air, fire, earth, and water. He explained the nature of the Universe by the interaction of love and strife on the elements.

▷ Anaximenes
The Greek philosopher Anaximenes believed that air was the most basic element of the Universe. The last of the important philosophers of Miletus, Anaximenes helped the transition from a mythological explanation of the world to a scientific one.

IGNIS
Acutus·Tenuis.
ac Mobilis

QVA
assa·Mobi=
et Obtusa.

AER
Mobilis·Acu-
tus·Crassus

TERRA
Crassa·Ob=
tusa·et Immobilis

Creatura

Solidata

**" To the elements it came
from Everything will return.
Our bodies to earth,
Our blood to water..."**

MATTHEW ARNOLD, "EMPEDOCLES ON ETNA," 1852

FIRE

Hot — Dry

AIR — EARTH

Wet — Cold

WATER

Classical elements and their properties
This diagram shows two squares, one inside the other. The corners of the larger square represent the classical elements, and the corners of the smaller square their properties.

AFTER »

The ancient Greek notion of four basic elements dominated Western scientific thought for more than 2,000 years.

ARISTOTLE'S FIFTH ELEMENT
Aristotle 36–37 » reasoned that the heavens were composed of a **fifth element**, called **ether** (from a Greek word meaning "to glow"), which was **perfect, eternal, and incorruptible**. The remaining four were **confined to Earth**, and given the qualities of hot, cold, wet, or dry.

THE WORKINGS OF THE UNIVERSE

ELEMENTS IN MODERN FORM
The ancient Arab alchemists believed all metals were composed of two elements, **sulfur and mercury**. The Persians added **salt**. The 16th-century Swiss alchemist **Paracelsus** 70–71 » combined the four Greek and three Arab elements. In 1661 the Irish chemist **Robert Boyle** 98–99 » coined the **first modern definition of an element**: a substance that can not be broken down into simpler substances.

SEE ALSO »
pp.232–33 THE PERIODIC TABLE

Early Medicine and Surgery

The origins of medicine lie with the origins of civilization itself. People in early civilizations across the world attempted to explain the reasons for disease and create treatments. China has a strong tradition in using herbs in medicine and many of the earliest ideas still resonate today.

« BEFORE

Prehistoric medicine, that is to say medicine before the written word, was probably characterized by a combination of primitive first aid and a belief in supernatural spirits.

TRIBAL MEDICINE

Clues about the **earliest medical practices** come from anthropological observations of indigenous peoples alive today and indicate that prehistoric societies probably used **herbs to treat simple ailments**. These would have been developed by a tribal shaman or medicine man.

FIRE AND METALWORKING

Controlled fire that would have been used for sterilization purposes and cooking food may also have been used by early man for closing wounds. In addition, **mineral and metalworking tools «« 18–19** would have been used in crude surgical procedures. Some blades were made out of obsidian, a glass found in volcanic rock, and were remarkably sharp. The earliest surgical procedures included **skull trepanning**, which consisted of holes being bored into a person's skull in the hope of relieving headaches and epilepsy. Healed skulls (see right) indicate that some individuals even survived this practice.

SURGICAL BLADE

Medical customs used to be handed down by word of mouth, but organized medicine began properly with the written word. The earliest known medical texts date from around 2000 BCE in China and Egypt.

Origins of medicine and surgery

Medicine had almost certainly been established in China by the middle of the 3rd millennium BCE, when the Yellow Emperor purportedly composed the *Huangdi Neijing* or *Basic Questions of Internal Medicine*. This document (much expanded 3,000 years later) formed the basis of traditional Chinese medicine, which became mainstream throughout much of Asia. Practitioners diagnosed and treated disorders based on the interplay between humans and their environment, using techniques such as meditation and acupuncture.

In the same millennium, in *c.*2600 BCE in ancient Egypt, Imhotep (a great polymath and architect of the pyramids) was revered as a god of medicine and healing. The *Edwin Smith Papyrus* from *c.*1700 BCE is arguably based on Imhotep's texts. The world's oldest surgical document, it is remarkably lacking in magical thinking with reference to diagnosis, treatment, and prognosis of disease.

Two centuries later in *c.*1500 BCE, a Babylonian text, the *Diagnostic Handbook* refers, perhaps, to the earliest

Earliest acupuncture
The first acupuncture needles date from 3000 BCE in China. Charts such as this identified points of the body where needles would bring maximum benefit.

physician—it suggests that a figure called Urlugaledin practiced primitive surgery way back in 4000 BCE.

The roots of the Ayurvedic (meaning "Life of Science") system of medicine are found in the Indus Valley even further back, possibly as early as 9000 BCE. This forms the basis of traditional Indian medicine today. It advocated a healthy lifestyle that prescribed herbalism, massage, and yoga. Its written records, such as the *Sushruta-samhita* text on surgery, appeared later, from *c.*500 BCE. This particular text refers to invasive practices that took place, including plastic surgery, cataract surgery, and even cesarean sections.

Ancient Greek medicine

In 700 BCE the first Greek medical school opened at Cnidus. Ancient Greek medicine, like that in Egypt and India, placed emphasis on control of diet, lifestyle, and hygiene.

It was 300 years later that Hippocrates established his own medical school on Kos. He described many diseases for the first time and introduced lasting

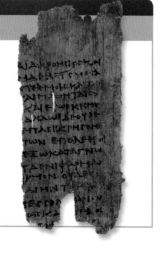

Skull trepanning
Holes were bored into patients' skulls in one of the earliest known surgical treatments dating back to 40,000 BCE.

medical terms, such as "acute" for an illness that is sharp and brief, and "chronic" for one that builds up slowly over time.

The Hippocratic School rejected supernatural causes in favor of seeking the physical causes of disease. The school emphasized care and prognosis, and encouraged thorough case studies, making it the forerunner of clinical medicine. Hippocrates also championed the idea of

BREAKTHROUGH

HIPPOCRATIC OATH

It is tradition that all physicians take the Hippocratic Oath, a guiding set of duties, which formed part of the *Hippocratic Corpus*, the texts from the Hippocratic School of ancient Greece. It is generally thought that Hippocrates himself wrote the oath, but it is likely that many contributed to the text. In the original version (see right), the oath swearer includes a debt of gratitude to their medical teacher, a promise that they will live a "pure" life and that they will preserve the confidentiality of the people in their care. Today the oath has been updated and modified in certain countries, for example, with the omission of clauses that forbid pregnancy termination.

PHYSICIAN (980–1037 CE)

AVICENNA

Avicenna (also known as Ibn Seena) was a prolific Persian scholar who made important contributions in medicine, chemistry, astronomy, mathematics, psychology, and geology. His major work *The Canon of Medicine* became a standard text in European universities. He pioneered many medical practices, including quarantine to control the spread of disease (see pp.242–43), and clinical trials in the experimental use of drugs.

Medical pioneers
These three medical pioneers (who could never have met) were masters of their craft: Galen of ancient Rome, Avicenna from Persia, and Hippocrates of ancient Greece.

AVICENA · ỤPOCRATES

AFTER

After the Middle Ages, the theory and practice of medicine advanced rapidly, but not always to the real benefit of the patient.

BLOODLETTING
This practice involved **withdrawing significant amounts of blood** for therapeutic reasons. William Harvey's circulatory theory discredited the practice in the 17th century. Today blood withdrawal is done chiefly for blood analysis or transfusion.

BLOODLETTING

REFINEMENT OF SURGICAL TOOLS
As **surgical techniques were refined**, tools evolved accordingly. During the Renaissance, saws for amputation were developed, but it was not until the discovery of stainless steel in the early 20th century that noncorrosive surgical tools were available for the first time.

SEE ALSO »
pp.70–71 RENAISSANCE MEDICINE AND SURGERY
pp.228–29 SAFER SURGERY
pp.312–13 THE DEVELOPMENT OF MEDICINES
pp.412–13 DISEASE CHALLENGES

humorism, which stated that the body contained four basic humors (fluids): black bile, phlegm, yellow bile, and blood. Moods and illnesses were attributed to imbalances of the humors. Hippocrates' student, the Greek physician Galen (129–c.216 CE), reinforced this idea with the belief that blood was continually made in the body and could stagnate. This encouraged the dubious practice of bloodletting, which involved the withdrawal of a large volume of blood in an effort to redress any imbalance in the humors. It wasn't until 1543 that a

Egyptian surgical instruments
Evidence suggests that bronze and copper instruments such as these were used in surgical procedures in ancient Egypt, such as drainage of sites of inflammation.

Flemish anatomist, Andreas Vesalius (see pp.72–73), disproved many of Galen's theories.

The birth of scientific medicine
In the 5th century CE, as Greek scholars fled Byzantine persecution and settled in Persia, the Academy of Gundishapur emerged as a center for medical study. It later became the first teaching hospital.

The golden age of Islam (c.700 CE to 1200 CE) saw the very first pharmacies and free public hospitals in Baghdad. Medieval Islam also produced medical treatises from scholars such as Avicenna (see right). During this era, Islamic thinkers, including Avicenna, started to introduce experimental methods into their study of medicine, which had a lasting impact on the discipline.

The First Astronomers

To our ancestors, the sky and Earth, the humans and gods, and the animals and plants, were all parts of one, interacting environment. The Sun, Moon, and stars provided heat and light, allowing living creatures to thrive. Their regular patterns of movement and change meant they could be used as compasses, clocks, and calendars.

The perceived "usefulness" of the stars and planets varied from place to place within the ancient world. People living by large rivers, such as the Euphrates and the Nile in the

« BEFORE

In the prehistorical period the night sky was thought to shed meaning and significance on human endeavor. Stories were woven around the recognized constellations.

EVIDENCE FROM CAVE PAINTINGS

Constellations and star groups were often pictured on the walls of caves occupied by our prehistoric ancestors. They noticed that certain stars and groups (such as Sirius and Orion's Belt) always rose and set at the same places on the horizon, and that risings that occurred just before dawn always happened in the same seasons of the year. Such observations led to the development of the first calendars.

Egyptian calendar
The world's oldest known astronomical representation lies in the Tomb of Senenmut, in Deir el-Bahri, Egypt. It dates back to 1500 BCE and depicts the decanal stars, which marked the ten-day Egyptian week.

Phases of the Moon
Every 29.5 days the Moon goes through a complete set of phases from full, through a new Moon (when it is dark), back to full. The shape appears to change as different parts of the illuminated Moon face Earth.

Middle East, were more influenced by rain and river levels than by the seasons. Farther from the equator, temperature and daylight variations had a greater effect on human, animal, and plant life. The critical hunting and gathering seasons were ruled by a solar calendar, since the Sun's movement was recognized as causing the spring, summer, fall, and winter cycle. Its overwhelming importance led people to believe that it needed to be observed and worshipped.

The importance of the Moon
In Mesolithic times, around 10,000 years ago, many people lived close to the water. In coastal areas, people noticed that tides were particularly high or low at certain times of the month and were connected with the lunar phases. These tides are now known as spring and neap tides (see pp.108–109) and would have affected fishing, shell collecting, and even opportunities for transportation.

The full Moon, acting as a "night light," would have affected how people lived, just as their daily tasks would have been governed by the number of hours of daylight. Their days were delineated by sunrise, noon, and sunset; and their activities determined by both a solar and a lunar calendar.

Finding patterns
A long time ago people observing the phases of the Moon would have noticed that a regular pattern occurred every 15, 30, 44, 59, and 74 days. The interval between successive full moons —29.5 days—became well known, establishing the concept that would later become known as a "month."

Importantly, the positions of heavenly bodies are predictable; both the Sun and the Moon rise in the east at regular times. Unlike both the vagaries of the weather and human relationships,

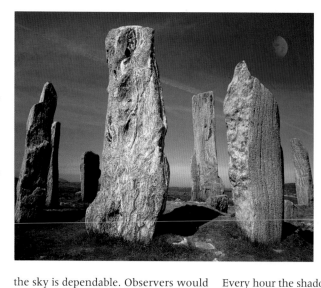

Astronomical calendar
Some of the Callanish standing stones in Scotland (c.3000–2500 BCE) seem to have been set up to record the Metonic Cycle of 18.61 years (see p.29).

the sky is dependable. Observers would soon realize that the time between one midsummer and another was 365 days.

Using the Sun as a clock
At midsummer the Sun reaches its maximum noontime elevation outside the tropics. At midwinter, 183 days later, the Sun is at its lowest in the noon sky. The idea of using the Sun's movement as a clock began with observations of

how a shadow cast by a vertical object points (in the Northern Hemisphere) toward the western horizon at sunrise and due north at noon. Every hour the shadow moves farther around, acting as a sundial. Similarly, most stars travel around the sky. An imagined line from any of these stars to the celestial pole acts like the hand of a gigantic clock.

Early astronomical advances took place in Babylon and Greece in the 700 years BCE, »

BREAKTHROUGH
MARKING TIME

A vertical obelisk, like this 3,300-year-old granite monolith at Luxor in Egypt, was used as a clock. The movement of its shadow tracks the hours of the day, and the length of its shadow marks the progress of the year. At noon the shadow is cast due north, allowing the obelisk to act as a compass as well as a clock. The four sides of the obelisk's tapered stone shaft are embellished with hieroglyphs, typically praising Ra, the Sun god. The top tenth of the shaft was covered in gold to reflect the Sun's rays.

| WAXING CRESCENT | FIRST QUARTER | WAXING GIBBOUS | FULL MOON | WANING GIBBOUS | THIRD QUARTER | WANING CRESCENT |

Pole star
A photograph exposed all night shows the stars spinning around the pole of the sky. The elevation of the center of the circle above the horizon is equal to the latitude of the observing site.

ASTRONOMER (190–120 BCE)

HIPPARCHUS

Hipparchus was a Greek astronomer who accurately mapped the positions of around 1,000 stars and divided them into six categories of brightness. He measured the year length to an accuracy of 6.5 minutes and, by comparing his observations with those made centuries before, realized that the direction of the spin axis of the Earth was changing. This caused the position of the Sun at the time of the equinox to move. He knew the seasons (intervals between solstices and equinoxes) were not of equal length and so calculated that the Earth was not at the center of the Sun's orbit.

>> spurred on by planetary astrology. It was believed that the positions of the planets, the Sun, Moon, and stars, and the times of their risings and settings, influenced life on Earth and foretold the future.

Accurate observations of planetary positions over long periods of time were recorded on Babylonian clay tablets and used to predict the planets' future movement. In Greece, Hipparchus (see above) wanted to improve their accuracy, so he measured the position and brilliance of around 1,000 major stars. This led to the theory of spherical trigonometry, and the acceptance of the "degree" as a unit of angle.

By the 2nd century CE the Greek astronomer Ptolemy had divided the Mediterranean sky into 48 constellations in his famous treatise, *Almagest*. He created tables of such positional accuracy that now—some 2,000 years later—they can still be used to detect the movement of stars that were previously thought to be fixed. Angles that had previously been estimated using the extended hand were now carefully measured with an adjustable cross staff or celestial compass.

Movements of the planets

The future movements of the planets were predicted using complicated Earth-centered models in which the planets moved at constant velocities around a series of theoretical circles. In the 6th century BCE the Greek astronomer

Ptolemy's cosmos
Ptolemy's vision of the Universe with the known planets on spinning crystal spheres centered on Earth. Their order was determined according to their speed, relative to the fixed stars on the outermost sphere. This view of the Universe was upheld for 1,400 years.

Pythagoras believed that the Earth was spherical. In the 3rd century BCE Aristarchus asserted that the Earth revolves around the Sun, but this theory was not widely accepted at the time. He also used careful eclipse timings to calculate that the Moon was about 60 Earth radii away (within 1 percent of today's value). Around the same time, Eratosthenes accurately measured the earth's radius and the tilt of the spin axis (see pp.38–39).

Marking time

Time was measured during the day using the Sun's shadow, and sundials became common. During the night people relied on the use of stellar positions, and primitive water, sand glass, and candle clocks were used to keep check.

Chinese constellations
This North Polar sky map was found in Dunhuang, China, and is thought to date from c.705 CE. The constellation of Ursa Major can clearly be seen at the bottom.

This was important because many prayers had to be made at specific times of the day. Religious observance also meant that the Moon and the Sun were closely observed. Once the major religions were established many societies used both a lunar and a solar calendar to accurately set the dates of festivals. Even today, the Christian festival Easter is set on the first Sunday after the first full Moon after the spring equinox.

Measuring the year

The length of the tropical year—the time interval between successive equinoxes—was estimated using the daily lengths of noontime shadows. The year length could also be found by

Farnese Atlas
This 2nd-century CE sculpture of the kneeling god Atlas includes the oldest known representation of the celestial globe. The sphere depicts 41 of the 48 classical Greek constellations described by Ptolemy.

" Astronomy **compels the soul** to **look upward** and leads us from **this world** to another."

PLATO, GREEK PHILOSOPHER, c.400 BCE

Astrolabe
This astrolabe, used to measure time and position, was produced in the late 15th century. If the latitude was known, the altitudes of the Sun or certain stars could be set to give the time. If the time was known, the Sun and star positions could be set to establish the latitude.

Rete
The rete is a cut-out plate that sits on top of the latitude tympan and rotates over it. This enables the user to line up the star pointers (on the rete) with the night sky for the particular latitude shown on the tympan beneath it.

Mater
The base plate is called a mater. Its outer rim is marked with a degree scale.

Tympan
A rotating plate, or tympan, sits on top of the mater. The celestial sphere for one particular latitude is mapped onto this plate.

Ecliptic ring
The ecliptic ring is part of the rete. It shows the annual path of the Sun across the sky.

Ulugh Beg
Islamic astronomer Ulugh Beg built the largest observatory of the time at Samarkand (in present-day Uzbekistan) in the 1420s. Working with his student Ali al-Qushji and others, Beg made major contributions to cataloging star positions. Ali al-Qushji also provided empirical evidence of Earth's rotation.

AFTER

More accurate observations of the motion of the planets paved the way for Isaac Newton and Johannes Kepler to formulate laws of universal gravitation and planetary motion.

INCAN EQUINOXES
The mysterious **Intihuatana stone** at Machu Picchu, dating from the 1400s, might have been a seat for a **priest-astronomer**, who observed the rising or setting of the Sun in certain valleys, with the timing of such events shaping the planting season and the harvest.

INCAN INTIHUATANA

HELIOCENTRIC VIEW
The Polish astronomer **Nicolaus Copernicus** (1473–1543) started modern astronomy in 1543 by proposing a **heliocentric system 74–75 »**, where the planets revolved around the Sun.

THE TELESCOPE
In 1609 **Galileo** turned a **telescope** to the sky, and began to **study** the **planets** and the **Moon** in greater detail. Astonishing new discoveries followed at an amazing rate.

SEE ALSO »
pp.74–75 THE SUN-CENTERED UNIVERSE
pp.76–79 PLANETARY MOTION

measuring the time taken for the Earth to orbit the Sun using fixed stars to measure the movement. The dawn rising of stars such as Sirius or star groups such as the Pleiades were used to do this. A breakthrough came when the Greek astronomer Meton (born c.460 BCE) discovered that 19 solar years was nearly equivalent in length to 235 lunar months. This became known as the Metonic Cycle, and it was used for setting the dates of religious festivals, and for predicting the times of solar and lunar eclipses.

The Julian calendar was introduced by the Roman Emperor Julius Caesar in 45 BCE. The year was taken to be 365.25 days long, and a leap year was added every four years. The inaccuracy was only one day in every 128 years, and this was corrected by Pope Gregory XIII in 1582 CE. The lunar calendar—with its month of 29.5 days—does not fit well with the solar calendar—whose year has 365.25 days. Chinese knowledge of astronomy had developed since the 12th century BCE but along independent lines. Chinese maps of the constellations (left) show a detailed knowledge of the night sky.

The astrolabe
The latitudes of places on Earth could be measured using the elevation of the Sun. This was important in Islam for determining the direction of Mecca, and led to the development of the astrolabe (above).

Ancient Number Systems

From the time of the earliest civilizations various signs and symbols have been used to represent numbers. As recording and calculating became ever more complex, several numerical systems developed, eventually leading to the numerals and decimal number system we use today.

The first way of recording numbers was probably making crude tally marks on stones or sticks, or, simply, piles of pebbles. Various numerical systems slowly evolved in the different cultures of the ancient world. Some proved inadequate for complex calculation, but others formed the basis of the system we are familiar with today.

One of the first ideas to emerge was the concept of counting in groups—what modern mathematicians call systems to a base. For example, eggs are still often counted in dozens, which can be seen as counting to

Ancient Egyptian scribe
In ancient Egypt scribes had detailed knowledge of writing and mathematics using hieroglyphs and, later, hieratics.

base 12. However, for most purposes, calculations now are usually done in a decimal system (that is, to base 10).

Cuneiform and hieroglyphs

The decimal system was not universal in ancient cultures. The Babylonians, in around 3000 BCE, worked to base 60, and the legacy of this system can still be seen in the way we measure time in hours and minutes, and angles in degrees. The Babylonians inscribed numbers on clay tablets with the end of a pointed stick, which left behind a wedge-shaped (cuneiform) mark. These cuneiform symbols were the first true numerals.

In central America, between the 3rd and 9th century CE, the Mayans developed a sophisticated system based on 5 and 20, with symbols that included zero—a concept that did not appear in other cultures until many centuries later. The ancient Egyptians, working to base 10, assigned separate hieroglyphs for the numbers one, 10, 100, 1,000, 10,000, and 1 million. Over time, these were simplified into simple brush strokes known as hieratics. The ancient Chinese also worked in a decimal system, with characters representing the numbers one to ten, and multiples of ten, a hundred, and so on. As well as this system of writing numerals, they used a set of small sticks called counting rods arranged on a board that was divided into rows and columns for performing calculations.

❮❮ BEFORE

The most ancient methods of representing numbers probably involved counting out small objects, such as stones, making marks—tally marks—on sticks or clay, or using counting rods.

TALLY MARKS
A simple **way of recording numbers**, tally marks were used, for instance, in counting the number of days that elapse or the number of animals in a herd. These were sometimes **scratched** on stones or sticks or impressed on clay tablets (a precursor of the **Babylonian cuneiform numerals**).

ISHANGO TALLY BONE, AFRICA

COUNTING RODS
The system of using an object, such as a pebble, bone, or stick **to represent each unit** evolved into counting rods, which were used in ancient China. In the **counting rod system**, digits are represented by the number of rods. This system also enabled **fractions**, and positive and **negative numbers** to be represented.

❮❮ SEE ALSO
pp.26–29 **THE FIRST ASTRONOMERS**

Algorist and abacist
This woodprint is an allegorical representation of the fierce competition in medieval Europe between rival groups of mathematicians. Abacists (right) used a traditional abacus for calculation, while algorists (left) did their calculations on paper using the new numerals in a positional system.

NUMBER SYMBOLS

Each of the ancient civilizations developed its own system of numeric notation. Some notation systems, such as the Mayan, had little or no influence on the evolution of modern numerals, and others, as in the case of Roman numerals, even hindered progress toward a universal positional notation system.

Babylonian	Ancient Egyptian	Ancient Greek	Ancient Roman	Ancient Chinese	Mayan	Modern Hindu–Arabic
𒁹	I	α	I	一	•	1
𒈫	II	β	II	二	••	2
𒐈	III	γ	III	三	•••	3
	II / II	δ	IV	四	••••	4
	III / II	ε	V	五	—	5
	III / III	ς	VI	六	•—	6
	IIII / III	ζ	VII	七	••—	7
	IIII / IIII	η	VIII	八	•••—	8
	III / III / III	θ	IX	九	••••—	9
∢	∩	ι	X	十	══	10

Ancient Greek and Roman numerals

With the rise of Greek civilization and its interest in mathematics came the practice of using letters of the alphabet as numerals. The alphabet was also used by the Romans. In their system I represented one, and was simply repeated for two (II) and three (III); other letters were used for five (V), ten (X), fifty (L), one hundred (C), five hundred (D) and one thousand (M). Numbers in between were expressed by repeating these symbols, as in XXX for thirty, and simply adding progressively smaller numerals, for example CCLXVII for 267. If a numeral preceded a larger one, however, this indicated that it should be subtracted from it, so that IV was four (five minus one).

Hindu–Arabic numeral system

The Roman system was difficult to use for calculation but still persisted in Europe until medieval times and the arrival of the Hindu–Arabic numerals and a decimal system in positional notation. This system had its roots in the

Roman numerals
Even today Roman numerals are still used in some contexts, such as copyright dates and on the dials of some clocks and watches.

numerals of the Indus Valley civilization contemporary with the Babylonians, which evolved into a recognizable ancestor of our modern numerals by the 3rd century BCE. It was further refined by Indian mathematicians, such

Mayan calendar

The Mayans used dots and bars to represent numbers, as can be seen on this calendar from the early 13th century Dresden Codex. Their number system was based on the numbers 20 and 5 and included a symbol for zero—commonly represented by a shell—which enabled them to write very large numbers.

MATHEMATICIAN (c.596–668 CE)

BRAHMAGUPTA

Born in India in c.596 CE, Brahmagupta was a mathematician and astronomer. He was the first to consider zero as a number. At the age of 30, he wrote a text called *Brahmasphutasiddhānta*, which is the earliest evidence of written rules for doing arithmetic with negative numbers and zero. He created several formulae and theories throughout his lifetime, and while they caused initial controversy, many still hold up more than a thousand years later.

as Brahmagupta (600 BCE), and spread to Persia and the Middle East, where it was adopted by Islamic scholars. It was at around this time, in the 9th century, that a symbol for zero was adopted. Previously in calculations, a gap was left as a placeholder in the appropriate column, but this led to ambiguity: out of context, it was practically impossible to distinguish between numbers such as 10, 20, 30, and 400 and simple 1, 2, 3, and 4. The gap was first replaced by a small dot, but this then evolved into the 0 symbol of Hindu–Arabic numerals. This symbol is still used today, although since the advent of computers, Ø is sometimes used for zero to distinguish it from the capital letter O.

Translations of Islamic texts in the 12th century brought the Hindu–Arabic system (often referred to simply as "Arabic numbers") to Europe, where it gradually replaced Roman numerals. The ease with which the Hindu–Arabic system enabled calculations to be done, and the fact that it allowed numbers to be written down unambiguously, ensured its status as the universal mathematical language that has been used, with minor additions and modifications, to the present day.

AFTER

The replacing of Roman numerals with Hindu–Arabic numbers in Europe made calculation simpler and helped accelerate progress in mathematics.

MATHEMATICAL SYMBOLS

From medieval times, various symbols were introduced as abbreviations for **verbal instructions** in mathematical problems. The first to appear were symbols for operations such as + (**plus**) and – (**minus**), and the = (**equals**) signs, followed were followed by conventions such as the superscript 2 for squares and √ √for roots. Letters of the Greek alphabet were also used as symbols, for example π for the ratio between a circle's diameter

and circumference, and later typographical symbols such as ! (factorial), ∞ (infinity), and ≈ (approximation) became part of a universally understood shorthand.

EXPONENTIAL NOTATION

One drawback of the positional notation system is that very large and very small numbers are difficult to read, especially when they include a long series of zeroes. This problem is overcome by the use of exponential notation, in which **large numbers** are **expressed as $a \times 10^b$** (a times 10 to the power of b); for example, 100 is 1×10^2.

SEE ALSO »
pp.48–49 ALGEBRA

MATHEMATICIAN AND PHILOSOPHER **Born c.569** BCE Died c.495 BCE

Pythagoras

"Number is the **ruler** of **forms** and ideas."

PYTHAGORAS, FROM "THE LIFE OF PYTHAGORAS"
BY IAMBLICHUS OF CHALCIS, c.300 CE

Pythagoras, the most famous mathematician of the ancient world, is remembered now for Pythagoras's Theorem. The Babylonians and others were familiar with its principles about a thousand years earlier, but Pythagoras proved it to be true; indeed, he introduced the concept of proof, now fundamental to mathematics.

Unclear beginnings
Pythagoras was such a towering figure that all sorts of stories were made up about him, and there is no way of proving which of them are true. He is known to have been born on the Greek island of Samos some time around 569 BCE. He may have traveled to Miletus; he may have been taught by Thales. It seems likely that he visited Egypt and Babylon, where priests may have taught the geometrical principles that underlie his theorem.

Pythagoras's school
Returning in about 520 BCE, Pythagoras journeyed in 518 BCE to Croton, in present-day Calabria, Italy, where he founded a school that was partly

The Tetractys
This triangular arrangement, known as the Tetractys, was a secret symbol of worship for the Pythagorean school.

mathematical, partly religious, and in part mystical. The mathematicians of the school regarded their life as a kind of exile, believing that the soul could leave the body, and that they would be reincarnated.

The school attracted about 2,600 young men, and probably women, who divided into two groups: an inner circle consisting of mathematicians, and their listeners. The mathematicians lived by a set of curious rules, were strict vegetarians, wore distinctive clothes, and went around barefoot. They also had a secret sign, the pentagram, which is a five-pointed star within a pentagon.

The school members lived their lives by mathematics, for Pythagoras thought that numbers were the essence of being. Odd numbers were thought to be male; even, female. For their arithmetical investigations they probably used pebbles in the

The school of Pythagoras
In southern Italy Pythagoras founded a mystical brotherhood, where "All is number." The mathematicians lived permanently in this unusual establishment, while listeners were permitted to attend during the day.

$$a^2 + b^2 = c^2$$

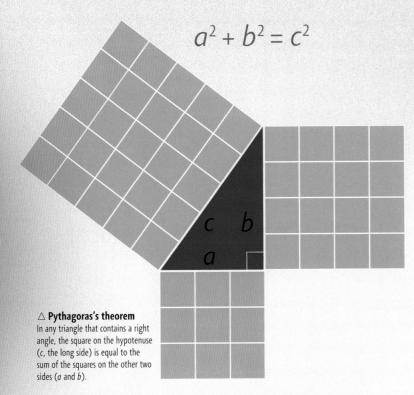

△ **Pythagoras's theorem**
In any triangle that contains a right angle, the square on the hypotenuse (c, the long side) is equal to the sum of the squares on the other two sides (a and b).

sand; in this way they learned about triangle numbers and square numbers. Especially important was the triangle with four pebbles on each side—the triangular number 10, which the

Mathematics of music
According to legend, Pythagoras experimented with simple instruments and invented the musical scale, based on mathematical principles.

◁ **Pythagoras of Samos**
The most famous mathematician of the ancient world, Pythagoras was one of the first to establish the notion that Earth is a sphere. He founded a curious school or brotherhood for whom the study of mathematics was tantamount to a religion.

school called the tetractys (see opposite). The school held the tetractys to have many meanings: for example, its rows of one, two, three, and four points were held to represent dimensions, from zero to three respectively.

Pythagoras's Theorem

When he found a proof for the theorem that bears his name, Pythagoras is said to have sacrificed an ox at the school to celebrate. The theorem states that in any right-angled triangle the square on the hypotenuse (the long side) is equal to the sum of the squares on the other two sides (see above). In other words, if the sides of a right-angled triangle are of lengths a, b, and c, and c is the longest side, then $a^2 + b^2 = c^2$. There is an infinite number of integral solutions to this equation—values for a, b, and c, which are all whole numbers. The simplest examples of these "Pythagorean triples" are (3, 4, 5) and (5, 12, 13).

Music and astronomy

Pythagoras enjoyed music. The story goes that he was walking past a blacksmith's shop and was intrigued by the notes as various hammers rang on the anvil. He investigated the notes made by plucking stretched strings, and so mathematically invented the musical scale. While this story is improbable, as he almost certainly studied music in Egypt, he may have experimented with stretched strings and gone on to formalize the musical scale mathematically.

Pythagoras believed that everything could be described in whole numbers, but in a right-angled triangle whose short sides are both one unit long, the length of the hypotenuse is given by the square root of two. Hippasus, one of the mathematicians in the school, managed to prove geometrically that the square root of two is an irrational number; it cannot be represented by a ratio, or fraction, of the form p/q, where p and q are whole numbers. Some say that Hippasus was thrown overboard and drowned; others that Pythagoras was so upset that he committed suicide. Whatever the truth of that, the square root of two is indeed irrational: written as a decimal it starts 1.4142135623730950488… but then goes on for ever, without any pattern.

Pythagoras's legacy

In his work on whole, triangle, and square numbers, Pythagoras founded number theory, a current branch of mathematics that deals with whole numbers. Of more immediate influence, he introduced the concept of proof. Euclid's *Elements*, written in 300 BCE, is entirely about the subject of proof and became the most influential mathematical text in the world.

Harmony of the spheres
Pythagoras believed that the relationship between the planets in some way reflected his musical scale. Robert Fludd (1574–1637) produced this interpretation of Pythagoras's "harmony of the spheres."

TIMELINE

STATUE OF PYTHAGORAS AT PYTHAGORION, SAMOS, GREECE

- **c.569 BCE** Pythagoras is born in Samos, on the island of Samos; his father was Mnesarchos, his mother Pythais. The town has since been renamed Pythagorion in his honor.

- **c.550 BCE** Goes to Miletus and is taught by Thales, one of the earliest Greek philosophers, an astronomer who seems to have predicted an eclipse of the Sun in 585 BCE, and his pupil Anaximander, who was interested in cosmology and geometry.

- **c.535 BCE** On Thales's advice he goes to Egypt, where there is already a sizeable community of people from Samos; they even have their own temple in Naucratis. Pythagoras studies astronomy and geometry.

- **c.525 BCE** Reputedly taken prisoner by Cambyses II, the King of Persia, and taken to Babylon, where he studies arithmetic, music, and other disciplines with the scholars.

- **c.520 BCE** Returns to Samos, where, after a visit to Crete to study its legal system, he forms a school. He is not treated well by the Samians, however, and travels on to mainland Greece, and from there to southern Italy.

- **c.518 BCE** Settles in Croton, a Greek seaport in southern Italy, where he founds a school or brotherhood devoted to the study of mathematics, but also including a medical school. The Pythagoreans are sworn to silence and secrecy and are bound by a set of curious rules. In particular they are not allowed to eat meat, fish, or beans, nor to drink wine. They are not allowed to wear woolen clothes, because wool comes from animals. This may be connected with the fact that Pythagoras believes in reincarnation, and is worried about being reincarnated as an animal.

- **c.510 BCE** The unusual brotherhood attracts hostility and distrust and is threatened with violence; Pythagoras escapes to Metapontion, another Greek city in southern Italy.

- **c.495 BCE** Dies in Metapontion.

BEFORE

Mathematics before the ancient Greeks was largely unsystematized and emphasized practical applications.

EARLY MATHEMATICS
Mathematics in ancient Egypt was focused on practical issues, such as **counting ‹‹30–31**, **calendrical calculations** to help predict the flooding of the Nile, and **simple geometry** for dividing up land and **building large structures** such as the **pyramids**. Ancient **Chinese mathematics**, which developed independently, was concerned with many of the same subjects, and similarly **emphasized practicality**.

PYRAMIDS AT GIZA, EGYPT

‹‹ SEE ALSO
pp.30–31 ANCIENT NUMBER SYSTEMS

The five regular polyhedrons
Also called Platonic solids, the five convex regular polyhedrons were first defined by Pythagoras. They are the only three-dimensional shapes whose faces are regular polygons and meet at equal angles.

Regular tetrahedron
The regular tetrahedron is composed of equilateral triangles. It has four triangular faces, six edges, and four vertices.

Cube
The cube is composed of squares. It has six square faces, 12 edges, and eight vertices.

Regular octahedron
The regular octahedron is composed of equilateral triangles. It has eight triangular faces, 12 edges, and six vertices.

Regular dodecahedron
The regular dodecahedron is composed of regular pentagons. It has 12 pentagonal faces, 30 edges, and 20 vertices.

Regular icosahedron
The regular icosahedron is composed of equilateral triangles. It has 20 triangular faces, 30 edges, and 12 vertices.

MATHEMATICIAN (C.325–265 BCE)
EUCLID

One of the foremost ancient Greek mathematicians, Euclid may have studied at Plato's Academy and certainly taught at the Library of Alexandria during the reign of Ptolemy I. Although little else is known about his life, his work was widely translated and is well known, especially his major treatise, *The Elements*. As well as establishing basic principles of geometry with his axioms, Euclid wrote about number theory, and also on the subjects of optics, mechanics, and music.

Greek Mathematics and Geometry

The mathematical thinking of the ancient Greeks marked a turning point in the development of the subject. Building on the empirical discoveries of the Babylonians and Egyptians, Greek mathematicians instigated a more scientific approach that is still the basis of mathematics today.

The big innovation that distinguished ancient Greek from previous mathematics was the development of deductive logic, and with it the concept of proof. Where the Egyptians were content to accept that, for example, certain dimensions produced a right-angled triangle, early Greek mathematicians sought to understand the underlying laws of these discoveries, and applied logic to prove their theories.

Early Greek mathematics
The earliest known of these Greek mathematicians was Thales of Miletus, who probably acquired much of his practical knowledge from his travels to Egypt. Because he was an engineer, most of his studies were in geometry, which he is credited with introducing

EUCLID'S FIVE AXIOMS

1 Any two points can be connected by one and only one straight line.

2 Any line segment is contained in a full, infinitely long line.

3 Given a point and a line segment starting at that point, there is a circle that has the point as its center and the line segment as a radius.

4 All right angles are equal to one another.

5 Given a line and a point that is not on the line, there is only one line through the point that never meets the first line.

to Greece. Geometry—at that time, the study of lengths, areas, and volumes—therefore became the foundation for subsequent Greek mathematics.

Thales found new applications for his geometry: he calculated the height of a pyramid by measuring the length of its shadow and comparing this with the ratio of the length of his own shadow (measured at the same time) to his height. Theories attributed to him confirm his analytical as well as practical appreciation of geometry, and also established the principle of deductive proof in mathematics.

It is thought that Thales taught Anaximander, the philosopher who first asserted a scientific rather than supernatural order to the Universe and whose students included Pythagoras (see pp.32–33). From his school in Croton, Pythagoras taught that everything could be explained by mathematics, an idea that influenced later thinkers including Plato and Aristotle (see pp.36–37). Using Thales's notion of deductive logic, Pythagoras (or members of his school, the Pythagoreans) provided proofs for a number of theorems. These included Pythagoras's famous theorem for right-angled triangles—that the square of the hypotenuse (long side) of a right-angled triangle equals the sum of the squares of the other two sides.

Pythagoras's disciples included Hippasus, who discovered that the square root of 2 was what he called an "irrational" number—one that cannot be written as a simple fraction—and among later followers of his ideas was Hippocrates of Chios, writer of the first geometry textbooks to state Pythagorean theorems and proofs.

Mathematicians of this period were fascinated by how many geometric problems could be solved using only a straight edge and compasses. They found solutions for bisecting an angle, drawing parallel lines, and more. But three problems remained unsolved: trisecting an angle, doubling a cube, and squaring a circle (making a square with the same area as a given circle).

Euclid's influence

Following the death of Alexander the Great in 323 BCE, the Greek empire seemed to be on the wane, but Greek mathematics flourished. The first, and perhaps greatest, mathematician of this period was Euclid (see opposite). His contribution was enormous, but at its heart was the notion of axioms, self-evident truths from which deductions could be made. In his treatise *The Elements*, probably the

IN PRACTICE

NAVIGATION

The advances made by the ancient Greeks in geometry and mathematics were particularly useful for navigation, especially when combined with astronomical observations. By applying the principles of trigonometry (the study of triangles) to their observations, sailors were able to navigate by the stars with greater accuracy than before. For example, by measuring the angle between a celestial object, such as a star, and the horizon using a sextant, latitude can be determined.

Irrational numbers
Applying Pythagoras's theorem to a right-angled triangle with sides of one unit, Hippasus found that the hypotenuse (long side) was equal to the square root of two, which was what he termed an irrational number—a number that cannot be written as a simple fraction.

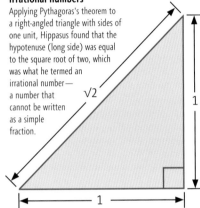

$\sqrt{2}$

1

1

most influential mathematics textbook ever written, Euclid laid down five axioms (see opposite) and proceeded from them to provide proofs for his mathematical theorems.

Building on Euclid's foundations, Archimedes proved that the area of a circle is pi (π) times the square of its radius, and managed to calculate π (approximately 3.14) to a remarkable

degree of accuracy. Diophantus of Alexandria took Euclid's ideas to the next stage in his work on equations, pioneering algebra (see pp.48–49). A particular subject of interest was conic sections, the shapes made by slicing through a cone, such as parabolas and ellipses. Foremost in this field was Apollonius of Perga, who used conic sections to explain the apparent motion of the planets.

Mathematics and astronomy

Astronomy dominated the last period of ancient Greek mathematics. Rather than simply mapping celestial objects, ancient Greek astronomers used geometric principles to understand their motions. The first comprehensive explanation of the motions of the planets was made by Ptolemy in the 2nd century CE. He devised a model that mapped their movements as seen from Earth. His treatise the *Almagest* (or "*Great Compilation*") was as influential to astronomy (see pp.28–29) as Euclid's *Elements* had been to geometry.

AFTER

Works of the Greek mathematicians were translated by Indian and Islamic scholars, influencing subsequent developments.

CLASSICAL INDIAN MATHEMATICS
From around 400 CE the era of classical **Indian mathematics** followed on from both the Greek and Chinese traditions. Mathematicians such as **Aryabhata** and **Varahamihira** built on their discoveries, introducing **innovations** including a **symbol for zero** and **decimal numbers**.

ISLAMIC MATHEMATICS
As the Islamic empire spread from about 700 CE **scholars translated** the works of ancient Greek mathematicians. **Islamic mathematicians** developed their work to make further **discoveries in geometry**, and in the emerging fields of **algebra 48–49 »** and **number theory**.

MATHEMATICS IN MEDIEVAL EUROPE
From about the 12th century contact with Islamic scholars stimulated **interest in mathematics in Europe**, but there were **no major discoveries** until **Fibonacci 62–63 »** in the 13th century.

MEDIEVAL GEOMETRY LESSON

SEE ALSO »
pp.48–49 ALGEBRA
pp.62–63 EAST MEETS WEST
pp.136–37 NAVIGATING THE OCEANS

INVENTION

SIEVE OF ERATOSTHENES

The sieve of Eratosthenes is a technique for finding prime numbers—whole numbers divisible only by themselves and by 1 (which, by definition, is not a prime). It consists of filtering out non-primes from a list of all numbers from 2 up to any top limit.

For example, to find all primes to 50:
1. List all numbers from 2 to 50.
2. Circle 2 then cross out all multiples of 2 (4, 6, 8, etc.).
3. Circle the next number that is not crossed out then cross out all multiples of that number.
4. Repeat step 3 until the end of the list is reached. The primes are the numbers circled.

Circled numbers are primes
Crossed-out numbers are non-primes

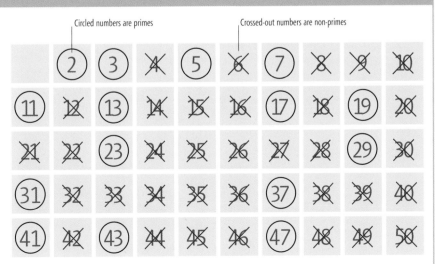

PHILOSOPHER AND SCIENTIST **Born 384** BCE **Died 322** BCE

Aristotle

"The **whole** is more than the **sum** of its parts."

ARISTOTLE, FROM "METAPHYSICS," 335–323 BCE

One of the greatest of the Greek philosophers, Aristotle laid the foundations for scientific thought in the Islamic and Christian worlds. His studies of various subjects, including physics, mathematics, logic, and biology, have influenced many disciplines.

Aristotle was born in Stageira, Macedon, the son of a doctor, Nicomachus. Appointed personal physician to the King of Macedon, Nicomachus probably took young Aristotle with him, beginning a close association with the royal family.

Aristotle was orphaned at the age of 10 and brought up by his uncle, who complemented his early medical education with lessons in rhetoric and poetry. At the age of 17 he went to Athens to become a student at Plato's Academy. Aristotle spent 20 years at the academy, becoming one of a line of great Greek philosophers: Socrates had been Plato's mentor, and now Aristotle was an obvious choice as Plato's successor as head of the Academy.

After Plato's Academy

Denied the job for political reasons, Aristotle left the Academy for Assos in Asia Minor (modern Turkey), where Hermias of Atarneus had assembled some philosophers. While at Assos, Aristotle explored the wildlife of nearby Lesbos, returning to the study of biology and zoology he had started with his father and beginning work on a systematic classification of living things. Assos was then invaded by the Persians;

▷ **Aristotle**
This bronze statue of Aristotle, created in 1915 by Cipri Adolf Bermann (1862–1942), is placed opposite one of Homer, also by Bermann, outside an entrance of the University of Freiburg in Breisgau, Germany.

ΑΡΙΣΤΟΤΕΛΗ

Aristotle fled, first to Lesbos, and then to Macedon at the invitation of King Philip. One of his pupils at the Macedonian court was Philip's son Alexander, who, after he succeeded the throne, wanted to send Aristotle to Athens in an ambassadorial role.

The Lyceum and the "Peripatetics"

When Aristotle did not receive the headship of the Academy in Athens, Alexander encouraged him to set up his own establishment in competition: the Lyceum. This offered a broader range of disciplines than the Academy, with an emphasis on the study of nature as well as traditional philosophy, politics, and rhetoric. The Lyceum attracted students who were called the "Peripatetics" after the colonnades (*peripatoi*) of the school; the name may also derive from Aristotle's habit of walking while lecturing.

Aristotle prepared his lecture notes in the form of treatises, and these survived to become his most important

Aristotle tutoring Alexander
Aristotle taught at the Macedonian court before setting up the Lyceum in Athens. Among his pupils there was King Philip's 13-year-old son Alexander, later Alexander the Great, in whom he inspired a great love of learning.

works. In particular, he developed his "analytics," the system of deductive logic that underpins Western logic and the scientific method today. His teaching on the subjects of "natural philosophy" (physics), mathematics, and zoology showed that he regarded these just as highly as the philosophical subjects such as metaphysics, ethics, and politics covered by his treatises.

Aristotle's heyday at the Lyceum was sadly cut short by political wrangling after Alexander died. Athens once again became fiercely anti-Macedonian, and Aristotle's connection with the deceased king made his position untenable. At the same time he was denounced for refusing to accept royalty as divine and was forced to leave Athens once more.

Aristotle's legacy

Aristotle died of natural causes shortly after leaving the Lyceum. The works he had published during his lifetime are now lost, but much of his teaching was passed down by his students and a collection of writings was published posthumously. Translations into Arabic were later made by Islamic scholars, and subsequently into Latin by medieval European translators, preserving his extraordinary breadth of knowledge and clarity of intellect for posterity.

> "All men by nature desire knowledge."
>
> ARISTOTLE, FROM "METAPHYSICS," 335–323 BCE

▽ **The School of Athens**
Raphael's fresco *The School of Athens*, painted in the Vatican, Rome, depicts many of the philosophers and scientists of ancient Greece. Deep in discussion at its center are Plato and his pupil, the young Aristotle.

STATUE OF ARISTOTLE AT STAGEIRA

Ancient Ideas of the World

Over the course of several centuries BCE a succession of philosophers, mathematicians, and early geographers sought to understand the basic shape and size of our planet, as well as defining concepts such as latitude and climate zones. The conclusions that they reached demonstrated a high level of sophistication and accuracy.

BEFORE

Clues to what people's ideas about the world may have been over 3,000 years ago come from examination of a small number of ancient texts and maps.

EARLY LOCAL MAPPING

Long before they had any idea of the overall shape of Earth or the distribution of its landmasses, people were making primitive maps of their local areas. The **oldest known local map**, which was discovered in Mezhirich, Ukraine in 1966, is thought to date from approximately 10,000 BCE. Inscribed on a **mammoth tusk**, it shows several dwellings along the banks of a river.

FOUR CORNERS OF THE EARTH

A common ancient theme was that **Earth was flat** and that it took the **form of a circle or square**, which was thought to be surmounted by a hemispherical sky. In ancient Egypt, which prevailed for nearly 3,000 years BCE, the sky was believed to be like a tent canopy, which stretched between great mountains that stood at the four corners of Earth.

MATHEMATICIAN (276–194 BCE)

ERATOSTHENES

The Greek scholar Eratosthenes was born in Cyrene, in North Africa. As a young man he studied in Athens and Alexandria, becoming an expert in many different fields, including mathematics and astronomy. At the age of 40 he was appointed chief librarian at the library of Alexandria, the greatest center of learning in the ancient world. In addition to his first accurate determination of Earth's circumference (see right), Eratosthenes drew important early world maps, calculated the distance from Earth to the Sun, and made a key contribution to the mathematical theory of prime numbers. He also coined the term "geography" and was, arguably, the first geographer. Tragically, toward the end of his life he became blind and subsequently starved himself to death.

Greek and Roman scholars, including the first geographers, developed new concepts about the world, some of which are now known to be correct. Remarkably, some of them appear to have been based purely on philosophical or aesthetic grounds rather than scientific observation. Until around 2,500 years ago, the prevailing belief in the Mediterranean region and in nearby Mesopotamia was that these lands occupied the upper flat surface of a drum- or coin-shaped disk. Earth's surface at this time was thought to consist mainly of land, with a sea (the Mediterranean) at its center, and an ocean surrounding the land, extending to the edges of the disk.

Babylonian map
Part of this clay tablet (c.600 BCE) shows inhabited lands encircled by an ocean. It is considered to be the oldest known world map.

A spherical Earth

Over a period of more than two centuries (c.550–300 BCE), Greek scholars made some key advances in understanding the world. The most important of these ideas, first advocated by the mathematician Pythagoras around 550 BCE, was that Earth is a sphere. Pythagoras reasoned that this shape is the most harmonious geometric solid and, therefore, the shape the gods would have chosen for the world when they created it. Accepting the possibility of a spherical Earth, around 360 BCE the philosopher Plato suggested that there was likely to be another landmass directly on the opposite side from the lands of the Mediterranean, which he called the antipodes. This idea, however, was based purely on a Greek belief in symmetry.

Aristotle (c.384–322 BCE) was the first scholar to put forward scientific evidence that Earth is a sphere. He argued that travelers to southern lands could see stars in the night sky that were hidden below the horizon for those living farther north, and this could only be explained if Earth's surface was curved. In addition, he pointed out that during lunar eclipses the shadow of Earth on the Moon has a curved edge. Aristotle's arguments were convincing and, by c.330 BCE, most educated people in the classical world had accepted that Earth was a sphere and not flat.

Earth's circumference

The next challenge was determining how big this sphere was—the circumference of the Earth. Scholars realized they should be able to figure out the Earth's circumference. First, they needed to find two

Evidence for a spherical earth
One of Aristotle's arguments for a spherical earth was that its shadow, seen here cast on the Moon during a lunar eclipse, has a curved edge.

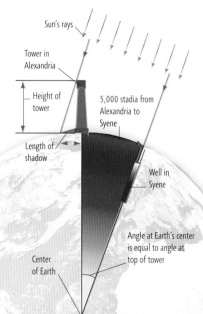

Sun's rays

Tower in Alexandria

Height of tower

5,000 stadia from Alexandria to Syene

Length of shadow

Well in Syene

Angle at Earth's center is equal to angle at top of tower

Center of Earth

Measuring Earth's circumference
By examining how the Sun's rays reached Earth at Alexandria and Syene, Eratosthenes figured out the difference in their latitudes (7° 12′), divided this into 360° (50), and multiplied it by the north–south distance between them (5,000 stadia) to give him the circumference of Earth (250,000 stadia).

AFTER

Improvements in gear technology since 1500 have included new types of gears and more precise, harder-wearing metal gears.

FROM ROTARY TO STRAIGHT LINE
In the 16th century the Italian doctor and mathematician **Girolamo Cardano invented the Cardan gear**. It provided a means of converting rotary motion into back-and-forth, straight-line motion by means of a cog rotating inside a large-toothed wheel.

19TH-CENTURY COTTON MILL

INDUSTRIAL REVOLUTION
The invention of **steam engines 132–33 »** and later **electric motors 168–69 »** during the Industrial Revolution of the 18th and 19th centuries led to an explosion in the use of metal gearing and further developments in the science of gear design.

SEE ALSO »
pp.50–51 WATER AND WIND POWER
pp.120–21 MEASURING TIME
pp.168–69 THE ELECTRIC MOTOR

Leonardo's designs
This drawing shows Leonardo da Vinci's design for a two-wheeled hoist with a caged gear from 1503, with the assembled design (above left) and an exploded view (above right). Although probably never built, Leonardo's hoist could have been used for raising or lowering a weight. It incorporated safety ratchets to prevent the weight falling freely.

(died *c.*220 BCE) described the use of simple gears in water-raising devices. By around 100 BCE, the Greeks were using metal gears with wedge-shaped teeth to form complex calculating devices and astronomical calendars. We know this from the discovery of a device known as the Antikythera mechanism (see opposite), which is the oldest-known complex geared machine. Constructed around 100 BCE, it was found in a shipwreck off the Greek island of Antikythera in 1900. The mechanism contained over 30 cogs and was designed to calculate astronomical positions.

In China, artifacts and written records indicate that gears were being used in watermills, crossbow-arming mechanisms, and chain pumps during the period from 200 BCE to 250 CE.

Middle Eastern technology
Whatever advances in gear technology were developed by the ancient Greeks, they were largely lost from Europe during the 1st millennium CE, and transferred to the Islamic world.

Several examples are known of geared calendars and astronomical devices from the Middle East dating from the 5th to 13th centuries CE. Parts have been found of a geared sundial calendar from around the 5th or 6th century that displayed the positions of the Sun and Moon in the zodiac. An early Persian astrolabe (see p.29) dating from the 13th century is the oldest geared machine in existence still in a complete state.

Reintroduction to Europe
The knowledge of mechanical gearing that had been kept alive in the Islamic world was reintroduced to Europe via the Islamic conquest of Spain in the 12th century. It was most evident from the sudden advances in clockmaking. These included water-driven clocks that for the first time incorporated epicyclical or planetary gears (one or more small outer gears, revolving about a larger central gear).

In non-Islamic Europe, sophisticated gears for transmitting high torque first appeared in astronomical clocks in the 14th century, such as the "Astrarium" made by Giovanni de Dondi of Padua in Italy in around 1365.

Artifacts, writings, and drawings by luminaries such as Leonardo da Vinci and Albrecht Dürer indicate that all the main types of gears were known and used in Europe by the 15th century.

IN PRACTICE

DIFFERENTIAL GEARING

Differential gearing allows power to be transmitted equally between the two wheels of a moving vehicle—such as a car—while allowing them to take paths of different lengths, and therefore turn at different speeds. This is necessary because the outside wheel of a car has farther to go when the car turns a corner. The differential consists of sets of bevel gears (see opposite) and pinions attached by a carrier to the crown wheel. When input torque (rotational force) is applied to the crown wheel, it turns the carrier, providing torque to the side gears. These drive the wheels via half-shafts. On a straight, the planet pinions do not rotate, and the wheels turn at the same rate. During a turn, the planet pinions rotate and, as a result, the side gears and wheels revolve at different speeds.

VEHICLE DRIVEN STRAIGHT

Fast — Fast

VEHICLE DRIVEN AROUND CORNER

Faster — Slower

Crown wheel

Carrier

Planet pinions (red)

Side gears

Back wheel

Rear axle

Drive shaft from the engine

Direction of travel

"Eureka!"

Many a great leap forward in science has come when a scientist has a sudden flash of inspiration. The first recorded case occurred 2,250 years ago in Syracuse and completely revolutionized the way people thought about measuring volume. The scientist in question was Archimedes, one of the finest minds of the time. His discovery came when trying to solve a tricky question of potential fraud.

In 240 BCE, Syracuse, on the island of Sicily, was ruled by the tyrant Hieron II, who ordered an elaborate new crown as a tribute to the gods. It was to be made from pure gold, which the king weighed out very precisely. When it arrived, however, he suspected that the crown maker had cheated by mixing some cheaper metal with the gold. The king was furious at the thought of being tricked, and yet did not know how to prove it. To investigate, he summoned Archimedes.

Archimedes was perhaps the finest mathematician in the ancient world, and also a skilled engineer. He was still worrying about the question posed by the king when he went for one of his rare baths in the public bath house. He knew the weight of the crown because he had put it on his scale at home. If he could measure the volume of the irregularly shaped object, he could figure out its density—the weight per unit volume. Archimedes knew that metals have different densities. For gold it would be about 19 g/cm^3, while other common metals are much less dense: lead 11, copper 9,

silver 10. So from the density he would know immediately whether the gold was pure.

Archimedes' dilemma was how to measure the volume of the crown. Then as he sank down into the bath, some water slopped over the side, and Archimedes had his flash of inspiration and utterly changed the way people thought about measuring volume. His body sinking into the bath had pushed some water out of it—and Archimedes realized that any "body" pushed into water would displace its own volume of water. Overjoyed, he jumped out of the bath, shouting "Eureka," and ran all the way home, stark naked. He filled a bucket with water and lowered the king's crown in on a piece of string, slopping some water over the side. Then he pulled the crown out and measured how much water he needed to refill the bucket exactly to the brim—and that was the volume of the crown. From this he calculated the density, which turned out to be much less than that of pure gold; so the king really had been cheated. But, what mattered most was this new way to measure volume.

A new breed of scientist
The story of Greek mathematician and inventor Archimedes' discovery of how to measure the volume of irregular objects is depicted in this 16th-century hand-colored woodcut.

"Eureka! I have found it."

ATTRIBUTED TO ARCHIMEDES

For thousands of years people have found increasingly ingenious solutions to the challenges of transporting people and materials over water.

ARISTOTLE

The great Greek philosopher **Aristotle «** 36–37 was a **keen observer** of science and technology. He noted that some boats that floated fully laden at sea sank when sailed up river. This is because fresh water is **less dense** than sea water, and is therefore less buoyant.

EARLY BOATS

The first **floating platforms** were probably rafts of logs, while some cultures made **kuphars** (reed boats) or stretched animal skins over frames to make small craft such as **coracles**.

KUPHAR MADE WITH WOVEN REEDS

"EUREKA!"

The Greek scientist and mathematician Archimedes **«** 44–45 is said to have shouted "Eureka!" ("I have found it!") when he noticed how his body **displaced** water from his bath. His understanding of the relationship between weight, volume, and water forms the basis of **Archimedes' Principle**.

Floating and Sinking

Whether things float or sink in water has always been important to people who live near rivers, lakes, and the sea. The ancient Greek inventor Archimedes was probably the first person to put buoyancy on a mathematical footing, and shipbuilders have made use of his ideas to manipulate buoyancy ever since.

The science of buoyancy

Items are buoyant if their weight (which includes the air they contain) equals or is less than the water displaced—the "upthrust." Ship designers can calculate the type and size of craft needed to carry heavy loads while remaining buoyant.

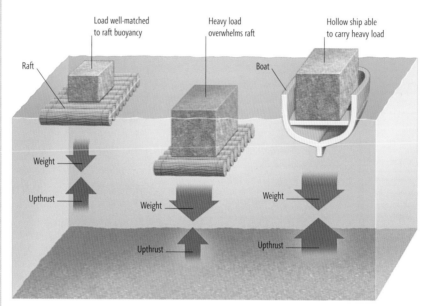

Load well-matched to raft buoyancy

Heavy load overwhelms raft

Hollow ship able to carry heavy load

Raft

Boat

Weight

Upthrust

Weight

Upthrust

Weight

Upthrust

Weight

Upthrust

A buoyant raft
The wooden raft can carry a small load provided the upthrust is greater than the combined weight of the raft and the load.

A sinking raft
With a heavier load the combined weight is greater than the upthrust; the water flows over the side, and the raft sinks.

A buoyant ship
Because a boat is hollow and deep, it displaces much more water than a raft, and the greater resulting upthrust can support the heavier load.

F or thousands of years people have wanted to travel over water, either in peace or in pursuit of war, and this has driven the gradual development of craft that operate either on the surface of water or under it.

Buoyancy

Buoyancy is a force that pushes things upward; it holds up things that float. The ancient Greek mathematician Archimedes realized that objects immersed in water displace some of it, and that this displacement is what causes the upthrust.

Archimedes' Principle states that any object immersed in water will experience an upthrust equal to the weight of water displaced. One consequence of this is that any object less dense than water will float, because if the object were to be submerged, it would displace a weight of water greater than

its own weight—and this upthrust would push it back up to the surface.

The density of an object is its mass per unit volume, normally measured in grams per cubic centimeter (g/cm^3) or kilograms per cubic meter (kg/m^3). Things that "feel" heavy for their size —such as lumps of metal or stone— have high density; while things that

"feel" light—such as table-tennis balls or cork—have a low density.

The density of water

Pure water has a density of $1g/cm^3$; anything with a higher density than this will sink in water, and anything with a lower density will float. Cork floats high in the water, since it needs to displace only one quarter of its volume of water to feel an upthrust equal to its own weight. People and

> **ARCHIMEDES' PRINCIPLE** states that the upthrust on an object in water is equal to the weight of the water that the object displaces.

icebergs float low in the water because their densities are about $0.9g/cm^3$. Most kinds of wood have densities less than 1, so most wood floats. Boats float mainly because they are hollow and full of air.

Another consequence of Archimedes' Principle is that although dense objects sink, they feel lighter under water because they experience some upthrust. Marine mammals such as whales can grow larger than land mammals because water supports their weight; land animals could never grow muscles and bones strong enough to hold so much weight.

In pure water, no inanimate object can float halfway between the bottom and the surface. Any object with a density of 1.001 or more will sink; anything with a density of 0.999 or less will float. Fish have to keep swimming to maintain a constant depth.

Cork has a structure of light molecules surrounding pockets of air

Water has a constant density

Only by swimming gently can the goldfish maintain its depth

The metal bolt is made of heavy atoms tightly packed together

Will it sink or will it float?

Water is incompressible, so the density of fresh water is the same at all depths. Objects that are less dense than water will float, while denser items will sink. Seawater is more dense because it also contains salt.

PLIMSOLL LINE

All large ships have to be marked with an International Load Line, or "Plimsoll Line" (after the British politician, Samuel Plimsoll, who helped introduce the law in 1876). This line is painted across a circle on a ship's hull, marking the position of the waterline when the ship is carrying the maximum safe load. The density of water varies with salinity and temperature, so various waterlines are marked: TF—Tropical Fresh Water; F—Fresh Water; T—Tropical Sea Water; S—Summer Temperate Sea Water; and W—Winter Temperate Sea Water.

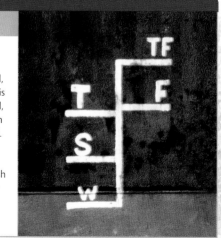

Submarine control

Submarines can be made to sink or rise to the surface through the use of their ballast tanks. Situated between the inner and outer hull, the tanks can be filled with air from compressed air tanks to decrease the vessel's density and make it rise; or with water to increase the vessel's density and make it sink.

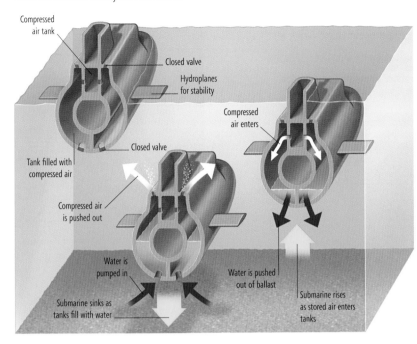

Compressed air tank

Closed valve

Hydroplanes for stability

Closed valve

Tank filled with compressed air

Compressed air enters

Compressed air is pushed out

Water is pumped in

Water is pushed out of ballast

Submarine sinks as tanks fill with water

Submarine rises as stored air enters tanks

1 Submarine afloat
When the ballast tanks are filled with air, the submarine's overall density is less than the water surrounding it, so it is able to float in the water.

2 Submarine submerged
For a dive, the ballast tanks are filled with water. This increases the submarine's weight, which become greater than the upthrust.

3 Submarine surfacing
Blowing compressed air into the ballast tanks drives out the water and lowers the weight of the submarine, so the upthrust takes it back to the surface.

Controlling density

The first boats were probably simple rafts of wooden poles tied together, but then people found they could make lightweight boats by stretching animal skins over wooden frames. Wooden boats were the next step up, and ranged from the hollowed-out logs of Polynesian canoes to the birch-bark canoes of Canada, and the northern European "clinker-built" ships, which used overlapping wooden planks nailed to sturdy wooden frames. At the end of the 18th century the first iron boats appeared, and today the vast majority of large ships are made of steel.

All of these designs for boats and ships depend on the fact that the space inside the hull is filled with air, and the overall density of the vessel is equal to the total weight divided by the total volume—in other words, shipbuilders can control the density, assuming that water is unable to get inside. However, even huge ships can sink. The RMS *Titanic*, the largest passenger ship in the world in 1912, sank within three hours because the iceberg it collided with sliced through the ship's hull and allowed five of the buoyancy compartments to fill with water.

Working with buoyancy

Boat and ship designers have learned to manipulate buoyancy, both in the materials and shapes of hulls and through the use of buoyancy tanks, while submarines are able to vary their buoyancy at will (see left).

Some marine creatures have gas bladders which fulfill the same function; they are able to fill them with oxygen to increase their buoyancy. But both supertankers and goldfish are subject to Archimedes' Principle—the upthrust is always equal to the weight of the fluid displaced.

Boats for hunting
The Palestrina mosaic from the 1st century BCE is evidence of the ways in which mastery of the water developed. It shows Romans fishing or hunting with spears on the Egyptian Nile on a large oar-propelled boat, possibly specially built for the purpose.

AFTER

Great efforts have been made to reduce the number of deaths by drowning. Knowledge of density and buoyancy has influenced the design of boats and led to the development of personal flotation devices.

FLOTATION CHAMBERS
Many recreational boats now have internal **flotation chambers** full of air, which prevent them from sinking even if they are swamped by water.

FLOTATION JACKET

BUOYANCY AIDS
Life jackets have **saved many lives** by keeping their wearers afloat and their faces well above water. Buoyancy aids or life jackets are routinely made available not only to sailors, but also to all **passengers in commercial aircraft**. These buoyancy aids are designed to be easy to put on, secure, and durable, and they carry whistles and lights for attracting attention.

SEE ALSO ≫
pp.144–45 LIQUIDS UNDER PRESSURE

Algebra

Early mathematicians lacked symbols for representing mathematical expressions and operations (such as addition and subtraction), which hindered progress significantly. With the introduction of symbols, algebra developed as a separate branch of mathematics and also led to major advances in other areas of the subject.

Ancient Babylonian, Egyptian, and Greek mathematicians managed to describe and solve mathematical problems by expressing them in words and numbers (rhetorical algebra), but it was a long-winded process. A system was needed in which the answer being sought—the unknown—could be expressed in terms of the information known about it.

This remained a verbal process until the 9th century, when the Persian mathematician Al-Khwarizmi laid the foundations of what we now call algebra by setting out systematic and logical rules for solving equations. Later mathematicians built on his work by introducing symbols for operations, such as +, −, and =, and using the shorthand *x* to represent an unknown and other letters for variables. The invention of specific mathematical

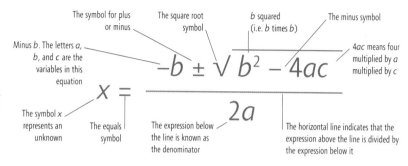

The symbol for plus or minus — The square root symbol — *b* squared (i.e. *b* times *b*) — The minus symbol

Minus *b*. The letters *a*, *b*, and *c* are the variables in this equation

4*ac* means four multiplied by *a* multiplied by *c*

$$x = \frac{-b \pm \sqrt{b^2 - 4ac}}{2a}$$

The symbol *x* represents an unknown — The equals symbol — The expression below the line is known as the denominator — The horizontal line indicates that the expression above the line is divided by the expression below it

An algebraic equation
The general formula for solving a quadratic equation $ax^2 + bx + c = 0$ shows some of the common mathematical symbols used in algebra.

signs and symbols simplified solving problems and established algebra as a field in its own right. It enabled people to conceptualize abstract ideas and to encapsulate general rules to solve problems in other disciplines, such as astronomy, physics, and economics.

Algebraic equations

Many problems can be expressed algebraically in the form of an equation, and unknown quantities can be calculated by applying the relevant rules of algebra.

As a simple example, suppose you paid a total of $28.75 for some stationery, and you know that the price included 15 percent tax. How much was the stationery before tax? To solve the problem algebraically, let the answer be *x*. Then *x* plus 15 percent of *x* is $28.75. In symbols, this is represented as $x + 0.15x = \$28.75$, or $1.15x = \$28.75$. Therefore the price before tax, $x = \$28.75 \div 1.15 = \$25.00$.

It is also possible to use algebra to solve problems in which there are two (or more) unknowns, as in the following example. Two donkeys are walking along, each carrying bales of hay. The first donkey says to the other one: "If I had one of your bales, we would have equal numbers." The second donkey replies: "If I had one of your bales, I would have twice as many as you." How many bales was each donkey carrying? To solve the problem, let *x* be the number of bales carried by the first donkey and *y*, the number carried by the second donkey. Then if one bale moves from donkey two to donkey one, donkey one has x + 1, donkey two has y − 1, and we know

▷ **Signs, symbols, and numbers**
A blackboard covered in algebraic expressions has become an immediately recognizable trademark of academic mathematics. Even when the equations are long and complex and contain unfamiliar symbols, the appropriate rules of algebra still apply.

« **BEFORE**

Before the invention of algebraic symbols, problems in algebra were stated in a verbal form, known as rhetorical algebra.

ALGEBRA IN THE ANCIENT WORLD:
From sources such as the Rhind papyrus we know that the **ancient Egyptians** could **solve linear equations** with several variables as long ago as 2000 BCE. The **Babylonians** of the same period were even more advanced, and **developed methods of solving quadratic** and possibly **cubic equations**. A Babylonian clay tablet (the Plimpton 322 tablet), dating from around 1600–1900 BCE, shows quadratic equations now known as Pythagorean triples.

RHIND PAPYRUS

MATHEMATICIAN AND SCIENTIST (c.700–c.850)

AL-KHWARIZMI

Although Abu Ja'far Muhammad ibn Musa Al-Khwarizmi was the foremost Persian mathematician, astronomer, and geographer of his time, little is known of his life. He was probably born in Khwarizm, in present-day Uzbekistan, but spent most of his adult life as a scholar at the House of Wisdom (a library and intellectual center) in Baghdad. Sometimes known as the "father" of algebra, he established algebra as a distinct branch of mathematics, and translations of his writings later brought the idea to the West, along with Hindu–Arabic numerals. The word "algebra" is derived from *al-jabr*, part of the title of his book on the subject, *Hisab al-jabr wa'l-muqabala*.

$$= 5 - 6x - 3x^2$$

$$6x + 4y = 8 \qquad y =$$

$$)^2 \qquad \frac{(y-8)^2}{64} =$$

$$x^2 - 2y + 4x - 4 =$$

$$< 0 \quad x^2 + (6-x)^2 =$$

$$x^2 - 6x + 8 =$$

$$\geq 0 \quad (x-2)(x-4) =$$

$$x =$$

$$> 0 \quad 1.96 = 0.5070($$

$$t \approx 8.1h$$

Types of algebraic equations

The most common type of equation, called a polynomial equation, includes variables raised to different powers. These equations can be classified according to the highest powers they contain. For example, in quadratic equations, at least one variable is squared (raised to the power of two).

Type of equation	Common mathematical form	Description/uses
Linear	$ax + by + c = 0$	No variable is raised to a power greater than one. Used in simple problems of addition, multiplication, and division.
Quadratic	$ax^2 + bx + c = 0$	At least one of the variables is squared. Used in calculations involving area.
Cubic	$ax^3 + bx^2 + cx + d = 0$	At least one of the variables is cubed. Used in calculations involving volume.
Quartic	$ax^4 + bx^3 + cx^2 + dx + e = 0$	At least one of the variables is to the power of four. Used in computer graphics.
Quintic	$ax^5 + bx^4 + cx^3 + dx^2 + ex + f = 0$	At least one of the variables is to the power of five. Solution of quintic equations was important to the development of abstract algebra.

they both now have the same number of bales, so $x + 1 = y - 1$. Adding 1 to each side gives $x + 2 = y$, or $y = x + 2$. Meanwhile, if one bale moves from donkey one to donkey two, then donkey two has $y + 1$ bales, which is twice as many as donkey one has $(x - 1)$. In other words, $y + 1 = 2(x - 1)$, which is the same as $y + 1 = 2x - 2$. Subtracting 1 from each side of the equation gives $y = 2x - 3$. We already know that $y = x + 2$, so we can write $x + 2 = 2x - 3$. Subtract x from each side to give $2 = x - 3$. Then add 3 to each side to give $x = 5$. And because $y = 2x - 3$, $y = 10 - 3 = 7$. So the answer is that donkey one had 5 bales and donkey two had 7 bales.

The same general principles can be applied to more complex equations. However, it is not always possible to explicitly solve a given equation, although it is usually possible to obtain an approximate solution to a high degree of precision.

AFTER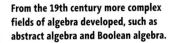

From the 19th century more complex fields of algebra developed, such as abstract algebra and Boolean algebra.

BOOLEAN ALGEBRA
The philosopher and mathematician **George Boole** (1815–64) brought together **algebra and logic** in his book *An Investigation into the Laws of Thought* in 1854. In Boolean algebra, **arithmetical operations** (such as addition and multiplication) are **replaced with logical operations** such as AND, OR, and NOT, and numbers are dealt with in a binary form. Largely overlooked at the time, **Boolean algebra** was later used in logic gates, which form the basis of **computer design**.

FERMAT'S LAST THEOREM
In the margin of a book the 17th-century mathematician **Pierre de Fermat** wrote "It is **impossible to separate** a cube into two cubes, or a fourth power into two fourth powers, or in general, **any power higher than the second into two like powers**." Known as Fermat's last theorem, this conjecture had confounded mathematicians for about 300 years until, using modern algebraic techniques, **Andrew Wiles** found a **proof in 1995.**

IN PRACTICE

COMPOUND INTEREST

Algebraic equations can be used to solve a huge variety of practical problems, including compound interest, which can be calculated using the equation:

$$A = P(1 + i)^n$$

where A is amount of money accumulated, including interest
P is the amount of money deposited (or borrowed)
i is the interest rate per period (e.g. 5% per year, expressed as a decimal—0.05 per year)
n is the number of periods for which the money is invested or borrowed

So if $100 is invested for 2 years at an annual interest rate of 5% (0.05), the amount after 2 years is given by:

$$A = 100 \times (1 + 0.05)^2$$

which gives the answer $110.25.

ANDREW WILES IN FRONT OF PART OF HIS PROOF OF FERMAT'S LAST THEOREM

SEE ALSO >>
pp.62–63 EAST MEETS WEST
pp.378–79 THE INTERNET

‹‹ BEFORE

People's need for water led to some of the most ingenious and long-lasting engineering in the ancient world.

ARCHIMEDEAN SCREW

One early device for **raising water** to **irrigate fields** was the "Archimedes screw" ‹‹ 40–41, which was attributed to **Archimedes** although it may have been in use since the 7th century BCE. It consists of an inclined spiral screw in a chute or cylinder, which scoops up water at its lower end, then carries it up as the screw is rotated.

EARLY IRRIGATION

Agriculture developed in the ancient world in areas where there was sufficient rainfall, and along the banks of large rivers, such as the Nile and the Euphrates. In order to **expand the area** available for **growing crops** in arid climates, **irrigation ditches** were dug to divert water from these rivers, in one of the earliest examples of **engineering** work.

TRANSPORTING WATER

As irrigation technology became more sophisticated, the irrigation channels developed into more **permanent structures**. Special water tunnels and bridges called **aqueducts** were built as early as the 7th century BCE, to carry water across long distances. These were perfected by the ancient Romans, who used them to supply both agricultural land and towns and cities.

IRRIGATION ALLOWED WATER CONTROL

‹‹ SEE ALSO
pp.20–21 EVOLUTION OF THE WHEEL
pp.40–41 SIMPLE MACHINES
pp.42–43 HOW GEARS WORK

Wind power from windmills
Windmills were a common sight throughout the world until they were gradually replaced by steam, combustion, and electric-powered engines. Post mills such as these could be turned to face the prevailing wind, their rotating sails driving machines to grind corn or, more rarely, to pump water for irrigation.

Water was an essential ingredient in early civilization: it was necessary for agriculture and for the growing towns. Early engineering focused on its transportation, and the flow of water in a river was soon recognized as a potential source of energy. By the time of the ancient Greeks, water was being used to drive simple machines. Wind power was also exploited in the ancient world by sailing boats.

Harnessing power

By the Middle Ages water and wind power were widely used to drive machinery. Water wheels—probably the earliest source of mechanical energy—powered the mills for grinding grain; later they were also used for sawing wood and clothmaking. They became more sophisticated in design: vertically mounted wheels were made possible by a system of gears that transferred power to the mill. Overshot wheels—where water flows over the top of the wheel—began to replace the less efficient undershot design, which relies on a fast supply of water under the wheel (see opposite).

The first windmills were built in the 7th century, in Seistan, Persia. Unlike the familiar modern windmill, these had sails mounted on a vertical shaft: wooden blades, sometimes covered with cloth, were mounted on arms to catch the wind. The sails were attached to a vertical axis driving a grinding stone.

A global trend

The idea soon spread to China and across the world. Windmills like those below soon began to appear in Europe in the 12th century; the sails were set at a slight angle to convert the

Hurricane power
The destructive power of the wind is seen at its most dramatic in tropical storms; hurricanes in the Atlantic and typhoons in the Pacific cause terrible damage on land and at sea every year. The enormous potential of wind power has been harnessed in minor ways since medieval times, and scientists today are researching the best ways of using this power to generate electricity.

Water and Wind Power

As early civilizations searched for alternatives to animal and slave power, two natural sources of energy were widely harnessed: water and wind. These remained the principal means of driving machinery until the discovery of steam and electrical power, and today they are seen as important sources of renewable energy.

BREAKTHROUGH

IMPROVEMENTS IN SAIL SHAPE

The rectangular sails of the earliest sailing ships were rigged perpendicular to the hull, efficiently catching the wind and propelling the boat forward when traveling in the direction of the wind, but they were not easy to handle against the wind. They were eventually replaced by triangular sails, which allowed the boats to sail closer to the wind.

Rectangular sails
These were hung from yards (crossbars), with one face to the wind. The yards could turn slightly around the mast.

Triangular sails
Fit lengthwise onto the ship on poles known as gaffs, triangular sails gave ships improved maneuverability.

Water and wind are the focus of today's scientific research into clean, renewable, and efficient forms of energy.

WIND TURBINES

Windmill technology is now being used to **generate electricity**. Wind turbines 424–25 » use huge propellers mounted on pillars to gather wind energy, which is converted into electricity. The turbines are often grouped in **wind farms**.

WAVE AND TIDAL POWER

Experimental plants are being developed to generate cost-effective power from the **motion of sea waves**. The changing levels and currents of tides are also being exploited by dams across **tidal estuaries**, while underwater turbines offer a more reliable potential than wind power.

11 percent of the US's energy came from renewable sources in 2019—surpassing coal for the first time since before 1885.

HYDROELECTRIC POWER

Up to 20 percent of the world's electricity is produced from hydropower. This **harnesses the energy** of water, under **gravitational** force, to drive a turbine connected to an electrical generator. Most hydropower plants are built as a **dam across a river**, creating a reservoir with a constant store of water whose release through the **turbine** can be controlled according to demand and the availability of water.

horizontal force of the wind into rotational movement. Later windmills were built around a solid vertical post (hence, the name "post mills") around which they could be turned so that the sails faced into the wind. A system of gears derived from the design of vertical water wheels transferred the motion of the sails to the grinding wheels inside the mill.

Later medieval windmills improved upon this design; the tower mill needed only the cap carrying the sails

to be turned into the wind. Windmills such as these were in common use even after the invention of steam engines, and only declined with the advent of the internal combustion engine.

Understanding hydraulics

Wind and water power were also used in sawmills, clothmaking plants, and smelting metal furnaces. In the Islamic world the principle of the water wheel was developed into the *noria*, an undershot water wheel incorporating

containers, which raised water for irrigation. In medieval China windmills inspired by the Persian design were used to pump water. Hydraulic power was widely used in water-powered machines throughout China and the Islamic world.

Interest in water and wind power diminished during the 19th century as new forms of energy were exploited, but these natural powers are exciting interest again today in the search for renewable energy sources to replace non-renewable carbon-based fossil fuels. Wave power, tidal power, wind turbines, and hydroelectric plants are all increasingly being used around the world to generate electricity.

Chinese astronomical clock
This mechanized armillary sphere and celestial globe demonstrates the technological sophistication of 11th-century China. Originally 37 ft (11.25 m) high, the water-powered clock also drove the sphere at the top of the tower, showing the positions of stars and planets.

◁ **Overshot water wheel**
Water from a reservoir is directed through a channel on to the paddles or blades on the rim of a wheel, making it rotate. More efficient than the undershot water wheel, the overshot uses not only the flow of water, but also the weight of water carried in paddles on one side of the wheel.

The force of the water and gravity combine to turn the wheel

The wheel is turned purely by the force of the current beneath it

Medieval irrigation
Widely used in the medieval Islamic world, *norias* use an undershot water wheel to raise water up to an aqueduct, where it can then be directed to a town or farmland. As the *noria* is rotated by the current, containers around its rim scoop water from the river and carry it up to the raised aqueduct.

ITAIPU HYDROELECTRIC DAM, BRAZIL

SEE ALSO »
pp.132–33 STEAM POWER TO STEAM ENGINE
pp.252–53 GENERATING ELECTRICITY
pp.422–23 GLOBAL WARMING

▷ **Undershot water wheel**
The earliest water wheels were slightly submerged in a stream or river, in an arrangement known as undershot. The flow of water acts on paddles at the bottom of the wheel to cause it to rotate, providing power for a mill.

« BEFORE

The inspiration for alchemists of the Middle Ages and the Renaissance went back to ancient Egypt and ancient Greece.

METALLURGY AND MYSTICISM

In the ancient world mystical significance was given to crafts, especially the craft of metalworking, where apparently "magical" transformations occurred in the smelting process. The occult learning of craftsmen was shared with the priestly class in cultures such as ancient Egypt, where the allegorical writings of Zosimos formed the basis for the "science" of alchemy.

THE HERMETIC WRITINGS OF THOTH

Medieval alchemists based many of their ideas on what they believed to be the writings of the Egyptian god Thoth, which were in fact Hellenistic texts. Ancient Greeks had integrated the figure of Thoth with their god Hermes into one, Hermes Trismegitus. The *Hermetica*, supposedly the writings of this hybrid god, proposed a number of "Hermetic Arts," of which alchemy was considered to be one.

THOTH

« SEE ALSO
pp.18–19 EARLY METALWORKERS

Alchemy

Although regarded sceptically today as a mere pseudoscience, alchemy was an important and respected branch of learning from the time of the ancient Greeks to the Renaissance, and many of its techniques and discoveries formed the basis of the modern sciences of chemistry, biochemistry, and pharmacology.

The central principle of alchemy, transmutation of matter, became the aim of philosophers in Hellenistic Greece, China, and India at much the same time, around the 3rd century BCE. Each of these cultures had sophisticated techniques of metal-working, often imbued with mystical significance. The processes of melting and mixing metals, essentially changing their character, became symbolic of the philosophical and religious ideas of the time. Alchemists saw them as the key to three distinct goals: the transmutation of base metal into gold; the discovery of a panacea or universal medicine; and creation of an elixir of life granting immortality.

From the empirical findings of these early metallurgists, which included the effects of fire on metals and elements such as mercury, sulfur, and arsenic, alchemy developed into a methodical discipline, involving experiments and the naming and classification of substances. Each of these was seen to have individual properties, which alchemists linked to mythical and astrological ideas. In addition, Greek alchemists proposed the possibility of creating a substance, the "philosopher's stone," which was thought to be both a panacea and a means of transforming metals to gold.

Chinese and Indian alchemy

While Greek alchemists placed emphasis on the "perfect" metals—gold and silver—and the search for the philosopher's stone, in China the primary aim of alchemical experiments was to find the "Grand Elixir of

Islamic alchemy

Rhazes, the most influential of the Islamic alchemists, discovered and classified a large number of natural substances in experiments using sophisticated apparatus invented by other Islamic alchemists.

COPPER CRYSTALS

METAL LIME

LIME

COPPER SAFFRON

URINE

Alchemical symbols
Alchemy's links with astrology and other occult beliefs were reflected in the symbols chosen to represent different materials, emphasizing their mystical significance.

Immortality." This quest was intimately associated with the search for the spiritual purity of Taoism and the martial arts. Discoveries of the medicinal uses of cinnabar (mercuric sulfide) and more importantly the mixture of sulfur, charcoal, and saltpeter (potassium nitrate) to make gunpowder (see pp.56–57), were in fact by-products of this experimentation.

Similarly, Indian alchemy of the same period stressed its pharmaceutical applications and was linked with Ayurvedic medicine and the Dharmic religions, even though its roots lay in the metallurgy of Indian craftsmen. In the course of their experiments, all three cultures developed techniques such as heating, burning, and most importantly distillation, which were later taken up by Persian and Arabic scientists in what was to be a golden age of alchemy in the Islamic world.

The influence of Islam
The Islamic empire, which spread across the Middle East in the 8th century, boasted remarkable cultural and scientific achievements. Building

their origins in names given at this time: alchemy is from the Arabic *al-kimia*, "the art of transformation," and terms such as alkali, alkane, and alkaloid also have Arabic roots.

The pursuit of gold
The great body of learning of the Islamic scientists, including alchemy, came to Europe in the 12th century through translations of Arabic and Greek texts. Medieval scholars such as the British Roger Bacon (see pp.56–57) took an avid interest in the art of alchemy, which fit well with the mystical Christianity of the time. The ability to transmute base metal to gold was also an attractive prospect in an increasingly commercial Europe, as was the possibility of eternal life. During the Renaissance, the emphasis in alchemy shifted back to the search for a panacea, with the Swiss Phillip von Hohenheim (1493–1541), who later took the name Paracelsus (see pp.70–71), leading the way in finding medicinal uses for chemicals. But with the advent of the so-called Scientific Revolution, from around the middle

Alchemist processes (c.1470)
Alchemists tried many different processes, most of them involving heat or fire, in their attempts to make gold, and by chance made important chemical discoveries.

Distillation by alembic
Invented by Geber in the 8th century, the alembic consists of a flask which is heated and from which vapor rises to be condensed and so transferred into a second flask. This woodcut shows a pair of intertwined alembics in action.

> " The **alchemists** in their **search** for **gold** discovered many other things of greater **value**."
>
> ARTHUR SCHOPENHAUER, GERMAN PHILOSOPHER, (1788–1860)

on the knowledge that had been passed down from the ancient Greek and oriental experimenters, alchemists in Persia such as Geber, Rhazes, and Avicenna treated the subject with a more rigorous scientific approach to experimentation. They refined and perfected the techniques of filtration, crystallization, and distillation using apparatus such as the alembic (a still for distillation invented by Geber), which allowed them to produce several new substances. Most important among the discoveries were alcohol and the mineral acids (including hydrochloric, sulfuric, and nitric acids); they also found that a mixture of concentrated nitric and hydrochloric acids subsequently called *aqua regia* could dissolve gold—which for an alchemist had mystical significance.

Perhaps the most methodical of these Islamic alchemists was Rhazes. By documenting all his experiments, he was able to name and classify the many new substances that he and his predecessors had produced for the first time, almost free of mystical references. Many of the words used in chemistry today have

of the 16th century, alchemy itself fell out of favor despite some notable followers in the scientific community, such as Robert Boyle (see pp.56–57) and Isaac Newton (see pp.110–11).

The contribution of alchemy to scientific knowledge had, however, been huge. Stripped of its occult associations by the Enlightenment, it evolved into the science of chemistry as we know it today.

AFTER

As the science of chemistry emerged in the 17th century, ideas about the elements began to change.

PHLOGISTON THEORY
In 1667 the German scientist Johann Joachim Becher, put forward a theory that **in addition to the four classical elements, there was a fifth**, which he called phlogiston. This new element was supposed to exist inside combustible materials, and was released during burning. Although the theory was soon discredited, it had prominent adherents including the British Joseph Priestley 146–47 **》**, who was credited with **the discovery of oxygen**.

CHEMICAL REVOLUTION
The so-called **Chemical Revolution**, spearheaded by early chemists including the French **Antoine Lavoisier**, dismissed the phlogiston theory and its continued adherence to the notion of the classical elements. Lavoisier recognized and named hydrogen and oxygen, and made the first steps toward a complete list of the elements.

SEE ALSO **》**
pp.138–139 THE NATURE OF MATTER

BREAKTHROUGH

DISCOVERING PHOSPHORUS

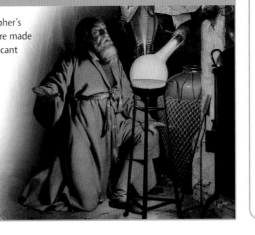

In the search for the mythical philosopher's stone, many important discoveries were made by alchemists. One of the most significant of these was made in 1669 by a merchant and alchemist from Hamburg, Germany, named Hennig Brand. By boiling and concentrating urine he managed to produce a substance that glowed with a pale green light, which he named phosphorus (from the Greek for "light-bearing"). This was a major turning point in the evolution of alchemy into chemistry.

ASTRONOMER **Born 78** CE **Died 139** CE

Zhang Heng

"**The Sun** is like **fire**, and the **Moon** like **water**. The fire gives out **light** and the water **reflects it**."

ZHANG HENG, FROM "LING XIAN" (MYSTICAL LAWS), 120 CE

Despite his modest and unambitious nature, Zhang Heng became an important official in the court of the Eastern Han emperors. He was also an influential astronomer, mathematician, inventor, and poet. His inventions and writings are among the most important of his time, and as a result Zhang Heng is considered one of the most significant figures in the field of astronomy.

Emergence of a poet

Zhang was born in Nanyang, China—where his grandfather had been a governor—and showed a talent for poetry and literature from an early age. While studying in the provincial capital, Luoyang, he established a reputation as a literary scholar as well as a poet. He was offered several posts at the Imperial court, but turned these down to continue his literary career until eventually accepting a minor post in the civil service in Nanyang. It seemed that Zhang, while in his twenties, was destined for a career as an administrator and poet.

Astronomer at the Imperial court

At the age of 30 Zhang took an interest in science. He devoted his time to studying astronomy and mathematics, and was soon as well known for his writings on these subjects as he had been for his poetry. As word spread of his expertise, he was once again invited to serve at the Imperial court in Luoyang, and this time he accepted. After a few years, he was promoted to the post of chief astronomer—an important position. The Chinese at that time devoutly believed in the astrological influence of the stars over their fortunes, but Zhang's interest was more scientific; he made detailed observations of the night sky, mapping

Armillary sphere

An invention that originated in China, the armillary sphere was improved to a high degree of sophistication by Zhang Heng. Beginning with a bamboo model, he went on to construct a bronze globe that showed the positions of all the known stars. The globe's movement of precisely one rotation each year was powered by a water clock positioned beneath it.

Lunar eclipse
From his observations of the night sky, Zhang suggested that moonlight was simply reflected sunlight and that that lunar eclipses resulted from Earth's shadow falling across the Moon. A crater on the Moon's dark side is named after him.

INVENTION

ZHANG HENG'S SEISMOMETER

Zhang Heng invented this device for detecting distant earthquakes in 132 CE. It consisted of an egg-shaped copper container, inside which was a sensitive pendulum. In the event of an earthquake, even at a great distance, the pendulum moved a mechanism that released a ball from the mouth of one of eight dragon heads fitted to the container. The ball would fall into the mouth of one of eight surrounding copper frogs, indicating the direction of the epicenter of the earthquake.

all the stars and planets. He stated in his treatise, *Ling Xian* (Mystical Laws): "North and south of the equator there are 124 groups that are always brightly shining. 320 stars can be named. There are in all 2,500, not including those which the sailors observe. Of the very faint stars there are 11,520."

He also came to some innovative conclusions about the nature of the Universe, observing:

"The sky is like a hen's egg, and is as round as a crossbow pellet, the earth is like the yolk of the egg, lying alone at the center. The sky is large and the earth small."

Zhang, unlike his predecessors, conceived a spherical Universe with Earth as a sphere at its center, and assumed that the planets were also spherical, as well as the Moon, which he controversially suggested was not in itself a source of light, but merely reflected the light of the Sun.

Mathematics

Zhang's mathematical work was no less innovatory. He studied "magic squares" of numbers—in which the rows, columns, and diagonals add up to the same number—and tried to find a method of calculating a value for pi (π), which until that time had been achieved only by practical measurement rather than theoretical calculation. In one of his mathematical treatises he provided a geometrical solution to the problem using circles drawn inside and around squares. It

Zhang Heng and his seismometer
Mainly remembered today as the inventor of the seismometer, Zhang Heng was also a gifted astronomer, mathematician, poet, and literary scholar.

gave $\pi = \sqrt{10}$ (approximately 3.162), and working from this he calculated the volume of a sphere.

Unfortunately, his work was not always appreciated by the court. Among the court ministers there were power struggles; Zhang took no interest in these but became embroiled when he suggested alterations to the calendar based on his astronomical data. The idea conflicted with the traditional astrological basis of the calendar, causing a controversy that denied him any further promotion.

Later achievements

Zhang nevertheless went on to devise some remarkable inventions. The first was a water-powered armillary sphere (see opposite), achieved through his detailed study of the night sky and thorough knowledge of the mechanics needed to make it rotate exactly once a year. Perhaps more famous was his seismometer (see above), which sensed distant earthquakes and indicated the direction from which they had been detected; presented to the court in 132 CE, it was not proved to be accurate until 138 CE.

Zhang died the next year, leaving a body of work that included some of the finest poetry of the age. His treatises on astronomy and mathematics were significant advances, and he is regarded as probably the most important Chinese thinker of his time. He is commemorated by his writings and inventions, and in the naming of later discoveries in his honor. These include the lunar crater Chang Heng, the asteroid 1802 Zhang Heng, and the mineral Zhanghengite.

Figurine of Zhang Heng
In China Zhang is commemorated by portraits, statues, and numerous porcelain figurines. A portrait of Zhang was featured on a Chinese postage stamp in 1955.

TIMELINE

78 CE Born in Xi'e in Nanyang Commandery (north of modern Nanyang City, Henan province), China.

88 CE His father dies, and he is brought up by his mother and grandmother.

94 CE Already an accomplished writer at the age of 16, he leaves Nanyang to study in Chang'an and Luoyang, the Han capital. He soon becomes well known for his literary writings, and has several works published.

101 CE Returns to Xi'e and works under Governor Bao De, with rank of officer of merit, as an archivist and writer of official documents.

106 CE Emperor He dies and is succeeded by the infant Emperor Shang, then less than one year old. When Shang also dies later that year, 12-year-old An is declared emperor.

108 CE Commences his studies in astronomy, and begins to work on treatises on astronomy and mathematics.

112 CE His reputation as an astronomer and mathematician spreads, and he is summoned to Luoyang to work for the Imperial Secretariat in the court of Emperor An.

115 CE Appointed chief astronomer to the Imperial court, and is made responsible for managing national documents and editing national history.

123 CE Introduces reforms to the calendar to bring it into line with his astronomical observations, but his suggestions cause controversy and he loses favor in the court.

125 CE Emperor An dies, at only 32, leaving a corrupt and failing government. He is succeeded by his son Shun, who is only 10 years old when he is declared emperor.

126 CE Re-appointed as chief astronomer by Emperor Shun, and is later given the rank of palace attendant to the emperor.

132 CE Presents his design for a seismometer to the court.

136 CE Amid increasing political rivalry in the court, Zhang retires from office under Emperor Shun, and is appointed chancellor of Hejian (in modern Hebei).

138 CE Zhang's seismometer detects an earthquake to the west of Luoyang, which is confirmed by messengers arriving a few days later.

138 CE Retires completely and returns to Nanyang, but is called back to Luoyang after only a few months to serve in the Imperial Secretariat.

139 CE Dies in Luoyang, at 61. His body is taken to be buried in his home town of Xi'e.

TOMB OF ZHANG HENG

« BEFORE

So terrifying and devastating is fire that it has been used widely as a weapon for at least 3,000 years.

SECRET WEAPON

"Greek fire" was a secret incendiary mixture brought to Constantinople by **Kallinikos** in 673 CE. It probably included **naphtha, or rock-oil**, a black, sticky material that oozes out of the ground in the region between the Black Sea and the Caspian Sea. This may have been distilled, and mixed with **sulfur or resin**. The material could be catapulted in tubs or pumped through blowpipes; it floated on water, and was easy to light, but extremely **hard to put out**, except, allegedly, with urine, vinegar, or sand. It was a major factor in the **defense of the Byzantine Empire** for 800 years.

FLAMING ARROWS

Arrows tipped with **burning oil-soaked rags** were fired at the wooden walls and buildings of **Athens** in 480 BCE. More than 2,200 years later, in 1806, **William Congreve** used his **rockets** to attack a French fleet in the harbor at **Boulogne**. The rockets all swerved off course, with the result that no ship was hit, but most of the town was **destroyed by fire**.

▷ **Early Chinese rocket**
The Chinese used bamboo for their rocket tubes; a fire stick equipped with a nozzle and a guide stick to steady the direction made a powerful weapon.

Gunpowder and Fire Weapons

Invented by the Chinese in the 9th century, gunpowder rapidly became a crucial factor in battle. The secret spread slowly to the West and reached Europe by the 13th century. Incendiaries, materials that readily set fire to enemy property, predated gunpowder, and are still in use today.

For hundreds of years gunpowder was a vital ingredient for such military hardware as bombs and guns, and in peacetime for mining and demolition. Gunpowder has its own built-in supply of oxygen, so that it burns extremely rapidly; in burning it produces a large quantity of gas. If it is ignited on an open surface it goes off with a fiery whoosh, but if it is contained then the gas generates enormous pressure in the container, and the result is an explosion.

Invention of gunpowder

Gunpowder was invented in the 9th century by Chinese alchemists or monks looking for the elixir of life. They had used saltpeter (potassium nitrate) in various medicinal compounds for hundreds of years, and probably came across this highly inflammable material by chance. A Taoist text, which is thought to date from the mid-800s, includes the passage. "Some have heated together sulfur, realgar [an arsenic sulfide mineral], and saltpeter with honey; smoke and flames result, so that their hands and faces have been burned, and even the whole house where they were working burned down."

Popular legend has it that the Chinese used gunpowder for fireworks, as they do today, but in reality they quickly harnessed it for military purposes. Fire sticks were bamboo tubes filled with gunpowder, lit, and pointed at the enemy at close range. Primitive guns were fire sticks with pebbles packed in front of the gunpowder. Rockets were fire sticks with a nozzle and a guide stick. Bombs were simply gunpowder in containers.

Arrival in Europe

Knowledge of gunpowder reached the Arab world in the 13th century, and soon traveled through Europe, possibly via a German monk and alchemist called Berthold Schwarz, but certainly through an English monk in Oxford, Roger Bacon (see opposite).

Bacon wrote down what appears to be the formula of gunpowder in Latin and in code: "*Luru Vopo Vir Can Utriet.*" This can be interpreted as a mixture of charcoal, sulfur, and saltpeter, but it is so low in saltpeter that it would only crackle and fizz; it would not explode.

Real gunpowder, or "black powder," is a mixture of the three ingredients in the proportions 15 percent charcoal, 10 percent sulfur, and 75 percent saltpeter. These should ideally be ground together to a fine powder so that they are intimately mixed (a very dangerous procedure). When burned, gunpowder produces large amounts of gas, including carbon dioxide and sulfur dioxide.

Roughly speaking, a teaspoonful of gunpowder will burn to produce about 5 liters of hot gas in perhaps one hundredth of a second. If the gunpowder is contained this will create an enormous pressure surge, which may be used to cause an explosion, drive a bullet from a gun or a ball from a cannon, or push a rocket into the air.

Old-fashioned black powder produces lots of smoke, and because it burns rather than detonating is classified as a low explosive. However, it is still used in fireworks, where its behavior can be tailored to specific requirements, such as directed bursts, time delays, and

Berthold Schwarz

Commemorated by a statue in Freiburg, Germany, Schwarz is credited with inventing gunpowder and the gun in 1353. He may have been a Franciscan monk and an alchemist, but some authorities say he never existed and was merely a legendary character.

Charcoal 15 %   Saltpeter 75 %
Sulfur 10 %

The components of gunpowder
Effective gunpowder contains 75 percent saltpeter (potassium nitrate), to provide the oxygen for the combustion; 10 percent sulfur; and 15 percent carbon. Sodium nitrate is a cheaper alternative to potassium nitrate but it tends to get damp.

TIMELINE

- **965** Abu Ali al-Hasan ibn al-Haytham (the name is Latinized as Alhazen or Alhacen) is born in Basra, then part of Buyid Persia, now in Iraq.

- **1011** Goes to Egypt to dam the Nile, having proposed this to Caliph Al-Hakim bi-Amr Allah in ignorance of the river's size. Realizing he cannot dam the Nile and fearing summary execution for the failure, he pretends to go insane.

- **1011–21** Placed under house arrest by the Caliph, he uses the time to continue his researches and writes the *Book of Optics* during his ten years of captivity.

- **1021** Caliph Al-Hakim bi-Amr Allah dies.

- **1021** Alhazen is thought to have traveled to Spain in this year to further his knowledge of science with the Spanish Moors.

- **1025–28** At some time in this period Alhazen publishes *Doubts Concerning Ptolemy* or *Aporias Against Ptolemy*, in which he identifies flaws and contradictions in Ptolemy's writings on science, astronomy, and mathematics.

- **1027** In order to devote himself to his scientific researches, Alhazen resigns from his position as a civil servant.

until the light reaches the retina. Alhazen experimented with bulls' eyes and showed that the light enters a small hole at the front, the pupil, and is focused by a lens onto the sensitive surface, the retina, at the back of the eye.

Investigations into light

Alhazen went on to experiment with mirrors and lenses, but perhaps the most fundamental thing he did was to show that light travels in straight lines. He probably inferred

Diagram of the human eye
This diagram from Alhazen's *Book of Optics* is remarkably accurate and shows the cornea, aqueous humor, vitreous humor, and optic nerve as they are called today.

this from the fact that solid objects cast sharp shadows in sunlight, but he proved it with the aid of the *camera obscura*, or "dark room."

Arranging for three lamps to be burning outside, he showed that three images appeared on the opposite wall inside. When he intercepted one of the paths with his hand, the corresponding image disappeared. He could "catch" the image anywhere on a straight line between the hole and the wall, and therefore the light that formed the image must be traveling along that straight line. He wrote, "The entering light will be clearly observable in the dust which fills the air." Further, he made a portable *camera obscura* in the form of a tent, and took it around on his travels. No doubt the moving images must have astonished all those who had the chance to see them.

Alhazen's groundbreaking *Book of Optics* remained the definitive book on the subject until the time of Isaac Newton, 650 years later. He also wrote

extensively on other subjects, ranging from astronomy to psychology. He tried to explain why the Sun and Moon look bigger when they are near the horizon than when they are high in the sky. He showed that twilight ends when the Sun is 19 degrees below the horizon, and estimated the height of Earth's atmosphere as 34 miles (55 km); in fact, three-quarters of the atmosphere lies within 7 miles (11 km). He studied the attraction between masses, and the acceleration of masses due to gravity. He even suggested that a body will stay at rest or move at a constant speed unless acted on by an external force; 650 years later this came to be called Newton's first law of motion (see pp.104–105). In addition to

The setting Sun
Alhazen suggested that the Sun and Moon appear to be bigger when they are near the horizon because when they are high in the sky there is no foreground with which to compare them.

these achievements, Alhazen was a pioneer of the scientific method, the basis of scientific research today.

The scientific method

The method or process begins with an observation, which is formulated into a statement of the problem to be solved. A suggested solution follows, called the working hypothesis. This is tested by experiment, ideally predicting the result expected. If successful, it can be refined into a theory; if not, another working hypothesis may be required and further experiments carried out to test it. When the problem is resolved, the work is written up and published.

Alhazen died in Cairo in 1039. His many and varied achievements in science are commemorated by a crater on the Moon, an asteroid, and by his portrait on Iraqi banknotes.

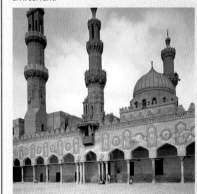

AL-AZHAR MOSQUE AND UNIVERSITY, CAIRO

- **1037** Almost certainly returns to Egypt, attending the Al-Azhar University and the Dar Al-Hekma (House of Knowledge) in Cairo. The latter's library is reputedly second only to that of Baghdad.

- **1038** Publishes *The Model of the Motions of Each of the Seven Planets*, the first model of the Solar System since Ptolemy to be based on radical ideas. Alhazen supports earlier Greek and Indian suggestions that Earth spins on an axis through its poles.

ALHAZEN ON IRAQI 10,000-DINAR BANKNOTE

- **1039** Dies in Cairo at 74 years of age. In the course of his life he wrote more than 200 manuscripts, of which the modern world has knowledge of nearly 100 scientific works. About 50 of these are about mathematics, 23 on astronomy, 14 on optics, and others on medicine. Numerous further works are lost.

BREAKTHROUGH

SIMPLE LENSES

Alhazen investigated the optical properties of glass, and discovered that the ratio between the angles of incidence and refraction varies; the actual relationship was discovered by Snell 600 years later (see p.116–17). Alhazen studied mirrors and simple lenses, investigating their magnifying power. When used to view a close object, a convex lens produces a magnified real image; a concave lens produces a diminished virtual image.

Convex lens
A convex lens is thicker in the middle than at the outside edge, and has the effect of converging incoming parallel rays of light and bringing them to a point of focus, where a real image is formed.

Concave lens
A concave lens is thinner in the middle than at the edge, and has the effect of diverging incoming parallel rays of light. A virtual image is formed at the point from which the rays appear to diverge.

« BEFORE

Our view of the material world is thousands of years old, but the modern Western scientific method has its origins in ancient Greece.

EARLY INDIAN SCIENCES

Scientific thought may have arisen as a necessity in India, where **medicine, metallurgy**, building, and crafts demanded an increasingly sophisticated understanding of the **physical processes** involved.

EARLY EASTERN SCIENCE

Scholars in China, Japan, Korea, and Vietnam in the 13th and 14th centuries BCE explained nature in terms of **numbers, yin** and **yang**, the five **elements**, and other intellectual constructs.

HELLENISTIC GREEK SCIENCE

Divisions of knowledge into subjects such as **mathematics and philosophy** were less distinct around 600 BCE. Thinkers such as **Thales «** 22–23 and **Pythagoras «** 32–33 began to observe and experiment, laying the basis of the modern **scientific method**. Much of what they discovered in the way of mathematics was included in **Euclid**'s *Elements*, written in c.300 BCE. At the same time, the Greeks drew ideas from other peoples, including the **Mesopotamians, Babylonians**, and **Egyptians**.

« SEE ALSO
PP.38–39 ANCIENT IDEAS OF THE WORLD
PP.52–53 ALCHEMY

East Meets West

After the collapse of the Roman Empire, science disappeared in the West. More alert to the achievements of the Greeks was the Islamic world, where many works were translated into Arabic, stimulating original thought in myriad areas. When scholars began to translate Islamic works into Latin after 1100 CE, Europe was transformed.

Although Hellenistic Greece was the scene of an extraordinary flowering of philosophy and science between 600 and 300 BCE, that event did not occur in a vacuum—astronomy and mathematics were being studied in India, Persia, and among the Sabaeans in eastern Syria. After the fall of the Roman Empire in the 5th century CE, Western science all but disappeared, but in the East scholars, particularly the Nestorians of Syria, had established centers of Greek learning, from where Greek ideas were disseminated in Latin, Greek, Syriac, and Persian. At the same time, ideas were migrating westward from China, India, and Persia, creating fertile conditions for debate in centers such as Byzantium.

Between 622 and 750 CE Arab tribes, united under Islam, gained control of Syria, Persia, Egypt, Spain, and parts of Africa. There developed a renewed fascination with learning, and works of Greek philosophy, mathematics, and science were rapidly translated into Arabic; these were to become the basis of the Arabic sciences.

An important new development was the Arab emphasis on using experiment, the empirical test, to try to establish scientific truth. In the 10th century there began an era of original research that lasted until the 13th century, producing leaps of knowledge in astronomy,

An Islamic view of the medieval world
This map was created by the Muslim geographer Al-Idrisi in the mid-12th century. South is at the top of the map, and the Islamic empire is positioned in the center between the East and West. Modern-day Mongolia lies at the bottom left.

△ **Widely influential book**
In the 12th century the Italian scholar Gerard of Cremona translated into Latin (see above) the work *Book of Medicine for Mansur* by Abu al-Razi, known in the West as Rhazes or Rasis. Written in 903 CE for Mansur, the Persian prince, and divided into nine chapters, the work was highly influential in both the Islamic world and the West.

◁ *The Canon of Medicine*
Printed in Rome in 1593, *The Canon of Medicine* was one of the first Arabic works to be published in Europe. It was written as the *Qanun* by the great scholar Avicenna (980–1037) (see pp.52–53). *The Canon* was translated into Latin by Gerard of Cremona in the 12th century.

LEONARDO FIBONACCI

The son of an Italian trader based in the port of Bugia, Algeria, Pisa-born Fibonacci learned the Hindu–Arabic numeral system while helping his father. He went on to study with Arab mathematicians. In 1202 he published *Liber Abaci*, in which he explained how using the digits 0–9 and place value (how the place of each digit has a value 10 times the place to its right) could benefit commercial activities such as bookkeeping. His book was hugely influential in Europe. Later works included *Practica Geometriae*, in which he addressed Arab geometry and trigonometry.

space were described in mathematical terms by Nicole Oresme, who based his work above all on Platonic concepts.

The Renaissance

During the European Renaissance, from the 14th to the 17th century, thinkers responded to Greek and Islamic ideas with new developments. Among them were the cubic and quartic equations of Tartaglia, Cardano, and Ferrari; Stevin's decimal fractions; Viète's algebraic formulas; advances by Copernicus (see pp.74–75) and Galileo (see pp.82–83) in astronomy; and gains in physics, chemistry, and medicine.

▷ **Nicole Oresme and King Charles V**
A 14th-century vellum manuscript shows the philosopher, bishop, polymath, and popularizer of science, Nicole Oresme, presenting his work to King Charles V of France. Beneath the two illustrations is the text of "Human Happiness" from Aristotle's *Ethics*.

AFTER

mathematics, and medicine. Arab alchemy was doomed in its bid to transform base metals to gold but its intense researches yielded empirical physical, chemical, and biological information. Important, too, were the instruments and apparatus introduced by the Arabs for scientific experiments, and the notion of the laboratory as a place dedicated to such research.

Translation into Latin

In Europe, scholars such as Boëthius and Cassiodorus in the 6th century had tried to preserve Greek learning, although many of their books and manuscripts were lost. From the 11th century, however, Arabic scientific works began to be translated into Latin by scholars such as Gerard of Cremona and Adelard of Bath. The ideas they contained, both early Greek and Arab, had a profound effect on European thought.

Direct contact with Arabic culture was also bringing rewards in the West. In the 13th century Fibonacci, also known as the Leonardo of Pisa (see above), founded his very influential approach

to numbers and mathematics on his tuition as a youth in Algeria; his work included the Fibonacci series for calculating the offspring of two rabbits: 0, 1, 1, 2, 3, 5, 8, 13, 21, and so on, each figure being the sum of the preceding two figures. Fibonacci drew on not only Euclid (see pp.34–35) and Arab ideas but also Diophantus of Alexandria, who is credited as the "father of algebra."

In the field of astronomy, the recovery of Aristotle's beliefs about the Universe challenged medieval thinking that had become mired in superstition. The Arabs, too, had been preoccupied by astronomy and astrology; in the 1080s, observers in Toledo and Córdoba, Spain, had compiled tables of the stars.

The infinity of space

In the medieval period, God and the heavens were seen as inextricable, yet Arabic texts discuss an infinity of empty space. In the 14th century Bradwardine of Oxford used mathematics to attempt to prove God's presence in the infinity of space, and theories of motion within

In many branches of science, proposals first made in the ancient world, East and West, had to wait centuries before full development.

CALCULUS

In the 5th century CE **Zu Chongzhi** used calculus to find the volume of a sphere; 1,000 years later, Bonaventura Cavalieri formulated what is known as **Cavalieri's principle** to achieve exactly the same end. **Pierre de Fermat** applied calculus 102–103 ❯❯ to curved lines. **Isaac Newton** subsequently applied calculus to **physics** in general, while his contemporary **Gottfried Leibniz** devised much of the **notation** used in calculus.

COPERNICUS AND GALILEO

For **scientific understanding** to advance, it was sometimes critical for **religious misconceptions** to be overturned. **Copernicus** observed the stars and concluded that Earth **revolves** around the **Sun** 74–75 ❯❯, while **rotating** about its axis. **Galileo** 82–83 ❯❯ agreed after observing moons crossing the face of **Jupiter**. Both men were thought

heretical, yet the notion of such a **heliocentric** solar system was first presented 1800 years before Galileo by the Greek astronomer **Aristarchus of Samos**, in c.250 BCE.

HELIOCENTRIC UNIVERSE

SEE ALSO ❯❯

pp.68–69 BIRTH OF EXPERIMENTAL SCIENCE

2

RENAISSANCE AND ENLIGHTENMENT

1500 – 1700

From 1500, the reemergence of classical and Islamic scientific ideas in Europe's Renaissance galvanized scientific thinking. Brilliant minds such as Galileo and Newton learned about the world by observation and experiment, and thinkers such as Descartes ushered in a new age of Enlightenment governed by the power of human reason.

RENAISSANCE AND ENLIGHTENMENT 1500–1700

1500

1550

c.1503
Leonardo da Vinci creates a design for a two-wheeled hoist with a caged gear.

1525
Galen's works on human anatomy published in the original Greek.

1526
Paracelsus pioneers the use of minerals and poisons in medicine.

Paracelsus ▶

1530
Father of botany, Otto Brunfels publishes the first part of *Herbarum Vivae Eicones* with realistic illustrations by Hans Weiditz.

1535
Berengario da Carpi produces *Anatomia Carpi*, the first anatomical text with illustrations based on dissections of human bodies.

1537
Ambroise Paré revives the practice of ligature for gunshot wounds.

1541
Giambattista Canano publishes illustrations of each muscle and its relation with the bones.

1542
Leonhart Fuchs publishes *Historia Stirpium*, naming about 500 plant species.

1543
Andreas Vesalius publishes *De Humani Corporis Fabrica*, with illustrations made at dissections. Copernicus asserts that Earth revolves around the Sun.

▲ Illustration from *De Humani Corporis Fabrica*

1546
Girolamo Fracastoro proposes that diseases are contagious and are spread by seedlike entities. Georgius Agricola publishes *On the Nature of Fossils*.

▼ Copernicus's Universe

1551
Leonard Digges invents the theodolite, used in surveying to measure angles.

1553
Miguel Seveto states that blood travels from one side of the heart to the other through the lungs.

1557
The equals sign = introduced in mathematics.

1559
Matteo Realdo Colombo states that the pulmonary vein carries blood and not air as previously believed.

c.1562
Gabriel Fallopio describes the ovaries, the uterus, and the tubes connecting them.

1564
Julius Caesar Aranzi discovers that the blood of a mother and fetus do not mix.

▼ The heart and the pulmonary artery

1572
Tycho reports sighting a new star, revolutionizing the idea that the heavens never change.

▼ Tycho's sextant

1577
Geronimo Mercuriali suggests that disease is spread by flies.

1579
Fabricius gives the first accurate description of the valves in the veins.

1583
Galileo discovers that the oscillations of a swinging pendulum of fixed length take the same amount of time regardless of their amplitude. Felix Platter proposes that the retina is the part of the eye stimulated by light.

1586
Simon Stevin shows that water pressure increases with depth.

1590
Hans and Zacharias Janssen make the first compound microscope.

Robert Hooke's ▶ compound microscope

1593
The Canon of Medicine is one of the first Arabic works published in Europe.

▼ *The Canon of Medicine*

The Renaissance placed humanity at the center of things; in Italy, where it began, anatomists such as Vesalius made the first systematic studies of the human body. As Renaissance ideas spread across Europe, the emerging belief in the power of human reason gave new impetus to mathematics and its application to science. And as the conviction grew that the best way to learn about the world is to study it directly, the invention of the telescope and microscope helped reveal things too distant or too tiny to see with the naked eye. The telescope enabled Galileo to confirm Copernicus's insight that Earth is not the center of the Universe, while Leeuwenhoek and Hooke were startled by the teeming life under their microscopes.

1600

1650

1600
William Gilbert publishes *On the Magnet, Magnetic Bodies and the Great Magnet of the Earth*, proving that planet Earth itself is a huge magnet.

William Gilbert ▶

1605
Michal Sedziwój proposes the existence of a "food of life" in the air.

1609
Galileo makes his first telescope. Johannes Kepler publishes his first two laws of planetary motion. (The third one is published in 1619.)

Galileo's telescope ▶

1610
Galileo discovers the mountains on the Moon, that the Milky Way consists of innumerable stars, Jupiter's four largest moons, the phases of Venus, and sunspots.

▼ Galileo's drawings of the Moon

1614
Logarithms devised and published by John Napier.

1621
The Doctrine of Signatures advocates the medicinal use of plants that resemble the human body.

▼ Napier's bones; an early calculating tool

1622
William Oughtred invents the slide rule.

1628
William Harvey proposes that blood is pumped around the body by the heart.

1630
Pierre de Fermat and Blaise Pascal develop a probability theory.

1631
Vernier scale invented. Pierre Gassendi publishes the first data on the transit of Mercury.

c.1632
Galileo proposes the concept of inertia.

1637
René Descartes states that the heart is just a mechanical pump.

1643
Evangelista Torricelli creates a vacuum.

▼ Evangelista Torricelli

1648
Johan Baptista van Helmont measures the water taken up by a tree.

1649
Pierre Gassendi states that the properties of atoms depend on their shape and that they may be joined together to form molecules.

1651
William Harvey states that young animals develop from eggs.

1653
Blaise Pascal publishes *On the Equilibrium of Liquids*, setting out Pascal's law of pressure, the basis of all hydraulics.

1654
Otto von Guericke gives a public demonstration of the vacuum.

1657
Christiaan Huygens builds the first mechanical clock with a pendulum.

1658
Jan Swammerdam observes red blood cells under a microscope.

1660
Robert Boyle discovers that air is necessary to carry sound, feed a flame, and sustain life.

1661
Richard Towneley and Henry Power note air pressure decreases with altitude. Marcello Malpighi discovers capillaries. Robert Boyle publishes *The Sceptical Chymist*, trying to bring scientific ideas to alchemy.

c.1665
Calculus developed by Isaac Newton and Gottfried Leibniz. Robert Hooke publishes *Micrographia*, and uses the term "cell" to describe structures seen in a slice of cork.

▲ Hooke's air pump

▼ Illustration from Hooke's *Micrographia*

1669
Nicolas Steno proposes that fossils are the remains of extinct animals. Hennig Brand makes phosphorus, isolating a new element.

1671
Newton builds the first reflecting telescope.

1675
Newton publishes *Hypothesis of Light*. Marcello Malpighi publishes *Anatome Plantarum*.

Newton's telescope ▶

1676
Olaus Roemer calculates the speed of light. Robert Hooke publishes Hooke's law. Antonie van Leeuwenhoek observes single-celled organisms under a microscope.

Spring demonstrating ▶ Hooke's Law

1681
In *Sacred Theory of the Earth*, Thomas Burnet suggests that Earth contains subterranean voids full of water.

1684
Isaac Newton writes a paper about the orbits of comets.

1690
Christiaan Huygens publishes *Treatise on Light*. Paul Hermann divides seed plants into angiosperms and gymnosperms.

1691
Jacob Bernoulli introduces polar coordinates.

▲ Isaac Newton

1687
Isaac Newton's *Principia* published, stating Newton's laws of motion and the law of universal gravitation.

« **BEFORE**

Ancient Greek and early Islamic scientists and philosophers asked many questions about the nature of the Universe.

ANCIENT GREEKS

Philosophers in ancient Greece believed that knowledge would be found through argument and reason. The idea of **performing experiments** rarely occurred to them.

PLATO AND ARISTOTLE

Plato argued that if enough men of sufficient intelligence thrashed out an idea for long enough, they would reach the "truth" about any subject. **Aristotle**, a student of Plato's, allowed that science could be studied but denied any place for experimentation.

ISLAMIC SCIENTISTS

For about 800 years after the great Greek philosophers, scientific study was led by the Arabs, notably **Alhazen**, **Al-Jahiz**, and **Al-Farabi**. While there was an emphasis on **investigation**, the discipline was still hampered by mysticism.

« SEE ALSO
pp.36–37 ARISTOTLE
pp.52–53 ALCHEMY

Until the mid-15th century, the ideas that governed European learning were a mixture of beliefs handed down from ancient Greece and the teachings of the Catholic Church. Then, with the start of the Renaissance, scholars began to break away from traditional methods, proclaiming their belief in reason as the measure of all things.

The Scientific Revolution

Scientists explored the world around them in a systematic way: careful observations led them to construct hypotheses, which they then tested through experiment. The first great thinker of the modern era to use this scientific method was Galileo Galilei (see pp.82–83). Among his discoveries were the principle of the pendulum (see p.83) and the uniform acceleration of falling bodies. To investigate falling, Galileo did not drop objects directly, since their motion is too fast to follow. Instead, he carried out series of experiments rolling a bronze ball down a wooden slope, thus slowing its rate of fall to measurable speeds. The popular stories of him dropping objects from the tower of Pisa are untrue, although similar experiments had been carried out by Flemish scientist Simon Stevin,

Novum Organum **(1620)**
The title page of Francis Bacon's book shows people sailing through the Pillars of Hercules—the limits of knowledge—into the future.

who in 1586 dropped lead balls from the church tower in Delft).

Galileo also refined the telescope, which helped him gather evidence to show that Earth was not the center of the Universe, but moved around the the Sun. This clashed with the accepted wisdom of theologians. He was pressed by the Church to retract, standing trial in 1633 for heresy. Found guilty, he was forced to issue a public denial.

Galileo was part of a new wave of scientists using experimentation to make sense of the world around them. Englishman William Gilbert, one of Queen Elizabeth I's physicians, studied the effects of magnetism (see pp.80–81). He determined that the ordering of the heavens was not mystical: the magnetic nature of "attraction" held the planets on their course. William Harvey developed ingeniously simple experiments to study the circulation of the blood (see pp.90–91). His ideas were considered so heretical that he had his work published by an obscure press in Germany.

A new optimism

Despite the opposition of the Church, scientific thinking continued to flourish and men developed a new optimism about what could be achieved by it.

The French mathematician and philosopher René Descartes (see pp. 102–03) was an admirer of Galileo. He claimed that he had learned nothing in school that wasn't subject to doubt, and stated that knowledge had to be "clear and distinct." When he came to write his *Regulae* (Rules) in 1628, and then the *Discours de la Methode* (Discourse on Method) in 1637, his purpose was to

Birth of Experimental Science

Science as we know it today is no more than 500 years old. It is firmly based on certain rules of procedure that scientists must follow to obtain accurate knowledge. These rules were formulated during a revolution in scientific thinking—the birth of experimental science—and are upheld today by the world's scientific societies.

▽ **Recording observations**
Good science needs precise recording of information. These notes by the French mathematician and physicist, André-Marie Ampère, concern the magnetic properties of electrical current.

▷ **Testing theories**
Scientists must design experiments that are open—not to prove a preconceived theory but if possible to disprove it. Objectivity and repeatability are vital.

set down a method not simply for solving mathematical or physics problems, but for reaching the truth about everything. He started with the first principle, "I think, therefore I am," and believed he could use this principle to construct a new body of knowledge.

A contemporary of Descartes, the English philosopher Francis Bacon (see opposite), was equally contemptuous of false knowledge, or what he called "idols," based only on a few untested observations. Like Descartes, Bacon demanded a new standard of precision. He proposed a scientific method that required the patient accumulation of data in three steps. The scientist makes his observations, then formulates a theory that might explain them, and finally he tests his theory by rigorous experiment.

System of logic

In 1620 Bacon published *Novum Organum*, or "New Instrument," in which he set out his belief in

AFTER »

Scientists still rely on the scientific method as a regular and generally accepted procedure for conducting research.

DESIGNING EXPERIMENTS
Scientists must first form a **hypothesis**, and then look for evidence to support it. The method requires **careful recording of data** and other information. **Experiments** must be performed under **controlled conditions**, and must be repeatable by others.

PEER REVIEWS
Scientists often hold conferences with their peers to **report on their findings**. The results are passed on for analysis. Only when the findings have been **rigorously tested** will they be **published**, usually in a scientific journal.

SEE ALSO »
pp.82–83 GALILEO GALILEI
pp.90–91 CIRCULATION OF THE BLOOD

> **"Many** will pass through and **knowledge** shall be **increased**."
>
> FRANCIS BACON FROM "NOVUM ORGANUM," 1620

was restored to the monarchy in 1660 they became the Royal Society of London for the Improvement of Natural Knowledge, generally abbreviated to Royal Society. Robert Hooke was appointed the first curator of experiments, which meant that he was expected to give practical demonstrations at every meeting. The English example was swiftly copied on the continent. In 1666 Louis XIV accepted the founding of the French Royal Academy of Sciences, and by 1700 similar organizations were established in Naples and Berlin. In turn they spawned many societies that are still in existence today.

These scientific societies provided not only a focus and a meeting place for discussion, by which scientific ideas thrive, but also a conduit for information. When Volta invented the electric battery in 1799, he wrote from Italy to Sir Joseph Banks, the President of the Royal Society in London, and told him about it. Banks immediately wrote to all those who he thought would be interested, and within weeks batteries were being made all over Britain, with the result that studies in electricity gathered rapid momentum. Even the Napoleonic wars were no barrier to such dissemination of information through the Royal Society.

---

this scientific method. His work had a direct influence on scientists Robert Boyle, Isaac Newton, and Robert Hooke.
One of the greatest legacies of Bacon's philosophy, which he stressed in his *Novum Organum*, was the concept of establishing an organization within which scientific methods and research

could be shared and explored. By the end of the 17th century this had become a reality.

Scientific societies
The first scientific society that published any proceedings was the *Accademia dei Lincei* (Academy

Birth of scientific societies
The first scientific societies were founded in the 17th century. This engraving depicts King Louis XIV of France visiting the Academy of Sciences in Paris.

of Lynxes), so called because of the members' sharp observational skills. Founded by Duke Frederico Cesi in 1603 in Rome, Italy, it welcomed Galileo into its ranks in 1610 following his astronomical discoveries, but then got into trouble with the Church and closed after the Duke died. The *Accademia degl' Investiganti* (Academy of Investigators) was founded in Naples in 1650, and the *Accademia del Cimento* (Academy of Experiment) in Florence in 1657, although it ended 10 years later.
Also during the 1650s a group of "natural philosophers" had begun meeting in England, sometimes at Gresham College in London, and later in the rooms of Robert Boyle in Oxford. They called themselves the "Invisible College," but when Charles II

PHILOSOPHER AND POLITICIAN (1561–1626)

FRANCIS BACON

Although not a scientist, Bacon established the principles of the Scientific (or Inductive) Method, based on observation and experiment, which has been a basic tool of scientists ever since. Bacon rose to be Lord Chancellor under James I, but in 1621 he was convicted of bribery and driven out of public life. He then devoted himself to writing. He died after catching a chill while stuffing a chicken carcass with snow, to see if the meat would be preserved.

Renaissance Medicine and Surgery

The 14th century saw a flowering of the arts and sciences around the world, and a new intellectual freedom that produced an upwelling and dissemination of new ideas as never before. Medicine at this time grew out of a fusion of ideas from ancient Greece and the Middle East. However, a combination of science and supernatural belief meant that trained physicians worked alongside practitioners of folk medicine.

« **BEFORE**

Previous knowledge passed on from the ancient world had a huge impact on the development of medicine and surgery.

DISEASE AND HYGIENE
The 13th-century Italian surgeon **Gugliemo da Saliceto** recognized that **good hygiene** was vital to health, and suggested that pus formation was bad for wounds (indicating an infection).

PHARMACOPEIAS
Some of the first books on the preparation of medicines, called **pharmacopeias**, were written in antiquity and added to by physicians and alchemists of the Middle Ages. In ancient Greece, **Dioscorides** wrote a text that **influenced pharmacological thinking** for 1,600 years.

« SEE ALSO
pp.24–25 EARLY MEDICINE AND SURGERY

From the 14th century Europe saw great advances in understanding the human body and the scientific basis of medicine — including surgery, anesthesia, and pharmacy. The new universities at Paris, Bologna, and Padua became key centers of learning and trained some of the finest physicians of the time.

Early textbooks
This 1512 book included an illustration of phrenology or craniology — the study of the shape of the skull.

Health and disease
Disease is a disorder of the body, and in the 17th century a new science of disease, pathology, was becoming a discipline in its own right.

The Christian world still tended to see disease as God's punishment, but scientists were starting to seek natural causes. Prompted by the devastation of plague and the experiences of foreign travel, they began developing theories to explain the spread of contagious diseases. Geronimo Mercuriali, an Italian physician, suggested that disease was spread by flies, and Gabriel Soares de Souza, a Portuguese explorer in Brazil, proposed that flies transmitted "poison" by sucking it from the sores of their victims. The first specialist of tropical medicine was the Jewish physician Garcia de Orta. He left Portugal, perhaps fleeing the Inquisition, to settle and practice in India, where he studied the effects of cholera. Increasingly, biological factors were being blamed for disease, even

though the details were misguided. In 1600 mass hysteria in Italy brought on by fear of spider bites spawned a "therapeutic" dance, called the Tarantella.

Balance and lifestyle
The ancient Greeks had proposed that good health relied upon maintaining a balance between the four body fluids, or humors: blood, black bile, yellow bile, and phlegm. The idea that disease resulted from an imbalance of body constituents was central to later medicine, too. It was championed by Paracelsus, a Swiss physician–alchemist who considered that the imbalance in the body associated with disease was caused by outside influence—"poison." For Paracelsus, disease could be treated by finding another external substance that would redress the imbalance caused by the poison, and so bring about a cure. He believed that there was a remedy for every disease in the minerals and herbs

AFTER ≫

◁ Surgery in progress
This 1563 woodcut depicting a surgical operation on a man's head is from one of the original works of Paracelsus (see below).

"The dose makes the **poison**."
ATTRIBUTED TO PARACELSUS,
PHYSICIAN–ALCHEMIST

of the natural environment. Pharmacopeias—catalogs of medicines—had existed in ancient China and Mesopotamia. However, Paracelsus was perhaps one of the first scholars to put forward a theory that explained the role of drugs in curing disease. More importantly, Paracelsus recognized the significance of dose. In fact, he articulated something that is still a principle of pharmacology today: "All things are poison and nothing is without poison, only the dose permits something not to be poisonous." Even something as seemingly innocuous as water can harm if taken in sufficiently large quantities. Substances such as arsenic are simply more toxic because they are lethal at much lower doses.

Many physicians thought they knew the appropriate cures for various diseases. However,

in spite of the increasing influence of experimental medicine, some ideas were less than scientific. The Doctrine of Signatures, which probably originated in ancient Greece and was popularized by the German mystic Jakob Böhme, became especially influential. Böhme suggested that cures for disorders could be found by looking for the Creator's "signature" in the

natural world. According to the Doctrine, plants that resembled the human body, or parts of it, harbored cures for disease—such as mandrake roots, thought to resemble human limbs.

While many of these teachings were at best ineffectual and at worst harmful, the herbalist origins of pharmacology

Doctrine of Signatures
The Doctrine advocated the medicinal use of plants with a similarity to the human body. These illustrations depict a plant with roots that mimic the shape of a hand (left) and flowers resembling the eye (below).

▽ Tongue depressor
By the 17th century medical instruments such as this one were routinely used by physicians as part of their clinical examinations.

did produce some things of genuine worth. For example, the painkilling properties of willow bark eventually led to the production of aspirin, while digitalins extracted from foxglove would later be used to treat heart problems.

Advances in surgery
Surgery would always be risky as long as there was pain, blood loss, and infection. Nevertheless, the Italian surgeon Theodoric Borgognoni refined a mixture of opiate-based substances that formed the basis of anesthesia (see pp.228–29) until the 19th century. In the mid 16th century, the French surgeon Ambroise Paré improved techniques of ligature (tying limbs or blood vessels to reduce blood loss) during amputations, which had been developed in the 10th century by Al-Zahrawi, a physician in Spain. Paré also revived an ancient Roman potion of turpentine, egg yolk, and oil extracts from roses for healing wounds, which proved more effective than the prevailing treatment of pouring boiling oil on the injury. Despite these efforts, many patients died after surgery as a result of a factor that was to evade even the finest minds of the time: infection.

Ambroise Paré treating a wounded soldier
Paré, a French pioneer of surgical techniques, became skilled in treating firearm wounds suffered by soldiers in battle. His techniques used ligatures to staunch blood flow and poultices to heal wounds.

The modern study of biochemistry grew out of European medicine between the 14th and 17th centuries. Scientists began to understand the chemical effects of drugs on the body with greater precision.

SAFER DRUGS
In the 17th century **Sylvius**, a medical professor, suggested that all processes of life and disease were based on **chemical actions**, ushering a new age of biochemical study. Most **traditional therapies**, such as using mercury to treat syphilis, had been developed by **trial and error**, without knowing how or why they worked. In time, **modern pharmacology** would replace these remedies with safer, more effective treatments.

OCCUPATIONAL DISEASES
When **Paracelsus** wrote a treatise on the illnesses of metalworkers, he sparked an interest in **occupational diseases**. With the rise of anatomical pathology, physicians began to see links between **some diseases and particular occupations**, such as the high levels of lung disease in miners.

SEE ALSO ≫
pp.228–29 SAFER SURGERY
pp.312–13 THE DEVELOPMENT OF MEDICINES
pp.406–07 MODERN SURGICAL PROCEDURES

METALWORKERS

ALCHEMIST (1493–1541)
PARACELSUS

Paracelsus was born in Switzerland but traveled widely in Europe, Africa, and the Middle East. His many interests included alchemy, astrology, medicine, psychology, and toxicology. A prolific writer, Paracelsus had a reputation for being arrogant. Born Philipp von Hohenheim, he adopted the name Paracelsus later in his life: it means "equal to or greater than Celsus" (a Roman scholar of medicine). In 1526 he recorded his discovery of zinc.

The **Human Body Revealed**

Rejecting ideas of old, European scientists of the 14th–16th centuries explored the body as geographers explored the world, dissecting the human form with a high degree of expertise to discover complex anatomical systems.

Anatomists have been dissecting the body to learn about its structure since the days of ancient Greece and Egypt.

Anatomy by direct observation

The world's first public dissection took place around 1315: Mondino de Liuzzi, an Italian Professor of Medicine, used the body of an executed criminal. European laws concerning human dissection at this time varied and scholars relied heavily on the availability of executed criminals' bodies. Liuzzi's purpose was to discover, to educate, and

Understanding the human body
This Italian anatomical wax figure was produced in the 18th century, when scientists had an accurate understanding of the position of most internal organs. Here, the small intestine has been removed to show the rectum.

even to entertain. His publication, *Anothomia,* set out a basic body plan, but it supported old Greek errors too, especially those of the physician Galen 129–200 CE (see pp.24–25), the established authority of anatomy at that time.

Denying the wisdom of antiquity was risky because the Catholic Church used to condemn challengers as traitors, but many remained undaunted. In 1535 the Italian physician Jacopo Berengario da Carpi produced *Anatomia Carpi,* the first anatomical text to be extensively illustrated and based on dissections

Right lung

Heart

Liver

Gall bladder

Stomach

Large intestine

Rectum

‹‹ BEFORE

Many early cultures regarded the human body as sacred and would only dissect animals. This eventually changed, first under Egyptian law and later under Christian law.

EARLY DISSECTIONS

The first anatomical text, written in the 6th century BCE by ancient Greek theorist **Alcmaeon**, is believed to have been based on **animal dissection**. Some 300 years later, pagan Rome forbade human dissection, which meant that Galen, the city's most famous physician, **proposed many mistaken theories** about human anatomy based on dissecting animals.

GALEN DISCREDITED

Human dissections were sanctioned in medieval Islam and **Avenzoar** was thought to be the first physician to perform **autopsies** in the 12th century CE. Many Arabic scholars doubted the

FOUR HUMORS

Greek idea that disease was caused by an imbalance in the four humors (body fluids) and **disagreed with Galen's anatomical theories**, such as the belief that blood was continually made in the body. However, they had little impact on Renaissance Europe where Galenic wisdom dominated.

‹‹ SEE ALSO
PP.24–25 EARLY MEDICINE AND SURGERY

Vesalius's muscle man
This illustration from the key work on human anatomy *De Humani Corporis Fabrica,* by the Flemish-born physician Andreas Vesalius, demonstrates his detailed understanding of the shapes of muscles.

of human bodies. Carpi's work undermined Galen's theories, including the description of a network of blood vessels below the brain, which were found not to be present in humans. In doing so Carpi was heralding a golden age of anatomy that was dominated by four men: Eustachi, Colombo, Fallopio, and Vesalius.

In the 16th century Italian anatomist Bartolomeo Eustachi made accurate diagrams of almost all the main organs, though his findings were not published until 1714. Italian surgeon Matteo Colombo studied blood circulation at the University of Padua and went on to become possibly the first Renaissance physiologist. Like others who followed him, he experimented on animals to answer his questions. Gabriele Fallopio, an anatomist from Italy, studied the bones, muscles, and reproductive organs, describing the tubes that link the ovaries to the uterus, which now bear his name.

However, the most famous anatomist of the 16th century was Flemish-born Andreas Vesalius (see below)—a pioneer for understanding anatomy in medicine. His superbly illustrated *De Humani Corporis Fabrica* from 1543 documented observations made from human dissections, and his work finally disproved many of Galen's theories.

Microscopic anatomy

By the 1600s anatomists saw that organs were made of different types of tissue, such as muscle, bone, and lining. The body was defined as an organized structure—tissues made up organs, and organs interacted in systems. Around this time a new type of anatomist—microanatomist—was routinely using the microscope to look at the body in even closer detail.

Many organ specialists of this period were physicians, eager to improve their diagnoses and treatments. One such was Marcello Malpighi, an Italian doctor, who scrutinized the skin and found it to be a multilayered organ— the Malpighian layer has dividing cells that generate the layers above it and bears the doctor's name today.

The science of pathology

Scientists also sought to find out more about diseased anatomy. In 1761, Italian anatomist Giovanni Morgagni published a treatise based on his observations of autopsies to study the causes of disease. It was this that established anatomical pathology as a science. With the rise of medical teaching and as executions became scarcer, demand for bodies exceeded supply. In 1832 the Anatomy Act was introduced in England, allowing licensed anatomists to dissect unclaimed and donated bodies. By the 1800s, when English surgeon Henry Gray wrote *Anatomy*, physicians such as Carl Ernst Bock, a German professor, routinely performed autopsies, and dissection was mainly used to discover cause of death.

Peripheral nerve in arm

Left lung

Diaphragm

Liver

Kidney

Fallopian tube

Ovary

Bladder

Anatomical illustrations
Vesalius's *De Humani Corporis Fabrica* was illustrated with a ground-breaking style by artists who were present at dissections.

> " I profess to **learn** and to **teach anatomy** not from books but **from dissections**."
>
> WILLIAM HARVEY, ENGLISH PHYSICIAN, 1628

From the 1700s, anatomists were looking at bodies in new ways, using new methods.

COMPARATIVE ANATOMY

Anatomical studies comparing species underscored **humans' place in the animal kingdom**. In 1698 British scientist **Edward Tyson** dissected a chimpanzee and found it was closer to humans than to monkeys. In the mid-18th century Dutch physician **Petrus Camper** showed that the orangutan was a different species and not another type of human.

ANATOMY FOR EXHIBITION

Anatomists developed techniques for preservation and **display of body parts**. Today, collections are used for research and teaching. New interest in anatomy came with the *Body Worlds* exhibits of German anatomist Gunther von Hagens in the 1990s.

HUMAN BRAIN SPECIMEN

IMAGING TECHNIQUES

Modern anatomy uses techniques to create images of the body without dissection. **X-rays** can reveal bone fractures and lung tumors, while magnetic resonance imaging (**MRI**) and **ultrasound** can show soft tissues.

SEE ALSO »
pp.294–95 THE DISCOVERY OF X-RAYS
pp.408–09 BODY IMAGING
pp.448–61 REFERENCE: BIOLOGY

ANATOMIST (1514–1564)

ANDREAS VESALIUS

Vesalius was born in Brussels, studied medicine in Paris, and later moved to Italy, becoming Chair of Surgery and Anatomy at the University of Padua. He dissected executed criminals to create anatomical drawings, correcting errors of earlier anatomists, including Galen. In 1544 he became court physician for Charles V, the Holy Roman Emperor. He died after being shipwrecked on the Greek island of Zakynthos.

IN PRACTICE

DISSECTION TECHNIQUES

When a thin section of an organ is taken, it reveals a great deal about the arrangement of the microscopic tissues that make up the organ. There are two techniques that are used for dissecting and examining an organ. A longitudinal section is sliced parallel to the long axis of the organ, whereas a transverse section is taken at right angles to it.

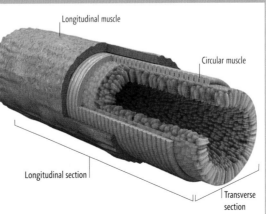

Longitudinal muscle

Circular muscle

Longitudinal section

Transverse section

PLANISPHÆRIVM
Sive
VNIVERSI TO:
EX HYPO:
COPERNI
PLANO

The Sun-Centered Universe

The ancients believed that the sun, moon, and stars orbited around the Earth, and—to the casual observer—this still feels true today. But in 1543 a Polish scholar named Nicolaus Copernicus suggested otherwise, placing the sun firmly at the center of the universe.

Almost 1,800 years before Copernicus, Aristarchus of Samos suggested that the Sun was the center of the Universe. But this view, together with his assertion that the Universe was therefore much larger than anyone had imagined, was ignored. The generally accepted view was that Earth lay at the center of the Universe, and this belief became crystallized in Ptolemy's great work *Almagest*, a 13-book treatise written in the 1st century CE. It was to remain the accepted model for more than a thousand years, until after the death of Copernicus in 1543.

Copernicus was a keen astronomer who came to the conclusion that it was unlikely that thousands of stars were circling rapidly around the earth every 24 hours. Apart from anything else, the distant ones would have to travel at impossible speeds. Surely it was much more probable that they were all stationary, while the Earth was spinning?

He then developed his own theory of the Universe, in which Earth and all the other planets orbit around the Sun. He described these planetary motions in his book, *De revolutionibus orbium coelestium* (*On the Revolutions of Heavenly Bodies*). Copernicus labored over his theory for more than 30 years but hesitated to publish it. He was dissatisfied with some mathematical details. He also feared criticism, as he explained in a preface dedicating the book to Pope Paul III. Saying that the Earth was moving would offend religious views as well as defy common sense. His work was finally published in 1543, the year he died.

Almost certainly unknown to Copernicus, a Lutheran scholar, Andreas Osiander, added an unsigned note to his work. He suggested that the theory was a mathematical device, not the literal truth. This did not reflect Copernicus's own views but helped prevent censure by the church. Almost a century passed before trouble with the church erupted, when Galileo confirmed the Earth's motion (see pp.82–83).

In fact the Sun is the center of our solar system, but not of our galaxy—the Milky Way (see pp.340–41)—nor the Universe. Nevertheless, Copernicus's theory was essentially correct, and it revolutionized the way people thought about the Universe.

The Copernican cosmos
This 1660 painting by Andreas Cellarius shows "The system of the entire created Universe according to Copernicus," including the six known planets and the moons of Jupiter, which were discovered in 1610.

> **"** Finally we shall place the **Sun** himself at the center of the **Universe**. **"**
>
> COPERNICUS, c.1540

« BEFORE

Modeling the paths, speeds, and brightness of planets had spurred astronomic endeavor for two millennia.

ANCIENT IDEAS ON PLANETS
Ancient Greek astronomers « 28–29 noticed that five bright "stars" were not "fixed" but **wandered about the ecliptic zodiac** of the sky. Two of these—**Mercury and Venus**— stayed relatively close to the Sun, while the other three—**Mars, Jupiter, and Saturn**—did not.

OLD ECLIPSES
Around 500 BCE Babylonian astronomers noted an **eclipse cycle known as the Saros**. They realized that 18 years, 11 days, and 8 hours after an eclipse of the Sun or Moon, these bodies, together with Earth, returned to similar geometries and **another eclipse** took place.

CHARTING THE STARS
One of the oldest maps of the stars is Chinese. **Su Song's** *Celestial Atlas* dates back to 1092 CE, and marks the position of **more than 1,350 stars**, including the correct position of the Pole Star.

« SEE ALSO
pp.26–29 THE FIRST ASTRONOMERS
pp.74–75 THE SUN-CENTERED UNIVERSE

During the course of the 16th century, the Copernican hypothesis that the Sun lies at the center of the planetary system became more widely accepted. If true, this also meant that Earth moved across its orbit every six months, and so the nearby stars should be seen to move against the background of more distant stars.

Astronomers wanted to detect this movement, so they began to make more accurate instruments. They were correct in their assumption, but unfortunately the huge distances involved—even the nearest bright stars are at least a quarter of a million times farther away from Earth than the Sun—meant that the movement was too small for their instruments to detect.

Mapping the skies
Tycho (1546–1601) was among Europe's earliest great observers. This Danish nobleman turned his back on a political career to focus on astronomy. The sight of the 1560 solar eclipse hooked his interest, and when astronomers failed to predict accurately the conjunction of Jupiter and Saturn three years later, he decided to devote his life to astronomy. In 1572 he reported a supernova (see right), and became famous.

Tycho observing the skies ▷
A 16th-century depiction of Tycho in his observatory on the Danish island of Hven. He sits beside his mural quadrant—a huge brass arc used for measuring celestial positions.

BREAKTHROUGH
TYCHO'S SUPERNOVA

Observing the constellation of Cassiopeia in mid-November 1572, Tycho noticed that a new star had appeared. Rivaling Venus in brightness, the new star remained visible until March 1574. Its appearance revolutionized astronomy by destroying the stale Aristotelian view that the heavens never changed. The exploding star is about 7,500 light-years away, and it is shown below as it appeared 400 years after its discovery.

Planetary Motion

The Ptolemaic system, which placed Earth at the center of the Universe, had held sway for more than 1,500 years, from the ancient Greek era to the 16th century. But then one astronomer accurately mapped the celestial sky and carefully charted planetary motion, laying foundations for the laws of planetary motion.

Tycho's sextant
This large floor-mounted wood and brass instrument used two sighting arms to accurately measure the distance between planets and stars that were closer than 60 degrees.

The Danish king, Frederick II, gave Tycho the small island of Hven and provided him with funding for his work. Tycho built two great observatories there—Uraniborg and Stjerneborg—equipping them with a series of extremely well-made and stable quadrants and sextants for measuring celestial positions. These were used with great care to minimize experimental errors.

Tycho showed that the comet of 1577 was farther away than the Moon, measured the length of the year to an accuracy of one second, and proved that the rate of precession (a change in direction of Earth's rotational axis) was not irregular. His greatest legacy was a long, accurate series of observations of the positions of the planets. These were consistently accurate to within a minute of arc (equal to $^1/_{60}$ of one degree).

Joining forces
Tycho's quarrelsome nature led him to leave Denmark for Bohemia, in what is now the Czech

> " Man makes a **great fuss** about **this planet**, which is only a **ball bearing** in the hub of the **Universe**."
>
> CHRISTOPHER MORLEY, AMERICAN AUTHOR, 1890–1957

JOHANNES KEPLER

The German astronomer Johannes Kepler was assistant to Tycho and Imperial Mathematician to Emperor Rudolf. His belief in the ancient philosophical concept of the music of the spheres, used to describe proportions of the solar system and sometimes credited to Pythagoras, led to Kepler's own ideas of ordered geometric spacing of planetary orbits, as explained in his *Mysterium Cosmographicum*. He also wrote three more very influential books: *Astronomia Nova*, *Epitome Astronomia Copernicanae*, and *De Harmonices Mundi*. He is best known for the three laws of planetary motion.

Planet moves faster when it is nearer to the Sun

The sun at one focus of the elliptical orbit

Both purple regions have the same area, with the planet crossing both in equal time

Orbiting planet

Planet moves more slowly when it is further from the sun

Elliptical planetary orbit

The first of Kepler's laws of planetary motion shows that the orbit of a planet forms an ellipse around the sun. The second law shows that a planet moves faster when closer to the sun and slower when further away.

Republic, in 1597. Johannes Kepler also moved to Prague in 1600, suffering from religious persecution, and became his assistant. Unfortunately Tycho died just one year later, but Kepler inherited his observations and used them to work out the true nature of the planetary orbits. He was a great mathematician and developed three laws of planetary motion still in use today.

Kepler's three laws

The first law states that every planet follows an elliptical orbit with the sun at a focus. The second law concerns the speed of a planet around its orbit and notes that an imaginary line between the planet and the sun would, in any given time, sweep out the same area (shown above).

Both these laws were published in *Astronomia Nova* (1609). His third law was in the last chapter of *De Harmonices Mundi* (1619). It states that the square of the time a planet takes to go around its orbit is proportional to the cube of the

orbit's semi-major axis (half the longest diameter of the planet's orbit). But not all of Kepler's ideas were correct. One of his theories stated that planets were swept along by radiated magnetic influence from the sun; later disproved by the Newtonian theory of gravitation (see pp.106–107). »

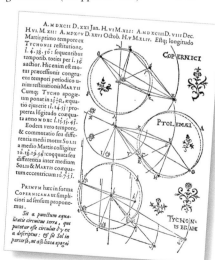

Kepler's calculations in *Astronomia Nova* This book demonstrates Kepler's mathematical workings, as he abandoned the Copernican and Ptolemaic views on the circularity of planetary orbits.

BREAKTHROUGH

PLANETS IN TRANSIT

A transit occurs when one astronomical body travels across the face of another. Only three major bodies pass between Earth and the sun: Mercury, Venus, and the moon. During a transit of Venus, the planet moves directly between Earth and the sun, appearing as a dot against the sun's disk. Venus transits the sun four times every 243 years, so this is one of the rarest predictable astronomical phenomena. This image shows the movement of Venus over a five-hour period during the 2004 transit.

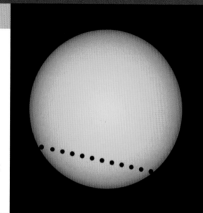

» Kepler's third law revealed the relative sizes of the planetary orbits but not their absolute sizes. The ancient Greeks had estimated that the sun was 1150 Earth-radii away, but Kepler's Mars calculations convinced him that the real value was far greater. There was much confusion in the 17th century—the 1687 and 1713 editions of Newton's *Principia* show a doubling of his estimate. Newton increased the estimated mass of the sun by an astonishing factor of eight.

In 1716 Edmond Halley suggested a way of measuring the distance from Earth to the sun, using the timing of the transits of either Mercury or Venus, measured by observers over a wide range of latitudes. This method was used in 1769 to observe the Venus transit. Today we know that the sun is about 93,000,000 miles (150,000,000 km), or 23,455 Earth-radii, away.

A day on Mars

The 1609 introduction of the telescope (see pp.84–85) revolutionized planetary astronomy. It led to the discovery of the moons of Jupiter, and revealed the

250 MILLION YEARS
The time it takes for the sun, and our solar system, to orbit once around the Milky Way, traveling at 155 miles (250 km) a second.

planets to be physical disks with observable surface features, rather than bright points of light. The sizes, spin rates, and spin-axis orientations could all now be measured. It was realized, for example, that Mars had a "day" that was 24 hours, 37 minutes long, and a spin-axis inclination of 24°, which implied that Mars had seasons, just like Earth. Saturn's inclination was 26.7°, which led to changes in the appearance of its rings.

Calculating the Earth-path

Around 135 BCE, Hipparchus (see p.28) had measured planetary positions to an accuracy of 60 minutes of arc. By 1586 Tycho had improved this to 1 minute, by 1700 Flamsteed to 0.3 minutes and by 1856, Argelander was working to an accuracy of 0.015 minutes. This

increasing precision led to a greater understanding of the stability of the solar system and the actions of one planet on another. It also improved our knowledge of the lunar orbit and our ability to predict eclipses. During the solar eclipse of 1715 Edmond Halley predicted its path across the Earth to an accuracy of less than half a mile (1 km).

Predicting eclipses

Eclipses themselves had always been hard to predict: Thales of Miletus (640–546 BCE) was only able to estimate their timing to within a year. Luckily, the eclipse he did predict passed over the River Halys in Turkey during a war between the Lydians and the Medes, and the soldiers on both sides were so awestruck that they immediately lowered their weapons and declared peace. The eclipse allows us to date the battle precisely, to May 28, 585 BCE, demonstrating the importance of these events to astronomers and historians.

Solar eclipses occur only when the new moon is close to a point (the node) at which the moon's orbit crosses Earth's. If

London solar eclipse
This print from 1748 was published to inform Londoners of the cause and appearance of a solar eclipse that would be visible from the capital on July 14, of that year.

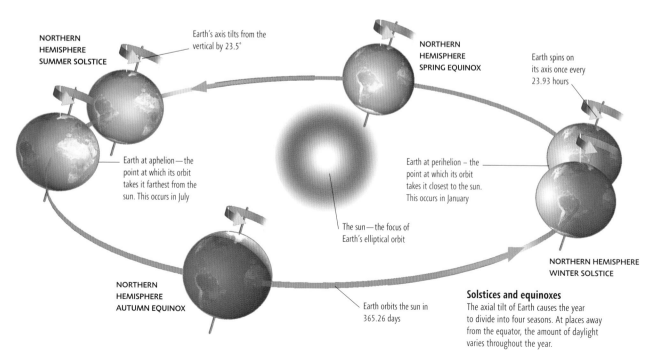

NORTHERN HEMISPHERE SUMMER SOLSTICE

Earth's axis tilts from the vertical by 23.5°

NORTHERN HEMISPHERE SPRING EQUINOX

Earth spins on its axis once every 23.93 hours

Earth at aphelion—the point at which its orbit takes it farthest from the sun. This occurs in July

Earth at perihelion – the point at which its orbit takes it closest to the sun. This occurs in January

The sun—the focus of Earth's elliptical orbit

NORTHERN HEMISPHERE WINTER SOLSTICE

NORTHERN HEMISPHERE AUTUMN EQUINOX

Earth orbits the sun in 365.26 days

Solstices and equinoxes
The axial tilt of Earth causes the year to divide into four seasons. At places away from the equator, the amount of daylight varies throughout the year.

PERIGEE The point on the moon's orbit where it is closest to Earth.

APOGEE The point on the moon's orbit where it is furthest from Earth.

the moon is near perigee (at its closest to Earth), its shadow cone intersects Earth's surface. Anyone standing in that shadow experiences a brief solar eclipse. The shadow can be as wide as 135 miles (217 km) across, sweeping Earth at a speed of 1,056 mph (1,700 kph). Totality can last for as long as 7.7 minutes.

Even though a total solar eclipse occurs somewhere on the Earth about every 18 months, at any fixed locality, one is visible on average only every 370 years. This rarity has given them a special place in mythology and folklore.

Eclipse science
When the moon completely covers the solar disk, the sun's coronal atmosphere springs into view. The corona is a gaseous region around the sun, extending more than 600,000 miles (1,000,000 km) from its surface, with temperatures exceeding one million degrees Celsius. Closer to the solar limb (the apparent edge of the sun in the sky) you can see large arches of red gas confined by magnetic fields: these are the solar prominences (cooler clouds of gas within the hotter coronal ones).

Kepler correctly identified these as solar phenomena, not lunar volcanic activity, as previously thought. But it took another 200 years for astronomers to realize that the size of the corona varied with the phase of the solar cycle.

The interest in stellar bodies, rather than planetary ones, increased hugely during the 19th century as telescopes improved.

STELLAR DISTANCE
Astronomy was revolutionized by the measurement of the **distance to the star 61 Cygni** in 1838. Stellar energy outputs could then be calculated, and stellar life cycles debated.

EINSTEIN
The stellar energy problem was solved in essence when **Albert Einstein** realized that $E = mc^2$ in 1905. **The general theory of relativity 304–05 »** also explained the extent to which starlight bends as it passes close to the sun.

SOLAR CORONA SCIENCE
Eclipses are now observed mainly for pleasure. Coronal observations are obtained using spacecraft **396–97 »**, which can view the **solar outer atmosphere** continuously, not just during the brief occasions of an eclipse.

SOLAR CORONA

SEE ALSO »
pp.182–83 THE SOLAR SYSTEM
pp.394–95 INSIDE THE SOLAR SYSTEM

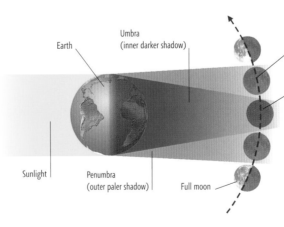

Only a slight darkening of the moon occurs in the light outer shadow.

The moon is darkest within the umbra.

Earth
Umbra (inner darker shadow)

Lunar eclipse
If the full moon is near an orbital node, it can pass through Earth's shadow cone and be eclipsed. People inhabiting the half of Earth that faces the moon would see the moon as darkened.

Sunlight | Penumbra (outer paler shadow) | Full moon

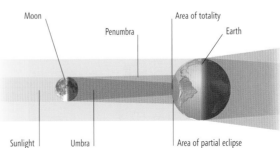

Moon | Area of totality | Earth
Penumbra

Solar eclipse
The shadow cast by the new moon just intersects Earth's surface. People standing in this small shadow region would see the solar disk totally covered by the moon.

Sunlight | Umbra | Area of partial eclipse

Solar wind
A stream of high-energy, electrically charged particles emitted by the Sun

Earth's magnetic field
This deflects the solar wind.

Earth's magnetic axis
Geographical north does not coincide with magnetic north.

Earth's magnetosphere
Earth's magnetic field stretches tens of thousands of miles into space. It protects our planet from the solar wind, directing its particles down toward the poles, where their energy causes air molecules to glow: these are the beautiful light displays known as the auroras (Northern and Southern Lights).

Geographic North Pole

Inner Van Allen belt

The Van Allen belts
Two doughnut-shaped regions (an inner and outer one) of charged particles held in place by Earth's magnetic field

Geographic South Pole

Elongated field
The magnetic field is elongated on the side opposite the Sun, stretching away like a tail from Earth.

"Squashed" magnetic field
Earth's magnetic field is squashed on the side closest to the Sun.

Magnetopause
The boundary where the magnetosphere meets the solar wind

Magnetic field lines
Within Earth's magnetosphere

Bow shock
Solar wind particles are deflected around the magnetosphere like water around the bow of a ship.

<< BEFORE

People first became aware of magnetism through finding naturally magnetic rocks.

ACCIDENTAL DISCOVERY
No one is sure of the true origin of the name "magnet." One of the most popular stories centers on a shepherd in Magnesia, Greece, who found nails sticking to a black rock. Until the 16th century, **magnetic rocks** were thought to have magical powers, which could be destroyed by garlic or diamonds.

THE WORK OF PEREGRINUS
French scholar Pierre de Maricourt (also known as Peter Peregrinus) was the first person to record the basic facts of magnetism, noting them in a letter written in 1269. He wrote in detail about magnetic poles, forces, and **compass** needles.

<< SEE ALSO
pp.18–19
EARLY METALWORKERS

Magnetic Fields

The area of influence of a magnet is called a magnetic field. Earth itself has a vast magnetic field emanating from its core and extending far out into space. The fields around much smaller, more familiar magnetized objects follow similar "lines of force" flowing from opposite poles.

Although magnetic fields are invisible, they can be visualized using tiny pieces of iron called filings. When iron filings are sprinkled around a magnet, they immediately arrange themselves along the field's "lines of force," illustrating the magnetic field.

Lodestone
Lodestones are a form of magnetite that were recognized as adopting a north–south alignment, and exercising repulsion and attraction. The ancient Chinese called them *tzhu shih* ("loving stones"), because they liked to "kiss."

Every magnet has two ends, or poles: one that is north seeking (N), and one that is south seeking (S). If you put together two magnets with like poles facing each other (N–N or S–S), they push each other apart, while if you hold the unlike poles facing each other (N–S), they pull together.

Materials that can be magnetized, such as iron, are composed of tiny regions called domains, each of which also has a north and south pole. When iron is unmagnetized, the domains are randomly oriented within the metal. But when the metal is magnetized, the domains all align—forming dense lines with all north-seeking ends pointing in

the same direction. If a metal paper clip is held near to a magnet, for instance, it becomes magnetized—its domains align, and it is pulled toward the magnet.

Earth's magnetism
Any magnet that is free to move and carefully balanced on a frictionless bearing will swivel so that the north-seeking end always points northward and the south-seeking end always points southward (a compass needle is just a small magnet on a pivot). This is because Earth itself has a magnetic field, caused by the powerful electric currents that run through the liquid metal of its outer core. This magnetic field—the magnetosphere

Creating a magnet
A piece of iron or steel can be made into a magnet by rubbing it repeatedly with another magnet.

Domains randomly oriented

Bar magnet is rubbed over surface of steel

Domains become aligned in the magnetized metal

Unmagnetized steel bar
Here the tiny regions of magnetism—domains—are randomly oriented. The overall magnetic field is zero.

Magnetized steel bar
Rubbing steel with a magnet causes the domains to line up in the same direction. The metal is now a magnet, even after the other magnet is taken away.

ENGLISH PHYSICIAN AND SCIENTIST 1544–1603

WILLIAM GILBERT

English natural philosopher William Gilbert graduated as a medical doctor, and began practicing in London in 1577. By 1599 he was the president of the College of Physicians. In 1600 he published his principal work, *On the Magnet, Magnetic Bodies, and the Great Magnet of the Earth*, which had taken 17 years to write. He proved that magnetism could attract through solid objects, not just air. He became a Royal Physician in 1601, but died—possibly of the plague—two years later. A second work, *A New Philosophy of our Sublunar World*, was published posthumously.

—extends into space, where it protects Earth from a stream of harmful particles emitted by the Sun called the solar wind.

Shifting magnetic fields
Although compasses point northward, they do not point to Earth's geographical north pole, defined by the planet's axis of spin. Compass needles actually point toward "magnetic north," which at present lies a few hundred miles away from the geographical north pole.

Earth's magnetic north pole gradually shifts position, as the planet's liquid outer core is free to swirl somewhat separately from the rest of the planet's spin. Occasionally, Earth's magnetic field flips—possibly as a result of turbulence in the liquid metals—so that the magnetic north ends up near the geographical south pole (in Antarctica). This "field reversal" occurs at irregular intervals, from a few hundreds of thousands to many millions of years.

Discovering magnetism
Magnetism was first recorded as a phenomenon around 600 BCE, when people realized that some minerals had magnetic properties. Naturally occurring forms of magnetite, known as lodestones (meaning "leading stones"), were observed always to align themselves in a north–south direction, and were used in the earliest compasses, from around 1000 CE onward.

The first person to study magnetism in a systematic way was English philosopher William Gilbert. In 1600, he published the results of his investigations, and suggested that our planet acts as a magnet. He also investigated static electricity (see pp.158–59). Like everyone else at the time, he considered electricity and magnetism to be completely separate. But in 1820, Danish natural philosopher Hans Christian

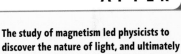

◁ **Magnetite compass**
Early compasses had a piece of magnetite attached to the bottom of a movable dial, with the north-seeking pole of the magnetite aligned with the North pointer on the dial.

AFTER ≫

The study of magnetism led physicists to discover the nature of light, and ultimately radio waves, X-rays, and infrared radiation.

MAGNETISM AND ELECTRICITY
English scientist Michael Faraday 170–71 ≫ discovered **electromagnetic induction** (the inverse of Ørsted's discovery) and showed that light is affected by magnetic fields.

ELECTROMAGNETIC RADIATION
In the 1860s Scottish theoretical physicist James Clerk Maxwell 167 ≫ produced a set of theorems—**Maxwell's Equations**—that unified the sciences of magnetism and electricity, and predicted the existence of electromagnetic waves.

SEE ALSO ≫
pp.166–67 ELECTROMAGNETISM
pp.170–71 MICHAEL FARADAY

Ørsted discovered that electric currents produce magnetic fields when the needle of a compass is deflected by a current going through a nearby wire. Physicists quickly came to realize that magnetism and electricity were very closely interrelated, and not separate.

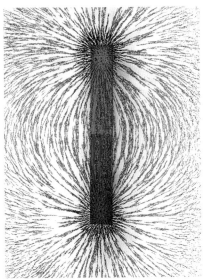

Iron filings around a magnetic field
A magnetic field can be made visible by placing iron filings around a magnet. Each tiny piece of iron lines up with the field, along the "lines of force." The closer each line is to the next, the stronger the force of the field.

Magnetic forces
A magnet always attracts unmagnetized objects made of iron or steel. But two magnets close together can either attract or repel, depending on whether like or non-like poles are placed nearer to each other.

Attraction and repulsion
Two poles of opposite types—north/south—always attract each other.

South-seeking pole

North-seeking pole

Magnetic field lines

North-seeking pole

North-seeking pole

Force direction
Attraction runs toward the south-seeking pole

ATTRACTION

Like poles repel
Two poles of the same type—north/north or south/south—always repel each other.

REPULSION

MATHEMATICIAN AND ASTRONOMER   **Born 1564** Died 1642

Galileo Galilei

"**Science** is written in the language of **mathematics**."

GALILEO GALILEI

Galileo has been called the "father of modern science." He was one of the first thinkers to challenge the doctrines of Aristotle (see pp.36–37) and test his ideas by experiment. Galileo applied mathematics to the real world. He got into trouble with the church over his views on the solar system, and he was eventually summoned for trial in Rome and put under house arrest.

From medicine to mathematics

As a teenager Galileo was tempted to go into a monastery, but his father persuaded him to study medicine; he therefore attended medical school at the University of Pisa. Galileo did not complete his medical degree; instead, he spent most of his time studying mathematics. He went on to teach that subject and later became professor of mathematics at the University of Pisa.

While at the university, Galileo experimented with pendulums, studied the fall of objects, and devoted himself to astronomy. Sitting in the cathedral, he noticed the great bronze lamp swinging to and fro on its chain in

Galileo observing a lamp in Pisa cathedral
Watching the swinging of the great bronze lamp in the *duomo* (cathedral) in Pisa, and timing it against his pulse, Galileo realized that a pendulum makes a simple and reliable timekeeper.

the draft from the door. He timed the swings, using his pulse, and was surprised to find that each swing took the same time, regardless of whether the lamp swung only a few inches or much more. Realizing that the pendulum is a regular natural timekeeper, he invented a little pulse meter (see left), and toward the end of his life he also designed a pendulum clock.

While at Pisa, Galileo performed

PULSE METER

While he was a medical student in Pisa, Galileo invented a simple pulse meter—a pendulum on a stand some 8 in (20 cm) high, constructed so that the length of the string was variable. The doctor would feel the patient's pulse, adjust the length of the string until the swing of the pendulum was just in time with the pulse, and then read the pulse rate off the scale on the stand, which probably just said "slow," "medium," and "fast." The medical authorities stole the idea, and Galileo received no credit for it.

Trial for heresy
Galileo's discovery of the moons of Jupiter was praised by the church, but when he suggested that Earth orbits the Sun, he was accused of heresy and summoned for trial by the Inquisition in Rome.

Galileo Galilei
Seen here in a 1636 portrait by the Flemish painter Justus Sustermans, Galileo is credited as the originator of modern science. He made fundamental contributions to mathematics, physics, and astronomy.

various experiments in motion and inertia (see pp.112–13). He challenged Aristotle's assertion that heavy objects fall faster than light ones, in proportion to their weight. According to legend, Galileo tested the theory by dropping balls of various weights from the Leaning Tower of Pisa. In reality he tested Aristotle's theory by rolling balls down an inclined plane and ascertaining their speed of descent (see pp.104–105).

In 1592 Galileo moved north to Padua, and settled into a long-term relationship with Marina Gamba; the couple never married, but had three children. His two daughters were sent to a convent; his son, Vincenzo, was selfish and often a trial to his father.

Further researches

In 1602 Galileo returned to the study of falling objects and trajectories; in due course he showed that a projectile, such as a cannonball, will follow a path close to a parabola as it is acted upon by gravity.

In 1609 Galileo heard about the spyglass invented by the Dutch spectacle maker Hans Lippershey. Although they never met, Galileo constructed his own instrument, and then increased the magnification from three to 30 times, creating the first astronomical telescope. Galileo made some money, or obtained some promise of favors, by offering spyglasses to the military, and then used his own to look at the heavens. First

he noted that the Moon was not the smooth and perfect sphere that Aristotle had claimed, but had a rough and irregular surface, with mountains and craters.

On January 7, 1610, Galileo looked at Jupiter. He could see colored bands on the planet, but there were also curious points of light in line—two on the right of the planet and one on the left. He thought they were distant stars, but on the following night all three seemed to have moved to the left of Jupiter. On January 13, he saw one more point of light. He then realized that they were not stars at all, but four small worlds—moons—orbiting the planet. This was a shock. Everyone "knew" that Earth was the center of the Universe, but here were moons orbiting Jupiter, not Earth. The moons, named Io, Europa, Ganymede, and Callisto, are now referred to as Galilean moons in his honor.

Punishment for heresy

The Jesuits in Rome were fascinated when Galileo showed them the moons of Jupiter; he was welcomed with enthusiasm by the cardinals and even by the Pope. However, these moons prompted Galileo to turn toward the idea that the Sun was the center of the Universe, as proposed by Copernicus in 1543 (see pp.74–75). This in turn meant that Earth must move around the Sun, which directly contradicted the Bible.

Galileo was sentenced to life imprisonment by the Inquisition in Rome, although this was reduced to permanent house arrest at his villa in Arcetri. He died there in 1642. In 1992, 359 years after his imprisonment for heresy, Galileo was pardoned by Pope John Paul II.

Pisa's leaning tower
According to legend, Galileo dropped balls of various weights from the Leaning Tower of Pisa to disprove Aristotle's assertion that heavy objects fall more quickly than light ones.

TIMELINE

1564 Galileo di Vincenzo Bonaiuti de Galilei is born in Pisa, Italy, the eldest child of musician Vincenzo and Giulia Ammannati.

1581 Encouraged by his father, he enrolls to study medicine at the University of Pisa, and invents a pulse meter. Later he switches to the study of mathematics.

1585 Begins teaching mathematics in Florence and then in Siena.

1589 Becomes professor of mathematics at the University of Pisa.

1590 Begins study of motion, rolling balls down inclined planes and measuring their velocity and acceleration.

1592 Moves to the University of Padua as professor of mathematics; the post is much better paid than the one in Pisa. He teaches geometry, astronomy, and mechanics, and describes this as the happiest time of his life.

1600 His first daughter, Virginia, is born to Marina Gamba.

1601 His second daughter, Livia, is born. She and Virginia are to enter a convent in 1614, remaining there for the rest of their lives.

1606 His son, Vincenzo, is born.

1609 Hears about the spyglass invented by Hans Lippershey, makes his own, improves it, and sells it to the militia.

GALILEO'S TELESCOPE

1610 Looks at the Moon and then at Jupiter, discovering four moons. He publishes the results in *Sidereus Nuncius*, or *Starry Messenger*.

1611 Travels to Rome to demonstrate his telescopic discoveries and to show the Jesuits the moons of Jupiter. He is made a member of the Accademia dei Lincei, one of the world's first scientific societies.

1614 His views on the motion of Earth in a sun-centered universe are denounced as dangerous and close to heresy by Father Tommaso Caccini.

1623 Publishes his second book, *The Assayer*.

1632 Publishes his third book, *Dialogue*, in Florence. He is summoned to appear before the Holy Office in Rome.

1633 Undergoes a papal trial in which he is accused of heresy. He is forced to recant and deny his own teachings and is then confined to house arrest in his villa at Arcetri.

1638 Loses his sight. Proposes a method for measuring the speed of light.

1641 Devises an escapement mechanism to govern the timekeeping of a pendulum clock; the idea is taken up 15 years later by Christiaan Huygens.

1642 Dies at Arcetri.

GALILEO'S TOMB IN FLORENCE, ITALY

‹‹ BEFORE

For thousands of years astronomers had observed and recorded heavenly bodies. By the 2nd century BCE the effect of lenses on light had been documented in the East.

VIEWING BY EYE

The **Babylonians** (c.1800 BCE) named the constellations, while the ancient **Chinese** recorded supernovae and comets from the 4th century BCE. **Arabian and European astronomers ‹‹** 26–29 made catalogs of the stars as long ago as the 2nd century BCE.

EARLY WORK ON OPTICS

In about 1010 Arab mathematician **Alhazen** working in Egypt, laid down the basic principles of optics, describing how lenses work in his *Book of Optics*.

‹‹ SEE ALSO
pp.60–61 **ALHAZEN**
pp.82–83 **GALILEO GALILEI**

Galileo's drawings of the Moon
Galileo observed that the Moon was not the smooth sphere described by Aristotle—instead, it was covered with mountains, valleys, and craters. His illustrations of the Moon were published in *Sidereus Nuncius* in 1610.

Exploring the Skies

People have always been fascinated by the stars and planets in the night sky, but until 1609 they had to rely on just the naked eye to make them out. The invention of the telescope opened up entirely new views into the universe.

The first telescope may have been made by English polymath Leonard Digges, around 1551, but there is no evidence that he used it to study the stars. Because his work was not widely known, the credit for inventing the telescope usually goes to the Dutch.

Early designs

The first designs for a telescope were produced by a Dutch glasses maker called Hans Lippershey. He discovered that if two lenses were installed in a tube, distant objects could be made to appear closer. The Italian scientist Galileo Galilei heard about Lippershey's telescope; he made his own versions, improved the magnification to 30 times, and sold the idea to the local militia. Then, during the last month of 1609, he turned his telescope toward the sky and was the first person ever to see an upclose view of the Moon. He also discovered four of Jupiter's satellites, and his instrument was the first to be given the name "telescope." In 1611 the astronomer Johannes Kepler improved on Galileo's design by using a convex lens for the ocular (eyepiece) instead of a concave one. This gave him a wider field of view and potentially higher magnification. One problem, however,

140

TIMES was the approximate magnification of Newton's telescope, even though it was only about 8 in(20 cm) long, with a 1¼-in (3-cm) mirror.

was that to achieve high magnification the objective (the lens at the front) needed a long focal length. As a result telescopes grew ever longer—some reaching 33 ft (10 m).

Refracting and reflecting light

Another problem associated with these first telescopes came from the way in which the lenses work. In 1671 Newton wrote to the Royal Society about his experiments with prisms, his explanation of the colors of the spectrum, and his theories about the nature of light. Newton explained that lenses, like prisms, refract (bend) light as it passes through, and refract blue light more than red; that is why a prism produces a spectrum. When light was refracted in a "refracting" telescope it caused all the stars to have colored fringes around them.

Almost in passing, Newton described his invention of a reflecting telescope, which used a mirror rather than lenses to focus the light. Mirrors do not refract light, and all colors are reflected equally; so a reflecting telescope produces no colored fringes. A Scottish astronomer, James Gregory, had described a similar instrument in 1663, but could not find anyone to grind the mirror for him. Newton was lucky—he managed to get his design made, and presented it to the Royal Society.

THE ACHROMATIC LENS

In the mid-18th century Chester Moore Hall and John Dollond independently solved the problem, highlighted by Isaac Newton, of chromatic aberration and the resulting color fringes around stars viewed through refracting telescopes. Hall and Dollond made lenses from a sandwich of two different materials: crown glass and flint glass. These have different refractive indices—meaning that they bend light waves to different extents. The resulting combined lens focused light of different wavelengths (and colors) at the same point, creating a sharp image with no fringe.

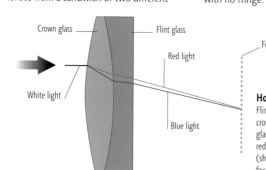

Crown glass — Flint glass
Focal plane
Red light
White light
Blue light

How it works
Flint glass bends the light more than crown glass. When the two types of glass are combined in a single lens, red (longer-wavelength) and blue (shorter-wavelength) light are focused at the same point.

△ How a reflecting telescope works
The primary mirror reflects the incoming light back up the tube. The secondary mirror in a Newtonian reflector is at 45° and reflects the light through the side of the tube to the eyepiece; in a Gregorian reflector it reflects the light back down the tube through a hole in the primary to an eyepiece behind.

Light enters telescope
Light collected by objective lens
Eyepiece
Secondary flat diagonal mirror reflects light to the eyepiece
Concave primary mirror

Convex objective lens refracts light rays to a focus
Refracted light
Focal point of light rays is one factor in determining length of telescope

△ How a refracting telescope works
Rays of light entering the objective lens at the top of the tube are brought to a focus near the lower end of the tube. They then pass through the eyepiece to the observer who sees a clearer, magnified, virtual image.

Eyepiece magnifies image

Eyepiece
The horizontal eyepiece made the telescope easier to use than refracting telescopes, whose users had to peer awkwardly upward at the sky.

▽ **Telescope development**
Technological advances allowed telescopes to become more and more powerful.

1611 Kepler introduces the convex eyepiece, improving clarity, magnification, and field of view.

1763 The apochromatic lens, or triple achromat, focuses red, blue, and green light together.

1857 Advances in chemistry allow the manufacture of silver-coated mirrors, which are more reflective.

1990s Adaptive optics are used to remove atmospheric distortion.

1600	1700	1800	1900

1609 Galileo makes a refracting telescope, with a concave eyepiece.

1663 The reflector telescope is described by James Gregory and later built by Isaac Newton.

1733 The invention of the achromatic lens allows red and blue light to be focused together.

1932 Aluminized mirrors are introduced; these are cheaper to produce and more durable.

1990 The Hubble Telescope is launched into Earth orbit.

Focus adjustment control
By turning this screw, the user can shorten or lengthen the telescope tube.

△ **Paranal Observatory, Chile**
The Very Large Telescope (VLT) at the Paranal Observatory comprises four units that can be used alone or combined to make one of the world's most powerful telescopes.

◁ **The remains of a star**
This picture, taken by the VLT, shows the remains of a star after a massive supernova explosion. This structure, called the Crab Nebula, lies in the constellation Taurus.

Sliding focus
The telescope tube was made of leather, in two parts, so that one could slide over the other to adjust the focus.

A replica of it is shown here. They were delighted, and invited him to become a fellow.

Further developments
In the centuries that followed, telescopes were used not only for star gazing but also for spying on enemy forces and for navigation at sea. Both refracting and reflecting telescopes were built with increasing precision. Today almost all the major astronomical telescopes are Gregorian reflectors—even the Isaac Newton Telescope on the island of La Palma.

Replica of Newton's telescope
The Royal Society in London keeps a replica of Newton's original reflecting telescope. It had a 1¼-in (3-cm) mirror and was surprisingly compact and simple, but it changed the way astronomers study the stars. Most modern telescopes are reflectors.

Telescopes have grown considerably since Newton's instrument with its 1¼ in (3 cm) mirror. One example is the Very Large Telescope (VLT) at Cerro Paranal in Chile. This array is actually a combination of four main telescopes, each with a mirror 27 ft (8.2 m) in diameter. The point of such huge mirrors is to capture more light and to increase the resolution of the image. Double the diameter of the mirror and you can collect four times as much light, and so see objects that are much fainter, which would be invisible with a smaller instrument. The power of telescopes is also enhanced by the use of electronic detectors in their imaging systems. From the 1970s, conventional photography was replaced by the use of sensitive charge-coupled devices (CCDs).

After its light-gathering capacity, the next most important feature of a telescope is its resolving power; that is, its ability to pick up fine details and produce sharp images. A modern large telescope could form a sharp image of a dime from several miles or kilometers away.

Magnification is much less important; indeed, many astronomical observations have been made with ordinary binoculars, which are, in fact, mini-telescopes.

AFTER

Today, telescopes observe from space as well as from Earth. Some are designed to detect invisible radiation, such as radio waves.

GATHERING RADIO WAVES

RADIO TELESCOPES
Grote Reber built the first **radio telescope** in 1937. It worked like a reflecting telescope, but picked up **radio waves** rather than light waves.

THE HUBBLE TELESCOPE
Launched in 1990, and last visited by astronauts in 2009, the Hubble Space Telescope operates in space, **above the atmosphere**. It is due to be replaced by the James Webb Space Telescope.

SEE ALSO »
pp.110–11 Isaac Newton
pp.116–17 Splitting and Bending Light
pp.396–97 Space Probes and Telescopes

Motion, Inertia, and Friction

One of the great scientific breakthroughs of the 17th century was to understand how and why things move and stop moving. The key to the problem is that objects have inertia, which means that they slow down and stop only when something, typically friction, forces them to do so.

BEFORE

For a long time, even thinkers such as Aristotle were misled about the nature of motion by common-sense experience.

EXTERNAL FORCES
Aristotle argued that things only keep moving because they are **continually pushed or pulled** by an **external force** and slow down as soon as the force is removed, just as a cart slows down when the ox stops pulling.

" …A body in motion can **maintain** this motion only if it remains **in contact** with a **mover**…"

ARISTOTLE, 384–322 BCE

ANTIPERISTASIS
Aristotle stated that an arrow keeps flying after it has left the bow because of **a natural reaction** or **antiperistasis** with the air, which generates **eddies** or **vibrations** that drive it on.

« SEE ALSO
pp.36–37 ARISTOTLE

For almost 2,000 years people subscribed to Aristotle's assertion that, unless things are continually propelled onward by some force, they soon slow down naturally to rest. It seemed only common sense. Yet there were problems with this view. How, for example, does an arrow continue in its flight long after it has left the bow?

In the Middle Ages the Islamic scholar Avicenna and the French priest Jean Buridan spoke of moving objects having "impetus." At first sight, this idea seems close to the modern idea of momentum—the natural tendency of an object in motion to keep on moving — but for Avicenna and Buridan, impetus was an active internal motive force pushing things onward against the natural tendency to slow down.

Galileo's experiments
The great insight that Italian scientist Galileo Galilei achieved was to realize that common sense is not always

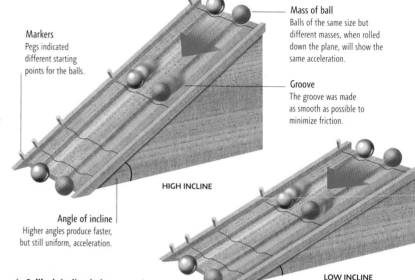

Markers
Pegs indicated different starting points for the balls.

Mass of ball
Balls of the same size but different masses, when rolled down the plane, will show the same acceleration.

Groove
The groove was made as smooth as possible to minimize friction.

HIGH INCLINE

Angle of incline
Higher angles produce faster, but still uniform, acceleration.

LOW INCLINE

△ **Galileo's inclined plane experiment**
By rolling balls down sloping wooden grooves and timing them as they traveled over different distances, Galileo proved that the acceleration at a given angle was uniform, no matter how far the ball traveled.

enough to explain different phenomena—careful observation and experiment are also needed to help people arrive at the truth.

During the 1630s, while under house arrest for championing Nicolaus Copernicus's heliocentric view of the Universe (see pp.74–75), Galileo carried out a series of experiments, the most famous of which involved rolling balls down a pair of inclined wooden grooves.

Rolling a ball down a slope allowed Galileo to slow the acceleration of

natural tendency to slow down at all. The opposite is true: a moving object will keep moving at the same speed unless something slows it down, and that slowing force is usually friction.

In fact, as Galileo realized, there is no real difference between an object moving at a steady rate and one that is not moving at all—both objects are unaffected by forces. A force is needed either to make the object speed up or slow down or to start it moving. This is the principle of inertia—the tendency

BICYCLE V–BRAKES

Cable acts on end of short lever called a caliper when lever on handlebars is squeezed

Brakes work by exploiting friction to counteract inertia. When a cyclist brakes, the extra friction as the brake pads are squeezed on the wheel rim slows the rotation of the wheel and reduces the bicycle's forward inertia.

Calipers are pivoted so that tension of cable squeezes brake pads against wheel

Brake pads are squeezed against rim, increasing friction

Brake pads are made of a rubberized material that causes maximum friction with minimum damage to wheel rim

Inertia in action
When a horse stops suddenly in front of a fence, inertia will often propel the rider over the horse's head—as in this case, where the rider is being thrown right over the jump.

falling objects to a measurable rate. By timing how long the ball took to travel a particular distance down the slope, he could figure out its acceleration, and by altering the slope, he could change the acceleration. To study deceleration, he rolled the ball down one slope and allowed it to roll up a slope opposite. Galileo had no clock, but he devised ingenious ways to time events, such as using his pulse or weighing how much water spurted from an urn as the ball traveled a certain distance.

Discovering inertia

It took Galileo to figure out that the role played by friction in slowing things down is crucial. Things don't have a

Galileo's demonstration
Galileo used his inclined plane experiment to demonstrate the law of falling bodies. He apparently exhibited it to groups of clergy and aristocrats, as shown in this 1839 fresco by Florentine painter Giuseppe Bezzuoli.

of an object to resist a change in motion, whether it is moving steadily or stationary.

Galileo also saw that it is impossible to distinguish between a steadily moving object and a still one without some point of reference; this idea formed the foundation of Einstein's special theory of relativity (see p.302) in the early 20th century. Galileo further realized that the rate of acceleration depends on the strength of the force and the heaviness, or mass, of the accelerating object.

Projectiles

The concept of inertia solved the problem of the arrow. An arrow keeps flying while inertia is dominant, but gravity is continually pulling it down. Galileo showed that the combination of inertia and gravity would mean that the path of a projectile, such as an arrow, through the air must be a curve called a

parabola (see below). The effect of drag, or air resistance, is to slow the arrow down, so that it falls short.

Galileo's ideas enabled a giant leap in the understanding of force and motion, and laid the groundwork for Isaac Newton (see p.104) to bring them together in his three laws of motion. Newton's first two laws are, in fact, restatements of Galileo's discovery of inertia and his finding that acceleration depends on the relative size of the

3.2 SECONDS

The time it would take for a ball or body to fall to the ground from the top of the leaning tower of Pisa.

force and the object—or, rather, the object's mass. In this way, Galileo's discoveries underpin our knowledge of virtually every movement in the Universe, from the motion of atoms to the rotation of galaxies.

The flight of projectiles
The discovery of inertia enabled Galileo to explain the trajectories of objects such as cannonballs. Their flight traces a parabola. The higher the angle at which a cannonball is fired (up to a maximum of 45°), the farther the ball travels.

[Graph: Height (meters) vs Distance (meters), showing parabolic trajectories]

AFTER >>

The concept of inertia is central to our understanding of motion.

STREAMLINING FOR SPEED
A race car relies on friction to make its wheels turn, but is held back by friction in its moving parts and by friction with the air, known as **drag**. To minimize drag, designers **streamline** the car's body so that it slips through the air with the **minimum of disturbance to air molecules.**

FORMULA ONE RACING CAR

SWIMSUITS
Champion **swimmers** are **held back by drag** as they race through the water. Swimsuit manufacturers use supercomputers in the **design of swimsuits** to minimize the **"skin friction."**

SEE ALSO >>
pp.104–05 NEWTON'S LAWS OF MOTION
pp.110–11 ISAAC NEWTON
pp.112–13 SPEED AND VELOCITY

‹‹ BEFORE

An ancient yet sophisticated system, used in China from the 5th century BCE, involved "counting rods" (*suan chou*).

CHINESE COUNTING RODS
Counting rods were usually made from lengths of bamboo. They were used according to the **base 10 number system ‹‹** 30–31. To form numbers, rods were set out in groups vertically or horizontally. The rods could be used for basic arithmetic, finding **square and cube roots**, and solving **complex equations. Negative numbers** could be formed by using black rods or laying rods diagonally across numbers.

‹‹ SEE ALSO
pp.30–31 ANCIENT NUMBER SYSTEMS

Methods of Calculating

Arithmetical calculating—adding, subtracting, multiplying, and dividing numbers—has always been important in commerce, industry, and everyday life. The development of more effective calculation methods, and even of mechanical calculating aids, enabled huge advances in scientific and practical knowledge.

In the civilizations of the ancient world, traders learned through practice the basic addition, subtraction, multiplication, and division that they required for their business, but any calculations more complicated than these required the work of an expert mathematician.

" ... by **shortening** the **labors** [Napier's logarithms] **doubled** the **life** of the astronomer."

PIERRE-SIMON LAPLACE (1749–1827), FRENCH MATHEMATICIAN AND ASTRONOMER

Early methods of calculating
One way to make calculation easier was to use mechanical aids. The abacus, for example, was a visual counting aid that first appeared in Mesopotamia in 2700–2300 BCE, and became

The abacus
The abacus as we know it, a frame with beads on wires, originated in China. Different forms developed in Japan, Korea, and Russia, and are still sometimes used in stores and schools. Shown here is a 16th-century Russian abacus.

MATHEMATICIAN (1550–1617)

JOHN NAPIER

Born in Edinburgh, Scotland, John Napier entered St. Andrews University aged 13, but it is thought that he acquired most of his mathematical learning while studying abroad in France and Italy. In addition to inventing logarithms and creating "Napier's bones," Napier popularized the use of the decimal point and introduced decimal notation for fractions. His powers of invention went beyond mathematics—he also experimented with the use of fertilizers to enrich farmland and designed machines for warfare.

widely used in the ancient world for calculating with large numbers in trade and engineering. The earliest versions were boards or tables with grooves, or even lines in the sand, into which people placed pebbles. (The Latin name for a pebble, *calculus*, has given us our word "calculate.") In China, people used counting rods on a board arranged in rows and columns. These devices were adequate for simple, quick calculation until a good system of writing numbers was formulated.

The development of new techniques also made calculation easier and more accurate. A breakthrough came when Hindu–Arabic numerals and a place-value system of notation (see p.31) were adopted in the West, replacing the old Roman numerals. Because digits were arranged in columns according to their value in the decimal system, addition and subtraction were made more straightforward. However,

multiplication and division were still difficult, labor-intensive jobs. Advances in the sciences, especially astronomy and navigation, demanded ever more complex calculations, and mathematicians sought ways to lessen this work. One method devised in the 16th century was prosthaphaeresis, in which formulas from trigonometry were used for multiplication and division, but this was short-lived due to its complexity and inaccuracy. Then a discovery was made that would revolutionize calculation.

Development of logarithms
In 1614 Scottish mathematician John Napier (see left) published *Mirifici logarithmorum canonis descriptio* ("Description of the Marvellous Rule of Logarithms"), in which he explained the principle of calculation using logarithms. What Napier had found was that any number can be expressed as a power of 10, which he called a logarithm, and that when the logarithms of two numbers were added together the result was the logarithm of the product of those two numbers (the numbers multiplied together). Similarly, division could be achieved by subtracting the logarithm of one number from that of another.

With his friend Henry Briggs, Napier worked out logarithms of numbers to the base 10, which Briggs published as a set of tables. Using these tables, it was simple to look up the logarithms of

numbers, add or subtract them, and consult a table of antilogarithms to find the result. Napier and Briggs also discussed logarithms to a base other than 10 and discovered the natural

"Napier's bones"
In this late-17th century set, the "bones" are cylinders marked with 0 to 9 and columns of multiples of the same. You turned the cylinders then read off the digits along a specific row.

Pascal's calculator
This replica of Blaise Pascal's original calculator shows the dials, labeled for units, tens, hundreds, thousands, and so on. Numbers are "dialed in" as on an old-fashioned telephone. The windows at the top show black figures for addition and red for subtraction.

Different scales are indicated by letters

1 on sliding center scale is aligned with first number to be multiplied on lower scale

Hairline on cursor is slid to second number to be multiplied by on sliding center scale. Reading down the hairline to the lower scale gives the answer

Slide rule

To use a slide rule, a mark on the sliding central scale is aligned with a mark on either the top or bottom fixed scale, and the sliding cursor moved along to read off the answer. Here a simple multiplication is shown.

logarithm (see right), also called the Napieran logarithm, which became important in later mathematical work.

Logarithms made it easier for users to multiply, divide, work out square and cube roots, perform trigonometrical functions, and represent data distributed over a large range of values.

Mechanical calculating tools

It was a small step from the tables of logarithms and antilogarithms to mechanical devices based on similar principles. In 1617 Napier invented a kind of abacus called "Napier's rods." This comprised a set of rods inscribed with columns of numbers; the user arranged the rods on a special board, then read along one of the resulting rows of numbers, adding or subtracting digits to perform multiplication or division. Variations on this device soon appeared, usually in the form of rotating wooden or ivory cylinders ("Napier's bones") in a box. Another calculating machine was invented by French mathematician Blaise Pascal in 1645, using an elaborate mechanism of rotating wheels to perform additions and subtractions (see opposite).

Most important of all, however, was an ingenious and relatively simple device—the slide rule. Invented in 1622 by William Oughtred, it had logarithmic scales (in which the gradations become closer together as the value increases) inscribed on a set of connected rulers: the user slid the rulers to realign the scales, then read off the answer to a calculation. Slide rules were so effective that scientists, engineers, and technicians continued to use them until the advent of the electronic calculator in the 1970s.

NATURAL LOGARITHMS

Natural logarithms are logarithms to the base *e*. Discovered by John Napier and Henry Briggs, their significance was only recognized by Nicholas Mercator in 1688. The number *e* is irrational (it cannot be written as a simple fraction) and has been calculated to millions of decimal places; to five places, it is 2.71828. It provides a "baseline" for complex calculations such as the rate of decay of a radioactive isotope (see pp.298–99). The number also appears elsewhere in nature, in objects or systems that undergo a constant, exponential rate of growth, such as a spiraling Nautilus shell (right).

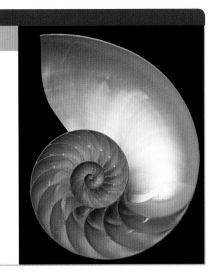

AFTER

Further advances in calculating techniques made mathematics an increasingly powerful tool in all areas of science.

CALCULATING MACHINES

Tables of logarithms continued to be used for calculations, not just in science but in everyday fields such as navigation and banking. There was, however, a constant worry about **inaccuracies** in the figures. This prompted inventor **Charles Babbage 174 ≫** to design his **Difference Engine** and **Analytical Engine**, to perform these calculations **mechanically** and print the results. In this way, he hoped to prevent errors.

MATHEMATICAL NOTATION

Swiss scientist **Leonhard Euler** was one of the greatest mathematicians. His innovations included **new symbols** for major concepts, such as Σ for "sum," *i* for the square root of −1, and *e* for the base of natural logarithms. He also found a rule known as Euler's identity, which is $e^{i\pi} + 1 = 0$. This simple formula involves the five most fundamental quantities in math.

SEE ALSO ≫
pp.102–03 GRAPHS AND COORDINATES
pp.174–75 CALCULATING AND COMPUTING

LEONHARD EULER

Ancient Egyptians never dissected humans and so proposed the theory that arteries carried air, an idea that persisted for centuries.

CHANNELS OF BLOOD AND AIR

In the 3rd century BCE ancient Greek physicians **Herophilus** and **Erasistratus** dissected blood vessels and saw heart valves. They distinguished **arteries** (which carry blood away from the heart) from **veins** (which carry blood toward the heart) by observing that it was only the severed arteries of a living pig that pulsed blood. However, in dead bodies, arteries empty into veins, so (like the Egyptians) they thought that **intact arteries carried air**.

BLOOD FLOW

In the 2nd century CE Roman physician **Galen** knew that arteries carried blood, but **proposed erroneous theories**, which endured for centuries. For example, he thought that arterial blood was made **in the heart** and venous blood made **in the liver**, and that the body consumed both.

GALEN

« SEE ALSO
pp.72–73 THE HUMAN BODY REVEALED

PHYSICIAN (1578-1657)

WILLIAM HARVEY

William Harvey was born in England and studied at the University of Padua. He had a deep conviction in the importance of experimental science and a particular interest in blood. In 1628 he published the results of 30 years of inquiry that established that blood circulated around the body through the heart's pumping action. His reputation as an accomplished doctor meant that he was favored in high quarters: he was physician to both King James I and King Charles I. He also tended the victims of the English Civil War.

By the 18th century scientists were faced with the challenge of reassessing blood production and blood flow in the body. They set out to disprove the established idea that large amounts of blood were constantly produced in the body in order to put forward the theory that we all possess a fixed quantity of blood, which is pumped around the body by the heart.

Blood in circulation

In 1553 a Spanish theologian named Michael Servetus proposed that blood circulated through lungs (pulmonary circulation). But his bold assertion conflicted with ancient Greek anatomy (see BEFORE) and so cost him his life. He was burned at the stake as a heretic. In fact, the medieval Arabic physician Ibn al-Nafis had said the same as Servetus back in the 12th century, but his theory was lost,

or ignored. Proof of their theory of pulmonary circulation came in 1559 with the experiments of Italian physician Matteo Colombo, who found that the pulmonary vein (which leads from the lungs to the heart) carried blood and not air as was previously believed. This was confirmed in 1564 by Italian anatomist Julius Aranzi, who also discovered that the blood of a mother and her fetus do not mix.

The pumping heart

Once scientists had ascertained that blood circulated around the body, their next challenge was to determine how this circulation took place. The ancient Greeks had thought that the heart worked by suction and that the pulse was due to arteries pumping blood. But it wasn't until the mid-16th century that Colombo suggested that the heart itself was the pump.

In 17th-century England the physician William Harvey (see left) had a special interest in blood and was a profound believer in the importance of experiments in science. In 1628 he established once and for all that blood flow was entirely circulatory. He experimented on animals to quantify the daily volume of blood pumped by the heart. He realized that such a great amount could not be continually made,

and so deduced that a fixed volume was circulated instead. He confirmed Colombo's idea of the heart as a pump, but went further in describing that blood flows away from the heart in arteries and returns to the heart in veins. He proposed that as well as pulmonary circulation there was another, more extensive, one through the rest of the body—the systemic circulation. Although some

Renaissance scholars, such as French anatomist Jean Riolan the Younger, insisted on adhering to older Galenic ideas, by 1700 Harvey's doctrine was established science.

In the late 17th century Danish polymath and physician Nicolas Steno found that the heart wall was composed of muscle; he was also the first to show that the heart consists of two relatively independent pumps. Around the same time, French anatomist Raymond Vieussens examined the left ventricle in detail, noting its extra thick muscle, necessary to pump blood all around the body.

Studying blood flow and valves
William Harvey found that blood pooled in the surface veins of a ligatured arm, and that valves—marked here by swellings—stopped blood inside from flowing back toward the hand.

Capillaries fill a gap

There remained one part of the story of circulation to complete: how blood passed from arteries to veins. Harvey predicted that blood passed through minute vessels too small to see and, in 1661, Italian doctor Marcello Malpighi discovered these microscopic tubes, called capillaries, in a frog's lung. Capillaries permeate all body tissues, bringing food and oxygen to respiring cells and removing waste carbon dioxide.

Life-giving blood

Malpighi may have been the first microscopist to see red blood cells, and English physician John Mayow

went as far as suggesting that blood carried oxygen from the lungs. However, it was many years before scientists could prove that it was hemoglobin (a protein respiratory pigment) in the red blood cells that carried oxygen from the lungs, where blood is oxygenated, to the body tissues (see pp.270–71).

Circulation of the Blood

Blood is a life-giving source on which all cells depend. It is pumped around the body through a complex system of vessels by the heart muscle. But although blood is such an obvious commodity in the body, its function and means of passage were not fully understood until the end of the Renaissance.

Capillary wall · Red blood cell

△ **Red blood cells**
These cells contain hemoglobin, which delivers oxygen around the body. Their disk form bends as they squeeze through tiny capillaries, bringing the oxygen close to capillary walls so it can be released into other cells.

Arteriole | Body cells | Capillary

◁ **Capillaries supply blood to organs**
As blood flows through capillaries as shown here, nutrients and oxygen pass from the blood across the ultra-thin capillary walls into neighboring cells. Carbon dioxide and metabolic wastes diffuse inward.

Aorta
The body's main artery, the aorta carries oxygenated blood away from the heart.

◁ **The human heart**
A wall (septum) divides the heart: deoxygenated blood from the venae cavae enters the right side and is pumped to the lungs, where it is oxygenated; the oxygenated blood enters the left side of the heart via the pulmonary vein and is pumped out through the aorta to the rest of the body.

Pulmonary artery
Artery that carries deoxygenated blood from the right ventricle to the lungs.

Semilunar valves
As the ventricles relax, these valves close to prevent backflow of blood from the arteries.

Pulmonary veins
Veins that carry oxygenated blood to the left atrium from the lungs.

Superior vena cava
Vein that carries deoxygenated blood to the right atrium from the upper body.

Right atrium
Thin-walled chamber that pumps blood into the right ventricle below.

Tricuspid valve
As the ventricles contract, this valve closes to prevent backflow of blood into the right atrium.

Right ventricle
Chamber with a thick wall to pump blood to the lungs.

Inferior vena cava
A large vein that carries deoxygenated blood to the right atrium from the lower body.

Left atrium
Chamber with a thin wall to pump blood into the left ventricle.

Descending aorta
Carries oxygenated blood away from the heart to the body tissues.

Left ventricle
Chamber with an extra thick wall to pump blood all around the body.

Mitral valve
As the ventricles contract, this valve closes to prevent backflow of blood into left atrium.

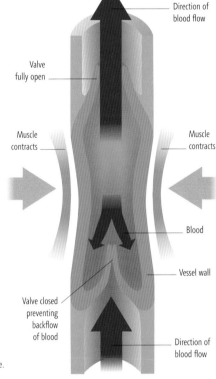

Direction of blood flow

Valve fully open

Muscle contracts

Muscle contracts

Blood

Vessel wall

Valve closed preventing backflow of blood

Direction of blood flow

△ **How valves in veins work**
Big veins run between limb muscles, which help squeeze deoxygenated blood upward against gravity. Valves within the veins prevent backflow as the blood returns to the heart. Portuguese physician, Amato Lusitano, was the first person to discover the role of valves in the mid-1500s by showing that air could be blown only one way through a dissected vein.

AFTER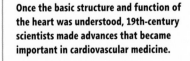

Once the basic structure and function of the heart was understood, 19th-century scientists made advances that became important in cardiovascular medicine.

THE ARTIFICIAL PACEMAKER
In 1839 a Czech anatomist called **Jan Purkinje** discovered specialized muscle fibers (known as Purkinje fibers) in the heart. The heartbeat is coordinated by the fibers as they carry an **electrical impulse** that causes the muscle of the heart's ventricles to contract. Malfunctions of this system are today treated with artificial pacemakers, which **stimulate the heart** with an implanted electrode. Pacemakers are being tested that are no larger than a vitamin pill and can be inserted into the heart via a leg catheter.

SEE ALSO >>
pp.228–29 SAFER SURGERY
pp.406–07 MODERN SURGICAL PROCEDURES
pp.448–49 REFERENCE: BIOLOGY

How the heart beats
The cardiac cycle is the series of heart movements associated with a single heart beat. Contraction of heart muscle is called systole, relaxation is called diastole. Blood always flows from high to low pressure, with valves preventing backflow.

Right and left atria contract

Oxygenated blood

Deoxygenated blood

Ventricles filled to capacity

Atrial systole
As the thin walls of the atria contract, they pump blood the short distance into the ventricles, ensuring they are filled to capacity with blood.

Semilunar valves open

Tricuspid and mitral valves forced shut

Right ventricle contracts

Left ventricle contracts

Ventricular systole
The thick walls of ventricles pump blood out of the heart; at the same time the triscupid and mitral valves shut, causing the first heart sound.

Semilunar valves close

Tricuspid and mitral valves open

Ventricles relaxed

Diastole
As the whole heart relaxes, blood flows from the veins into the atria and then partly fills the ventricles, causing the semilunar valves to shut, which produces the second heart sound.

PHILOSOPHER AND POLYMATH **Born 1635** Died 1703

Robert Hooke

"The science of **Nature**…should return to the plainness and **soundness** of observations."

ROBERT HOOKE, "MICROGRAPHIA," 1665

Robert Hooke has been described as "England's Leonardo." An inventor of instruments such as the compound microscope, he was the scientist who helped form the Royal Society, as well as Christopher Wren's equal partner in London's reconstruction after the Great Fire of 1666. His *Micrographia* was so enthralling that Samuel Pepys recorded in his diary how he read it until 2 AM, describing it as "the most ingenious book that ever I read in my life."

Hooke came from a modest home on the Isle of Wight, where he was born in 1635, the youngest child of John Hooke, the curate of All Saints Church in Freshwater. A sickly child, he was too poorly to attend school at first, but was very smart, with an aptitude for making things. Given an old brass clock to play with, he took it apart to see how it kept time, and made a wooden copy of it that actually worked. After his father died in 1648, Hooke left the island with an inheritance of £50 to attend Westminster School, London, where he studied music and became an excellent draftsman.

The first professional scientist

At age 18 Hooke gained a place at Oxford University as a choral scholar, although he was really studying what we would now call science. By this time he had developed a curvature of the spine that gave him a pronounced stoop, probably due to a condition called Scheurmann's kyphosis, aggravated by poor diet as a child.

Title page from *Micrographia*
Robert Hooke's *Micrographia* astonished readers such as Samuel Pepys with its detailed drawings of tiny creatures, but the work also dealt with the nature of gravity, fossils, craters on the Moon, and many other topics.

This foldout, detailed illustration of the underside of a louse is typical of Hooke's illustrations.

INVENTION

COMPOUND MICROSCOPE

While lenses had previously been used together for magnification, Hooke invented the first practical compound microscope ("compound" meaning it has more than one lens). He used an oil lamp as a light source and a concave mirror to direct the light onto the object via a lens. Inside the microscope itself were four extendable concentric tubes containing an object lens, a middle lens, and the eyepiece. To focus, Hooke raised and lowered the entire tube to change the distance between the objective lens and the object. Fine adjustments were made by turning the end of the tube.

Lacking money to support himself, Hooke worked as a part-time assistant of an Oxford doctor, Thomas Willis, who had his own chemical laboratory, and then got a similar job with the great Robert Boyle. It worked out so well that he never completed his degree, becoming instead the first full-time, paid, professional scientist.

Boyle was a wealthy aristocrat, but he always treated Hooke as a scientific equal, not as a servant. While they were working together, Hooke discovered the law that relates the pressure and volume of a gas. Boyle wrote about this in a book published in 1662, and was careful always to say that it was Hooke's discovery. But because the discovery appeared in Boyle's book, to this day it is known as "Boyle's Law," although Hooke did get his own name on the law that describes how springs stretch.

Birth of the Royal Society

In the same year, 1660, that Charles II was restored to the throne, a group of men, including Robert Boyle, formed a society "for the promoting of Experimentall Philosophy" under the King's patronage. This became the Royal Society. Hooke became its first "Curator of Experiments." His job was to demonstrate and test new discoveries, either those made by the Fellows of the Society, or by himself, or reported from other places.

In his own right, Hooke worked a lot on optics; his book *Micrographia* was published in 1665. Containing many engravings of creatures he had seen under the microscope, it was the first truly popular book about science.

London Monument

As City Surveyor, Hooke designed the Monument to the Great Fire of London, a hollow pillar 202ft (61.5m) high, although Wren is unjustly credited for it. The pillar was completed at the end of 1676.

astronomer and a Fellow of the Royal Society, as well as being an architect) were given the job of planning the reconstruction of the city. Hooke now had a salary as a City Surveyor, with the opportunity to accept commissions. Hooke was Wren's equal partner in this work and many of the "Wren" churches in the City of London were actually designed by Hooke.

Hooke knew poverty for many years, but when he died in 1703 he left a fortune of £8,000 in a great iron chest, earned through his many achievements.

BREAKTHROUGH

HOOKE'S LAW

Hooke's Law describes the behavior of a stretched spring or other elastic object. It states "As the extension, so the force" — the amount of stretch is proportional to the force pulling on the spring. When the force is doubled, the spring stretches twice as much; when it is removed, the object reverts to its original size. This is only true for any elastic object up to its "elastic limit." If stretched beyond this limit, the springy object will never return to its original size, and may break.

Hooke's life was greatly changed after the Great Fire of 1666 destroyed much of London. He and Wren (who was an

MERCHANT (1632–1723)

ANTONY VAN LEEUWENHŒK

Van Leeuwenhoek first used a magnifying glass during his apprenticeship with a textile merchant in Amsterdam, Holland, and later made his own microscopes. He was not trained in science and never wrote a book, but academic organizations published his findings. The Royal Society was initially sceptical about his discovery of single-celled organisms, but he proved their existence in 1680, so securing his reputation.

≪ BEFORE

The idea that a microscopic world existed beyond the naked eye is an ancient one, but it wasn't until the first lenses were made that this world could be seen.

THE FIRST MAGNIFIERS
The writings of **Seneca the Younger** of ancient Rome include perhaps one of the earliest explicit references to a **device for magnification**: "letters, however small and indistinct, are seen enlarged and more clearly through a globe or glass filled with water." In the Middle Ages, **Alhazen's** *Book of Optics* ≪ 60–61 was the foundation of an area of science that led to the **invention of eyeglasses** in 13th-century Italy, and the **microscope** in 16th-century Holland.

MAGNIFYING GLASS

≪ SEE ALSO
pp.72–73 THE HUMAN BODY REVEALED
pp.84–85 EXPLORING THE SKIES

Microscopic illustrations
Robert Hooke's *Micrographia*, published by the Royal Society, illustrated microscopic life with a degree of detail not seen before. It is known for its foldout plates, such as this example depicting a flea, but Hooke also included scientific ideas about light and astronomy.

Microscopic Life

The invention of the microscope opened up a new world of discovery for biologists of the Renaissance. Not only could they study anatomy in far more detail than they had previously thought possible, but they found whole communities of new life forms in a single drop of pond water.

Studies of medieval scholars were limited by simple magnifying glasses. Microscopy opened up life in miniature and eventually led to the development of ideas critical to biology.

First views of a miniature world
The identity of the first inventor of the microscope is debatable, but the Dutch glasses makers Hans and Sacharias Jansen are likely contenders. Around 1590 they discovered that a hollow tube containing two lenses, one behind the other, could produce more enlarged images than a simple magnifying glass. They had made a compound microscope. In 1609 Italian astronomer Galileo (see pp.82–83), inspired by the invention of the spyglass by Dutch spectacle maker Hans Lippershey, constructed his own instrument for observing the skies and increased the magnification from three to 30 times. Around this time German botanist Giovanni Faber coined the word "microscope"—from the Greek words *micron* (small) and *skopein* (to look at)—to distinguish it from the telescopes that had been developed using similar technology (see pp.84–85). It was in 1665 that English philosopher Robert Hooke (see pp.92–93) coined the term "cell" to describe the microscopic units he found in a section of cork. In the same year Hooke published his book *Micrographia*, an exquisitely illustrated miscellany of the microscopic world (see below).

Early simple microscope
Van Leeuwenhoek fixed a tiny spherical lens in the hole in a brass plate. His specimen was held behind the lens on the tip of a needle, its position adjusted by turning a screw to focus the image.

The father of microbiology
After seeing *Micrographia* Dutch merchant Antony van Leeuwenhoek developed a technique of making lenses that would have allowed for more cost-effective production of his microscopes without compromising on the quality of the lens. However, he kept his lens-making technique a secret, and it was 200 years before comparable models were made. His microscope had a single spherical lens, made by drawing out a thread

Basic modern microscope

Light reflected from a mirror passes through the specimen and then through different lenses in the microscope's tube to produce a magnified image. Focusing is achieved by knobs on the side of the instrument. Pictured here is a simple instrument, commonly used in schools.

Eye
The viewer sees a focused image.

Eyepiece lens x10
Together with objective lenses, total possible magnifications are x400, x100, and x40.

Fine focus
Function that allows small movement of the tube for use at high magnifications.

Coarse focus
Function that allows large movement of the tube for use at low magnifications.

Tube
Light passes through here from the objective lens to the eyepiece lens.

Revolving nosepiece
Piece that holds different objective lenses above the specimen.

Stage
Part of the microscope that supports the specimen.

Objective lens
Magnifications are one of three: x40, x10, and x4.

Specimen
A specimen would be placed on a glass slide held over a hole in the stage by clips.

Light
Light is reflected up to the specimen by the surface of the mirror.

Pivot
Allows microscope to be angled for ease of use.

Base
The base is weighted to provide stability.

of glass, melting and rounding off its tip, and filing one side flat once it had cooled. He showed that pond water was teeming with miniature life forms, which he called "animalcules." He was seeing single-celled microbes, and—unlike Hooke's static cork cells—they were buzzing with life. Light microscopes rely on light rays being directed through a very thin transparent specimen. Leeuwenhoek's lenses magnified up to 270 times and he was the first to see some of the smallest things that can be viewed by a light microscope, including human sperm and even, possibly, bacteria.

Improving the view

The light microscope was routinely used in science after the Renaissance. Today, liquids, such as blood or pond water are drawn out into a fine film, while solid objects, such as wood or stems, must be cut into slices of around 0.02 mm thick, using an instrument called a microtome.

In the mid-19th century, German physicist Ernst Abbe made a breakthrough in understanding the critical position of the lenses in the microscope for sharper image quality. But the light microscope is limited by its resolution due to visible light available. Two objects closer than half the wavelength of light (around 0.0004mm) appear as one, which means that certain tiny particles, such as viruses, are entirely beyond the microscope's resolution. Three centuries after Hooke and Leeuwenhoek, the electron microscope (see right) solved this problem and produced images of the miniature world that they would scarcely have thought possible.

Staining cells

These blood cells have been stained and photographed through a microscope to reveal their structures more clearly. Here, white blood cells have been stained purple from a patient with myeloid leukemia, a disease characterized by the large numbers of white blood cells.

AFTER

Emerging theories dominated a new kind of biology, which relied on the microscope.

MICROSCOPIC THEORIES
In 1850 the **microscope** was key to uncovering the theory that all living organisms were **composed of cells 214–15 »**. In the 1800s microbiologists used microscopes to show that **decomposition** could be caused by **microorganisms 242–43 »**.

ELECTRON MICROSCOPE
First built by German physicist **Ernst Ruska** in 1931, the electron microscope resolves **more detail** than the light microscope since **electrons** have a **smaller wavelength** than light.

SEE ALSO »
pp.214–15 ANIMAL AND PLANT CELLS
pp.392–93 NANOTECHNOLOGIES
pp.408–09 BODY IMAGING

Microscopic findings

Early microscopes could magnify 30 times, revealing objects around 0.1 mm in size, such as plant cells. Advances in design allowed greater magnifications, so revealing smaller objects inside cells.

1674 Antony van Leeuwenhoek observes "giant coiled" chloroplasts in *Spirogyra*; length around 60 micrometers.

1890 Richard Altmann publishes description of mitochondria (energy-generating cell structures); length 1 micrometer.

1876 Robert Koch observes rods of anthrax bacteria in blood of infected animals; length 3 micrometers.

mid-1950s George Palade observes ribosomes (protein-generating cell structures) with an electron microscope; diameter about 0.02 micrometer.

1650	1700	1750	1800	1850	1900	1950

1663 Robert Hooke observes cork cells; diameter 40 micrometers.

1683 Leeuwenhoek possibly observes bacteria; diameter 2 micrometers.

1676 Leeuwenhoek observes protozoa, such as *Amoeba* and *Paramecium*; diameter 50 micrometers.

1831 Robert Brown discovers nucleus in cells of plant leaves; length 40 micrometers.

1848–49 Wilhelm Hofmeister sees clusters of chromosomes in plant leaves coiled in nucleus; cluster diameter 2 micrometers.

1939 Gustav Kausche, Edgar Pfankuch, and Helmut Ruska view Tobacco Mosaic Virus using a transmission electron microscope; length 0.3 micrometer.

Discovery of the Vacuum

The ancient Greek philosophers were convinced that a vacuum was nothing, and therefore an impossibility: "Nature abhors a vacuum." In the 17th century, however, vacuums were discovered and their properties explored in detail. Scientists were quick to realize their potential value.

The Magdeburg hemispheres

In 1650 Otto von Guericke invented the air pump and four years later gave a public demonstration of the vacuum. Having pumped the air from two copper hemispheres, he showed that two teams of horses could not pull apart "nothing at all."

« BEFORE

Before the 17th century, people believed that there was no such thing as a vacuum.

"WHAT IS" AND "WHAT IS NOT"

The ancient Greek philosopher and astronomer **Thales of Miletus** said that everything in the Universe must be "Thing" or "No-Thing." A **vacuum** was "No-Thing" and therefore, by definition, did not exist.

FOUR ELEMENTS

The Greek philosopher **Empedocles** asserted that all matter is made of only **four elements**: earth, air, fire, and water, in different proportions. He ingeniously showed that **air is not nothing** by pushing an upturned bucket into the sea and showing that the inside did not all get wet.

ESSENTIAL PARTICLES

The Greek philosopher **Democritus** believed that everything is made of tiny particles (which he called "**atoms**," from the Greek word for "**indivisible**"). A **vacuum** would be "nothing"—an **absence of particles**. One of Democritus's successors, **Aristotle**, disagreed with the theory of atoms, stating that a **vacuum was impossible**, and so could not exist. "Nature abhors a vacuum" is an Aristotelian idea. This view was later supported by the Catholic Church, as they said the absence of everything implied the absence of God.

ARISTOTLE

A vacuum is literally nothing; it is empty space. For thousands of years philosophers thought such a thing could not exist, but in the early 1600s the Italian scientist Galileo Galilei (see pp.82–83) became involved. He had been told by pump makers that they could not pump water out of a well more than 30 ft (10 m) deep. Their pumps could push water higher than this, but they could not suck it up.

He discussed this observation with a pupil, Evangelista Torricelli, who suggested experimenting with mercury instead of water. Mercury is 13 times denser than water and might show the same phenomenon at a smaller scale, making it much easier to investigate.

Torricelli's experiments

Torricelli took a 3-ft (1-m) tube, closed at one end, and filled it completely with mercury. He dipped the open end in a bowl of mercury and raised the closed end. Suddenly—when the height of the mercury column reached about 30 in (76 cm)—a gap appeared in the tube above the mercury. It could not be air that had been dissolved, because when he lowered the end of the tube the gap closed immediately; air cannot dissolve so quickly. The gap reappeared when he raised the end of the tube again. The gap was a vacuum, and it became known as a Torricellian vacuum.

Torricelli realized that the mercury must be rising because of atmospheric pressure, which pressed down on the mercury in

BOTTOM OF MOUNTAIN

- Torricellian vacuum in tube above mercury
- Mercury column 28 in (71 cm) high
- Tall, thick-walled glass tube (closed at top)
- Atmospheric pressure pushes down on mercury surface
- Open end of tube immersed in dish of mercury

TOP OF MOUNTAIN

- Vacuum above mercury
- Mercury column 25 in (63 cm) high
- Atmospheric pressure lower at mountaintop

Pascal's experiments with atmospheric pressure
Blaise Pascal's brother-in-law Florin Périer measured a height of 28 in (71 cm) of mercury at the bottom of the 3,300-ft (1,000-m) mountain Puy de Dôme, then carried the apparatus up, and measured 25 in (63 cm) at the top.

PHYSICIST (1608–47)

EVANGELISTA TORRICELLI

Born in Faenza, Italy, of a poor family, Torricelli was fortunate in being able to study mathematics, and in becoming assistant to Benedetto Castelli, who taught at the University of Sapienza in Rome. Through Castelli, Torricelli made contact with his hero Galileo, and was taken on as his assistant in October 1641. When Galileo died three months later, Torricelli was appointed to succeed him as court mathematician to the Grand Duke Ferdinando II of Tuscany.

Creating the unbreakable seal
Otto von Guericke pumped the air out of two copper hemispheres of around 20 in (50 cm) diameter sealed with a greased leather ring. Because this created a vacuum, there was no air pushing outward from the insides of the sphere, but there was pressure (atmospheric) pushing on them from outside, which forced the spheres tightly together.

the bowl, forcing the heavy liquid up the tube. The atmospheric pressure could only push the mercury up by 30 in (76 cm), because this figure corresponds to normal atmospheric pressure. But water is 13 times less dense than mercury, so atmospheric pressure can push water up 13 times higher (about 33 ft/10 m).

Varying pressure
French philosopher Blaise Pascal predicted that the height of a mercury column would be lower on top of a mountain, where the pressure of the

atmosphere should be lower. Experiments by his brother-in-law, Florin Périer, in 1648, proved him right. He also predicted that the atmosphere must therefore gradually thin out, and at some height above the earth give way to empty space—in other words, a vacuum.

Robert Boyle and Robert Hooke repeated Torricelli's experiments in Oxford, UK, from the 1650s onward. They noticed that the height of the mercury column varied from time to time. At first they thought that it depended on the phases of the Moon, like the tides, but then realized that it depended on the weather: when the mercury was rising there was good weather to come; when it was falling there was bad weather on the way. In effect, they invented the barometer.

The air pump
By 1659 Hooke had made an effective air pump for Boyle, and they used it in a series of experiments. They discovered that a piece of red-hot iron inside could be seen undimmed—so light does travel through a vacuum—but a bell rung inside could not be heard, which shows that sound waves need air (or another medium) to carry them. They discovered that a candle would not burn in a vacuum, and also that small animals died if left inside the flask. Later Joseph Priestley and others showed that oxygen is necessary both for combustion and for breathing (see pp.146–47); the air pump removed the air, including the oxygen.

In practice, making a perfect vacuum is impossible. Von Guericke, Boyle, and Hooke probably reduced the air pressure by three-quarters but no more; even in deepest space there is on average about one hydrogen atom in every cubic yard (or cubic meter). The rules of quantum physics mean that even a perfect vacuum contains short-lived virtual particles.

Glass flask full of air

Flame is extinguished

Air is pumped out of flask

Flame is extinguished in a vacuum
A candle burns in a flask full of air, but when the air is pumped out the flame is extinguished, showing that air (or specifically, oxygen) is needed to support combustion.

Bell rings audibly in flask full of air

Air is pumped out of flask

Sound is muted in a vacuum
A bell ringing in a flask full of air can be clearly heard from outside the flask. But when the air is pumped out, no sounds escape the flask. This is because sound is carried by air, and cannot travel through a vacuum.

Bell still rings, but is now inaudible

AFTER

Scientists soon realized they could harness a vacuum's power.

ATMOSPHERIC STEAM ENGINES
The first effective **atmospheric engine** 130–31 ⟫ was built by **Thomas Newcomen** in 1712. By generating a vacuum, it used the power of atmospheric pressure.

THE LIGHT BULB
Joseph Wilson Swan demonstrated the first useful **light bulb** in the UK, in February 1879, just ahead of

Thomas Edison 264–65 ⟫ in the US. The bulb's vacuum ensured that the carbon filament did not burn out.

EXPERIMENTS IN A VACUUM
The largest vacuum system on Earth, with the volume of a cathedral, was built for the Large Hadron Collider 404–405 ⟫.

SEE ALSO ⟫
pp.98–99 **ROBERT BOYLE**
pp.292–93 **VACUUM TUBES**

CHEMIST AND PHYSICIST Born 1627 Died 1691

Robert Boyle

"There is a **spring**, or **elastical power**, in the air we live in."

ROBERT BOYLE, FROM "NEW EXPERIMENTS PHYSICO-MECHANICALL," 1660

Robert Boyle has been called the "father of chemistry"; he broke away from the old ideas of alchemy and defined what is meant by an element. With the help of Robert Hooke, he performed extensive research into the properties of the vacuum, and the relationship between pressure and volume in gases.

Robert was born in 1627 at Lismore Castle, in County Waterford, Ireland. He was the seventh son and 14th child of Richard Boyle, who had become the Earl of Cork in 1602 when he bought Lismore from the adventurer Sir Walter Raleigh (then in prison). Boyle watched as his father married off all his older siblings, mainly for financial or social advantage, and this

sorry spectacle may have put Robert off marriage. At the age of eight he was sent to Eton College, the well-known English boarding school. He loved the academic work but hated the compulsory games. Four years later he and his brother Francis were taken by a French tutor on a tour of Europe, winding up in Italy, where they spent the winter of 1641–42 in Florence. The great Galileo (see p.82–83) died there in 1642, and the public reaction inspired Boyle to pursue similar ideas.

Reduced circumstances

Returning to England in 1644 Boyle found the country in civil war, his father dead, and himself broke, since all the income that had come from the Irish estates had been stopped. He first settled at Stallbridge in Dorset, but visited his favorite sister, Katherine, in

> **BOYLE'S LAW** Robert Boyle referred to the relationship between pressure and volume as "Mr. Towneley's hypothesis."

Pendle Hill
On this hill in Lancashire, once famous for its witch trials, Richard Towneley and his physician Henry Power first noted, in 1661, that a trapped sample of air increased in volume as they went up the hill.

Gresham College
Founded in London in 1597 by Sir Thomas Gresham, Gresham College provided lectures that were free to the public. It became a meeting place of the "invisible college," and later of the Royal Society.

Boyle and Papin in Maiden Lane Laboratory
French scientist Denis Papin, a friend and colleague of Boyle's, invented some of the first steam engines. In this laboratory Boyle's technician, Ambrose Hanckwitz, was set up to make phosphorus.

London. Full of enthusiasm for science, the "new philosophy," Boyle met at Katherine's house a group of like-minded people who called themselves the "invisible college." They included John Wilkins, diarist John Evelyn, Robert Hooke, and William Petty, inventor of the catamaran. They met sometimes at Gresham College in London, and later in Boyle's rooms in Oxford—safely distant from Oliver Cromwell's base—and exchanged ideas by letter, or by sending one another books with notes in the margins. In 1652 Boyle visited Ireland, then torn apart by rebellion, and, with the help of William Petty, restored some order to his estates, which meant that money started to flow again. Thereafter he had a substantial income for life.

A scientific partnership

In 1654 Boyle hired the ingenious Robert Hooke as an assistant, and although Hooke was of a lower social class they seem to have worked together as equal partners. Boyle had lots of ideas, and Hooke was the right person to put them into practice. They investigated the Torricellian vacuum (see p.96), and it may well have been Boyle and Hooke who discovered that the height of Torricelli's column of mercury varied with the weather, and thus invented the barometer. When Hooke built an air pump, which they used and demonstrated extensively, it came to be called the Machina Boyleana.

Two years after Cromwell's death in 1658 Charles II returned from exile and was

"I could be content the world should think I had scarce looked upon any other book than that of nature."
ROBERT BOYLE

crowned king, and the pursuit of science was once again safe for gentlemen. In the same year the invisible college became the Royal Society of London for the Promotion of Natural Knowledge, abbreviated to the Royal Society. Boyle was one of the founding Fellows.

Discovery of "Boyle's Law"

In 1661 Boyle received a letter from Richard Towneley (see opposite), an astronomer and the first person in England to measure rainfall systematically. On April 27 he and his physician, Henry Power, had climbed Pendle Hill, 1,827 ft (557 m) high, near Towneley Hall in Lancashire, carrying a barometer. They found that the air pressure at the top of the hill was lower than at the bottom, and that trapped air occupied a greater volume when the pressure was lower. After getting Hooke to check the measurements, Boyle published the results in 1662, but then Power published the findings himself in 1663.

Although Boyle referred to the work as "Mr. Towneley's hypothesis" and said that Hooke had done the checking, the relationship is known as "Boyle's Law" (see p.93). In Europe the relation is often attributed to Edme Mariotte, but he did not publish it until 1676.

With the large income from his Irish estates, Boyle was able to do scientific research, publish his own books, and pay assistants, but he remained a modest man. He refused offers of a peerage and a bishopric, and turned down an invitation to become President of the Royal Society.

Air pump
Built by Hooke, this device was a lot like a bicycle pump working in reverse, pulling the air out of a flask. There were great technical difficulties in making airtight seals, however, and it is unlikely that a complete vacuum was ever achieved with it.

Robert Boyle
Johann Kersebook painted this portrait around 1689, when Boyle was 62, although he barely looks so old. Boyle was uncomfortable about having his portrait painted, which may be why he looks so awkward.

TIMELINE

January 25, 1627 Robert Boyle is born at Lismore Castle in County Waterford on the south coast of Ireland. He is the 14th child and the seventh son of the First Earl of Cork.

1635 Sent to Eton College, near Windsor, England.

1639 Travels to Europe with his brother Francis (who had been married only days before) and a French tutor. They go to France, settle in Geneva for a while, and then go on to Italy.

1641 Spends the winter in Florence, where the aging Galileo dies the following year. Boyle is inspired by the public interest in him to read Galileo's work and to follow in his footsteps.

1644 Returns to England and begins to work on the "new philosophy," namely the ideas of experimental science as advocated by Francis Bacon (see pp.68–69) some 20 years before. He becomes one of the leading exponents of science based on observation and facts, rather than mere guesswork.

1652 Visits Ireland, with William Petty, in order to recover his estates, ravaged by the Irish rebellion. Thereafter he has a substantial income.

1654 Moves to Oxford and rents rooms in Cross Hall on the High Street. A plaque now hangs there, on the outside wall of University College. He hires Robert Hooke as assistant.

1659 Boyle and Hooke make an air pump, following the example of Otto von Guericke (see p.97). They use it to carry out a series of experiments, showing that air is necessary to carry sound, to sustain a flame, and to sustain life.

1660 Publishes the results of his experiments with the air pump: *New Experiments Physico-Mechanicall, Touching the Spring of the Air, and its Effects*. The work is criticized by the English scientist Franciscus Linus. There is much philosophical argument about the nature of the vacuum (see p.96).

1661 Publishes *The Sceptical Chymist*, an attempt to bring scientific ideas to alchemy, but also the beginning of chemistry as we now understand it.

1662 In the second edition of *New Experiments Physico-Mechanicall*, Boyle publishes "Boyle's Law," although he does not claim credit for the idea for himself. The Royal Society receives its charter, and Boyle becomes a Fellow.

1668 Moves to London to live with his sister Katherine, who is 14 years older than he is, but his favorite sibling. He never marries, nor seems to have had a long-term partner.

1669 Publishes *Paradoxa Hydrostatica Novis Experimentis*, describing hydrostatical paradoxes.

1691 Katherine dies on December 24; Robert Boyle dies on December 30.

PARADOXA HYDROSTATICA

The **Behavior** of **Gases**

Before the 17th century no serious scientific work was done on gases—they were all thought to be "just air."
However, the pioneering work of scientists studying the weather led to the discovery that air and other gases
behave according to a set of mathematical laws. This led to a greater understanding of atoms and molecules.

« BEFORE

Gases were not studied in detail before the
17th century, although the ancient Greeks
included "air" as one of the four elements
from which everything is made.

EXPLOITING PRESSURE

In 9th century Baghdad the **Banu Musa brothers**
produced a book of **ingenious devices**, some of
which used clever control of air pressure—including
a pitcher that would not pour unless you removed
your thumb from a hole near the handle.

THE WORK OF TORRICELLI

Torricelli « 96–97, inspired by his mentor
Galileo « 82–83, investigated and discovered
the vacuum. His work inspired other scientists to
start experimenting on air held at low pressure.

« SEE ALSO
pp.44–45 "EUREKA"
pp.96–97 DISCOVERY OF THE VACUUM
pp.98–99 ROBERT BOYLE

During the 17th century science
flourished in England, and several
natural philosophers became interested
first in the behavior of the air and,
later, of other gases. The result of their
investigations was a set of "gas laws"
that mathematically describe the
interrelationships between pressure,
volume, and temperature.

Pressure and volume

Richard Towneley of Lancashire,
England was a keen meteorologist
interested in the characteristics of the
atmosphere. In 1661 he collected a
sample of "Valley Ayr" (probably
trapped by mercury in a J-tube) and
took it up Pendle Hill, where he found
that the volume of the air had increased.
Then he collected a sample of
"Mountain Ayr" and took it down the
hill, to find its volume had decreased.
  Knowing that the atmospheric
pressure must be lower at the top of
the hill, because of the experiments
done by Florin Périer in France 13 years

earlier, he realized that for a fixed
amount of air, lowering the pressure
increased the volume, and vice versa.
He told his friend Robert Boyle
(see pp.98–99) about this, who in
turn asked his colleague Robert Hooke
(see pp.92–93) to confirm the findings,
using the air pump that they had

1 IN (25 MM) is about the
distance mercury falls in
a barometer for every 1,000 ft
(300 m) increase in height,
reflecting the fall in air pressure.

developed to create a vacuum. In 1662
Boyle published the results, and the
relationship came to be called "Boyle's
Law," although Boyle himself referred
to it as "Mr. Towneley's hypothesis."

Bouncing gas

Robert Boyle thought air particles were
like tiny coils of wool, which could be
squashed, but would bounce back. In
fact, gases are made up of millions of

Single weight produces
pressure (P) in gas

Two weights produce
double
the pressure
(2P) in gas

Low pressure
in flask

High pressure
squeezes particles
into half the
original volume

Boyle's Law
The pressure and volume of a fixed mass of ideal
gas at constant temperature are inversely proportional
to one another. If the pressure is doubled, the volume
is halved. This relationship, which can be written as
the formula $PV = constant$, was first formulated by
Robert Boyle in 1662.

Two weights
produce equal
pressure (P)
in each flask

High pressure inside

Hot particles move faster and
expand volume while
maintaining pressure

Heat

Charles's Law
At constant pressure, the volume of a fixed
mass of ideal gas is directly proportional to its
absolute temperature. If the temperature is doubled,
the volume is doubled. This can be written as
$V/T = constant$. The relationship was formulated
by Jacques Charles in around 1787.

One weight produces
pressure (P) inside

Two weights produce
double the pressure
(2P) inside

Low pressure inside

Pressure increases because
hot particles move faster

Heat

Gay-Lussac's Law
For a fixed mass of ideal gas at constant volume,
the pressure is directly proportional to the absolute
temperature. If the absolute temperature is doubled,
the pressure is doubled. This can be written as
$P/T = constant$, and it was first published, along with
Charles's Law, by Joseph Gay-Lussac in 1802.

◁ **The dawn of hot-air ballooning**
The first hot-air balloon flight carrying people was made in France in 1783. Scientists soon realized that hot-air balloons could provide them with a platform for studying the atmosphere at various altitudes.

Air of equal density

Balloon steady
The air becomes gradually less dense with increasing altitude. When the air's density matches the balloon's density, the balloon stops climbing.

Air inside is less dense than the air outside

Direction of movement

Rising balloon
The air inside the balloon is at the same pressure as the atmosphere; when it is heated, its volume increases, and some of the balloon's air escapes. This lowers the average density of the balloon, and causes it to rise.

individual atoms or molecules, all flying about at great speed. The pressure on the walls of any container they are in is the same in all directions, and it is caused by these particles crashing into the walls and bouncing off them. If the volume of a container is halved, any particle inside will hit the walls twice as often, and so the pressure doubles.

The effect of temperature

In the 18th century French scientists Jacques Charles and Joseph Gay-Lussac investigated the effect of temperature on gases. They found that when a fixed amount of gas is heated and its pressure is kept constant, its volume increases. This became known as Charles's Law (see opposite). On the other hand, if its volume is kept constant, its pressure increases (Gay-Lussac's Law).

When a fixed amount of gas is heated, the particles travel much faster—their speed is directly related to their temperature—so they hit the walls of a container much harder. This maintains the pressure if the volume is allowed to expand, or increases it if the volume is kept constant. However, these theoretical gas laws are only approximate, because the atoms and molecules of real gases are all somewhat "sticky," which tends to decrease their volume and their pressure slightly.

Independent gases

One of the most enthusiastic early weather watchers was Englishman John Dalton, who realized that gases in a container do not sit in the bottom, as liquids do, but occupy the whole space. This means that if there are two gases in a container, both will occupy the entire volume, and their particles will behave independently. The total

pressure exerted by the gases is therefore the sum of the partial pressures of all the gases inside.

In 1858 the Italian chemist Stanislao Cannizzaro applied Dalton's Atomic Theory and Avogadro's Hypothesis (for both, see pp.138–39), and the experimental methods of Gay-Lussac to deduce relative atomic masses for

Pioneering the pneumatic tire
The first practical air-filled rubber tires were made by the Scot John Boyd Dunlop, who described them as "pneumatic." He made them in response to his son's complaint that a bicycle ride was "too bumpy."

various elements. Thus, what started for some scientists as an interest in the weather gradually became more mathematical and led to real insights into the behavior and properties of atoms and molecules.

AFTER

Understanding how gases behave started a whole train of inventions and discoveries in both science and technology, ranging from the gas laws to the steam engine and the household vacuum cleaner.

AIR COMPRESSION
Compressed air is used to drive brakes in trucks, to spray paint, and to power many **industrial machines**. It is also used in hovercraft and in divers' air tanks.

AIR PRESSURE
The discovery that air flows from areas of **high pressure** to areas of **low pressure**, causing winds and weather, was the **beginning of meteorology**.

COMPRESSED AIR TANK

SEE ALSO ››
pp.104–05 NEWTON'S LAWS OF MOTION
pp.132–33 STEAM POWER TO STEAM ENGINE
pp.138–39 THE NATURE OF MATTER

PHYSICIST AND CHEMIST (1778–1850)

JOSEPH GAY-LUSSAC

Frenchman Joseph Louis Gay-Lussac studied chemistry in Paris before working in Claude-Louis Berthollet's prestigious laboratory as a research assistant. In 1802 he published his groundbreaking work on the relationship between gas volume and temperature. Two years later he and a colleague took a hot-air balloon to a height of 21,000 ft (6,400 m) to collect samples of air in order to establish that the composition of the atmosphere remains the same with increasing altitude. Gay-Lussac died in Paris in 1850, and is buried in the Père Lachaise cemetery.

◀◀ BEFORE

The roots of the coordinate system lie far back in the geometry of ancient Greece and the mathematical brilliance of ancient Islam.

ANCIENT GREEK GEOMETRY

As long ago as the 3rd century BCE, ancient Greek mathematicians such as **Dicaearchus** and **Eratosthenes** were **pinpointing locations** on maps using **grids** that were **precursors** to the modern system of **latitude** and **longitude** ◀◀ 38–39.

KHAYYAM'S CUBICS

In the 11th century CE the Islamic scholar **Omar Khayyam** found an ingenious way of using **geometry to solve cubic equations** ◀◀ 48–49. "No attention should be paid," he wrote, "to the fact that **algebra and geometry** are different in appearance."

OMAR KHAYYAM

◀◀ SEE ALSO
pp.38–39 ANCIENT IDEAS OF THE WORLD
pp.48–49 ALGEBRA

Graphs and Coordinates

Graphs present statistical information visually so that anything from monthly rainfall to rising crime rates can be seen at a glance. However, underpinning graphs and coordinate systems is a fundamental link between algebra and geometry that places them at the heart of mathematics and science.

The connection between algebra and geometry was established in the 17th century, principally by two French mathematicians, Pierre de Fermat and René Descartes. What they realized is that algebra can be converted into geometry, and geometry into algebra, using coordinates—two or more numbers that pinpoint a location in relation to reference lines such as the axes of a graph. Through coordinates all geometry can be expressed as algebraic equations—and, in theory, every algebraic equation can be represented on a graph.

Development of coordinate systems

What interested Fermat and Descartes was curves, which are involved in everything from the trajectory of missiles to the orbits of planets. All but the simplest curves were hard to deal with using the basic geometric approach developed 2,000 years earlier by Euclid. Fermat and Descartes realized independently that the key to working with curves is coordinates.

Coordinates had been used in maps for thousands of years. In studying the geometry of curves in 1629, Fermat realized that a single equation could be used to pinpoint every point along the line of the curve according to its distance from two other fixed lines. In *La Géometrie* of 1637, the appendix to his great book *Discours de la Méthode*, Descartes expressed this idea more fully, describing two coordinate lines intersecting at right angles called the x and y axes. This system later became known as the Cartesian coordinate system, after Descartes.

The algebra of graphs

Coordinates make it possible to convert curves and shapes into equations, which can then be manipulated using algebra and, conversely, algebraic equations can be shown visually on a graph. Using Cartesian two-dimensional coordinates, a straight line can be described by the general equation $y = mx + c$, where y is how far up, x is how far along, m is the slope of the line, and c is where the line crosses the y axis (see above).

Equations can also be found for much more complex lines and shapes. In 1691 the Swiss mathematician Jacob Bernoulli introduced polar coordinates (see opposite). With polar coordinates, simple equations could describe curves as complex as the cycloid, the path traced by a point on the rim of a wheel rolling along the ground. Adding extra axes to the basic x and y ones or basing graphs on different shapes (such as a cylinder or sphere) allowed points to be specified in three or more dimensions, and equations could be created for shapes in multiple dimensions.

The connection between algebra and geometry opened up a whole new kind of mathematics called analytic geometry, which could analyze how things move and change. It also gave birth to calculus, created principally by the English scientist Isaac Newton (see pp.110–11) and the German philosopher and mathematician Gottfried Leibniz. Descartes saw his ideas as the key to a mechanical view of the universe, in which everything can be explained in terms of motion and extension (length, breadth, and height). However, using this model, it was possible to describe mathematically only a static universe, in which things either do not move or move at a constant speed. Calculus allowed a description of a more realistic, dynamic universe, still using coordinates and curves (see box, left).

The power of coordinates

Separately, algebra and geometry can deal well with static relationships but together—in analytic geometry and calculus—they can explore dynamic relationships, making them essential tools for the scientist today. Without this combination, key breakthroughs, such as Newton's law of gravity (see p.106) and quantum physics (see p.404), would not have been possible. Coordinates also have a huge number of practical uses. They are employed in everything from computer graphics to satellite positioning, and for plotting statistical data (see pp.196–97).

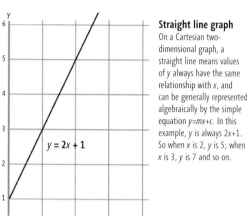

Straight line graph
On a Cartesian two-dimensional graph, a straight line means values of y always have the same relationship with x, and can be generally represented algebraically by the simple equation $y=mx+c$. In this example, y is always $2x+1$. So when x is 2, y is 5; when x is 3, y is 7 and so on.

$y = 2x + 1$

BREAKTHROUGH

CALCULUS

The development of calculus by Gottfried Leibniz and Isaac Newton answered two fundamental questions: how to figure out the tangent to a curve (which on a graph gives rates of change) and how to calculate the area under a curve (which gives quantities). For example, in a graph of speed against time, applying differential calculus (differentiation) gives the slope of the tangent at a point, which is the acceleration (rate of change of speed with time) at that point; applying integral calculus (integration) between two specified times gives the distance traveled in that period. Interestingly, differential and integral calculus are inverse processes.

Differentiating a curve's equation for a specified value of x gives the slope of the tangent at that point

Tangent

Here, the slope at $x = 3$ is 0.6, which means that for every 1 unit you go along, you will go 0.6 points up

Differential calculus

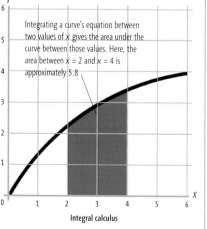

Integrating a curve's equation between two values of x gives the area under the curve between those values. Here, the area between $x = 2$ and $x = 4$ is approximately 5.8

Integral calculus

BEFORE

The fact that things fall to Earth was obvious to the earliest humans—but not why they fell.

ANCIENT GREEK THEORIES
Aristotle believed in the four elements of earth, air, fire, and water « 22–23, and thought that things want to return to where they belong; so stones fall toward Earth, and they accelerate because as they get closer to home they move more jubilantly.

GALILEO'S EXPERIMENTS
In the 16th century Galileo « 82–83 conducted experiments to try to find a mathematical

description of falling objects. He rolled balls down inclined planes rather than dropping them, so their rate of fall was slower, and measurable. But he never solved the mathematical equations, nor extended the idea to the heavens.

« SEE ALSO
pp.86–87
MOTION, INERTIA, AND FRICTION
pp.106–07 NEWTON'S IDEA OF GRAVITY

Gravitational Force

Why things fall to the ground has always been a mystery. Isaac Newton did not invent gravity, but he was the first to realize that things do not just fall, but are attracted to one another. He saw that this force affects everything in the Universe, from stars and planets to people and apples.

Gravity is a force of attraction between masses. It holds us on Earth and holds Earth together; it holds the solar system and galaxies together. It also determines weight and the speed at which things fall.

Newton's discovery
Legend has it that Newton witnessed the fall of an apple and spontaneously realized how gravity works, but it seems that in reality it was the result of years of deliberation and calculation. He was aware of the theories of the French philosopher Renée Descartes and the Italian scientist Galileo Galilei, and he knew about the German astronomer Johannes Kepler's laws of planetary motion (see p.77), but apparently he did not apply himself to the problem seriously until the 1680s, when he finally figured out the mathematics,

and explained the concept of universal gravity. This was "a drawing power" and Newton further recognized that "If matter thus draws matter; it must be in proportion of its quantity." This means the attraction is proportional to mass; larger objects exert a stronger attractive force. An astronaut on the Moon weighs only a sixth as much as at home, because the Moon is much less massive.

Weight and mass
A person's weight is the attractive force between the mass of the person and the mass of Earth. The mass of an object is the amount of matter in it: a 2.2 lb (1 kg) bag of flour contains 2.2 lb (1 kg) of flour, and this amount does not change whether the bag of flour is on the Moon or weightless in deep space. But the weight of that bag of flour is the force with which it

is pulled down by gravity—this would be 9.8 newtons on Earth, or about 1.6 newtons on the Moon. (A newton, usually expressed as a metric measure, is the force needed to accelerate a mass of 1 kg by 1 m per second every second.)

Fellow scientist Robert Hooke (see pp.92–93) had figured out that the force decreases with distance, and

Newton proved that this was in accordance with an inverse square law—that the strength of the force is inversely proportional to the square of the distance from the source of the force. Therefore an object twice as far away from the source has only one quarter of the attraction, while an object three times as far away has only one ninth of the attraction.

IN PRACTICE

SPRING TIDES AND NEAP TIDES

The moon pulls on the ocean facing it more strongly than it pulls on Earth because the ocean is closer than is the center of Earth. This draws up a bulge of water—high tide. A bulge also appears on the opposite side of Earth because the Moon pulls more strongly on

the center of Earth than it does on the ocean on the planet's far side. When the sun is in line with Earth and the moon, the pull is stronger, giving extreme spring tides. When the sun and moon are at right angles, the two pulls cancel each other to some extent, giving a neap tide.

The relationship between mass and distance
Gravitational attraction depends on the product of both objects' masses divided by the square of their separation. If the masses are doubled, the attraction is four times stronger; but if the distance is doubled, the attraction is four times weaker.

Attraction
Two bodies are attracted by a gravitational force.

Doubling the mass
If each body is twice the mass, then the gravitational force is four times as strong.

Increasing the distance
If the distance between the masses is doubled, the gravitational force is four times less.

SPRING TIDE   NEAP TIDE

Increased gravitational pull exaggerates high tides

Sun counteracts Moon's pull

Moon exerts pull on sea

High tides reduced

Saturn: the ringed planet
Saturn is the sixth planet from the sun, and the second most massive planet in our solar system. It is a gas giant, with no firm surface, orbited by at least 82 moons. Its main rings (below) measure around 124,000 miles (200,000 km) from side to side, but are only about 33 ft (10 m) thick. They consist mainly of millions of pieces of ice, all held in individual orbits by Saturn's gravity.

AFTER »

Since Einstein's groundbreaking theories, scientists have been attempting to learn more about the gravitational force, including the speed of gravity itself.

THE FOURTH DIMENSION
Einstein's theory of relativity 300–301 » introduced the new concept of **four-dimensional spacetime** and redefined gravity.

GRAVITATIONAL WAVES
According to Einstein, objects with mass moving around can cause **fluctuations or ripples** in spacetime 302–303 »; these gravitational waves were finally detected in 2016 by US-based researchers using a highly sensitive technique called laser interferometry.

THE SEARCH FOR THE GRAVITON
The **graviton** is a theoretical fundamental particle proposed by **quantum field theory** 314–17 ». It has no mass, nor any electrical or magnetic charge, but is thought to carry the **gravitational force**. The existence of gravitons has not yet been confirmed experimentally.

SEE ALSO »
pp.300–01 EINSTEIN'S EQUATION
pp.368–69 ARTIFICIAL SATELLITES
pp.404–405 GRAND UNIFIED THEORY

force, the weak nuclear interaction, and gravity. However, only gravity and electromagnetism work over long distances.

Spacetime distortion
Einstein had a different explanation, arguing that there is no attraction between masses. What happens is that objects with mass distort spacetime around themselves, just as weights placed on a rubber sheet make "dents" in it. Because an object that has mass makes a dent in spacetime, other objects with mass tend to "roll down" towards it. Einstein said that gravity acts at the speed of light. If this is true, something must be "carrying" the gravity—these particles, known as gravitons, are equivalent to the photons that carry electromagnetic force.

Newton's great leap forward was to realize that the force of gravity extended out into space, and that it controlled not only the fall of the apple in front of him, but also the orbit of the Moon. Today Earth's gravity holds the Moon, the International Space Station, and thousands of satellites in orbit.

Gravity is the weakest of the four forces that operate in the Universe: electromagnetism, the strong nuclear

Changing weight
Astronauts and spacecraft in orbit around Earth have the same mass as on the ground, but are weightless because they are in "free fall," with the centrifugal force counterbalancing the force of gravity.

6,000 MILLION
million million tons is the total mass of planet Earth, as first estimated by Isaac Newton, and as subsequently confirmed by Henry Cavendish's ingenious torsion balance experiment in 1797 (see below).

Michell-Cavendish experiment
Below is a modern representation of Henry Cavendish's experiment of 1797 in which he used a torsion balance apparatus, designed by John Michell, to measure the mass of Earth. From this scientists were able to calculate a value for the gravitational constant, G, which is part of Newton's Law of Gravity.

Screen

Mirror

Lead spheres

Light source (lamp)

Larger lead spheres attract the smaller ones

1 Two spheres are mounted at either end of a free-swinging beam that has a fixed mirror in the middle. When the rod is still, light shone at the mirror is reflected to the center of the screen.

Mirror moves with the rod, and deflected light beam indicates amount of attraction

2 Two large lead spheres are brought up close to the small ones, one in front and one behind, and their gravitational attraction on the small spheres turns the rod and mirror clockwise, moving the light to the right.

Light beam is deflected in the opposite direction

Larger lead spheres attract the smaller ones in a different direction

3 When the large spheres are introduced on the other side, the rod and mirror turn counterclockwise and the light moves to the left. Using the relative attraction of the large spheres and Earth, Earth's mass can be calculated.

PHYSICIST AND MATHEMATICIAN **Born 1642** Died 1727

Isaac Newton

"**Truth** is ever to be found in **simplicity**, and not in the **multiplicity** and **confusion** of things."

ISAAC NEWTON, FROM "RULES FOR METHODIZING THE APOCALYPSE"

Perhaps the greatest mathematician and scientist of all time, Isaac Newton was a contentious genius who made extraordinary strides in mathematics, optics, mechanics, and gravitational attraction. He invented calculus, he explained the colors of the rainbow, and he realized that the same gravitational force that gives us weight also controls the movements of the Moon and planets. In later life he became Master of the Mint, and President of the Royal Society.

Childhood without parents

Isaac Newton was born in 1642 at Woolsthorpe Manor, near Grantham, England. His father had died three months earlier, and when Newton was three his mother married a clergyman, leaving him in the care of her parents. She returned when her second husband died, but that was not until 1653; it is possible that Newton's difficult relationships may have begun with the absence of his parents. After

education at the King's School in Grantham and some time assisting on his mother's farm, Newton went to Cambridge University.

Works of genius

As an undergraduate he showed no great promise, but in 1665 the University was closed because the Plague was rife, and Newton returned home to Woolsthorpe, where, during the next 18 months, he produced his best work in mathematics and physics. Some time later, in 1671, he made a reflecting telescope, and presented it to the Royal Society.

His first scientific paper explored the nature of light. Robert Hooke (see pp.92–93) disagreed with some of Newton's conclusions in this paper. Unfortunately Newton took this as a personal affront, and vowed not to publish anything else about optics during Hooke's lifetime. He had disputes with others, too: with Gottfried Leibniz about who

Woolsthorpe Manor
Isaac Newton was born in this house in Woolsthorpe, near Grantham, England. His grandparents were entrusted with his care.

However, a sheet of polished metal is more convenient than a pool of water, and metal mirrors have long been used by men and women for personal grooming. Modern telescopes and domestic mirrors use a sheet of glass backed with a coating of silver or aluminum; the clean metal surface reflects the light, while the glass protects it from dirt and tarnishing.

Light-bending materials

When light passes from one material into another, for example from air into glass or water, it changes direction; light is refracted (bent) as it goes through the interface. This refraction makes ponds look shallower than they really are, and can make a spoon appear bent when it stands in a glass of water.

Light travels more slowly through denser materials, such as water and glass, than it does in air, and it refracts to minimize the time it takes to travel between two points. In around 1620 the Dutch scientist Willebrord Snell discovered that the amount of bending depends on the refractive indices of the two media. He created a mathematical formula, now known as Snell's Law (see p.473), that can be used to calculate the refraction of light when traveling between two materials of differing refractive index.

Splitting light

In 1666 Isaac Newton (see pp.110–11) acquired a glass prism. He cut a hole in his window shutters, so that a narrow beam of sunlight shone through, then he intercepted this beam with his prism so that a spectrum formed on the opposite wall. He noted that the spot

of light which came into the room spread out into a long patch of different colors: blue (which refracted the most) lay at one end and red (which refracted least) at the other. Newton showed that white sunlight is a mixture of all the

colors of the rainbow. It can be split into its many colors using a prism, while a second prism will recombine them to make white light again (see below). Newton himself claimed, probably for mystical reasons, that he could see seven colors—red, orange, yellow, green, blue, indigo, and violet—but few people can distinguish between indigo and violet.

IN PRACTICE

POLARIZATION OF LIGHT

Light consists of electromagnetic waves that oscillate at right angles to the direction of travel. The waves within a beam of light oscillate up-down, left-right, and all directions in between. Polarizing filters act by transmitting the waves oscillating in one direction only – for example up-down – and blocking waves at other orientations. In the diagram below, only vertical waves travel through the polarizing filter; others are absorbed. Photographers use polarizing filters to reduce surface reflection and to deepen the blue of the sky in an image.

Light beam | Polarizing filter | Vertical waves are transmitted through the filter | Photographic sensor | UNFILTERED IMAGE

POLARIZED IMAGE

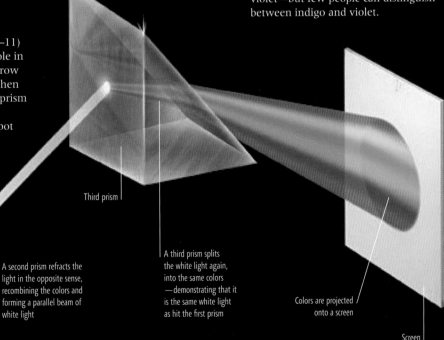

Second prism

A second prism refracts the light in the opposite sense, recombining the colors and forming a parallel beam of white light

Third prism

A third prism splits the white light again, into the same colors —demonstrating that it is the same white light as hit the first prism

Colors are projected onto a screen

Screen

"In the beginning of the year 1666… I procured me a **triangular glass–prisme**, to try therewith the celebrated phaenomena of colors."

ISAAC NEWTON, LETTER TO THE ROYAL SOCIETY, FEBRUARY 6, 1672

AFTER

Intricate optics—lenses and mirrors—are crucial components within many modern scientific instruments. Some optics are used with infrared and ultraviolet light.

VISIBLE AND INVISIBLE LIGHT

The light we see is only a small fraction of the **electromagnetic spectrum 258–59 》**, which ranges from gamma rays to radio waves. Most of these waves are invisible to the human eye.

SPECTROSCOPY

A crucial development of the splitting of white light into its constituent colors was the development of **spectroscopy 259 》**, which has become a major analytical tool. Scientists can identify many elements and compounds by looking at the **unique spectrum** of light they emit or absorb.

HOLOGRAPHY

In 1947 Dennis Gabor invented a way to make **pictures appear three-dimensional**. However, **holography** did not become a practical possibility until some ten years later when the laser had been invented.

APOLLO 11

The first astronauts to visit the Moon left a special mirror on the **lunar surface**: the "laser ranging retroreflector array." This has allowed scientists to discover that the **Moon is receding** from the Earth by $1^{1}/_{2}$ in (4 cm) each year.

SEE ALSO 》
pp.258–59 ELECTROMAGNETIC SPECTRUM
pp.300–01 EINSTEIN'S EQUATION
pp.302–03 THEORIES OF RELATIVITY
pp.402–403 GRAVITATIONAL WAVES

« BEFORE

Until 1577, comets were a complete mystery. Usually regarded with fear, a comet's appearance in the sky was rare and startling.

BAYEUX TAPESTRY EVIDENCE
An embroidered comet hovers above King Harold, noting **Comet Halley's 1066 appearance**, and is included as a symbol of death and disaster.

ENSISHEIM METEORITE
On November 7, 1492, **a meteorite landed near the church of Ensisheim**, Alsace, France. Some of the meteorite was taken by locals, but a large chunk of it is still exhibited in the village.

« SEE ALSO
pp.76–79 PLANETARY MOTION

Comets and Meteors

Huge numbers of comets were formed at the dawn of the solar system some 4.6 billion years ago. Comets are largely made up of rock, dust, and ice. Meteoroids are smaller fragments of rock in space. When they enter our atmosphere, they become meteors or "shooting stars."

Comet Churyumov–Gerasimenko's nucleus
The Rosetta craft reached Churyumov–Gerasimenko in 2014 and traveled with the comet for two years before it was deliberately crashed. It released a small probe, Philae, which landed on the comet's nucleus.

Comets are found mainly in the outer regions of the Solar System, beyond Neptune's orbit. Each one is an icy, dusty lump, miles across. When a comet's orbit brings it closer to the Sun than about the distance of Jupiter, solar radiation heats the surface of this solid core, or nucleus, enough to convert its ice into gas. The escaping gas begins to push away from the nucleus, until eventually the expanding gas and dust form a huge spherical cloud, called a coma, and two tails (see right). This is what we see in the night sky. The comet's tails extend away from the coma, in the opposite direction from the Sun. They can extend for many millions of miles into space.

A comet's life
A comet nucleus can range in size up to 25 miles across (40 km). Comet Halley, for example, has a nucleus that is around 9 miles (15 km) long and 5 miles (8 km) wide. Halley takes about 76 years to travel around its medium-sized orbit and loses about 6 ft (2 m) from its

surface every time it passes close to the Sun. This is how comets die: Halley will last around another 3,000 orbits.

Orbiting bodies
The major breakthrough in our understanding of comets came in 1687 when Isaac Newton showed how to calculate the orbit of a comet using three positional sightings taken a few weeks apart. He used the Great Comet of 1680 as an example. However, this comet had such an elongated orbit that it was impossible for Newton to calculate where it had come from and where

The comet's tails
As a comet nears the Sun, it develops two tails, which extend away from the Sun as the comet orbits. The curved tail is formed of dust, pushed away by solar radiation. The straight tail consists of ionized gas that has been blown away from the coma by the solar wind.

it was going. Newton passed his accumulated positional data on to his friend Edmond Halley, who then started to calculate more cometary orbits. By June 1696 Halley realized that the comets that had been seen in 1607 and 1682 had very similar orbits,

Tails are longest close to the Sun

Dust tail is curved

Gas tail is straight and narrow

Tails expand as they move toward the Sun

Perihelion (orbital point nearest the Sun)

Sun

Tails shorten as the comet travels away from the Sun

Aphelion (orbital point farthest from the Sun)

Naked nucleus

Uranus

Comet Swift-Tuttle (orbital period about 135 years)

Saturn

Comet Halley (orbital period about 76 years)

Mars

Earth

Sun

Jupiter

Comet Encke (orbital period 3.3 years)

Comet Tempel-Tuttle (orbital period 32.9 years)

Comet Hyakutake (orbital period about 30,000 years)

Comet Hale-Bopp (last orbit of 4,200 years; next orbit predicted to last 3,400 years)

Comet orbits
All the comets shown here pass very close to the Sun. Encke is a short-period comet that orbits in the plane of the Solar System. The others are intermediate- and long-period comets."

3

THE INDUSTRIAL REVOLUTION

1700–1890

As the Industrial Revolution began to transform the world with factories and urban sprawl, science truly came of age. It was no longer just the intellectual pastime of a few gentlemen "natural philosophers" but a career pursued by professionals in academic and scientific institutions.

THE INDUSTRIAL REVOLUTION

1700–1890

1700

1725

1701
Giacomo Pylarini gives the first smallpox inoculations in Europe.

1709
Daniel Gabriel Fahrenheit invents the alcohol thermometer.

1712
Thomas Newcomen builds the first atmospheric steam engine, to pump water out of mines.

1717
Sébastien Vaillant proposes that plants reproduce sexually and have male and female parts.

▼ Newcomen's engine

1727
Stephen Hales demonstrates that plants transpire in his essay *Vegetable staticks*.

1742
Anders Celsius develops the centigrade (Celsius) temperature scale.

1744
Pierre-Louis Maupertuis proposes the principle of least action.

1746
Leonhard Euler describes the refraction of light mathematically.

Stephen Hales ▶

1704
Isaac Newton publishes *Opticks*, exploring theories on light. Final volume of John Ray's *History of Plants* published.

1714
Bartolomeo Eustachi's *Tabulae Anatomicae* published posthumously, giving an accurate description of the human nerve map. Fahrenheit invents the mercury thermometer.

c.1730
Charles du Fay discovers that there are two types of electricity (positive and negative charges).

c.1740
Benjamin Franklin starts to keep regular weather records.

1733
Stephen Hales takes measurements of blood pressure.

1741
Stephen Hales presents the first ventilator to the Royal Society, UK.

1747
James Lind discovers that citrus fruits prevent scurvy.

◀ James Lind

▲ Isaac Newton

1705
Edmond Halley predicts that the comet of 1682 will reappear in 1758. The comet is later named after him.

1715
Before a solar eclipse, Edmond Halley predicts the path of totality across England to an accuracy of 30km (19 miles) and the time of the eclipse to within four minutes.

1745
Pieter van Musschenbroek invents the Leyden jar, in which a static electric charge can be stored. Pierre-Louis Maupertuis states that offspring are made from components derived from their parents' bodies.

▼ Comet in night sky

▼ London solar eclipse

1710
John Ray's *History of Insects* published posthumously.

1735
John Harrison produces his first marine chronometer. Carl Linnaeus publishes *Systema Naturae*, proposing a system for classifying the natural world.

▲ John Harrison's first chronometer

1749
The Comte de Buffon publishes the first of 44 volumes of *Histoire Naturelle*, covering the entirety of natural history.

◀ Leyden jar

The rise of industry spurred scientific discovery both directly and indirectly, as scientists explored the nature of heat and energy, and of chemicals and their processes, leading to the discovery of the laws of thermodynamics, the atomic theory, and the periodic table. The power promised by science became intoxicating, and no discovery was more exciting than that of electricity and its links with magnetism. Yet in some ways the most profoundly disturbing discoveries were made, as natural historians left the factories and cities behind to explore the natural world. They revealed the shocking truth that we humans are just recent arrivals in an ancient world that has seen many creatures, including dinosaurs, come and go.

1750

1775

1754
Joseph Black isolates carbon dioxide, which he calls "fixed air," and shows that air is a mixture of gases.

◀ Lightning strike

1758
Linnaeus extends the two-part system of classification to every species in the tenth edition of *Systema Naturae*.

1768
Joseph Banks sets sail for the South Seas (Tahiti) with Captain James Cook on HMS *Endeavour*.

▼ Harrison's H4 chronometer

Boulton & Watt ▶ steam engine

c.1787
Jacques Charles writes his law of ideal gas.

1796
Jean Senebier demonstrates that plants take in inorganic carbon dioxide. Edward Jenner performs the first vaccination. Pierre-Simon Laplace proposes his theory on the formation of the solar system.

1752
Benjamin Franklin and Thomas-François d'Alibard confirm that lightning bolts are huge electrical sparks.

1755
Immanuel Kant proposes that the solar system originated as a type of cloud around the Sun.

▼ Illustration of *Linnaea borealis*, the twinflower

1775
James Watt makes a steam engine with a separate condenser.

1780
Charles Coulomb formulates Coulomb's law.

1781
William Herschel discovers Uranus. Immanuel Kant's *Critique of Pure Reason* published.

1789
Antoine Lavoisier publishes *Traité Elementaire de Chimie*, the first modern chemistry textbook.

1790
Henry Cavendish determines the balance of gases in the atmosphere.

▼ Edward Jenner's first vaccination

1761
John Harrison makes his H4 marine chronometer, finally solving the problem of finding longitude at sea. Joseph Black discovers latent heat.

1770
Antoine Lavoisier establishes that food in the body releases heat in a similar way to combustion.

1776
Henry Cavendish discovers "inflammable air" —hydrogen.

1777
Nicolas Desmarest proposes that some rocks were formed from magma from volcanoes.

1783
First hot-air balloon flight to carry people. John Mitchell proposes the idea of black holes. Lazzaro Spallanzani proves stomach juice can liquefy meat.

▼ First balloon flight carrying people

1753
Carl Linnaeus produces *Species Plantarum*, classifying plants according to his binomial system (two-part names).

1756
Johann Lehmann publishes his study of rocks, describing sedimentary rock layers and the origin of mineral deposits.

1757
The sextant is invented.

1773–74
Joseph Priestley and Carl Wilhelm Scheele each independently isolate oxygen.

1778
Fossilized bones of a giant marine reptile, a mosasaur, found in the Netherlands. Joseph Bramah patents his improved flushing toilet.

▼ Joseph Priestley

1791
Luigi Galvani describes experiments on animal electricity.

1798
Henry Cavendish accurately calculates Earth's density to be five times that of water.

Sextant ▶

1763
Antoine Lavoisier measures daily air pressure, humidity, and wind speed in an attempt to predict the weather. Claudius Aymand performs the first appendectomy.

1779
Jan Ingenhousz describes the carbon cycle and respiration in plants. Antoine Lavoisier names oxygen and recognizes its role in combustion.

1785
James Hutton writes *Theory of the Earth*, explaining his ideas on rock formation.

1786
Caroline Herschel discovers Comet Herschel (C/1786 P1). She is the first woman to discover a comet.

1793
Christian Sprengel's treatise on flowers examines how they are pollinated. John Dalton publishes *Meteorological Observations and Essays*.

1799
Alessandro Volta makes the first battery. Richard Trevithick builds his first high-pressure engine. William Smith publishes the first geologic map.

1800

1825

1800
William Herschel discovers infrared light. Henry Maudslay invents a screw-cutting lathe.

1801
Giuseppe Piazzi finds the first known asteroid.

1802
Joseph Gay-Lussac formulates a law stating that the pressure of gas is proportional to the absolute temperature.

Joseph Gay-Lussac ▶

1804
Nicolas-Théodore de Saussure confirms photosynthesis as a chemical process.

▼ Section of leaf showing internal structure

1803
John Dalton proposes that atoms of different elements have different weights.

1808
John Dalton publishes *A New System of Chemical Philosophy*, suggesting that all elements are made of indivisible atoms.

Dalton's symbols ▶

1810
Humphry Davy invents the arc lamp and isolates and names chlorine.

1811
Amedeo Avogadro states that equal volumes of different gases contain the same number of molecules.

▼ John Dalton's models of atoms

1821
Michael Faraday builds the first electric motor.

Faraday's ▶ original experiment

1822
Charles Babbage proposes the first design for his Difference Engine—predecessor of the computer.

▼ Difference Engine

1820
Hans Christian Ørsted discovers that electric currents produce magnetic fields. Joseph Fourier proposes a formula for estimating the age of Earth.

1824
Joseph Fourier discovers what is later known as the greenhouse effect. Nicolas Carnot publishes Reflections on the Motive Power of Fire, in which he explains how heat can make things move.

1825
William Sturgeon invents the electromagnet. Jöns Berzelius publishes a list of the atomic weights of 43 elements.

1827
William Prout classifies nutrients in food categories. Robert Brown defines Brownian motion.

1828
Christian Ehrenberg observes bacteria.

1830
Charles Lyell publishes the first of three volumes of *Principles of Geology*, in which he argues that Earth must be several hundred million years old.

1831
Robert Brown identifies the nucleus of a cell. Michael Faraday discovers electromagnetic induction. HMS *Beagle* sails with Robert Fitzroy and Charles Darwin.

The Coriolis effect ▼

1833
Carl Friedrich Gauss and Wilhelm Weber build the first operational telegraph system. Daguerre produces the daguerrotype, an early form of photograph.

▲ Early telegraph machine

1835
Gustave Coriolis describes the movement of winds around the globe, now known as the Coriolis effect. Robert Anderson invents the first electrically powered carriage.

1837
Louis Agassiz proposes that Europe went through an ice age. Samuel Morse and Alfred Vail devise the Morse Code.

1840
William Fox Talbot develops the calotype, a form of photographic reproduction that allows multiple positive prints to be made from a single negative image.

1842
Justus von Liebig discovers body heat is produced by chemical processes. Christian Doppler predicts the Doppler effect, based on sound waves.

1843
Ada Lovelace writes the world's first computer program.

1845
William Armstrong invents the hydraulic crane.

▲ Anemometer

1846
The anemometer introduced. Ignaz Semmelweis tries to introduce antiseptic procedures into hospitals. Johann Galle discovers the planet Neptune.

▼ Neptune

1847
William Budd suggests that typhoid is transmitted through infected water. James Joule introduces the law of conservation of energy.

1848
William Thomson shows that there is an absolute zero of temperature. James Joule estimates the average speed at which gas molecules move.

"We must, however, acknowledge… that man, with all his **noble qualities**… still bears in his bodily frame the indelible stamp of his **lowly origin**."

CHARLES DARWIN, *The Descent of Man*, 1871

1850

1875

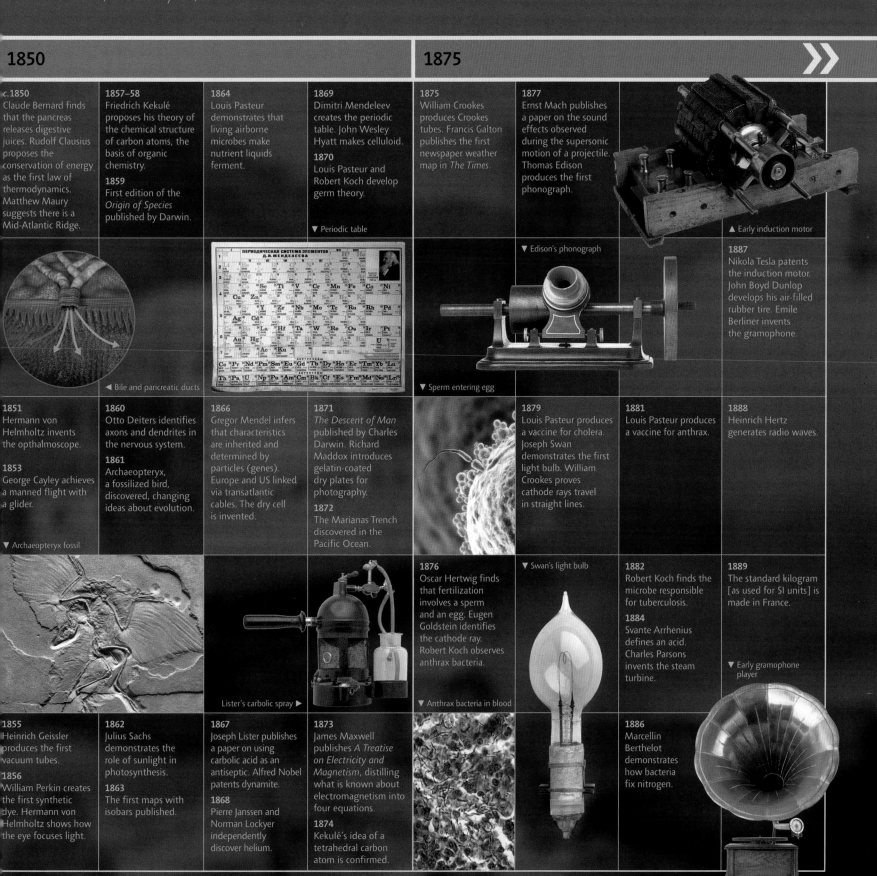

c.1850
Claude Bernard finds that the pancreas releases digestive juices. Rudolf Clausius proposes the conservation of energy as the first law of thermodynamics. Matthew Maury suggests there is a Mid-Atlantic Ridge.

1857–58
Friedrich Kekulé proposes his theory of the chemical structure of carbon atoms, the basis of organic chemistry.
1859
First edition of the *Origin of Species* published by Darwin.

1864
Louis Pasteur demonstrates that living airborne microbes make nutrient liquids ferment.

1869
Dimitri Mendeleev creates the periodic table. John Wesley Hyatt makes celluloid.
1870
Louis Pasteur and Robert Koch develop germ theory.

▼ Periodic table

1875
William Crookes produces Crookes tubes. Francis Galton publishes the first newspaper weather map in *The Times*.

1877
Ernst Mach publishes a paper on the sound effects observed during the supersonic motion of a projectile. Thomas Edison produces the first phonograph.

▲ Early induction motor

▼ Edison's phonograph

1887
Nikola Tesla patents the induction motor. John Boyd Dunlop develops his air-filled rubber tire. Emile Berliner invents the gramophone.

◄ Bile and pancreatic ducts

▼ Sperm entering egg

1851
Hermann von Helmholtz invents the opthalmoscope.
1853
George Cayley achieves a manned flight with a glider.

1860
Otto Deiters identifies axons and dendrites in the nervous system.
1861
Archaeopteryx, a fossilized bird, discovered, changing ideas about evolution.

1866
Gregor Mendel infers that characteristics are inherited and determined by particles (genes). Europe and US linked via transatlantic cables. The dry cell is invented.

1871
The Descent of Man published by Charles Darwin. Richard Maddox introduces gelatin-coated dry plates for photography.
1872
The Marianas Trench discovered in the Pacific Ocean.

1879
Louis Pasteur produces a vaccine for cholera. Joseph Swan demonstrates the first light bulb. William Crookes proves cathode rays travel in straight lines.

1881
Louis Pasteur produces a vaccine for anthrax.

1888
Heinrich Hertz generates radio waves.

▼ Archaeopteryx fossil

1876
Oscar Hertwig finds that fertilization involves a sperm and an egg. Eugen Goldstein identifies the cathode ray. Robert Koch observes anthrax bacteria.

▼ Swan's light bulb

1882
Robert Koch finds the microbe responsible for tuberculosis.
1884
Svante Arrhenius defines an acid. Charles Parsons invents the steam turbine.

1889
The standard kilogram [as used for SI units] is made in France.

▼ Early gramophone player

Lister's carbolic spray ►

▼ Anthrax bacteria in blood

1855
Heinrich Geissler produces the first vacuum tubes.
1856
William Perkin creates the first synthetic dye. Hermann von Helmholtz shows how the eye focuses light.

1862
Julius Sachs demonstrates the role of sunlight in photosynthesis.
1863
The first maps with isobars published.

1867
Joseph Lister publishes a paper on using carbolic acid as an antiseptic. Alfred Nobel patents dynamite.
1868
Pierre Janssen and Norman Lockyer independently discover helium.

1873
James Maxwell publishes *A Treatise on Electricity and Magnetism*, distilling what is known about electromagnetism into four equations.
1874
Kekulé's idea of a tetrahedral carbon atom is confirmed.

1886
Marcellin Berthelot demonstrates how bacteria fix nitrogen.

The Newcomen Engine

In 1712 Thomas Newcomen built an "atmospheric" engine to pump water up from the depths of a coal mine at Dudley in the West Midlands of England. This was the first time that reliable power had been made available when and where it was needed. The Newcomen engine was arguably the most important invention of all time—one that marked the beginning of the Industrial Revolution.

The demand for coal in England increased steadily from 1709, when Abraham Darby discovered how to smelt iron using coke rather than charcoal. By 1712 coal miners were digging deeper and deeper to supply the demand, and flooding of coal mines was an ever-increasing problem. Manual and horse-powered pumping was no longer enough. Then Newcomen, an ironmonger from Dartmouth in Devon—today he would be called an engineer—built the world's first successful engine to pump out the water.

The engine was a huge lumbering beast. Coal was burned to boil water and make steam, which was fed into a vertical cylinder; this forced a piston inside to rise. When the piston reached the top of the cylinder, cold water was squirted into the cylinder to condense the steam and create a partial vacuum. The pressure of the atmosphere then pushed the piston down, pulling down in turn a huge beam above, and so pulling up pump rods at the other end of the beam. The whole enormous engine was protected from the weather by an engine house some 33ft (10m) high.

Because they were inefficient, Newcomen engines were used mainly in coal mines, where there was a plentiful supply of fuel. Each engine cost around £1,000 to build—a vast sum of money in the early 18th century. Nevertheless, the engines were so important that some 75 of them were built in the 17 years before Newcomen died, and by the end of the century the figure had risen to more than 1,000. They were exported to Sweden, Germany, Belgium, France, Spain, and even to America. They were rugged and reliable, and some of them lasted for more than 100 years.

The first Newcomen engine had a brass cylinder 21in (53cm) in diameter and 8ft (2.4m) long; it ran at about 12 strokes a minute and could lift 10 gallons (45 liters) of water from 150ft (45m) down with each stroke. Later versions had much larger cylinders of cast iron, and so were more powerful. Improvements were made to the basic engine in the middle of the 18th century, and in the late 1770s James Watt's improved version with a separate condenser (see pp.132–33) moved the Industrial Revolution into the next phase.

Newcomen's historic engine
No Newcomen engine survives today, but the Black Country Living Museum at Dudley has a full-size working replica of the original 1712 engine. The replica is based on this early drawing, which shows all the main working parts, and is occasionally operated under steam for the benefit of the visiting public.

"In the whole history of technology it would be difficult to find a greater single advance than this, nor one with a greater significance for all humanity."
L.T.C. ROLT, FROM "THOMAS NEWCOMEN: THE PREHISTORY OF THE STEAM ENGINE," 1963

Development of steam transportation
James Watt and Richard Trevithick made the breakthroughs in steam engine technology that allowed a century of steam-powered transportation on both land and sea.

1765 James Watt conceives the idea of a separate condenser, which greatly improves the efficiency of steam engines.

1804 Richard Trevithick runs the world's first railroad locomotive, which pulls 10 tons of pig iron 9 miles (14.5 km) from the Pen-y-Daren ironworks in South Wales to the wharf at Abercynon.

1812 John Blenkinsop develops a rack railroad system in collaboration with Matthew Murray of Leeds Round Foundry—the Middleton Railway.

1830 George Stephenson builds the world's first passenger railroad, between Liverpool and Manchester. The first locomotive on it was his son Robert's *Rocket*.

1750	1775	1800	1825

Steam Power to Steam Locomotive

Thomas Newcomen invented his "atmospheric engine" in 1712, long before the start of the industrial revolution. However, by the end of the 18th century, it had been greatly improved by James Watt, whose more efficient double-acting steam engines were good enough to replace water power in factories. Watt's innovations also paved the way for steam-powered locomotion.

The discovery of the vacuum in the middle of the 17th century showed philosophers the immense power of atmospheric pressure, which pushes with a force of about one kilogram on every square centimeter (14 lb per square inch) of Earth's surface. One way to use this force was to create a vacuum by condensing steam, then allow the atmosphere to eliminate the vacuum. The Newcomen engine worked in this way, and for decades pumped water from coal mines.

Improving the efficiency
In 1764 a model Newcomen engine was brought for repair to a young Scottish engineer called James Watt (see p.149) at the University of Glasgow. After fixing it, he discovered that it needed loads of coal to run at all, because it wasted so much heat. Watt had heard about latent heat—the heat needed to turn water into steam—from his friend and mentor Joseph Black (see pp.148–49). He realized that the Newcomen engine was inefficient because for every stroke the entire cylinder had to be heated above boiling point to fill it with steam, then cooled below it for the steam to condense.

Watt pondered the problem for a few months, then one Sunday morning in May 1765 the answer came to him in a flash of inspiration as he was walking on Glasgow Green. He realized that it would be far better to keep the cylinder above boiling point all the time, and have a separate condenser.

In 1769 Watt went to London to acquire a patent, and when his wife died and his Scottish backer went bust he moved to Birmingham to go into partnership with Matthew Boulton. Boulton gave him money and skilled craftsmen as his assistants. Watt's engine was right at the edge of what was technically possible, but in 1775 he finally got it

to work. From then until about 1800 Boulton & Watt steam engines were the best in the world.

Watt made a double-acting engine by letting in steam at both ends of the cylinder, instead of steam at the bottom and air at the top. This was a real steam engine, as opposed to an atmospheric engine; both up and down were power strokes (see opposite). Installed with a flywheel it could drive machinery in carpenters' shops, mines, cotton mills, and factories of every kind.

The next big step forward was the high-pressure steam engine, pioneered by fiery Cornishman Richard Trevithick

in 1799. This used the force from the steam to drive the piston. In 1801 Trevithick built a steam carriage, but it quickly came to grief after getting into a rut and turning over. Then in 1804 he successfully ran another engine on wheels along a track. The age of steam transportation had begun.

« BEFORE

Before steam power, people had to rely on animal and human muscle power, and on windmills or water wheels.

POWER FROM WIND AND WATER
Windmills have long been used for grinding corn and raising water, but wind power is unreliable. **Water power** brought great concentrations of mills and factories to places with fast rivers—but not elsewhere.

HERO OF ALEXANDRIA
The Greek mathematician **Hero** (or Heron) lived in Alexandria in the 1st century BCE. An inventive man, he devised the first-ever vending machine, **primitive robots** powered by falling weights, and a wind-powered organ. But his most impressive invention was the **aeolipile**, the world's **first steam engine**. Steam from a boiler went up an arm into a sphere and hissed out of two tangential jets, spinning the sphere at speed. The device was for amusement only.

AEOLIPILE

« SEE ALSO
pp.50–51 WATER AND WIND POWER

Lots of coal
Early steam engines burned vast quantities of coal and so were initially used only at mines where coal was available as fuel.

1841 The railroad from London reaches Bristol, after which it is clear that local time differences and railroad timetables are not compatible, so London time, or railroad time, is adopted as standard.

1883 Mark Twain publishes *Life on the Mississippi*, detailing his experiences as a paddle steamer pilot in the years before and after the American Civil War. Steamboats and steamships gradually replace sailing vessels throughout the 19th century.

1897 The Stanley twins Francis and Freelan make the first Stanley Steamer, which goes on to become the world's most popular steam-powered car.

1850

1875

AFTER ≫

Internal combustion and electricity replaced steam for powering engines and machinery in the 20th century, but today we depend on steam turbines for most of our electricity.

INTERNAL COMBUSTION
In 1885 Karl Benz made the **Benz Patent Motorwagen**, the first ever motor vehicle to be powered by an internal combustion engine.

THE STEAM TURBINE
In 1884 **Charles Parsons** invented the **steam turbine**: a series of fans driven by super-high-pressure steam, like propellers in reverse. This is enormously powerful, and steam turbines now generate 75 percent of the world's electricity.

STEAM SPEED
In 2009 a British-built vehicle reached a speed of 139 mph (223 kph), setting a new record for steam-powered cars.

SEE ALSO ≫
pp.168–69 THE ELECTRIC MOTOR
pp.252–53 GENERATING ELECTRICITY
pp.254–55 THE INTERNAL COMBUSTION ENGINE

▽ How a double-acting cylinder works
Steam is exhausted to the separate condenser from the bottom of the cylinder, creating a partial vacuum inside. Steam from the boiler then enters the cylinder from the top end, allowing the piston to move down. The process is then reversed, so that both up and down are power strokes.

Steam inlet from boiler

Steam outlet to condenser

Piston moves down

Piston moves up

Steam outlet to condenser

Steam inlet from boiler

1 Down stroke. Steam from the boiler enters the cylinder at the top, allowing the piston to move down, while steam from the previous stroke escapes to the separate condenser.

2 Up stroke. Steam escapes from the top to the condenser, while steam from the boiler enters below, allowing the piston to rise.

Boulton and Watt engine
In this working model of a Boulton and Watt steam engine the double-acting cylinder is near the left side and the large flywheel on the right. The parallel motion above the cylinder was Watt's proudest invention.

Parallel motion
The parallel motion converts the vertical movement of the piston rod to the curved movement of the end of the beam.

Beam
The beam moves up and down in an arc.

Governor
The speed of the engine is controlled by a centrifugal governor, which cuts off the supply of steam if the engine goes too fast, and increases it if the engine goes too slow.

Connecting rod

Flywheel
Cranks or sun-and-planet gears convert the up-and-down motion of the beam to turn the flywheel.

Piston

Valve gear
The entry of steam into the cylinder is controlled by the valve gear.

Double–acting cylinder

Crankshaft

Cistern
Inside a cistern full of cold water are a condenser and a pump. The condenser cools the steam back into water, which is removed by the pump.

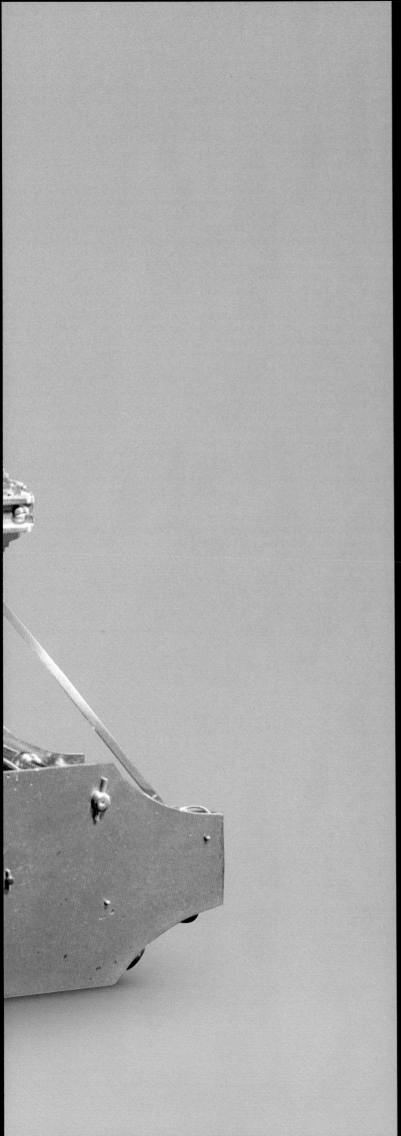

Harrison's Chronometer

In 1707, the British Navy suffered a humiliating disaster when a fleet ran aground on the Isles of Scilly, off southwest England, with the loss of more than 1,400 lives. The disaster spurred efforts to find a solution to the "longitude problem"—the accurate assessment of a ship's position at sea—and inspired one man to take engineering precision to a new level.

Longitude is a measurement of the position east or west of a fixed location, such as the home port of a vessel. Since the late 19th century this location has been along the prime meridian, at Greenwich, England. The key to finding longitude was deceptively simple. This is because a vessel's location on Earth's surface can be determined by the time at which it experiences local noon in relation to the time on a clock that has been set from noon at a fixed location.

In 1714 a prize of £20,000 was offered by the British parliament for a solution to the problem, and a "Board of Longitude" was established to sit in judgment on the ideas proposed.

Most of the suggestions offered to the Board were astronomical in nature, and all relied on precise observations that would require a steady deck, a fine telescope, and an experienced observer. But Yorkshire-born clockmaker John Harrison invented a mechanized alternative—a precise, robust, maritime timekeeping machine, or "chronometer," which was capable of keeping track of the time in the vessel's home port.

In 1730 Harrison traveled to London and presented the Astronomer Royal, Sir Edmond Halley, with plans for his revolutionary marine chronometer. Halley referred him to the renowned clockmaker George Graham, who personally financed the construction of Harrison's first prototype, now known as "H1," over the next five years.

While pendulum timepieces were unreliable timekeepers at sea because the pendulum would not always swing accurately, H1's main technological advantage was that its spring-driven mechanism was unaffected by the pitch and roll of a ship's deck. The clock would only run for a day at a time, but Harrison's later timepieces incorporated an ingenious mechanism that allowed them to continue running while being wound. H1 caused a sensation when it was presented to members of the Royal Society, and proved its accuracy at sea on its voyage from England to Portugal and back. Although Harrison was paid by the Board of Longitude and died a rich man, a bias against his mechanical solution meant that he never received the full amount of the prize.

Harrison's first chronometer
When H1 was put to a sea trial onboard HMS *Centurion* in 1736 it lost only a few seconds during the voyage to Lisbon, Portugal. Decorated with eight cherubs and four crowns, the brass timepiece is almost 2ft (0.6m) tall, wide, and deep, and weighs 75lb (34kg). The brass balls on the top of the clock are part of the oscillating mechanism that replaced the pendulum.

"One of the most **exquisite movements** ever made."

WILLIAM HOGARTH, ON H1, "ANALYSIS OF BEAUTY," 1753

BEFORE

Through the ages, people have devised a variety of methods for finding their way at sea.

USING STICK CHARTS

STICK CHART

Between 2300 BCE and 1500 CE, the Polynesians and Micronesians gradually **spread across a large part** of the Pacific, without any physical compass. Their methods included noting prevailing directions of waves and winds, and the use of **early types of maps** called **stick charts**.

HIPPARCHUS'S CONTRIBUTION

Although it had already been in use for decades, the system of **latitude** and **longitude** for locating a point on Earth's surface was formally described by the Greek astronomer Hipparchus in about 150 BCE ≪38–39. It has continued, almost unchanged, to the present day.

VIKING SUN COMPASSES

In the 8th and 9th centuries CE, the **Vikings** used **non-magnetic compasses**, based on the Sun's position, to provide crude **estimates of latitude** and the **direction of true north**. When they thought they were near land, they released ravens to find out the direction of the coast.

≪ SEE ALSO
pp.26–29 THE FIRST ASTRONOMERS
pp.134–35 HARRISON'S CHRONOMETER

In order to navigate the seas with any degree of certainty, seafarers needed to ascertain not only what direction they were sailing in, but also where they were at any one time in terms of latitude and longitude (see pp.38–39).

The magnetic compass

Until around 1100 CE the main way of navigating at sea was to stay in sight of land, or to observe the Sun and stars. The invention of the magnetic compass was a big step forward, since it allowed seafarers to orient themselves and steer a course even in cloudy weather or fog. The first definite reference to a magnetized device "for finding south"—a tiny iron fish in a water bowl—comes from China in about 1044. By 1120 this device had become a magnetized needle. These early Chinese compasses were "wet"—they consisted of needles floating in bowls of water. The true mariner's compass, using a pivoting needle in a dry box, was in use in Europe by c.1300.

By the end of the 15th century, seafarers realized that compass readings require a degree of correction. This is because changes in Earth's magnetic field cause the magnetic poles to wander relative to the geographical poles of Earth's rotation.

Determining latitude and longitude

Though useful, the compass was not much help when determining position at sea. To find out where they were, and to create maps, navigators

Mariner's compass
In this 16th-century compass, the magnetized needle is attached to the underside of a pivoting circular card, which displays the compass points.

needed to work out their latitude and longitude. Latitude is easy to determine by measuring the angle of elevation of the Sun or a particular star in relation to the horizon (see opposite). To make such a measurement, a variety of instruments such as the astrolabe, cross-staff, and backstaff were used until the mid-18th century when these became extinct following the invention of the sextant (see opposite).

Between the 15th and 18th centuries, although voyagers could determine latitude, they had only a vague idea of where they were in the East–West direction. This was because longitude was much harder to measure. Compasses gave virtually no clue, and instruments such as quadrants (predecessors of sextants) were also of no use on their own, since the celestial position and paths of the Sun and stars appear exactly the same at different longitudes. To find their way back to home ports after voyages to the Americas, European seafarers relied

Navigating the Oceans

Until instruments such as the magnetic compass came into use around 1300 CE, navigating the world's oceans was extremely difficult. Precise navigation only became possible several centuries later, following the invention of the sextant (an angle-measuring device) and an accurate seagoing clock.

Magnetic map
The lines on this chart, which dates from 1700, indicate the correction that had to be applied to compass readings to adjust for the difference between Earth's magnetic poles and the true (geographic) poles.

on sailing north or south until they reached the same latitude as their home port, then sailing directly east. However, this method resulted in frequent shipwrecks in storms or fog. By the 18th century, the military and economic importance of solving the longitude problem was a major concern for some European governments.

Solving the longitude problem

The scientists of the day already knew, in general terms, the solution. Ships' navigators needed a "universal clock," which would always tell them the exact time at a particular location, such as the navigator's home port. Wherever they went, they could then calculate their longitude by comparing the time on this universal clock to their local time (determinable using a sextant). Each hour of disparity would be equivalent to a 15-degree difference in longitude (see opposite). Two types of clock were possible. The first would be a physical clock on board a ship, which would accurately keep track of the time at the ship's home port—but such a clock, capable of coping with rough

AMORPHOUS SOLIDS

An amorphous solid is not crystalline, which is to say there is no regular repeating pattern to the position of the atoms (see p.142). Many substances form amorphous solids when they are cooled so rapidly that the particles do not have a chance to become arranged in an organized way. Glass is an amorphous solid—the common belief that it is really a supercooled liquid arises from the fact that its solidification from liquid form does not happen in the same abrupt way as the phase transition exhibited by other types of solids.

STAINED GLASS

the inter-particle attractions are negligible. As a result the particles are free to move independently of one another, so that neither the shape nor volume of a sample of gas is fixed.

Phase transitions

When matter changes from one state to another, the change is known as a "phase transition." For example, a solid melting, a liquid boiling, a vapor condensing into a liquid, and a liquid solidifying are all phase transitions. Metal becomes more malleable when heated so its atoms become more energized, but they are still fixed by bonds: heated metal is still solid. But when it finally melts its particles move freely in the new liquid phase: melting is the abrupt phase transition.

A pure substance melts or freezes, and boils or condenses, at characteristic temperatures. These temperatures can be used as a gauge of the purity of substances, since any contamination alters their values. For example, pure water freezes at 32°F (0°C), whereas the salt dissolved in seawater depresses its freezing point to about 28.4°F (−2°C).

In order for a solid to be melted into liquid, the particles need to gain sufficient energy to escape the bonds of their rigid confinement. They must acquire yet more energy to become gaseous. In each case the amount of

heat that is required to change the state (but without changing the temperature) is called the latent heat, and this varies from substance to substance (see pp.178–79).

A few substances have anomalous properties that mean phase transition occurs directly from solid to gas, with no liquid state in between: this is called sublimation; for example, dry ice (solid carbon dioxide) sublimes directly into gaseous carbon dioxide.

The fourth state of matter

In 1879 Sir William Crookes discovered the fourth state of matter by means of experiments on gases. He called it "radiant matter." Later, the English physicist J.J. Thomson showed that a proportion of the particles in the gas had lost their electrons, with the result that this state of matter was composed of flowing negative electrons and positive ions. It was named plasma in 1928 by US scientist Irving Langmuir.

Plasma resembles gas but differs from it in conducting electricity and responding strongly to magnetic fields, often forming beams or filaments when it does so. It is the most abundant state of matter in the Universe, being a component of stars—including the »

Rising temperature gives bound water molecules more energy, turning them to liquid

◁ **Sublimation**
Solid carbon dioxide (also known as dry ice) dropped into a tube of water turns instantly into gas. Dry ice is one of relatively few substances that sublime—change directly into gas when it is heated, bypassing a liquid phase.

Stream of light emitted from ionized gas

The bonding of water molecules keeps them in pools

Metal ball in the center of the sphere is charged with electricity

Water's changing states
As ice melts its molecules escape from their fixed positions in the ice crystals, but—unlike most other substances—come closer together. As a consequence ice, being less dense than water, floats (see p.142).

▷ **Liquid mercury**
Mercury is one of only two liquid elements (the other being bromine), and the only metal that is liquid at room temperature. Its high surface tension makes it form rolling spheres that do not wet surfaces as water droplets do.

Mercury drops on a surface are more spherical than water drops

Plasma globe
An electrically charged metal ball in a glass sphere filled with low-pressure gas discharges electricity through the sphere to the glass wall. This forms plasma by making particles of gas ionized along the path of discharge, emitting light in the process.

Malleable materials
Metals are highly malleable because their atoms can slide past one another so that the metal changes shape without fracturing. Reshaping by application of force is more easily done when the metal is red hot.

Mohs' Hardness Scale

Ten common minerals are listed in Mohs' Hardness Scale, which enables the hardness of a solid to be determined by finding the hardest mineral listed that is capable of scratching it. For example, if a solid is scratched by feldspar but not by apatite, its Mohs hardness is determined to be between 5 and 6.

Mineral	Mohs hardness
Talc (softest)	1
Gypsum	2
Calcite	3
Fluorite	4
Apatite	5
Feldspar	6
Quartz	7
Topaz	8
Corundum (e.g. ruby)	9
Diamond (hardest)	10

≫ Sun—and is formed naturally by lightning. In the home, artificial plasma is used in fluorescent lamps and TV plasma displays.

Hardness of solids
In many crystalline solids the particles naturally adopt a specific repeating pattern as a result of the highly organized ways in which the particles bond with their neighbors. The nature of the crystal arrangement has important consequences for the properties of the solid. Some solids—such as glass—lack a crystalline structure entirely and are called amorphous solids (see p.141).

In 1812 Friedrich Mohs came up with a scale of hardness based on the ability of one solid to scratch another. Mohs' Hardness Scale is still used by geologists today. Thomas Young later devised a way of measuring a property of solids which he described as elasticity. Some solids are highly elastic, meaning they can revert to their original shape when deformed, or plastic, meaning that they cannot. Young's expression of elasticity, published in 1807 and now known as Young's Modulus, has important applications in mechanical engineering since it quantifies how materials behave when under a load.

Ductility
Glass consists mainly of quartz sand (silica) with other additives, such as sodium carbonate, which lowers its melting point. When heated, glass is immensely ductile and can be stretched almost indefinitely; yet when cold it is brittle.

Solids also vary in their response to different types of stresses, which may be pulling (tensile) or pushing (compressive). Some are highly ductile, meaning that they can be pulled without breaking. Others are highly malleable and can be compressed without breaking; for example, copper and gold can be hammered without breaking whereas cast iron cannot.

Fluidity and wetness of liquids
When a solid melts to form a liquid the particles become separated from their neighbors in such a way that the volume usually increases. However, the melting of ice into water is a notable exception: ice shrinks to form water that has higher density, with the result that ice floats on water.

Water owes its bizarre property to its bonding. In ice, the molecules are held rigidly apart in an "open" lattice arrangement by hydrogen bonds (see pp.288–89). On melting at 32°F (0°C), the molecules come slightly closer together, and the lattice begins to disintegrate. As the temperature rises, this disintegration continues and is complete at 39.2°F (4°C), when the cold water has its maximum density. As the temperature rises more the density of water drops a little as the molecules move faster and farther apart, but its density in liquid form never becomes as low as that of ice.

A liquid's resistance to stress is called viscosity; the more viscous a fluid, the less easily it flows. The physical principles of viscosity were first formulated by Isaac Newton (see pp.110–11). Today we know that viscosity—just like the hardness of solids—depends upon the size of the molecules and forces between them. Water has very low viscosity, while in olive oil it is higher, and in syrup higher still. The lack of rigidity of liquids also means that many can be mixed, though certain mixtures—such as water and oil—will spontaneously separate, repelled by the intermolecular forces

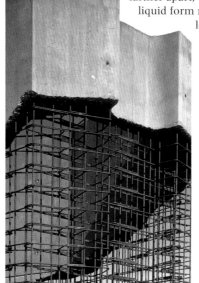

Reinforced concrete
Concrete on its own is a brittle material, but its strength can be improved for building purposes by incorporating bars of non-brittle steel.

that preferentially engage water with water and oil with water but keep the different molecules apart.

Intermolecular forces also account for the fact that liquid particles at the surface pull inward to create a "skin"—a phenomenon called surface tension. The cohesive molecular forces responsible for surface tension plus the repulsive forces that keep oil and water apart explains why perfect rolling drops of water form

PLASTICITY The characteristic of a solid to change shape without recovering.

ELASTICITY The capacity of a solid to change shape and recover.

MALLEABILITY The ability of a plastic solid to be reshaped without breaking.

DUCTILITY The capacity of a plastic solid to be stretched without breaking.

VISCOSITY The tendency of a liquid to resist flow.

on oily surfaces, such as a waxy leaf or a duck's back. However, on something that can be more easily wetted—such as clean glass—the attractive forces between the water and glass molecules are stronger than those between the individual water molecules. In 1804 Thomas Young explained that the attraction between water and glass is so strong that it enables water to creep up the sides of a narrow tube—a

Surface tension
Water molecules at the surface are under tension, being pulled inward by their neighbors. The result is a thin surface "skin" that can support the weight of certain insects with water-repelling hair on the ends of their legs, such as this pond skater.

« BEFORE

People probably observed the effects of static electricity in ancient times, by rubbing objects with fur, for example, and seeing lightning.

DISCOVERY OF STATIC

In 600 BCE the Greek philosopher **Thales of Miletus** noted that rubbing fur on the natural material **amber** (fossilized tree resin) gave the amber the ability to attract feathers and other small objects. Amber has an especially high **electrical resistance**; its own **electrons** are bound very tightly to their atoms, but amber does allow electrons from other objects to **migrate** onto it, so that it becomes **negatively charged** overall. The electrons in materials such as fur, feathers, hair, wool, silk, and paper are not so tightly bound, which allows them to have an **electrical attraction** to materials such as amber. The word for amber in Greek is "elektron," which gave us the word "**electricity.**" Its usage can be traced back to the British scientist **William Gilbert** « 81, who used the word "**electricus**" to describe the force that exists between charged objects.

AMBER ATTRACTING FEATHER

PHYSICIST (1736–1806)

CHARLES COULOMB

Like any other child of a wealthy French family in the 18th century, Charles-Augustin de Coulomb received a good education in literature, philosophy, and the sciences. He particularly excelled at mathematics, and after further study in applied math he became a qualified civil engineer. He worked for 20 years on large engineering projects for the French Crown, in several different countries. After winning a prize in a competition organized by the French Academy of Sciences, Coulomb became more interested in physics than in engineering. He made important contributions to the subject of friction, as well as static electricity.

Static Electricity

From the dramatic flash of a lightning bolt to the gentle, invisible force that makes a rubbed balloon stick to a wall, the effects of static electricity are everywhere. They are all due to a simple imbalance of electric charge, and the fact that opposite charges attract one another while like charges repel.

Electric charge comes in two varieties, either positive (+) or negative (−). All everyday objects contain these two charges, since they are all made up of atoms—and every normal atom contains both positively charged protons (within its nucleus) and negatively charged electrons (in the outer part of the atom).

Generally, the negative charges balance the positive ones; atoms are electrically neutral. But electrons can move from one atom to another in certain circumstances—such as when some materials are rubbed together—which confers an overall negative charge on the atoms they migrate to, and a net positive charge on the ones they leave. For example, when rubbing a balloon on a sweater, friction causes the negative charges to move from the sweater to the balloon, making it negatively charged.

Any two oppositely charged objects (+ and −) attract each other, and similarly a large imbalance of negative charges in one object will attract the positive charges of a second object. Sometimes, when two objects come near each other a stream of electric charges passes across the gap between the objects, causing a spark, which equalizes the imbalance of electric charges—as with lightning.

Experimenting with static

Investigations into static electricity became popular during the 17th century. At that time, scientists thought that heat, magnetism, and electricity were made of weightless, invisible fluids, like gases or liquids. When they charged objects with static electricity, they believed they were filling them with electrical fluid—in English the verb "charge" originally meant "fill."

Early experimenters were able to make small sparks by charging objects and placing them close to each other. In 1745 Dutch scientist Pieter van Musschenbroek invented a device called a Leyden jar, which made it possible to store static electric charge. In 1752 American statesman and scientist Benjamin Franklin proved that lightning bolts are just huge electrical

sparks. In his classic experiment, Franklin—after insulating himself to avoid electrocution—flew a kite up into a thundercloud and successfully charged up a Leyden jar by drawing electrical charge down the wet string from the cloud. He also made sparks fly from a metal key attached to the string.

Electric force fields

About 20 years later French physicist Charles-Augustin de Coulomb was measuring the forces between objects that were charged with static electricity, known as electrostatic forces. Coulomb formulated an equation that relates the strength of the force, amount of charge, and the distance between two charged objects. He discovered that doubling the distance between objects, for example, reduces the force to one quarter of its original magnitude. The unit of electric charge, the coulomb, is named after him.

Storing electric charge
The Leyden jar was the first capacitor, "storing" electricity between two electrodes.

Detecting static electricity
The gold-leaf electroscope, an instrument for detecting static electricity, relies on the fact that inside metals electrons are free to move around. A charged object just above an electroscope causes the metal parts to become charged, and this causes the foils to separate. The greater the charge, the higher the foil lifts.

1 Charge introduced
A positively charged resin rod is moved over the metal plate of the electroscope to induce a charge.

3 Negative charge
The metal plate becomes negatively charged due to the arrival of free electrons.

2 Movement of electrons
The free electrons from the gold-leaf foil pass up the metal rod to the metal plate.

Metal plate

Glass case protecting gold leaves from damp

Metal rod

Gold-leaf foil strips which hang together when uncharged

Escaped free electrons move to metal plate

Gold atom

Negatively charged electron

4 Gold leaves part
As the gold leaves become positively charged, they move apart.

BEFORE CHARGE INDUCED

AFTER CHARGE INDUCED

AFTER »

By the end of the 19th century physicists were able to make sense of static electricity and how it related to atoms.

THE CONNECTION WITH MAGNETISM
In the 17th century scientists believed **electrical** and **magnetic forces** were completely separate until, in the 1870s, British physicist **James Clerk Maxwell** proved they were both forms of a single phenomenon: **electromagnetism** 166–67 »

THE FORMS OF ELECTRIC FIELDS
In 1839 British scientist **Michael Faraday** 170–71 » visualized the shapes of electric fields and proved that **static** and **current electricity** were different aspects of the same phenomenon.

ATOMS AND IONS
The **electron** was discovered by physicist Joseph John (J.J.) Thomson in 1897. In due course **quantum physics** 314–17 » was able to work out the behavior of electrons in detail.

Christmas Lecture
Faraday gave 19 sets of Christmas Lectures in what is now called the Faraday Lecture Theatre at the Royal Institution in London. He originated the lectures to introduce complex scientific ideas to the public, especially to young people.

Electromagnetic rotation
On September 3, 1821, Faraday performed the experiment that was to result in the world's first electric motor. He wrote out the details in this laboratory notebook.

Michael Faraday
An intuitive physicist, Faraday had an exceptional gift for visualizing lines of magnetic force and how they interacted with electric current, when both phenomena were poorly understood. In addition, he was an exceptional experimental chemist.

Davy took him on, first as his secretary, and then as chemistry assistant. When, in 1813, Davy decided to make a tour of Europe with his new wife, a grand heiress named Jane Apreece, Faraday was asked to go with them as scientific assistant. Davy's valet did not wish to go, and Faraday was given his duties, too. Jane treated him as just another servant and he had a miserable time.

For five years after they returned, Faraday worked on various projects in chemistry. Then, in 1821, he started on what was to become his greatest work, in electricity and magnetism.

Visionary power
Lacking formal education, Faraday had almost no mathematics. Instead he had an extraordinary ability to visualize things. He looked at a magnet and could "see" the magnetic field lines sprouting from the ends and curling around from one pole to the other; he called them "lines of force," and we still use the expression today. This visionary power enabled him to make strides in several branches of chemistry and physics. His ideas were also crucial to physicist James Clerk Maxwell, who

was to continue in electromagnetism what Faraday left unfinished.

In 1820 the Danish scientist Hans Christian Ørsted showed that around a wire carrying a current there is a magnetic field, almost like a sleeve around an arm. Faraday talked about this with Davy and William Hyde Wollaston, and they speculated about whether this sort of magnetism could be used to generate motion.

Ground-breaking invention
On September 4, 1821, Faraday made the world's first electric motor—a wire dipping into a pool of mercury and revolving around a magnet. His invention paved the way for others to develop the idea (see pp.162–63). Faraday failed to acknowledge the contribution of Davy and Wollaston sufficiently, and they argued. Davy then gave Faraday a useless project investigating various types of glass, which in effect wasted six years of his life. Only when Davy died in 1829 was he able to return to electromagnetism; Faraday then discovered the process of electromagnetic induction, the basis of transformers and dynamos.

Later interests
Faraday's broad scientific curiosity was not confined to the laboratory. He investigated the spiritualist movement, which began in 1848, although he was convinced it was all nonsense. He devised experiments to show that, during table-tipping séances, the sitters themselves were unconsciously moving the table.

Michael Faraday also was a brilliant lecturer and demonstrator. He initiated the Christmas Lectures at the Royal Institution in 1825. And yet he was humble and self-effacing. He twice refused to be President of the Royal Society, and he even refused a knighthood for religious reasons, saying he preferred to remain plain Mr. Faraday to the end of his life.

BREAKTHROUGH
LAW OF INDUCTION

Faraday's law of induction, recognized independently in the US by Joseph Henry, is a basic law of electromagnetism and applies to electrical generators. The law states that the voltage induced in a coil of wire is proportional to the rate at which it cuts through a magnetic field, multiplied by the strength of the field. In practical terms, a higher voltage can be obtained by spinning a generator more quickly, by using a stronger permanent magnet, or by using more turns of wire in the input side of an induction ring or transformer.

FARADAY'S INDUCTOR

TIMELINE

- **1791** Michael Faraday is born on September 22 in Newington Butts, south London, England, the son of a poor Yorkshire blacksmith.

- **1805** He is apprenticed to a bookbinder, George Riebau, in Bloomsbury, London.

- **1812** Goes to a series of lectures given by Humphry Davy (see opposite) at the Royal Institution (RI). He sits in the seat above the clock, and is entranced. He then writes, illustrates, and binds a 300-page book based on his notes from the lectures, and sends it to Davy, asking for a job.

- **1813** Davy employs Faraday first as secretary and then as chemical assistant at the RI.

- **1813–1815** Davy takes Faraday as assistant on his honeymoon tour of Europe, during which he investigates volcanic action. Davy's valet refuses to go, so Faraday has to act as valet as well. Davy's wife treats him as one of the servants, and Faraday becomes so miserable that he is tempted to return home and give up science.

VOLTAMETER USED BY DAVY

- **1821** Marries Sarah Barnard on June 2. She is a fellow member of the Protestant Sandemanian sect, which opposes the Church of Scotland.

- **1821** Invents the first electric motor.

- **1823** Liquefies chlorine.

- **1824** Elected a Fellow of the Royal Society.

- **1825** Elected Director of the RI. Initiates the Christmas Lectures, which are still running today.

- **1827** Gives his first of 19 sets of Christmas Lectures; one of his best-known was called "The Chemical History of a Candle."

- **1829** Invents the induction ring—the basis of the transformer and the dynamo.

- **1831** Observes that a moving magnet induces an electric current.

- **1833** Elected Fullerian Professor of Chemistry at the RI.

- **1836** Inspired by Faraday, John Frederic Daniell succeeds in making the first battery (see pp.162–63) to produce electrical current over a long period.

DANIELL AND FARADAY

- **1847** In what could be called the birth of nanoscience (see pp.392–93), Faraday reports that the optical properties of gold colloids differ from those of the bulk metal.

- **1855** Writes a strong letter to *The Times* newspaper about the Great Stink, caused by foul pollution of the Thames River.

- **1867** Dies on August 25 at his grace-and-favor house in Hampton Court, Surrey.

Accurate Measurement

Nearly every scientific experiment involves some kind of measurement. Today scientists can measure to extreme accuracy, using a wide range of precise measuring devices. They use a logical and clearly defined system of units called the International System of Units, or SI units.

Accurate measurement in scientific investigations became much more important in the 17th century, when people began to use the scientific method (see pp.68–69). Experimenters became aware that only with reliable measurements could their experiments support or disprove their theories.

Along with the measurement of time (see pp.120–21), the most frequent and important measurements that scientists make are mass (or weight) and length (or distance).

Mass and weight

In everyday language, people confuse the terms "mass" and "weight"—both words seem to refer to how heavy something is. But to a scientist there is

an important difference: an object's mass is the amount of "stuff" it contains, while its weight is the force of gravity (see pp.108–109) on that object.

In 1678 the English scientist Robert Hooke (see pp.92–93) invented the spring balance, which measures an object's weight according to how much it extends a metal spring. Hooke's invention, however, is unreliable when measuring mass. This is because the extension of the spring depends on the force of gravity on the object being measured; gravitational pull varies with the irregular shape of our planet and the thickness of Earth's crust; it can be different at different locations.

In contrast, the lever-arm balance gives the same reading wherever you take it. It works by comparing standard weights or masses on one side of the pivot with the weight of an object on the other. A traditional balance would even give the same readings on the Moon, where an object's mass is unchanged, but its weight is only one-sixth of that on Earth.

Modern chemical balances are accurate to less than millionth of a gram (about a 30-millionth of an ounce), while a device called a mass spectrometer can select individual atoms by mass.

Length or distance

In 1631 French physicist Pierre Vernier invented the Vernier scale—an add-on to existing measuring devices to help scientists measure small distances very accurately. Vernier attached his scale to

« BEFORE

All the ancient civilizations had systems of weights and measures, which were essential for many different activities, including building, making items such as clothes, agriculture, and trade.

ANCIENT MEASUREMENTS

The units of measurement in early civilizations were often based on the **human body** or other natural quantities. For example, the cubit, which originated in ancient Egypt, was the **length of the forearm**, and in several civilizations the weight of precious stones was often given in terms of grains of wheat. However, human bodies and wheat grains vary; so it is impossible to measure consistently and accurately using these. As a result, even in ancient times governments and rulers made **"standard" weights and measures** against which people could compare their own measuring devices. Nevertheless, systems of weights and measures remained complicated and confusing, and the measurements people made were probably inaccurate.

ANCIENT ASSYRIAN WEIGHT

Pointer indicates object's weight of 7 lb (equivalent to 3.2 kg)

Rack-and-pinion links spring to pointer

Dial shows weight in pounds

Oval spring deforms into a more circular shape when a weight is hung on the hook

Spring balance

Inside this 18th-century spring balance is an oval spring, which deforms when an object is hung from the hook. The greater the mass of the object, the greater its weight and the more the spring deforms. As the spring deforms, a ratchet connected to it moves a pointer around a calibrated dial.

Weight (7 lb) suspended on a hook, which is linked to the spring by a metal bar

ENGINEER (1771–1831)

HENRY MAUDSLAY

English toolmaker and inventor Henry Maudslay gained early experience of engineering when he began filling artillery shells in the Royal Arsenal, Woolwich, at the age of just 12. Later he was apprenticed to the English lockmaker Joseph Bramah, who was obsessed with precision. In 1791 Maudslay set up on his own, and in 1800 invented a type of lathe that enabled standard screw threads to be mass produced. He is known as the "father" of precision engineering.

INVENTION

VERNIER SCALE

The Vernier scale was invented in 1631 by French mathematician Pierre Vernier. Here, a Vernier scale is fitted to a calliper, a device whose jaws close around an object (or open into a space to make internal measurements). In the example shown, the approximate reading on the fixed main scale is somewhere between 18 mm and 19 mm. The point at which the markings on the fixed scale and Vernier scale coincide give greater precision. The actual reading below is 18.64 mm.

Movable jaws for making internal measurements

Vernier scale (in inches)   Fixed scale (in inches)

Movable jaws for making external measurements

Vernier scale (in mm)   Fixed scale (in mm)

Reading on fixed main scale between 18 mm and 19 mm

Vernier scale and fixed scale coincide here, giving a reading of 18.64 mm

a quadrant—a device for measuring angles—but it was later used on the calliper, a device that enables dimensions of objects to be measured.

Around the same time, English astronomer and mathematician William Gascoigne invented another device that would revolutionize accuracy: the micrometer. Gascoigne's device was based around a screw: as you turn a screw, it advances very slowly, and incorporating this into a measuring device makes it possible to measure small distances accurately. Gascoigne used his micrometer in telescopes, but the principle was soon being used to engineer tools.

In 1800 Henry Maudslay invented the screw-cutting lathe, which made it possible to cut standard screw threads. Until then nuts and bolts were made in pairs, and only a matching pair could be guaranteed to work together, but Maudslay's invention enabled engineers to mass produce screws, nuts, and bolts.

Maudslay used his expertise with screw threads to produce an extremely accurate micrometer that was capable of measuring to an unprecedented accuracy of one ten-thousandth of an inch (about 0.0025 mm). One of Maudslay's ex-employees, Joseph Whitworth, did even more to promote precision in the 19th century. In 1833, having learned the technique from Maudslay, he produced a new way of making extremely flat surfaces. These "surface plates" are used to set up accurate measuring devices, and Whitworth's technique enabled him to make a measuring device that was accurate to one millionth of an inch (about 0.000025 mm).

SI units

By the end of the 19th century great accuracy was common in science and engineering. However, scientists were still using traditional systems of measurement, with the result that different systems were being used in different countries.

With science becoming increasingly international during the 20th century, it became important for there to be a single measurement system. The Système International, or SI units system (see pp.430–31), was introduced in 1960, enabling scientists to share the results of their measurements and calculations meaningfully in international journals.

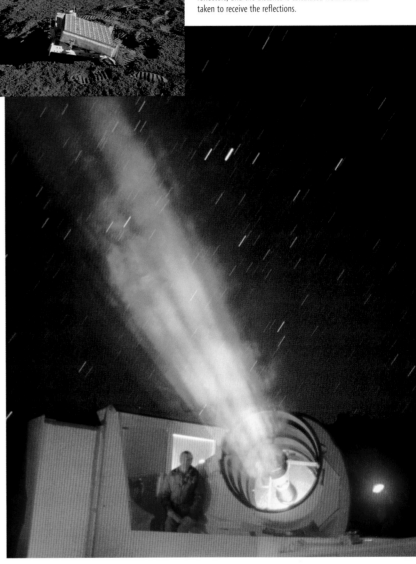

The measurement of the distance to the Moon is one of the most precise ever made, accurate to better than one part in ten billion. It is made possible by reflectors placed on the lunar surface (left) by US and Soviet missions. Laser beams from observatories on Earth (below) bounce off the reflectors, and the distance is calculated from the time taken to receive the reflections.

One kilogram
The kilogram was, for many years, defined in terms of a physical object—the International Prototype Kilogram (IPK), which was made in 1889. However, with the need for greater accuracy and repeatability, the definition of the kilogram is changing; it will soon be defined according to physical constants.

AFTER

Modern science and technology is so sophisticated that measurements of unprecedented precision are necessary.

EXTREME PRECISION

It is now possible to measure the distances **between individual atoms** using an instrument called an **atomic force microscope**. This works by measuring the tiny forces between atoms on the microscope's probe and those on a surface, enabling the **surface atoms to be mapped**. In 2016, scientists operating an instrument known as **LIGO** (based at two sites in the US) measured the **interference between laser beams** to detect the effects of **gravitational waves** for the first time, confirming predictions made in Albert Einstein's general theory of relativity a century earlier.

LIGO GRAVITATIONAL WAVE DETECTOR

Calculating and Computing

From the last years of the 18th century onward the demands of science and industry prompted the development of techniques that ultimately gave us the computer. These techniques included the forerunners of apparently modern concepts such as programs and computer languages.

As the Industrial Revolution spread across Europe, a demand arose for more automated machinery. At the same time scientific discovery called for increasingly complex calculations.

‹‹ BEFORE

For millennia, people have used mechanical aids for calculation. Tools such as the abacus, Napier's bones, and the slide rule are the direct ancestors of our computers.

LOGARITHMS AND NAPIER'S BONES

First set out by Scottish mathematician **John Napier**, **logarithms** make calculations easier by allowing **multiplication** and **division** to be **simplified** to addition and subtraction. Napier also invented **Napier's bones**, a set of **rods** or **cylinders** marked with **multiplication tables** for numbers 1 to 9, as a calculating tool.

NAPIER'S BONES

SLIDE RULE

The slide rule is a **mechanical computer** invented in 1622 by English clergyman **William Oughtred**. The device comprises two sliding bars marked with **logarithmic scales**, plus a marker, or **cursor**. Deceptively simple in appearance, **slide rules** enabled users to do highly complex **calculations**. They were essential in science and engineering until the 1970s—even being used in the **Apollo** missions.

‹‹ SEE ALSO

pp.30–31 ANCIENT NUMBER SYSTEMS
pp.88–89 METHODS OF CALCULATING

Although various calculating machines had been invented, they were limited in capability and were not always accurate. For large and difficult calculations logarithms provided the best solution. The main problem with logarithms, however, was that their use depended on sets of tables—which themselves had to be calculated. The invention of the slide rule provided an easy-to-use alternative to consulting tables, although it was not as accurate.

In 1792 Napoleon Bonaparte appointed the French mathematician Gaspard de Prony to compile a set of more comprehensive and accurate logarithmic tables. To do so, de Prony employed a hierarchy of workers. Mathematicians devised the formulas, assistants broke the calculations down

Ada Lovelace
The daughter of Romantic poet Lord Byron, Ada Lovelace was encouraged to pursue mathematics instead of literature. She worked closely with Charles Babbage on his Analytical Engine and recognized its potential beyond calculation, writing the world's first computer program in 1843.

into instructions for simple addition and subtraction, and these basic tasks were then carried out by workers known as "computers."

The **Nature** of **Heat**

People have known how to make things hot since they first tamed fire, but not why their methods worked. In the early 18th century a German scientist, Daniel Fahrenheit, invented a thermometer with a reproducible scale, and shortly afterward an American spy made a breakthrough in the understanding of heat itself.

BEFORE

The concepts of "hot" and "cold" were obvious to the ancients, but what caused things to turn from cold to hot was not clear.

FOUR ELEMENTS
The 5th-century BCE Greek philosopher **Empedocles** thought that everything was made from **four elements**: air, earth, water, and fire.

CALORIC THEORY
Antoine Lavoisier introduced the idea of a fluid named "**caloric.**" He claimed that hot items contained high amounts of caloric, which **flowed into cold things** that they touched.

« SEE ALSO
pp.16–17 FIRE POWER
pp.176–77 ENERGY CONVERSION

Conducting heat
An iron bar heated at one end glows red then orange, then yellow as it gets hotter. Heat energy flows along the bar as more atoms start to jiggle around. This type of heat transfer is called conduction.

Atoms move more vigorously as they gain heat energy from the fire

Atoms adjacent to the vigorously moving ones also start to move more actively as the heat energy flows between them

PHYSICIST (1753–1814)
COUNT RUMFORD

Benjamin Thompson spied for the British during the American Revolution and eventually fled to England. From there he went to Bavaria, reorganized the army, invented a stove and a coffeepot, and was made a Count of the Holy Roman Empire. Returning to England in 1799, he founded the Royal Institution; he travelled to Paris and married the widow of the guillotined Antoine Lavoisier.

Heat makes things get hotter; it raises their temperature. But the nature of heat puzzled not only the ancients but also the scientists of the 17th and 18th centuries. The solution came from the American-born physicist, Benjamin Thompson, Count Rumford.

At the end of the 18th century cannons were made by casting the metal in one solid piece and then drilling or boring out the barrel to make an accurate cylindrical bore. While he was making cannons in Bavaria, Rumford realized that not only did the barrel always become hot during the boring,

Convection currents
As water is heated, the warm water closest to the heat source becomes less dense and rises. It is replaced by cooler, denser water sinking down from above. The convection currents continue until the water boils.

but also that when the borer became blunt, the barrel became much hotter than it did when the borer was sharp.

This surprised him, because people then believed that heat was a fluid called caloric, which lay hidden within substances. The blunt borer was removing less iron, so it should be generating less heat—but instead it was generating more. Thompson came to the conclusion that heat is nothing to do with caloric, but is in fact a form of motion. The borer was causing vibration of the iron atoms—the longer he ground away with his blunt borer, the more he was making the particles vibrate.

Three types of heat transference
During the cannon boring, both cannon and borer became hot; heat was being transferred from one to the other. It is now known that heat can be transferred in three different ways: by radiation, conduction, and convection. Anyone lying on a beach can feel the radiant heat from the sun, traveling through space and the atmosphere to the skin. Conducted heat is soon felt by anyone putting the end of an iron poker into a fire—the fire end may get red hot, but

heat is also conducted along the poker. Convection is the movement of air or other fluids caused by heat. If a radiator is turned on at one end of a room, it heats the air in front of and above it. This hot air—less dense than cool air—rises and moves across the ceiling then slowly drifts down a wall as it cools. In due course, the whole room becomes warmed by convection.

Inventing the scale
In 1714, the German physicist Daniel Fahrenheit proposed the measurement of temperature along a scale that gave the freezing point of water at 32° and its boiling point at 212°. This scale has generally been superseded by that devised by Swedish scientist Anders Celsius in 1742. He originally suggested that the boiling point of water should be 0° and the freezing point 100°, but this was reversed after he died; the scale we use today has the freezing point of water at 0° C and its boiling point at 100° C.

In 1848 William Thomson (later Lord Kelvin) calculated a value of –273°C for "infinite cold"—the temperature at which all particle vibration ceases. He devised a new temperature scale based on this absolute value, which began at 0K (kelvin) and gave water's freezing point as 273K and boiling point as 373K.

Specific heat and thermal capacity
Some materials are easier to heat than others. The "specific heat" of a substance is the amount of heat needed to raise the temperature of 1 g of it by 1° C. Water has an unusually high specific heat, so it can carry more energy than other common liquids. The "thermal capacity" of an object is the amount of heat needed to raise its temperature by 1° C; a large pan of water takes much more heat to reach boiling point than a small pan. Thermal capacity depends on the mass of the object and on its specific heat. Joseph

Black (see pp.148–49) discovered that even after being heated to 100° C a pan of water needs yet more heat to make it boil. He called this heat "latent" (from the Latin, meaning "hidden"), because it seemed just to disappear.

AFTER

Heat measurement has produced surprising results, including the quantum revolution.

BLACK BODY RADIATION
A heated black body (perfect emitter) glows, and the higher the temperature the shorter the wavelength. The exact relationship could not be explained without **quantum theory 314–17 »**.

TEMPERATURE OF THE UNIVERSE
The temperature in space is not absolute zero, but 2.725K, due to the background radiation—the heat echo of the Big Bang 320–21 ».

ULTRA COLD ATOMS
When **atoms** are **cooled close to 0K**, they slow to a crawl and display **quantum mechanical properties**. In the future, ultracold atoms could be used to build quantum computers.

SEE ALSO »
pp.180–81 THERMODYNAMICS

IN PRACTICE
KEEPING COOL

Many mammals have fur, which traps air and is a good insulator. It keeps them warm in cold weather, but can be a problem in higher temperatures. To avoid overheating, some animals have big ears which radiate heat, while others animals pant heavily—the evaporation of water from their mouths and tongues takes away the latent heat. Kangaroos lick their forearms to make them wet, and this water also evaporates to help keep them cool.

Laws of Thermodynamics

The area of science known as thermodynamics originated in efforts to understand and improve the steam engine early in the 19th century. Thermodynamics is the study of heat and energy. Its findings are set out in four laws, which provide deep insight into the way the Universe works.

The origins of thermodynamics go back to 1824, when Nicolas Sadi Carnot, a French engineer, published an influential book, *Reflections on the Motive Power of Fire*. In this groundbreaking work Carnot explained how steam engines work as he understood it, and how heat can make things move.

Motive power

Carnot used the term "motive power" to describe the ability to do work or to make things move. For example, steam engines have the motive power to lift heavy weights or pull trains. The source of this motive power is heat.

Carnot's book explained how heat could drive any engine. Whenever heat flows from a hot to a cold place, some of the heat can be "extracted" to do work. Carnot proposed that the most efficient way to extract heat was via a cycle of changes in the temperature and pressure of a gas. Known as the

PHYSICIST AND MATHEMATICIAN (1847–1907)

WILLIAM THOMSON, LORD KELVIN

Born in Belfast and raised in Glasgow, William Thomson published more than 600 scientific papers during his life. He made important contributions in many areas of science, including electricity and chemistry. Thomson also invented a device that predicted ocean tides, and he supervised the laying of the first successful transatlantic cable in 1858. He became known as Lord Kelvin after he was given the title Baron Kelvin in 1892, in recognition of his accomplishments (the River Kelvin runs past Glasgow University, where he was a professor for 50 years).

Carnot cycle, this idea paved the way for the development of the internal combustion engine (see pp.254–55) toward the end of the 19th century. In his book, Carnot relied on the caloric theory—the (incorrect) idea that heat is

a physical substance. By 1830 he had rejected this notion, and realized that heat is related to the movement of the particles of matter: the hotter something is, the more vigorously its atoms and molecules move. This is the dynamic theory of heat ("dynamic" means changing or moving).

Conservation of energy

With the recognition that heat is related to the motion of atoms and molecules, scientists began to understand heat as a form of energy. When an object slows down because of friction, for example, the energy of the moving object is not lost—it is merely transferred as heat to the atoms and molecules of the object and its surroundings. In other words, the energy of motion, kinetic energy, changes to thermal (heat) energy.

During the 1840s scientists realized that energy can neither be created nor destroyed, an idea investigated by the English scientist James Joule and expressed mathematically by the German physicist Hermann von Helmholtz. By 1850 Rudolf Clausius, another German physicist, had suggested that this idea, called the conservation of energy, should be the first law of thermodynamics.

Entropy and absolute zero

Clausius suggested a second law, which concerns the fact that not all energy is usable. When a steam engine cools down, for example, the heat energy still exists, but it can no longer do any work since it has dissipated into the air. Heat produced by friction dissipates, too. The second law of thermodynamics states that the amount of unusable,

Bouncing ball

A bouncing ball rises a little less high with each bounce, because some of the ball's motion energy is shared out among the atoms and molecules of the ball and with those of the floor. Each impact also produces a tiny amount of sound energy.

« BEFORE

Benjamin Thompson's work in the late 18th and early 19th centuries was fundamental to the understanding of heat as energy.

CALORIC THEORY

During the 17th and 18th centuries, physicists believed that **heat was a physical substance**, which they called caloric. They suggested that **caloric passed from hot to cold objects**, explaining why hot things cool down. The word "caloric" comes from the Latin word for heat, *calor*, and is the origin of the word "calorie."

BORING CANNONS

The American-born English physicist **Benjamin Thompson** (Count Rumford) **threw the caloric theory into doubt** after observing the boring out

THOMPSON AT A CANNON WORKS

of cannons. He noticed that when a drilling machine bored into a cannon barrel the cannon grew extremely hot. The **heat was being produced by friction** between the drill and the cannon, not released from within the gun metal as the caloric theory suggested.

« SEE ALSO
pp.176–77 ENERGY CONVERSION
pp.178–79 THE NATURE OF HEAT

LAWS OF THERMODYNAMICS

0 The "zeroth law" states that if two thermodynamic systems are each in thermal equilibrium with a third, they will also be in thermal equilibrium with each other.

1 Energy can neither be created nor destroyed.

2 The total entropy of an isolated system increases over time.

3 There is a minimum temperature, at which the motion of the particles of matter would cease.

dissipated energy can only stay the same or increase; it can never decrease. In 1865 Clausius coined the term "entropy" to refer to this unusable energy. Entropy is therefore a measure of disorder in isolated systems, and as time passes the Universe is becoming increasingly disordered.

The laws of thermodynamics give scientists a better understanding of the universe, including the world of living things.

BLACK HOLES AND QUANTUM PHYSICS
The **mathematics of thermodynamics** play a **key role in modern physics**. For example, it helps physicists understand the relationship between **black holes** 330–31 » and **quantum physics** 314–17 ».

ENTROPY AND LIFE
The second law of thermodynamics states that the entropy of the Universe is increasing, but there are some systems in which **entropy is decreasing**. The 20th-century German physicist **Erwin Schrödinger** looked at the role of entropy in the processes of life. He realized that **living things produce order from disorder**, like "islands of negative entropy," although they need an input of energy to do so.

SEE ALSO »
pp.314–17 QUANTUM REVOLUTION
pp.328–31 THE LIFE CYCLE OF STARS

Ship's engine room
Here, seamen stoke the furnace in the engine room of the USS *Mississippi.* Heat from the fire boiled water into steam, which was used to turn the ship's propellers.

"Nothing in life is certain except death, taxes and the second law of thermodynamics."

SETH LLOYD, PROFESSOR OF MECHANICAL ENGINEERING AT MIT, 2004

The third law of thermodynamics concerns temperature. As an object cools, its atoms and molecules vibrate less energetically. Scottish physicist William Thomson—Lord Kelvin (see opposite)—realized that there must be a temperature at which all motion would cease. He called it "absolute zero," and he even figured out that its value would be −459.67° F (−273.15° C). The kelvin temperature scale (K), which scientists use today, is named after him. Absolute zero is 0 K.

In 1852 Thomson came up with the the idea of a device he called a "heat pump." Such a device would be able to move heat from a cold place to a hot one—the opposite of what occurs naturally. With his heat pump idea,

Thomson had outlined the principles behind the refrigerator and air conditioning. The first cooling device based on Thomson's ideas was built in 1855, by an Austrian mining engineer, Peter von Rittinger.

The "zeroth law"
The fourth law of thermodynamics is also known as the "zeroth law." (The name comes from the idea that it is more trivial than the first law and

should come before it.) It states that if two objects are in thermal equilibrium with a third object—in other words, at the same temperature—they will also be in thermal equilibrium with each other. This basically means that using the same thermometer to measure the temperature of different objects will give meaningful results. The term "zeroth law" was coined by the British physicist Ralph Fowler in the 1920s.

	Kelvin	Celsius	Fahrenheit
	373 K	100° C	212° F
	300 K	27° C	81° F
	273 K	0° C	32° F
	255 K	−18° C	0° F
	200 K	−73° C	−99° F
	100 K	−173° C	−279° F
Absolute zero	0 K	−273° C	−460° F

Temperature scales compared
The Kelvin scale takes absolute zero as its starting point. On the Celsius scale, zero (0° C) represents the freezing point of water, while the Fahrenheit scale starts from the freezing point of saturated salt water (0° F).

‹‹ BEFORE

Every ancient culture watched the Sun, Moon, and planets progress across the sky, assuming they revolved around the earth.

FROM EARTH- TO SUN-CENTERED
In 1543, Nicolas Copernicus described a **sun-centered planetary system ‹‹ 74–75**, overturning centuries of belief that the earth was at the center of the Universe. In the following century, Johannes Kepler and Isaac Newton gave this theory mathematical form and structure.

THE INVENTION OF THE TELESCOPE
Telescopic observations made from 1609 by Galileo Galilei **‹‹ 82–83** provided **proof of the heliocentric view** for the first time. Although Galileo did not invent the telescope, he used a description of the first telescope to make a much more powerful version, allowing the skies to be more thoroughly studied than ever before.

‹‹ SEE ALSO
pp.74–75 THE SUN-CENTERED UNIVERSE
pp.84–85 EXPLORING THE SKIES

Formation of the solar system
About 4.6 billion years ago, a vast spinning cloud of gas and dust contracted to form a central star: the sun. This was surrounded by a slowly rotating flattened disk of material. Over millions of years, tiny pieces fused and grew, until eventually eight large lumps remained; these were the planets. Rocky metallic material near the sun formed Mercury, Venus, Earth, and Mars. The four giants—Jupiter, Saturn, Uranus, and Neptune—formed at a greater and cooler distance out of snow, ice, rock, metal, and gas.

2 The sun takes shape
Gravity pulls material into the center of the nebula, where it becomes concentrated and heats up. The central young sun is surrounded by a flattening disk of material.

1 Solar nebula forms
Gas and dust collect together to make a huge cloud many times larger than the present-day solar system. The cloud spins and contracts.

4 Rocky planets
Close to the sun, where it is hot, rocky and metallic material forms planetesimals. These join together to form the rocky planets.

5 Giant planets
In the outer, colder regions of the disk the planetesimals are of rock, metal, snow, and ice. These form together into bodies which become the cores of the giant planets. These cores attract large amounts of gas.

6 Unused debris
Remaining chunks of dust, rock, and snow are drawn into the Sun and destroyed or thrown out of the planetary solar system. Others form the vast Oort Cloud of comets.

The Solar System

Earth is part of the solar system, which consists of the sun and the large number of objects that orbit around it. Five planets had been noticed in the skies since antiquity, but the invention of the telescope brought with it the discovery of new planets and asteroids, leading to theories about the origin of the solar system itself.

Jupiter
The largest planet and the fastest spinner. It rotates in less than 10 hours.

Asteroid Belt
This consists of small rocky objects never incorporated into planets. Ceres, the largest asteroid, was the first to be discovered.

The majority of the known objects within the solar system are too small and distant to be seen easily from Earth, and they remained undiscovered until the invention of the telescope in the early 17th century. By the end of the 19th century, astronomers had discovered two major planets, approximately 20 planetary moons, and more than 400 asteroids.

A larger universe
In the mid-18th century astronomers believed the solar system consisted of six planets: Mercury, Venus, Earth, Mars, Jupiter, and Saturn. Astronomers were convinced that Saturn marked the outer limit of the system. So it came as a complete surprise when a seventh planet, subsequently named Uranus, was discovered. Even more surprisingly,

this discovery meant that the known solar system was double the size anyone had previously thought.

New planets
The discovery of Uranus was made on March 13, 1781 by the British astronomer William Herschel. He was making a routine observation of the sky as part of a longer-term survey, when he noticed an unexpected object. He recorded this as "a nebulous star or perhaps a comet," but further observation confirmed that he had, in fact, discovered a planet.

Irregularities in Uranus' orbit led astronomers to the prediction of another, more distant planet. Urbain Le Verrier and John Adams used Newton's gravitational theory to pinpoint the planet's position, and on

September 23, 1846 Neptune was discovered by Johann Galle, using Le Verrier's calculations.

Meanwhile, others had discovered smaller but more numerous objects. In the early 17th century Johannes Kepler had come to the conclusion that there should be something between Mars and Jupiter. In the 1770s this idea was strengthened by the Titius-Bode Law, which states that the spacing of the planets follows a mathematical sequence. When the newly discovered Uranus fit into this theory, it confirmed the likelihood of its truth.

It was late 1800 before a group of astronomers, organized by Franz Xaver von Zach, and known popularly as the "Celestial Police," started to look in the Mars–Jupiter gap. In 1801 Giuseppe Piazzi found the first asteroid here as

Uranus
Uranus is tilted on its side.

The solar system
The solar system (not shown to scale, see p.432) has existed for 4.6 billion years. The sun is the most massive member; its gravity keeps the system together. All the other objects orbit around it, and rotate as they travel.

3 Planetesimals form
The disk of material continues to spin and flatten. Grains of material collide to form planetesimals: rocky objects that are the precursors of planets.

ASTRONOMER (1749–1827)

PIERRE-SIMON LAPLACE

This Frenchman was one of the most influential mathematicians and astronomers of the 18th and 19th centuries. His mathematical ability was recognized at an early age, and by the time he was 20 he was working with Jean d'Alembert, the famed mathematician. Laplace's scientific legacy includes his work on positional astronomy, the stability of the solar system, and the theory of its formation. He survived the French Revolution, and was made chancellor of the Senate by Napoleon.

The size and content of the known solar system continues to grow. We also know that our solar system is not unique; we have seen others in the process of formation.

OORT CLOUD OF COMETS
In the early 1930s **Ernst Opik** studied the paths of **long-period comets** (comets taking more than 200 years to return) and concluded there is a reservoir of them, far beyond Neptune. This is now known as the **Opik-Oort**, or Oort Cloud.

SPACECRAFT EXPLORATION
From the 1950s onward, robotic spacecraft **visited all seven planets other than Earth**, as well as the sun, planetary moons, comets, and asteroids. Between 1969 and 1972, 12 astronauts landed on the moon and returned to Earth.

PRE-PLANETARY DISKS
In 1983 the **infrared space telescope** IRAS provided the first evidence of a star with a flattened disk of pre-planetary material.

DATING THE MOON AND EARTH
In 1956, a study of iron meteorites and ocean sediments **dated Earth to about 4.55 billion years**. Dating of moon rocks brought to Earth in 1969–72 shows that **the moon is 4.527 billion years old**.

SEE ALSO »
pp.394–95 INSIDE THE SOLAR SYSTEM
pp.396–97 SPACEPROBES AND TELESCOPES

he was updating a star catalog. The asteroid—Ceres—was joined by three more in the next six years; within 70 years, 112 were known.

Theories of the system's birth
The excellence of William Herschel's telescopes and skills enabled him to discover two moons of Saturn and two of Uranus. He also produced extensive catalogues of star clusters and nebulae. The term "nebula" then referred to indistinct starlike objects that we now know to include a range of objects, such as galaxies, as well as true nebulae (clouds of gas and dust).

Such discoveries opened up new areas of research, such as the idea that the planets had formed from a nebulous ring of material that once surrounded the Sun. This proposal, made in 1796 by Pierre de Laplace, remains the basis of our understanding of the formation of the Solar System.

Mercury
The smallest planet and the closest to the Sun, Mercury also as the widest range of temperatures of any planet.

Earth
The only planet to have stable bodies of liquid water and to support life.

Mars
The outermost of the four rocky planets, Mars is rust-red in color.

Venus
Just a little smaller than Earth, Venus is shrouded by a thick and permanent atmosphere, and is the slowest-spinning planet.

Saturn
Surrounded by a spectacular system of rings, Saturn is the most distant planet visible to the naked eye.

Neptune
The most distant and coldest planet, Neptune is 30 times farther from the Sun than Earth is.

How Rocks Form

Until the mid-18th century few scientists had investigated the composition of rocks or tried to figure out how they had formed. By around 1770 two distinct theories of rock formation had been proposed, and by the beginning of the 19th century it had become clear that there were three main rock types.

Rocks are aggregates of minerals brought together by a variety of geological processes. It was not until the late 18th century that Scottish geologist James Hutton recognized that Earth's rocks consist of three main types linked by a cycle of processes of formation and alteration.

Neptunists versus plutonists

Before the 18th century Earth's rock layers were widely believed to have been deposited during the biblical flood. But, by the 1770s, two rival theories of rock formation became established. The German geologist Abraham Werner thought that an ocean had once covered the whole planet, and that granites and other hard rocks, which he believed to be the oldest (because in his local area they mostly lay at great depths) had formed as minerals crystallized out from this sea. As the fluid level lowered, continents began to emerge. Werner's theory, termed "neptunist" due to its emphasis on the ocean, was highly influential for a time.

Opposing Werner were vulcanists and plutonists, who looked to volcanoes and magma (molten rock from inside the planet) solidifying on top of, or beneath, Earth's surface as the main cause of rock formation. Thus, in 1777 French geologist Nicolas Desmarest proposed that columnlike basalt formations in the Auvergne in France and off Ireland (the Giant's Causeway) were the result of magma from local volcanoes crystallizing as it cooled.

Hutton's theories

James Hutton leaned toward the vulcanist and plutonist view. He was sure that some rocks were formed from magma—he saw fingers of granite cutting through sandstones in a way that indicated the granite had once been molten. And he identified some geological formations as rocks that had been altered by heat (metamorphic rocks), confirming the importance of heat in rock formation. However, he also accepted sedimentation (the process whereby solid matter is deposited from air, water, or glacial ice) as important. He thought that a high proportion of rocks originally formed from sediment at the bottom of the

Geological tools
A field geologist uses a rock hammer (right), and sometimes chisels (bottom center and left), or a pocket knife (top center) for collecting rock samples. Many of these tools haven't changed in centuries.

SEDIMENTARY PARTICLES

Deep burial, compaction, and cementation

Uplift, weathering, erosion, and sedimentation

Uplift, weathering, erosion, and sedimentation

≪ BEFORE

A small number of scientists before the 18th century helped lay the foundations of modern petrology (the study of rocks).

AVICENNA'S THEORY
11th-century Persian scholar **Avicenna** proposed that **rock formation** occurs through processes that have **operated uniformly throughout Earth's history**—uniformitarianism.

EARLY STUDIES OF ROCKS
Johann Lehmann's 1756 study of rocks in southern Germany was a pioneering work. In addition to **describing layers of sedimentary rock**, he explained the **origin of many mineral deposits** within them.

LEHMANN'S GEOLOGICAL DRAWINGS

≪ SEE ALSO
pp.38–39 ANCIENT IDEAS OF THE WORLD

Sedimentary rock layers
These multihued layers of sedimentary rock—sandstones and mudstones—are in Arizona. They were sedimented between 200 and 230 million years ago in an arid environment.

Giant's Causeway
In the late 18th century, this 60 million-year-old formation of basalt in Northern Ireland was the subject of a dispute: vulcanists (correctly) thought it had a volcanic origin; neptunists that it was altered sedimentary rock.

IGNEOUS ROCK

GEOLOGIST (1726–97)

JAMES HUTTON

Scottish geologist James Hutton qualified as a medical doctor at age 23, but then immersed himself in studying chemistry and experimental farming. His agricultural interests led him to start investigating soils and rocks, and from the 1760s he made a series of geological tours of Scotland. These led him to develop his theories of rock formation. He was often referred to as the father of modern geology, and his *Theory of the Earth* (1785) was highly influential, despite being somewhat unintelligible.

sea. Later, these rock layers were uplifted and eroded, sometimes tilted and deformed, subsided again, and had more sedimentary layers added. He believed that this cycle had occurred many times over in the past.

Three rock types

By the early 1800s it had become clear that there are three main rock types: igneous, sedimentary, and metamorphic. Igneous rocks are formed from cooling and crystallization of magma. These include granite, basalt, and obsidian. Sedimentary rocks, such as sandstone, shale, and limestone, are formed from compaction of sedimented particles of mineral or organic material, or precipitated chemicals. Metamorphic rocks, such as gneiss, schist, and marble, occur whenever existing rocks of any type become buried under mile-thick layers

of overlying rock or when rocks are affected by movements of Earth's crust (see pp.352–53). The pressure and heat in these circumstances cause physical and chemical changes, which result in the formation of completely new types of rock.

The rock cycle

Also arising from Hutton was the idea of a rock cycle. Thus, when any rock is exposed at the surface, it is weathered and eroded, turning it into sedimentary particles. Over time, these are buried, compacted, and cemented to make sedimentary rocks. Both these, and igneous rocks, may be heated and compressed at depth to form metamorphic rock, and any rock may melt and subsequently cool to create new igneous rock. Rock may also be uplifted and exposed to start a new cycle.

IN PRACTICE

OIL EXTRACTION

Some sedimentary rocks are oil-bearing—they can hold a mixture of hydrocarbon chemicals (crude oil, petroleum, or gas) in their pores. These substances come from the remains of organisms, which settled millions of years ago in the mud or sand at the bottom of a lake or sea. This material was then compacted into sedimentary rock and, as more rock layers formed above, pressure and heat altered the organic matter into petroleum. When searching for oil, geologists look for promising permeable rocks trapped between layers of impermeable rock. Oil rigs, such as this one in the North Sea, enable oil to be extracted from beneath the seabed.

> " A **stone** is **ingrained** with **geological** and **historical** memories."
>
> ANDY GOLDSWORTHY, BRITISH SCULPTOR, 1956

∇ **Rock cycle**
This illustrates how rocks are formed and altered by a variety of different processes to make new rock types.

Uplift, weathering, erosion, and sedimentation

Heat and pressure

METAMORPHIC ROCK

SEDIMENTARY ROCK

Melting

Heat and pressure

Cooling

Melting

MAGMA

Metamorphic rock
These crumpled layers of metamorphic rock, mainly schist, in the Sierra Nevada mountains of California, were originally deposited as sedimentary rock but were later altered by subjection to heat and pressure.

Igneous rocks from volcanoes
The cooling of lava flowing from volcanoes such as Arenal in Costa Rica (one of the most active in the world) produces igneous rocks including basalt and andesite.

AFTER

Over the past 200 years, the details of rock formation have been further elaborated by laboratory and field studies.

MICROSCOPY AND GEOLOGY

In the mid-19th century, the English geologist **H. C. Sorby,** known as the "father of microscopical petrography," developed the **microscopical** study of **ultrathin rock sections** in polarized light to understand rock composition and formation.

CROSS SECTION OF QUARTZ

PREDICTING MINERAL FORMATION

During the 1920s and '30s the Swiss–Norwegian geochemist **Victor Goldschmidt** developed methods for **predicting** what **minerals** would form in a rock given specific combinations of **chemical elements** and **geological conditions.**

PLATE TECTONICS

The plate tectonic theory of the 1960s modified ideas on rock formation. Spreading **mid-ocean ridges** 212–13 ❯❯ (undersea mountain ranges) emerged as sites of rock formation, and **subduction zones** 358–61 ❯❯ (where one section of Earth's crust moves below another) as regions where rocks melt to form magma.

SEE ALSO ❯❯
pp.356–57 PLATE TECTONICS

BEFORE ‹‹

The **Fossil Record**

Fossils record the history of life on Earth, revealing the remains (or evidence) of plants, animals, and other organisms preserved in rocks. Although the study of fossils, called paleontology, is a field of biology, its development has been closely linked to efforts to understand the history of Earth itself.

The study of fossils and what they reveal has a history going back over 2,500 years.

MARINE DISCOVERIES
The 6th-century BCE Greek philosopher **Xenophanes** described **imprints of fish and seashells** he had found **in rocks** far from the sea. He recognized these as animal remains that must once have been underwater.

RENAISSANCE VIEWS
In 16th-century Florence **Leonardo da Vinci** described **fossil shells** he had observed in the foothills of the Alps and proposed that they were the **remains of once-living organisms**.

FOSSIL DRAWINGS
Swiss naturalist **Konrad Gesner** produced the **first drawings of fossils** in his book compiled in 1558, although it is unclear whether he realized that they were the remains of animals.

‹‹ **SEE ALSO**
pp.184–85 **How Rocks Form**

Geological succession
Each geologic period has characteristic fossils, which lie in sedimentary rocks formed during that period. An example fossil from each period is shown in the sequence below, and the start date for each period is given.

Fossils have been the key to understanding extinct life forms and informing us about evolution. They have also helped us identify the geologic periods in Earth's history.

Reality of extinction
The scientific study of fossils (the preserved remains of living organisms) arguably started in the mid 17th century with English scientist Robert Hooke (see pp.92–93), who observed fossilized seashells and wood. He was the first to formulate a reasonably accurate explanation of how fossils form: they had been soaked with "water impregnated with stony and earthy particles." In 1667 Danish geologist Nicolas Steno deduced that objects known as tongue stones (found in some rocks in northern Italy) are, in fact,

fossilized shark teeth. In 1778 giant bones and a skull were found near the Meuse River in the Netherlands, later identified as the relics of a giant marine reptile, a mosasaur. Although many

found the concept of extinction difficult to accept on religious grounds, most scientists recognized that such fossils were the remains of once-living species that had died out. In the early 19th

PALEONTOLOGIST (1769–1832)

GEORGES CUVIER

Born in France, Georges Cuvier studied in Stuttgart, Germany, then made his reputation as a naturalist. By 1799 he was Professor of Natural History at the Collège de France in Paris. He had a brilliant ability to reconstruct animals and their lifestyles from their skeletal remains. In addition, he was appointed a state councilor, remaining in this position through the reigns of Napoleon and three successive kings of France.

Cambrian period
541 mya
Trilobites (*Paradoxides*) were marine arthropods that left many seabed fossils.

Ordovician period
485 mya
Didymograptus was a marine animal that left fossils in seabed shales.

Silurian period
444 mya
Monograptus was a graptolite, a fossil common in Silurian slates and shales.

Devonian period
419 mya
Brachiopods like this *Mucrospirifer* were hard-shelled marine animals.

Carboniferous period
359 mya
Plants like *Asterophyllites* grew in swampy areas and fossilized in coal layers.

Permian period 299 mya
Marine reptiles such as *Stereosternum* left fossils in some limestone shales formed on the seabed.

Fossilization
Animal and plant remains will be preserved as fossils only if they are protected from destruction by being quickly buried after death. Even then, usually only the hard parts of the organism will be fossilized.

Organism dies
To become a fossil, an organism must die in an environment (such as on the seabed or near a river mouth) where it will be buried by sediment before it disintegrates.

Skeleton buried
After burial, the soft parts of the organism soon decay and dissolve, leaving just a skeleton in the case of an animal and a cellular framework in the case of a plant.

Rock forms
As the skeleton or plant remains are compressed by overlying rock, organic substances are replaced by minerals, although the remains retain their shape.

Fossil revealed
Earth movements slowly lift up the sedimentary layers, and weathering and erosion then wear them down, eventually exposing the fossil at Earth's surface.

century French naturalist Georges Cuvier (see left) established the reality of extinction, noting that fossils bore no resemblance to anything living.

A series of fossil discoveries in England in the 1800s strengthened the case for extinction. Since many of the species found were restricted to specific rock layers, it seemed that they had existed for only a part of Earth's history.

Rock layers and evolution
In the early 19th century, knowledge of the fossil record led to the development of stratigraphy—the study of rock layers, or strata. British geologist William Smith used fossils to determine the relative ages of strata, and by 1880, nearly all the currently recognized geological periods had been named and linked to the succession of strata (see pp.190–91).

Growing paleontological activity in the mid-19th century saw the discovery of *Archaeopteryx*, a primitive bird, in 1861 (see pp.188–89). In the 1870s, the "bone wars" between Edward Drinker Cope and Othniel Marsh in their search for fossils in North America led to the discovery of many new dinosaurs, including the giant plant-eater *Diplodocus*.

The 1909 discovery of abundant and diverse fossils in the Burgess Shale of British Columbia, Canada, demonstrated an explosion of life in Cambrian times, around 530 million years ago. Decades later, there were finds of exceptionally

well-preserved dinosaurs in Mongolia and China. The 1946 discovery of the remains of soft-bodied organisms, known as the Ediacaran biota, in late Precambrian strata continue to puzzle paleontologists. From the 1970s, the recovery and analysis of chemical fossils in the form of organic molecules from the rock record revolutionized understanding of the early evolution of life and the inter-relationships of different plant and animal groups. Although complex molecules of DNA do not survive for long in the fossil record, their recovery has transformed understanding of recent human evolution.

Fossil hunter
The English paleontologist Mary Anning discovered several important fossils including those of ichthyosaurs and plesiosaurs. She contributed to many theories of the 19th century despite being excluded from scientific circles.

AFTER

Since the 1950s, there has been an ongoing search to find even older fossils.

PRECAMBRIAN FOSSILS
In 1954 American paleontologist **Elso Barghoorn** and geologist **Stanley Tyler** discovered **microfossils of bacteria** in a **2 billion-year-old** rock formation found in the Canadian Shield.

FOSSILS IN CHINA
Since the 1980s intriguing fossils **preserving soft parts** such as feathers have been found in China. They show that birds evolved from a group of feathered dinosaurs.

FEATHERED DINOSAUR

SEE ALSO >>
pp.188–89 FINDING *ARCHAEOPTERYX*
pp.190–91 DATING THE EARTH
pp.200–01 HOW EVOLUTION WORKS

BREAKTHROUGH
MASS EXTINCTION
Mass extinctions are sudden cataclysmic decreases in the number of living species on Earth. During the 19th century it became clear from the fossil record that a severe mass extinction happened at the end of the Cretaceous period when huge numbers of species died out, including the dinosaurs and, from the sea, ammonites.

Triassic period 252 mya
Dicellopyge was a freshwater fish that left fossils in rocks formed from lake or river sediments.

Jurassic period 201 mya
Ichthyosaurus was a streamlined marine reptile that is found as a fossil in limestone shales on the seabed.

Cretaceous period 145 mya
Sea urchins such as *Micraster* were marine animals of which the test, or shell, is a common fossil in chalk beds.

Tertiary period 66 mya
Heliobatis was a freshwater stingray and is found as a fossil in some rocks formed from lake sediments.

Quaternary period 2.6 mya
Hominins—modern humans and extinct relatives—have left fossils in locations such as rock formed from lakeside mud.

Finding *Archaeopteryx*

Discovery of a "missing link" is a rare event in the paleontological world. Few such finds have been more significant than *Archaeopteryx*, an extraordinary fossil skeleton that came to light in a Bavarian quarry in 1861. Combining the characteristics of a reptile with the feathered wings of a bird, the fossil was the first example of evolution in action. *Archaeopteryx* marks the midway stage between birds and reptiles.

In the 19th century the limestone quarries of Solnhofen, in southern Germany, yielded a number of interesting fossils. It was not unusual for the quarry workers to come across relics from Jurassic times within the 170-million-year-old rocks. One notable find in 1860 was a fossilized feather that predated any previously discovered fossil feathers by millions of years. But this was nothing when compared to the sensational discovery made the following year.

The splitting of a slab of rock exposed the almost entire skeleton of a small animal surrounded by the outlines of feathered wings, and at the time of this discovery any animal with feathers was classified as a bird. Named *Archaeopteryx* (meaning "ancient feather"), the fossil quickly acquired the status of "first bird," although it had some distinctly unbirdlike features. The long, bony tail and claws on the wing bones were definitely those of a reptile. After the discovery of the first specimen, a number of complete fossils of *Archaeopteryx* were found, which showed that this particular bird also had teeth.

Karl Haberlein, a local doctor and fossil collector, bought a fossil from the quarrymen and, scenting a large profit to be made from avid collectors, offered it for sale. The prize was bought for £700 by Dr. Richard Owen, superintendent of the British Museum's natural history collection.

Archaeopteryx was a priceless gift to the supporters of Darwinism. At the time of the fossil's discovery, Darwin's still-new theories on evolution (see pp.200–03) and the divergence of species were not universally accepted. His opponents had been demanding evidence that one life form could evolve into another—and here the evidence was, set in rock. Owen himself was no evolutionist, although he was correct in his understanding of *Archaeopteryx* as a bird at a primitive stage of development. But it was left to his Darwinist arch rival Thomas Henry Huxley to recognize *Archaeopteryx* for what it was—the intermediate stage between reptiles and modern birds.

Recently discovered Chinese fossils show that birds are effectively avian dinosaurs, having evolved from small feathered theropods.

Snapshot in time
This specimen of *Archaeopteryx* has been almost perfectly preserved in a slab of limestone. The imprints of feathers are clearly visible splaying out from the animal's wing bones and around the long, bony tail.

> "If **evolution has taken** place, there will its **mark be left**."
>
> THOMAS HUXLEY, "HOW PALAEONTOLOGY CAN HELP TO UNDERSTAND LIFE," 1860

Dating the Earth

From the mid 18th century, a growing body of scientific evidence made it clear that the earth was much older than previously believed. Discoveries in geology contributed to this revolution, but ultimately the key to dating the earth was supplied by advances in physics.

The first person to try to figure out Earth's age using a reasoned scientific approach is thought to have been the 18th-century French naturalist, the Comte de Buffon. He argued that Earth must have started as a hot, molten orb, and that its age could be calculated by figuring out how long it had taken to cool to its present state. He estimated Earth's age at about 75,000 years—far older than conventional opinion of the time. After pressure from the Church, Buffon was forced to retract this claim, but belief that the earth was just a few thousand years old was already fading. In 1788, the Scottish enlightenment scientist James Hutton concluded from his search for the most ancient strata that "we find no vestige of a beginning, no prospect of an end," implying that

BEFORE

Over recorded history, ideas about Earth's age have varied from literal interpretations of the Bible to philosophical thoughts.

ANCIENT BELIEFS
The Greek philosopher **Aristotle** (4th century BCE) thought **the world had always existed**. The Maya (2000 BCE–900 CE), of Central America, thought it was a **few thousand years old**.

NOAH'S ARK

IDEAS BASED ON THE BIBLE
In Judeo-Christian tradition, **the world is about 6,000 years old**, based on a literal interpretation of the biblical Old Testament and narratives such as that of Noah's Ark and the Flood. In 1644 **James Ussher**, Archbishop of Armagh, dated creation to the evening before October 23, 4004 BCE.

SEE ALSO
pp.186–87 THE FOSSIL RECORD
pp.184–85 HOW ROCKS FORM

Earth's age seemed inestimable and much greater than accepted at the time.

Rock layers and fossils
Observation of fossils and rock layers (see pp.186–87) led to an understanding of the sequence of rock deposition. Geologists were starting to realize that layers of sedimentary rock were laid down at different ages and that very thick layers of rock must have taken millions of years to deposit. By 1800 it was realized that rock layers from different sites could be regarded as the same age if they contained the same fossils. However, geologists still had only a vague idea of the absolute age of each rock layer.

One of the most important 19th-century believers in the view that Earth is extremely ancient was Charles Lyell (right). In the 1830s, his book *Principles of Geology* produced a convincing argument that Earth must be several hundred million years old. His ideas of gradual change influenced Charles Darwin's Theory of Evolution (see pp.200–01).

Cooling Earth and the salt clock
In 1862 a new figure entered the arena: the Irish-born physicist Lord Kelvin. Like Buffon, Kelvin maintained that it must be possible to calculate Earth's age from the time taken for it to cool from a molten state. Using new theories in physics, Kelvin first estimated that Earth must be between 24 and 400 million years old, but later refined his estimate down to 20–40 million years. His findings dismayed most geologists and biologists, few of whom thought that this time frame fit with their observations in the natural world.

In 1899 the Irish physicist John Joly suggested a novel approach to dating the Earth, called the salt clock solution. He argued that the age of Earth's oceans could be ascertained from the time it had taken for them to reach their current salinity. He came up with a range of 80–150 million years. But Joly hadn't realized that salt is also removed from the oceans by geological processes, and so his estimate was too low.

The radioactivity revolution
At the end of the 19th century a major advance occurred that proved key to establishing the age of the earth. This was the discovery of radioactivity in 1896 by the French scientist Henri Becquerel (see pp.302–03). One early finding was that

GEOLOGIST (1797–1875)
CHARLES LYELL

Born in Forfarshire, Scotland, Charles Lyell started his career as a lawyer, but soon turned to geology. In 1828 he became convinced of Earth's immense age from his examination of the enormously thick layers of volcanic rock produced by Mount Etna in Sicily. A great traveler and observer of natural phenomena, Lyell popularized the idea that geological history could be explained by the operation of the same slow, gradual processes that can be seen on Earth today. His *Principles of Geology* was one of the most influential scientific books of the Victorian era and, in 1848, he was knighted.

4.55 BILLION
years old is the scientifically accepted age of the Earth today.

Earth has an internal heat source—in the form of radioactive elements such as uranium. This finding cast doubt on Kelvin's calculations, which assumed that no such heat source existed.

More importantly, advances in the early 1900s demonstrated that radioactivity supplies a type of natural "clock" that can be used to establish the absolute ages of some rock types (see below). The American chemist Bertram Boltwood tried dating a variety of rock samples and came up with ages for the different samples ranging from 265 million to over 2 billion years old.

For a while, the scientific community was sceptical of the new technique of "radiometric dating," but following further isotope work by Alfred Nier in the US and British geologist Arthur Holmes the method came to be accepted as valid. Holmes's 1947 timescale of Earth's history is close to accepted modern values.

This was not quite the end of the story. Because Earth may have existed for some time before any rocks currently

> **"...we find no vestige of a beginning, no prospect of an end."**
>
> JAMES HUTTON, GEOLOGIST, 1788

BREAKTHROUGH
RADIOMETRIC ROCK DATING

In 1902 the New Zealand-born physicist Ernest Rutherford suggested that the rate at which radioactive material decays can be used to date rocks. Granite, for example, can be dated by comparing the concentration of the radioactive isotope, uranium-235, with its decay product, lead-207, in minerals within the rock. When granite forms, grains of the mineral zircon usually crystallize within it. Inside each grain of zircon, atoms of uranium-235 are trapped and decay into atoms of lead-207. The rate of this decay can be accurately predicted—every 704 million years half the remaining uranium atoms in the mineral grain decay into lead-207 atoms. The ratio of the two isotopes allows the rock to be dated.

Atom of uranium-235   Atom of lead-207   Increased ratio of lead-207 to uranium-235

Grain of zircon forms when granite cools

Radioactive decay process
Over time the ratio of uranium to lead atoms in zircon drops at a predictable rate.

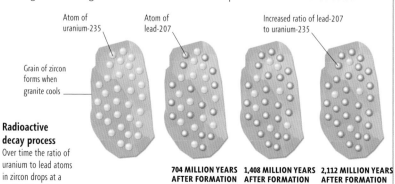

704 MILLION YEARS AFTER FORMATION   **1,408 MILLION YEARS AFTER FORMATION**   **2,112 MILLION YEARS AFTER FORMATION**

present in its crust came to be formed, scientists looked at materials, such as meteorites, that formed at the same time as Earth elsewhere in the solar system. In 1955 an American geochemist, Clair Patterson, used radiometric dating to date a piece of a meteorite whose chemical makeup matched Earth's. Patterson dated this rock at 4.55 billion years old, a figure that is still widely accepted as the true age of the Earth.

Geological timescale
This geological "clock" shows the start dates for the presently accepted divisions of the geological timescale (mya—million years ago). It is an updated version of a timescale proposed by the geologist Arthur Holmes in 1947 as a result of his work on radiometric rock dating.

CENOZOIC 66 MYA
CRETACEOUS 145 MYA
JURASSIC 201 MYA
TRIASSIC 252 MYA
PERMIAN 299 MYA
CARBONIFEROUS 359 MYA
DEVONIAN 419 MYA
SILURIAN 444 MYA
ORDOVICIAN 485 MYA
CAMBRIAN 541 MYA
ORIGIN OF EARTH 4,554 MYA
PRECAMBRIAN
1,138 MYA
3,415 MYA
2,277 MYA

Since the 1950s rock dating methods have increased in number and sophistication and in the range of materials that have been dated.

THE OLDEST MINERALS
In 1972 a rock found in the lunar highlands was brought back to Earth by the Apollo 17 astronauts, and has subsequently been dated at 4.5 billion years old. This is the **oldest moon rock** found so far. Its age meets with the currently accepted theory of how the Earth–Moon system formed.

ZIRCON CRYSTAL

DATING OF SEDIMENTARY ROCKS
Since the 1980s techniques have developed for direct **radiometric dating** of some sedimentary rocks, as well as igneous (volcanic) and metamorphic rocks. This is done by measuring the age of tiny crystals on mineral grains during the sedimentation process.

WORLD'S OLDEST ZIRCON CRYSTAL
In 2001 a zircon crystal was radiometrically dated at 4.4 billion years old. This crystal holds the current record for the **oldest known object** of terrestrial origin.

SEE ALSO ≫
pp.298–99 RADIATION AND RADIOACTIVITY

252 MYA PERMIAN PERIOD
260 MYA PERMIAN PERIOD
265 MYA PERMIAN PERIOD
300 MYA CARBONIFEROUS PERIOD
335 MYA CARBONIFEROUS PERIOD
359 MYA DEVONIAN PERIOD
515 MYA CAMBRIAN PERIOD
530 MYA CAMBRIAN PERIOD
541 MYA CAMBRIAN PERIOD
825–1,000 MYA PRECAMBRIAN
1,100–1,250 MYA PRECAMBRIAN
1,700–2,000 MYA PRECAMBRIAN

Dating rock sequences
The Grand Canyon in Arizona contains a remarkable sequence of rock layers. Their approximate ages have been established by dating to extend from 250 million years old (myo) to more than a billion years old, from near the end of the Permian period back to the Precambrian era.

BEFORE

Some explanations of the formation of common features, such as hills and valleys, were first put forward over 1,000 years ago.

THE EFFECTS OF WATER
The 11th-century Persian scholar **Avicenna** stated that land may be shaped by "the effect of water, which, cutting itself a new route, has **denuded the valleys**, the strata being of different kinds, some soft, some hard."

FOUNDATION OF GEOLOGY
After studying local landscapes, the Chinese scientist **Shen Kuo**, in his *Dream Pool Essays* of 1086, outlined the principles of **uplift, erosion,** and **sedimentation**.

VALLEY FORMATION
Danish scientist **Nicolas Steno** proposed in 1669 that some **valleys** in Italy may have originated from the **collapse** of sections **of crust into vast underground chambers**. Although this was wrong, the idea that **gravity** was the only force powerful enough to create mountains and valleys seemed rational then.

« SEE ALSO
pp.184–85 HOW ROCKS FORM

Shaping the Landscape

Earth's land surface has an extremely varied relief, with landscapes ranging from extensive flat plains, craggy peaks, and deep canyons to delicately sculpted valleys and hummocky hills. Explanations for landscape formation have varied from the effects of past catastrophes to the results of uplift, weathering, and erosion.

The formation of landscape involves three basic processes: uplift, weathering and erosion, and sedimentation. Land is built up through uplift, caused by plate movements (see pp.356–57) and volcanic activity. Uplifted land is then reduced by the gravity and climate-controlled processes of weathering and erosion. Weathering is the breakdown of rocks by physical or mechanical action (these include extremes of temperature which can break rocks apart), chemical attack, or biological activity (such as effects of lichens and plant roots). Erosion is the carrying away of the loosened material by wind, water, glaciers, and landslides.

Early explanations
From the 16th to the 18th centuries reports of explorers made it plain that Earth had undergone great physical changes in the past, which had altered its landscapes. Examples included evidence of massive historic landslides, and the existence of huge solidified sheets of volcanic lava in places such as Iceland and India. In America immense

canyons had been discovered, which had multiple layers of thick sediments forming their walls.

There were two main theories as to the cause of these phenomena. Most geologists thought that sudden

IN PRACTICE

UNRAVELING PAST CLIMATES

Some periods in Earth's past have had a colder or warmer climate than today, which has affected the shaping of the landscape. Evidence is found in the geographical distribution and chemical makeup of fossils, particularly of climate-sensitive species, such as beetles, some plants, and tiny marine organisms such as foraminifera (right). Other evidence comes from fossil tree rings and gases trapped in ice sheets.

Karst formations in China
This dramatic scenery in the vicinity of Guilin, China, known as a tower karst landscape, was produced by the chemical weathering and erosional effects of rainwater, which fell on what was originally a thick level bed of limestone.

Processes of weathering and erosion

Weathering and erosion are the main processes whereby rocks that are exposed to the air at Earth's surface are broken down and lowered. Four common types are shown here.

Wind erosion
Yardangs such as this one in the central Sahara are shaped by the cutting action of wind-blown sand, which is strongest near ground level.

Freeze-thaw weathering
If water gets into cracks in rocks and then freezes, it can split the rock as it expands to form ice. When it thaws, the rock may disintegrate or new cracks may appear.

Water erosion
Antelope Canyon, in Arizona, was sculpted by rainwater picking up sand in an upstream basin, then rushing into the canyon, forming "flowing" shapes in rocks.

Biological weathering
This occurs where plant roots penetrate and force apart cracks in rocks. Some plants also release organic acids, which dissolve the rock and widen the spaces.

catastrophic events, such as massive earthquakes, floods, and volcanic eruptions were responsible. This idea came to be known as catastrophism.

But, in his 1785 book, *Theory of the Earth*, Scottish scientist James Hutton outlined a different view. He proposed that most geological features could be explained by slow, gradual processes taking place over very long periods of time, such as weathering, erosion, and deposition (process whereby solid matter is deposited from air, water, or glacial ice). His philosophy came to be known as uniformitarianism (see p.184). It can be summarized by the phrase, "the present is the key to the past," which means that past geological changes are explainable by the same uniform events that can be seen operating today. During the 19th century uniformitarianism gained in popularity within the scientific community. It was supported, in particular, by Scottish geologist Charles Lyell. Today, it is still widely used to explain how Earth's landscapes evolved.

Wind erosion and deposition

Weathering and erosion by wind occurs mainly in desert regions with little or no water and vegetation. There are two main aspects to it. First is the actual transportation of material (grains of sand and dust) in the wind, either directly or by a process called saltation, in which flying and bouncing grains help move the heavier ones. In this way, a persistent wind can transport loose material over huge distances. Second is the abrasive effect of wind-blown particles sculpting rock formations into sometimes striking shapes. Where the wind drops, the particles are deposited to form sand dunes or loess (blankets of fine silt). The size of these deposits can be considerable, including sand dunes as high as 6,500 ft »

"**Running water** on the **land** has **ever had** the same **power** of **wear** and **transportation.**"

JAMES DANA, AMERICAN GEOLOGIST, 1867

Glacial erosion
Glaciers such as the Fox Glacier in New Zealand erode the bedrock in high mountainous areas by "plucking" (removing) material from it and, lower down, by grinding trapped rocks against it.

Ice becomes heavily fractured as glacier moves downhill

Bedrock

Boulder is "plucked" from bedrock after freezing to glacial ice

Trapped boulder scours bedrock as it is carried along by glacier

PLUCKED ROCK

Kettle lake
Small circular lake formed by ice blocks burying themselves in ground and melting as glacier retreats.

Glacier

Erratic
Glacier has carried this large boulder down mountain.

Erosion by a waterfall
A waterfall typically occurs where a layer of hard rock lies over a softer layer. As the soft rock is eroded, an overhang of hard rock develops. This eventually collapses due to lack of support and the waterfall erodes its way upstream leaving a steep-sided gorge, as in Lower Yellowstone Falls, Yellowstone National Park, Montana.

A layer of hard rock is undercut by erosion of softer rock

Waterfall

Soft rock is worn away by splashback and abrasive action of rock particles in water

A plunge pool is formed by force of water hitting soft bedrock at its base, and is deepened by stones rubbing against base

WATERFALL FORMATION

Gorge
Deep valley carved from the landscape by river or waterfall.

Floodplain
Flat region next to a river that has formed from sediment deposited by river.

Meander
Bend in river caused by erosion of one bank and deposition on opposite bank.

(2,000 m) in Peru, while in China there are accumulations of loess more than 300 ft (100 m) deep, covering hundreds of square miles.

Glacial erosion and retreat
As a glacier flows through mountains, it can reshape them, producing features such as cirques (hollowed-out depressions) and U-shaped valleys. At the lower end of a glacier, where it melts, features such as moraines (piles of rock debris), erratics (large boulders), and lakes can occur.

10 IN (25 CM) The rate at which the Himalayas, in Asia, are being eroded down every century. They are also being uplifted at about 24 in (60 cm) a century.

Retreating glaciers, such as those at the end of the last ice age 13,000 years ago, left other marks on the landscape. As ice melted, parts of the world depressed by the weight of the ice sheets, such as Scandinavia, Scotland, and Canada rebounded upward. In Washington State, a glacial lake (Lake Missoula) unleashed cataclysmic floods, creating deep, rough channels—the Channelled Scablands.

Water erosion and deposition
Erosion by water flowing downhill, in streams and rivers, is the main cause of landscape features such as gullies,

wadis (dried-up river channels), gorges, and V-shaped valleys. Even slow, smooth flows of water wash away small particles of rock, but where the flow is fast and turbulent, due to a steep gradient, it has much greater erosive power. Typically erosion is both downward, deepening the gorge or valley, and upward, extending the gorge or valley into the hillside. As water flow continues downstream, it picks up more and more solid material (usually sand and rock) and it is the abrasive effect of this material that causes most of a river's erosive power.

As a river reaches its base level (the lowest point to which it can flow) and its gradient decreases, the erosive activity switches from vertical to horizontal, the stream carries away rock and soil from its banks by friction, suction, and dissolution. This widens the valley floor and creates a floodplain. Where the river slows (as a result of widening, for example) and its carrying capacity decreases, sediment begins to be deposited, producing features such as point bars (ridges of sand or gravel on the inside bends of meandering rivers—see above

Erosion and deposition in drainage basin
From the glaciers that are the source of many rivers, along the course of the rivers themselves, and around the coast into which they discharge, a variety of erosional and depositional processes have their effects on the landscape, ranging from U-shaped glacial valleys and canyons, to floodplains and deltas.

right), deltas (triangular regions of deposited sediment), and sand bars (low, partially submerged ridges), where the river meets the sea.

A different pattern of erosion occurs when rainwater eats away at regions dominated by layers of limestone rock, producing karst landscapes (see main picture, pp.192–93). Rainwater trickles down along the vertical joints that divide limestone blocks. Since rainwater is usually slightly acidic, it dissolves the limestone, widening the joints, forming chimneys, and creating underground chambers linked by passages.

Coastal erosion and deposition
Coastal cliffs undergo continual weathering and erosion through the action of waves (see above right).

Weathering occurs through abrasion, as waves hurl beach material against the cliff. As waves impinge on a cliff they also compress air in cracks in rocks. As the air then decompresses it can shatter the rock. Depending on the cliff's rock composition, wave action may produce a cave, or the cliff above the eroded part may collapse.

Like coastal erosion, deposition is primarily caused by wave action though tidal currents also play a part. Sand and silt in the water may come from a river, eroded cliffs, or offshore. Waves that hit the coast at an angle can shift this material along shorelines, sometimes for long distances, by a mechanism called longshore drift. Where the sediment settles out and established his reputation back in England. During the voyage he read Charles Lyell's *Principles of Geology* 1830, which suggested that Earth had changed gradually over very long periods of time (see pp.190–91) and Darwin saw evidence for this all around him. But Lyell also suggested that species had originated in "centres of creation" and, based on observations on the distribution of species in places such as the Galapagos Islands, Darwin was harboring doubts about this view.

Developing his ideas

Returning to England in 1836 Darwin began to organize his specimens with the help of experts. British ornithologist John Gould pointed out that the slightly varying mockingbirds he had collected in the Galapagos Archipelago were actually three unique species individual to different islands. This, and many examples like it, served to foster Darwin's growing conviction that species were not fixed. After writing the journal of his voyage he began a secret notebook on the subject, known then as "transmutation."

In 1838 Darwin read an essay by British economist Thomas Malthus, that had a profound influence on his

Voyage of the *Beagle*
HMS *Beagle*'s task was to map southern coastlines, as seen here, to observe tides, currents, and trade winds, and make meteorological records. Charles Darwin spent almost five years on board as the expedition's naturalist.

Rural retreat: Down House
Darwin's seclusion in rural Kent, England meant that he was isolated from much of the furor caused by the publication of his evolutionary theory.

ideas. It made stark predictions about human overpopulation. Darwin began to consider whether the same theory would apply to plants and animals—that population growth would lead to competition over resources.

Substantiating his theory

In 1839 Darwin was elected Fellow of the Royal Society. He also married his cousin, Emma Wedgwood, and published his journal of the *Beagle's* voyage, which was popular and made him well known.

In the years that followed he settled with his growing family in Down House in Kent. He established a routine of prolific correspondence,

experimentation, and research on topics ranging from barnacles to the behavior of plants. He gathered facts about the variation that could be produced by animal breeders, such as pigeon fanciers, and explored human behavior too. During this time Darwin's Christian faith was dwindling—much to the consternation of his wife, Emma—and there can be little doubt that the death of his beloved

Darwin's tree of life
Darwin penned a sketch that illustrated his idea that life evolved and diversified like the branches of a tree.

> "No work on **Natural History** Science has made so **great** an **impression** on me."
>
> BRITISH BIOLOGIST THOMAS HENRY HUXLEY, DISCUSSING *ON THE ORIGIN OF SPECIES*, 1859

daughter Annie, from scarlet fever in 1851, was the final blow.

In July 1858 he produced a joint paper with British naturalist Alfred Russel Wallace (who had formed the same idea independently) describing the theory of evolution by natural selection (see pp.198–99). A year later the first edition of Darwin's "summary" of his evidence and ideas appeared in the famous *On the Origin of Species*.

Amid the furor that followed publication—condemned by many for contradicting biblical creation—Darwin continued with his work at Down House. He subsequently produced work that made it clear he considered humans to be part of this natural world too. When he died in 1882 he was honored with a state funeral and burial in Westminster Abbey in London.

(see pp.190–91) ... (see pp.198–99)

BEETLE SPECIMEN

DARWIN'S COMPASS

TIMELINE

- **February 12, 1809** Charles Robert Darwin is born in Shrewsbury, England, son of wealthy physician, Robert Darwin, and grandson of Erasmus Darwin and Josiah Wedgwood.

- **September 1818** Attends Shrewsbury School.

- **1825** Starts to study medicine at University of Edinburgh.

- **1827** Abandons medicine to train for the clergy at Christ's College, Cambridge University.

- **1831** Performs well in final examinations.

- **1832** Studies geology and natural history of South America and Galapagos Islands on board HMS *Beagle*.

- **October 2, 1836** Returns to England.

- **January 4, 1837** Publishes his first scientific paper on the geology of South America.

- **February 17, 1837** Elected to council of Geographical Society.

- **June 20, 1837** Finishes writing *Beagle* journal.

- **March 1838** Becomes Secretary of the Geological Society.

- **January 24, 1839** Elected Fellow of the Royal Society.

- **January 29, 1839** Marries his cousin, Emma Wedgwood.

- **May 1939** Publishes *Journal and Remarks* about *Beagle* voyage.

- **May 1842** Publishes *The Structure and Distribution of Coral Reefs*.

- **November 1842** Family moves to Down House in Kent.

- **1846** Starts a six-year study of barnacles, using the idea that all organisms are descended from a common ancestor.

- **1853** Awarded the Royal Society's Royal Medal for work on barnacles.

- **1856** Investigates whether eggs and seeds could survive sea water and spread over oceans.

- **July 1, 1858** Darwin-Wallace paper presented to the Linnean Society (British natural history).

- **Nov 22, 1859** First edition of *Origin of Species* released and sells out on the first day.

- **1868** Publishes *Variation of Plants and Animals under Domestication*, first part of his "big book."

- **1871** Publishes second volume, *The Descent of Man*, dealing with sexual selection.

- **1872** Publishes *The Expression of the Emotions in Man and Animals*, concerning human psychology; smaller books follow on insectivorous plants, climbing plants, and earthworms.

- **April 19, 1882** Darwin dies at Down House, at 73. He is given a state funeral.

Laws of Inheritance

Family resemblances have long been recognized, but a scientific explanation for them was slow in coming. Then in the middle of the 19th century an Augustinian monk began to breed peas and so discovered the answer to this mysterious question.

Most biological characteristics are inherited, and early efforts to explain how came from French philosopher Pierre Maupertius who concluded in 1745—as English naturalist Charles Darwin (see pp.201–202) did 100 years later—that offspring are made from components derived from every part of their parents' bodies. In a way they were correct: DNA in every cell of each parent determines the outcome in the progeny. But that alone could not explain why some traits skip generations.

Mendel's experiments

In 1866 an Austrian monk published the results of his experiments in breeding peas. Gregor Mendel (see below) had found that pea plant characteristics, such as flower color, were passed down in fixed proportions. He called this the Law of Segregation (see opposite). He deduced that each characteristic was determined by a particular particle (today called a gene) and that they exist in different variables, such as color. His theory suggested that characteristics could be inherited intact. This differed from popular notions of the time, which suggested that a purple-flowered plant crossed with a white-flowered plant, for example, would produce a blend of the color. Importantly, Mendel's work explained generation-skipping traits; these recessive traits can remain hidden in generations when "masked" by dominant traits.

Mendel also looked at what happened when

Rediscovering Mendel's laws
Geneticist William Bateson publicized Mendel's laws in his 1909 book *Mendel's Principles of Heredity*. Here, however, he shows how sweet peas do not follow the laws.

two characteristics, such as seed color and shape, were inherited together. This resulted in his Law of Independent Assortment (see opposite), which stated that different characteristics are inherited independently of one another. These findings established the distinction between phenotypes (the observable characteristics of organisms) and genotypes (the genetic characteristics of organisms that cannot be seen).

Human heredity

Pea plants clearly showed Mendel's patterns because any two parents can produce a very large number of

« BEFORE

Early attempts at figuring out meaningful explanations for inheritance were flawed by basic misconceptions.

EARLY THEORIES OF HEREDITY
Early understanding of heredity was hindered by a belief in preformation: **individuals were preformed in sex cells**. This was championed by scholars such as French philosopher René Descartes **«** 102–03 in the 17th century.

THE SCIENCE OF BREEDING PLANTS
Plant breeders had long since experimented with **selective breeding** of crops but it was 18th century German botanists Christian Sprengel and Joseph Kölreuter who established the role of **pollination** in plant breeding **«** 152–53.

« SEE ALSO
pp.152–53 PLANT LIFE CYCLES

Experimental organisms
Mendel examined seven characteristics of pea plants: flower position and color, pod shape and color, seed shape and color, and height. Peas were easy to grow, and—luckily for Mendel—the characteristics he selected for study were inherited in simple ways that he could analyze.

PRIEST AND SCIENTIST (1822–84)

GREGOR MENDEL

As a young Augustinian monk, Gregor Mendel studied at the University of Vienna, Austria, and then returned to his abbey in Brno (in what is now the Czech Republic) as a science teacher. His mathematical skills proved useful in establishing his laws of inheritance (see opposite) in his breeding experiments—between 1856 and 1868 he cultivated and tested more than 29,000 pea plants. His work was not recognized until 1900. His scientific work came to an end in 1868 when he became an abbot, and his papers were burned by his successor after his death.

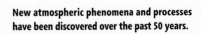

Layers of the atmosphere
Each layer is defined by the way temperature varies within it, and is separated by distinct boundaries.

Thermosphere
This layer of rarefied gas extends from about 55 miles (85 km) up to between 300 miles (500 km) and 600 miles (1,000 km) depending on solar activity. Much of the gas here is ionized (consists of charged particles) and this layer of ionized gas is known as the ionosphere. It is found in the thermosphere, and the exosphere higher up.

Mesosphere
The mesosphere extends from 30–50 miles (50–85 km) above the surface. Its boundary with the thermosphere (the mesopause) is the coldest part of Earth's atmosphere, with temperatures as low as −130° F (−90° C). Most meteors burn up here.

Stratosphere
This layer extends up to a height of about 31 miles (50 km). The ozone layer is found between 12 miles (20 km) and 25 miles (40 km).

Troposphere
All weather occurs in this layer, which extends up about 10 miles (16 km) at the equator and 5 miles (8 km) at the poles.

HEIGHT ABOVE SEA LEVEL

80 miles	130 km
74 miles	120 km
68 miles	110 km
62 miles	100 km
56 miles	90 km
50 miles	80 km
43 miles	70 km
37 miles	60 km
31 miles	50 km
25 miles	40 km
18 miles	30 km
12 miles	20 km
6 miles	10 km
Sea level	

MESOPAUSE

STRATOPAUSE

TROPOPAUSE

Aurora borealis

Meteor burning up as a result of air friction

Noctilucent clouds

Thin layer of ozone gas

AVERAGE TEMPERATURE

140° F	60° C
14° F	−10° C
−112° F	−80° C
−130° F	−90° C
−112° F	−80° C
−58° F	−50° C
−22° F	−30° C
14° F	−10° C
−4° F	−20° C
−40° F	−40° C
−76° F	−60° C
−76° F	−60° C
59° F	15° C

Aurora borealis
This occurs when charged solar particles collide with atoms and molecules in the thermosphere.

Noctilucent clouds
These ice crystal clouds, which form in the mesosphere, reflect light after the Sun has set.

Nacreous clouds
Forming in the lower stratosphere, these bright clouds, occasionally seen in polar regions, contain water and sometimes nitric acid.

Rainbows
Rainbows are caused by water droplets in the troposphere, which refract and reflect the light from the Sun.

AFTER »

New atmospheric phenomena and processes have been discovered over the past 50 years.

BOUNDARY WITH SPACE
There is **no clearly defined boundary** between the atmosphere and space, although in the 1950s, the International Aeronautic Federation accepted a **height of 62 miles (100 km)**, called the **Kármán line**, as a boundary above which the air is too thin for aeronautical purposes. The level of harmful radiation also increases above this line.

DISCOVERY OF OZONE DEPLETION
In the mid 1970s scientists warned that chemicals called chlorofluorocarbons (CFCs) were destroying Earth's ozone layer, resulting in increased ultraviolet radiation at the surface. In 1978 several countries banned the use of CFC-containing aerosol sprays. But, in the **mid 1980s, a seasonal ozone "hole"** (serious depletion) was detected above Antarctica. From 1987 an international treaty sharply reduced CFC production, which was phased out by 1996. By 2003 the **depletion of the ozone layer was slowing**, but the hole still reappears annually from September to December.

OZONE HOLE OVER ANTARCTICA

SEE ALSO »
pp.368–69 ARTIFICIAL SATELLITES
pp.422–23 GLOBAL WARMING

IN PRACTICE

LIGHTNING: AN ELECTRICAL DISCHARGE

The idea that lightning might be an electrical discharge was first suggested in the early 18th century and, in 1752, American Benjamin Franklin (pp.160–61) and Frenchman Thomas-François Dalibard confirmed this view. They noted that clouds are electrically charged: negatively at the base and positively on top. A bolt of lightning can carry a current of over 10,000 amps, generating temperatures higher than 36,000° F (20,000° C), which is hotter than the surface of the Sun.

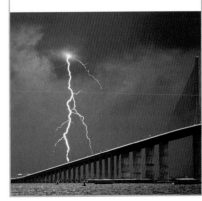

mainly to occupy the thermosphere. Later, an even higher layer, the exosphere, was defined, so rarefied that its gas particles hardly ever collide.

Atmospheric chemistry
In the 20th century scientists figured out how the composition and concentration of atmospheric gases, especially the increase of oxygen and the amount of carbon dioxide and other greenhouse gases, such as methane, has changed through Earth history, largely due to the activities of plants and other photosynthetic organisms. More recently, human-induced changes to atmospheric gaseous composition have promoted climate change. A related phenomenon is the ozone layer of the stratosphere, discovered by French physicists Charles Fabry and Henri Buisson in 1913. Its essential role in protecting life at Earth's surface was uncovered in 1930 by British physicist Sydney Chapman—molecules of ozone (O_3) and ordinary oxygen (O_2) are continuously converted into each other in a cyclical process that also absorbs harmful ultraviolet radiation and produces heat.

Scientists now study the chemistry of both the oceans and the atmosphere when trying to solve problems such as global warming.

BEFORE

People have contemplated oceanic phenomena ever since they first went to sea.

EARLY IDEAS ON TIDES
The 2nd-century BCE Babylonian astronomer **Seleucus** of Seleucia was the first to state that **tides** are linked to the **phases of the Moon**.

ORIGIN OF OCEAN BASINS
In 1644 the French philosopher **René Descartes** suggested that in the distant past, **parts of Earth's crust had collapsed** into underground chambers, which had been formed by ancient gaseous outpourings. He thought this had **created the ocean basins** with the remaining crust forming continents.

THE GULF STREAM
Originally discovered in the 16th century by Spanish sailing expeditions, the strong, warm ocean current known as the **Gulf Stream**, was **first studied and mapped** in the 18th century by **Benjamin Franklin**.

SEE ALSO
PP.108–09 GRAVITATIONAL FORCE
PP.136–37 NAVIGATING THE OCEANS
PP.206–07 ATMOSPHERIC MOVEMENT

Studying the Oceans

The roots of oceanography go back centuries. By around 1800 certain aspects—such as winds and tides—were already well understood because of their importance to trade and naval warfare. But it was not until later in the 19th century that marine science first began to take off as a recognized field of study.

From the early 1800s scientists began to gather immense amounts of data about marine life, paving the way for future knowledge about the importance of the planet's oceans. In the 1840s and '50s American scientist Matthew Maury (see opposite) began to analyze data on currents and the depths of ocean basins. Among other accomplishments, Maury noted that the Atlantic is shallower in the middle (the first indication of the mid-Atlantic ridge, see p.356). The lowest point in the oceans—the Mariana Trench of the western Pacific, at over 35,800ft (10,900m) deep—was discovered in 1872.

Ocean currents
Strong surface currents in the oceans, some warm, others cold, were known about from the 16th century onward.

HMS *Challenger*
The natural history laboratory on board HMS *Challenger*, where some 4,000 new animal and plant species were identified.

During the 19th century scientists gradually realized that many of these currents are linked in gyres—these are large-scale circular patterns of currents that operate over whole oceans (see left). In 1905 the Swedish oceanographer Walfrid Ekman

published a theory about the effect of prevailing winds blowing over ocean areas, which demonstrated that these winds were driving the gyres. Today it is known that ocean currents play a major role in affecting localized climate. Warm currents carry heat away from the tropics and subtropics, giving some temperate regions, such as the British Isles and Scandinavia, a warmer climate than they would otherwise

Westerly winds
Water pushed to right of wind direction in Northern Hemisphere
Northeast trade winds
Water pushed to left of wind direction in Southern Hemisphere
Southeast trade winds
Westerly winds

Pacific gyre formation
A gyre is a circular system of currents driven by winds whose effect is to push water at right angles to the wind direction.

KEY
■ Direction of wind
■ Direction of gyre

The great ocean conveyor
This continuous global flux of seawater is driven by water cooling becoming more dense in Arctic waters, then sinking to the seabed.

KEY
■ Warm surface current
■ Cold deep-ocean current

Reykjanes Basin (abyssal plain)
Atlantis Fracture Zone
Mid-Indian Ridge (mid-ocean spreading ridge)
Mid-Atlantic Ridge (mid-ocean spreading ridge and mountain range)
Angola Basin (abyssal plain)
Walvis Ridge (volcanic ridge caused by a hot spot)
Atlantic-Indian Ridge (mid-ocean spreading ridge)

Striated muscle
Contraction of striated muscle moves one bone with respect to another. The body has voluntary control over this—stimulation of the nervous system causes nerve impulses to travel via motor nerves to muscles.

Smooth muscle
Smooth muscle occurs in organs, such as the gut, and in blood vessels; contraction of this muscle is important in internal processes. It is stimulated by a part of the nervous system over which we have no conscious control.

that make up the skeleton are bone and cartilage. Bone contains important minerals for the body and is harder than cartilage because it contains calcium phosphate. A joint occurs where bone meets bone. Some joints, such as those in the skull, are fixed, but others allow movement of bones—synovial joints (see opposite). Some, such as the elbow and knee joints, can only move in one direction, acting like hinges. Others allow for more rotation, such as the ball-and-socket joints of the shoulder and hip.

Muscles
Muscles are fibrous tissues, that twitch when stimulated, pulling their ends together to create tension. Muscle

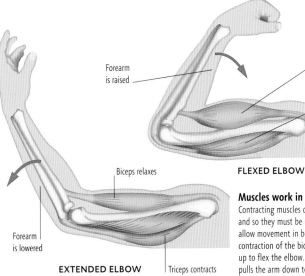

Forearm is raised

Biceps contracts

Triceps relaxes

FLEXED ELBOW

Forearm is lowered

Biceps relaxes

EXTENDED ELBOW   Triceps contracts

Muscles work in pairs
Contracting muscles can only pull (not push), and so they must be in pairs around joints to allow movement in both directions. Here, contraction of the biceps pulls the lower arm up to flex the elbow. Contraction of the triceps pulls the arm down to extend it.

contracts when it receives an impulse from the nervous system via nerves (see pp.220–21). This contraction is caused by the interaction of two muscle proteins: myosin, which was first isolated in 1864 by German physiologist Wilhelm Kühne, and actin, which was found by British physiologist William Halliburton in 1887. It was not until 1954 that two more British scientists, Andrew Huxley and Jean Hanson, showed how individual filaments of the two proteins slide over one another to contract muscle with energy provided by a substance called ATP (see right).

Voluntary contraction of muscles connected to the skeleton can move one bone in relation to another bone (see above). This type of muscle is sometimes found in pairs around joints, which allows movement in both directions (see above). Other muscles are found in the walls of organs, and their involuntary contraction is important in continuous vital functions, such as peristalsis—the pulsation of the gut (see pp.216–17).

Thin actin filaments

Thick myosin filament

RELAXED MUSCLE   **CONTRACTED MUSCLE**

How muscle contracts
In relaxed muscle, thick myosin filaments overlap slightly with thinner actin filaments. When the muscle is stimulated to contract, a cellular "fuel" called ATP enables the myosin to crawl along the actin filaments, overlapping further, and shortening the muscle.

"**Muscles** are in a most **intimate** and **peculiar** sense the **organs of the will**."

GRANVILLE STANLEY HALL, AMERICAN PSYCHOLOGIST, 1907

AFTER

Study of the chemistry of muscles has helped to understand neuromuscular diseases, offering hope for new treatments.

MUSCULAR DYSTROPHY
In 1987 American geneticist **Louis Kunkel** identified a protein called **dystrophin**, which **stabilizes muscle fibers**. Lack of it causes Duchenne muscular dystrophy. Research is focused on genetically modifying stem cells 390–91 ❯❯ so they can develop into normal working muscle fibers.

FRAGILE BONES
Osteoporosis is a bone disease where the thinning of bones increases risk of fracture. The American biologist **Fuller Albright** found that the disease is most common in **postmenopausal women** after their estrogen levels fall. This led to the development of **treatment with hormones**.

SEE ALSO ❯❯
pp.274–75 REGULATING THE BODY

Cranium encases the brain

Ribs form a cage to surround and protect heart and lungs

Vertebral column encases spinal cord

Endoskeleton
The internal endoskeleton of a mammal, such as this dog, is composed of bones articulated at joints. As well as providing a hard framework, bone also acts as a reservoir of calcium for the body and contains marrow for production of blood cells.

Limbs consist of upper parts (single limb bones) and lower parts (two parallel limb bones)

Exoskeleton
Many invertebrates have a hard casing around the body, with flexible material at joints. An exoskeleton restricts growth so it must be periodically molted.

Hydroskeleton
Many animals, such as this starfish, are supported by an internal skeleton of water. This is effective in invertebrate animals in aquatic environments.

BIOLOGIST (1729–99)

LAZARRO SPALLANZANI

Born in Italy, biologist Lazarro Spallanzani spent much of his career as Director of Natural History at the University of Pavia. He refuted the prevailing idea that life came spontaneously from inanimate matter rather than reproduction and used frogs to prove that eggs only develop into embryos after contact with semen. He was the first scientist to perform artificial insemination, on a dog. He also experimented with chemical digestion in the stomach (see p.217).

Conception to embryo
About once a month an egg is released from the ovary and funneled down through the fallopian tube, where, if sperm are present, fertilization will take place. Over the next few days, the fertilized egg divides into an embryo, initially just a ball of cells, which becomes implanted into the blood-rich lining of the wall of the uterus (womb).

2 Earliest embryo stage
Enclosed within a coat, the embryo forms when the zygote undergoes its initial cell cleavage (division) in the first 24 to 36 hours. The earliest divisions are not accompanied by cell growth, so individual cells shrink to the size of normal body cells.

Fallopian tube lining
The lining has a coating of hairlike cilia to waft the embryo down the tube.

First cleavage (division)

Nucleus of embryo cell

Egg cell
The egg cell contains 23 maternal chromosomes.

Fallopian tube

Follicle cell
A cell that helps nourish the egg when it is in the ovary.

Sperm tail
Part of the sperm that lashes to propel sperm toward the egg.

Sperm head
The head of the sperm contains 23 paternal chromosomes.

Acrosome
Tiny bag of chemicals in the sperm head that digests the egg coat to penetrate it.

1 Fertilization
Fertilization takes place in the upper third of the fallopian tube, and occurs when the sperm cell penetrates the much larger egg to form a zygote.

Human Reproduction

All human life starts as a fertilized egg. Unraveling the events that produce this egg and its subsequent development relied on the microscope and bold experiments. Today scientists are not only able to monitor the health of a baby in the uterus, but can even manipulate the entire reproductive process.

« BEFORE

Before fertilization was observed through a microscope, many scientists doubted the roles of eggs and sperm in reproduction.

LIFE FROM EGGS
William Harvey, discoverer of blood circulation and physician to King Charles I of England, studied the reproductive systems of deer in the royal parks, proposing in 1651 that young animals could only develop from eggs, refuting the theory of spontaneous generation (see above).

MYSTERY OF SPERM
In 1678 Dutch scientists Antony van Leeuwenhoek and Nicolas Hartsoeker made the first microscopic observations of sperm. Other microscopists of the period reported seeing tiny human beings (homunculi) inside sperm heads (since condemned as hoaxes), which fueled the idea of preformation—that embryos are preformed in sex organs.

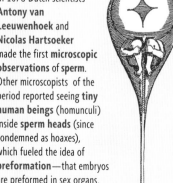

HOMUNCULUS IN SPERM

« SEE ALSO
pp.72–73 THE HUMAN BODY REVEALED

Sexual reproduction involves the fusion of the gametes (sperm and egg) in a process of fertilization to make a single cell called a zygote. Today we know how hormones control the monthly release of human eggs. But until the early 1800s, scientists were not even convinced that sperm was needed to make a baby.

Identifying the sperm and egg
In 1835 celebrated English zoologist Richard Owen classified the tadpolelike microbes in semen as parasites. In reality, they were sperm. A decade earlier, in France, physician Jean-Louis Prevost and chemist Jean-Baptiste Dumas had shown that sperm were

in the testes of many different kinds of animals. Later, in 1852, English physician Henry Nelson claimed to see fertilization down his microscope—in a worm called *Ascaris*—and others saw sperm penetrating eggs of sea urchins and starfishes. But all these animals naturally fertilized outside their bodies. Finding fertilization inside the body of a human being proved to be much more elusive.

Sperm are produced from the walls of tiny tubules in the testes in a process that involves meiosis (see pp.306–07)—a type of cell division that halves the genetic content of the cells to prevent genetic doubling at fertilization. Meiosis also happens in ovaries to produce

eggs. In 1826 a German embryologist Carl Ernst von Baer found eggs inside the ovaries of a dog. Swiss physicist Pierre Prévost and French chemist Jean Dumas studied the ovaries of mammals too. They recorded the growth of swellings in ovaries—the follicles that contain eggs—and noted the presence of embryos in the fallopian tubes after mating, suggesting fertilization happened there.

Understanding female hormones
Two of the hormones controlling reproduction (estrogen and progesterone) are produced in the ovaries themselves. In the late 1920s a group of American scientists found that

The menstrual cycle
The hormone FSH stimulates follicle growth (swelling in the ovary). Estrogen is released from the follicle and stimulates the uterine wall lining to thicken and fill with blood. The hormone LH stimulates ovulation (the release of an egg from the swollen follicle). Without fertilization the empty follicle (corpus luteum) shrinks, progesterone levels drop, and the uterus lining is shed in menstruation.

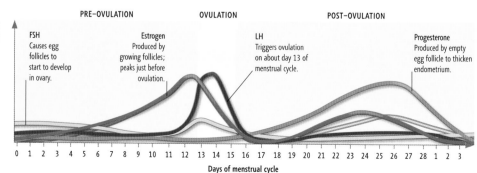

| PRE-OVULATION | OVULATION | POST-OVULATION |

FSH Causes egg follicles to start to develop in ovary.

Estrogen Produced by growing follicles; peaks just before ovulation.

LH Triggers ovulation on about day 13 of menstrual cycle.

Progesterone Produced by empty egg follicle to thicken endometrium.

0 1 2 3 4 5 6 7 8 9 10 11 12 13 14 15 16 17 18 19 20 21 22 23 24 25 26 27 28 1 2 3

Days of menstrual cycle

3 Morula stage
The term "morula" is used when the embryo is still a ball of cells with a protective coat. It leaves the fallopian tube to enter the uterus three to four days after fertilization.

Morula
A compact ball of 16 to 32 cells.

Coat around embryo

Fimbria
A fringe of tissue at the opening of the fallopian tube that catches and guides the released eggs from the ovary.

Ovary

Ovarian ligament

Blastocoele
A fluid-filled cavity.

Blastocyst

Inner cell mass
Mass of cells that will develop into the embryonic body.

Trophoblast
An outer layer of cells that branches into the uterus lining, forming the placenta.

Endometrium
Blood-filled uterus lining.

4 Blastocyst stage
Six days after fertilization, this structure has a fluid-filled hollow—a liquid formed from the embryonic cells. The inner mass of cells becomes the maturing embryo; the outer layer of cells will form part of the placenta as it implants in the uterus lining, two days later.

Maternal blood vessels

Ultrasound scan
Twelve weeks into pregnancy, the fetus is revealed by a technique that uses high-frequency sound waves, which reflect from tissue to give a three-dimensional image.

Mammalian in vitro (literally "in glass") fertilization (IVF)—in a hamster—was first achieved by Japanese clone specialists five years later. Then, in the 1960s, British physiologist Robert Edwards and physician Patrick Steptoe achieved IVF with human sex cells, heralding new hope for thousands of infertile couples (see pp.388–89).

Stem cells
By the mid 1800s, scientists had discovered that embryos develop from shapeless balls of cells. In 1892 German biologist Hans Driesch found that each cell in the ball had all the information needed to form any body part. Today, we call these stem cells.

Understanding the role of hormones in reproduction led to developments, which are considered routine today.

PREGNANCY TEST KITS
In the 1930s American scientist **Georgeanna Jones** discovered that a **hormone** called human chorionic gonadotropin (**hCG**) was produced by the **placenta**. Early pregnancy tests detected hCG in urine. Later this method was replaced by a more direct analysis of hCG concentration using an **antibody-binding reaction**, which became the basis of the home pregnancy test kit in the 1970s.

BIRTH CONTROL
Contraceptive methods involving suppositories or barriers can be traced back to ancient Egypt, but the development of a **contraceptive pill** only became possible with 20th-century **understanding** of **hormonal control** of the menstrual cycle. **"The Pill"** contains hormones that **suppress ovulation** (the release of an egg from the ovary), reducing the chance of fertilization.

CONTRACEPTIVE PILLS

SEE ALSO »
pp.348–49 THE GENETIC CODE
pp.386–87 GENETIC TECHNOLOGY
pp.388–89 IN VITRO FERTILIZATION (IVF)
pp.390–91 CLONING AND STEM CELLS

hormones from the pituitary gland, at the base of the brain (see p.275), also caused changes in the ovaries: follicle-stimulating hormone (FSH) triggers growth of the follicle and luteinizing hormone (LH) causes it to burst at ovulation (see opposite). Today, drugs that mimic FSH are important in treating infertility.

On the path to IVF
In 1951 Australian biologist Colin Austin and Chinese scientist Min Chueh Chang discovered that sperm cells need to be "activated" by chemicals before they can fertilize an egg—a process known as capacitation. Austin found how capacitated sperm penetrated the egg coat by releasing digestive enzymes from their heads.

20 MILLION
The average number of sperm cells in a milliliter of human semen (fluid produced by the testes).

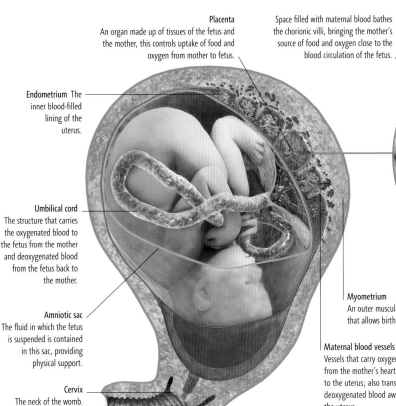

Placenta
An organ made up of tissues of the fetus and the mother, this controls uptake of food and oxygen from mother to fetus.

Space filled with maternal blood bathes the chorionic villi, bringing the mother's source of food and oxygen close to the blood circulation of the fetus.

Chorionic villus contains blood capillaries of the fetal umbilical blood circulation.

Endometrium The inner blood-filled lining of the uterus.

Umbilical cord
The structure that carries the oxygenated blood to the fetus from the mother and deoxygenated blood from the fetus back to the mother.

Amniotic sac
The fluid in which the fetus is suspended is contained in this sac, providing physical support.

Cervix
The neck of the womb.

Myometrium
An outer muscular layer of the uterus that allows birth contractions.

EXCHANGE ACROSS PLACENTA

Maternal blood vessels
Vessels that carry oxygenated blood from the mother's heart and lungs to the uterus; also transports deoxygenated blood away from the uterus.

Fetus in the uterus
Food and oxygen pass from mother to fetus along a vein in the umbilical cord, and waste passes back along an artery. This circulation stops at birth, when the baby's lungs and gut begin to function.

« BEFORE

Early surgeons attempted to address problems of blood loss and pain, but infection was a major issue.

HERBAL PAIN RELIEF
Early attempts to numb sensation used **plants with chemicals** that affect the nervous system — alkaloids. Poppies contain alkaloids called **opiates** and were used as anesthetics in Egypt and Asia. Islamic physicians pioneered the use of **inhalant anesthesia** with drug-soaked sponges, using a combination of different opiates.

OPIUM POPPY

BLOOD TRANSFUSIONS
Early **blood transfusions** were seen as a way of boosting "vitality." In 1665 an English physician Richard Lower used it to restore the health of dogs after substantial blood loss. Jean Denys, a French physician, tried to transfuse **sheep blood into humans**: animal blood was thought to be "impure of passion or vice." But a fatal incident led to its widespread condemnation.

« SEE ALSO
pp.24–25 EARLY MEDICINE AND SURGERY
pp.70–71 RENAISSANCE MEDICINE AND SURGERY

Safer Surgery

Surgery is a form of medicine concerned with treating injuries and diseases and curing life-threatening conditions. The history of surgery is a triumph of medical science over pain, blood loss, and infection. Some of the biggest surgical breakthroughs took place during the 19th century.

Surgery is highly invasive. It manipulates the body, exposing the patient to severe pain and blood loss in its aim to cure disease. And, for those who survive the ordeal, postoperative infection is an ever-present threat.

Reducing infection
At the start of the 19th century many physicians subscribed to the theory that disease was caused by "bad air"—the miasma theory. Florence Nightingale advocated a regimen of fresh air in her hospital wards. Some physicians advised improving hygiene, but they were largely ignored. Then in 1864 the French microbiologist Louis Pasteur (see p.244) demonstrated that organisms in the air caused nutrient liquids to ferment, and German physician Robert Koch found bacteria in animals

Lister's carbolic spray
This instrument was developed by Joseph Lister to deliver a fine spray of the antiseptic solution over the wound during surgery, drastically reducing the incidence of infection.

suffering from a disease called anthrax. This gave new focus to the control of infection (see pp.242–43). If bacteria—and other "germs" (pathogens)—could spread from person to person they could just as easily invade an open wound. But germs could be suppressed or killed by chemicals, and could not develop in a germ-free—sterile—environment.

Foremost in the campaign to eliminate microbes using chemicals was Joseph Lister (see opposite). He treated both wounds and surgical instruments with carbolic acid (also known as phenol—a chemical that had been used to deodorize sewage), establishing an antiseptic technique that significantly reduced infection.

Numbing the pain
Chemicals that deaden pain are called analgesics; those that eliminate all sensation are called anesthetics, a term coined by American physician Oliver Wendell Holmes in 1846. The earliest anesthetics were herbal and usually used on small areas of the body (locally). General (whole-body) anesthesia was too risky, since standardizing doses of plant extracts was difficult. The first recorded use of general anesthesia using plant chemicals was by Japanese surgeon Seishu Hanaoka in 1804. Soon after, the German pharmacist Friedrich Sertürner isolated morphine, the active ingredient of the opium poppy. Because of its purity, doses could be consistent.

BREAKTHROUGH

GERM THEORY OF DISEASE

Louis Pasteur demonstrated in 1864 that microorganisms did not arise spontaneously (as previously thought), but were present because of infection. Robert Koch applied this idea to explain the cause of infectious disease and, in 1890, arrived at a set of criteria known as Koch's Postulates that determine if a disease is caused by microbes. These microorganisms could be isolated from infected individuals and grown in a laboratory for identification, so that an appropriate treatment regimen could be devised. Today many disease-causing microbes have been identified.

JOSEPH LISTER

Joseph Lister was a skilled surgeon. He developed his ideas of sterile practice in surgery when working at Glasgow Royal Infirmary, Scotland. He promoted the use of antiseptics, recommended the use of gloves and the replacement of instruments with nonporous materials to minimize infection. Lister became President of the Royal Society in London but, in spite of his achievements, remained a modest character. He was honored with a funeral service in Westminster Abbey.

> " ...it is from the vitality of the **atmospheric particles** that all the **mischief arises**..."
>
> JOSEPH LISTER, *THE LANCET*, 1867

a hemorrhage after childbirth. The descriptions of patients' reactions afterward, which Blundell attributed to air bubbles in the blood, suggest that many had received incompatible blood. This issue was resolved in 1901 after Austrian biologist Karl Landsteiner discovered that there were four blood groups. Each group contains different antigens (chemicals that cause immune response). If compatible blood is transfused, see below left, immune response can be prevented.

Scientists also began to investigate the potential of gases as anesthetics. In 1800 English chemist Humphry Davy (see p.170) recorded that nitrous oxide (laughing gas) had pain-numbing qualities, and in 1846 American dentist William Morton used ether to anesthetize a patient. The effectiveness of the gas was proved at Massachusetts General Hospital in Boston, and its use spread through the US and to Europe.

Improvements in infection control
A medical team working in an operating room in the early 1900s at London's Charing Cross Hospital demonstrates that, although Lister's antiseptic principles were in practice, certain precautions—such as the use of face masks—were yet to be adopted.

The English surgeon James Simpson advocated the use of chloroform in 1847, but this gas posed a higher risk of overdose, so ether remained popular throughout the 19th century.

Dealing with blood loss
Bungled attempts at blood transfusion in the 17th century had meant that the practice had all but disappeared. A revival of interest came with the work of English physician James Blundell. In 1818 he performed the first human-to-human blood transfusion on a man with what would now be called stomach cancer; the man died 56 hours later. His first successful human blood transfusion was some years later in a woman suffering from

AFTER

The 20th century saw important scientific breakthroughs in anesthesia and hematology (the study of blood).

MODERN ANESTHESIA
In 1951 British chemist **C.W. Suckling** made the anesthetic **halothane**, a nonflammable replacement for ether. Later it was supplanted by less-toxic and faster acting drugs, often used in combination to produce more effective anaesthesia in patients.

BLOOD GROUPS AND BLOOD BANKS
By 1907 **routine blood-typing** for transfusions was recommended. World War I brought a new urgency for transfusions—and with it developments that enabled long-term storage of blood and blood products.

MODERN OPERATING ROOM

SEE ALSO »
pp.410–11 MODERN SURGICAL PROCEDURES

ABO blood groups
Safe blood transfusions are possible only when chemicals called antigens on the surface of donated red blood cells do not stimulate an immune reaction in the recipient. This reaction is mediated by antibodies dissolved in the blood. There are two main types of antigens. These can be present on their own (blood groups A and B), together (blood group AB), or both absent (blood group O).

A antigen
Anti-B antibody does not react

Blood group A
This blood cannot be donated to blood groups B or O because the A antigens react with their anti-A antibodies.

B antigen
Anti-A antibody does not react

Blood group B
This blood cannot be donated to blood groups A or O because the B antigens react with their anti-B antibodies.

B antigen
A antigen

Blood group AB
This can be donated only to a recipient of the same blood group, to avoid anti-A and anti-B antibodies in others.

Anti-A antibody does not react
Anti-B antibody does not react

Blood group O
This blood can be donated to anyone because the lack of antigens prevents the possibility of an immune response.

Mendeleev's Table

No chemistry textbook, classroom, or laboratory wall today is complete without a copy of the periodic table—recognized by anyone who has studied science as an exceedingly useful framework for classifying and comparing the elements. But what is perhaps more surprising is that the original version of the table was literally "dreamed up" by an eccentric Russian professor while taking an afternoon nap.

Over the first 65 years of the 19th century a bewildering variety of new chemical elements were discovered, many of them rare and obscure metals, but also substances that are well known today such as silicon and aluminum. By 1866 a total of 63 elements were known, but although these fell into a few obvious categories, such as metals and nonmetals, overall there seemed to be no rhyme or reason to them, nor indeed an explanation for why there were so many or why new ones kept turning up. In short, chemists at the time were in desperate need of a new way of understanding and classifying the elements.

By early 1869 a Russian chemist, Dmitri Mendeleev, had been trying to make sense of the elements for some months. He wondered whether if they were ordered by their atomic weights, or in some other way, a pattern might emerge. One February day, he became so engrossed with the problem that he canceled all engagements and began writing down the names of the elements on the backs of playing cards. He then tried arranging the cards, with the elements ordered by atomic weight, in a grid of rows and columns. This met with mixed success—there seemed to be some "periodicity," or recurring pattern, to the elements, but also many misalignments. In the midst of these efforts, he fell asleep and, while dreaming of playing patience, an idea came to him. What if spaces were left in his grid for elements that were undiscovered? When he awoke, Mendeleev found that a few adjustments to his grid produced a layout that worked brilliantly to show the relationships between all the known elements. It was Mendeleev's very own "Eureka" moment—the very first version of the periodic table had been born.

A table of elements
This version of the periodic table, with a photograph of Mendeleev (top right), comes from a textbook that was published in Russia during the Soviet era. It is somewhat different from the modern version of the table, particularly in its placing of the transition elements (the ones displayed here with black or dark blue element symbols).

> "It is the **function** of **science** to discover the **existence** of a general reign of **order in nature** and to find the **causes** governing this order."
>
> DMITRI MENDELEEV, LATE 19TH CENTURY

ПЕРИОДИЧЕСК[...]
Д. И.

	I	II	III
1	H		
2	Li 3 — 6,94₁ 2s¹ — ЛИТИЙ	Be 4 — 9,01218 2s² — БЕРИЛЛИЙ	5 2p¹
3	Na 11 — 22,98977 3s¹ — НАТРИЙ	Mg 12 — 24,305 3s² — МАГНИЙ	13 A[...] 26,9[...] 3p¹ АЛЮМИ[...]
4	K 19 — 39,09₈ 4s¹ — КАЛИЙ	Ca 20 — 40,08 4s² — КАЛЬЦИЙ	Sc 21 — 44,9559 3d¹ СКАНДИЙ
	29 Cu — 63,54₆ 3d¹⁰4s¹ — МЕДЬ	30 Zn — 65,38 4s² — ЦИНК	31 G[...] 4p¹ ГАЛ[...]
5	Rb 37 — 85,467₈ 5s¹ — РУБИДИЙ	Sr 38 — 87,62 5s² — СТРОНЦИЙ	Y 39 — 88,9059 4d¹ ИТТРИЙ
	47 Ag — 107,868 5s¹ — СЕРЕБРО	48 Cd — 112,40 5s² — КАДМИЙ	49 I[...] 5p¹ ИН[...]
6	Cs 55 — 132,9054 6s¹ — ЦЕЗИЙ	Ba 56 — 137,3₄ 6s² — БАРИЙ	La 57 — 138,905₅ 5d¹ ЛАНТАН *
	79 Au — 196,9665 5d¹⁰6s¹ — ЗОЛОТО	80 Hg — 200,5₉ 6s² — РТУТЬ	81 T[...] 6p¹ ТАЛ[...]
7	Fr 87 — [223] 7s¹ — ФРАНЦИЙ	Ra 88 — 226,0254 7s² — РАДИЙ	Ac 89 — [227] 6d¹ АКТИНИЙ **

Ce 58	Pr 59	Nd 60	Pm 61	S[...]
140,12 4f¹5d¹6s² — ЦЕРИЙ	140,9077 4f³6s² — ПРАЗЕОДИМ	144,2₄ 4f⁴6s² — НЕОДИМ	[145] 4f⁵6s² — ПРОМЕТИЙ	150, — САМ[...]

Th 90	Pa 91	U 92	Np 93	P[...]
232,0381 6d²7s² — ТОРИЙ	231,0359 5f²6d¹7s² — ПРОТАКТИНИЙ	238,02 5f³6d¹7s² — УРАН	237,0482 5f⁴6d¹7s² — НЕПТУНИЙ	[244] — ПЛУ[...]

СИСТЕМА ЭЛЕМЕНТОВ ДЕЛЕЕВА

VII	VIII
1 **H** $1s^1$ 1,0079 ВОДОРОД	2 **He** $1s^2$ 4,00260 ГЕЛИЙ

Периодический закон открыт Д.И.МЕНДЕЛЕЕВЫМ в 1869 году

V	V	VI	VII	VIII
7 **C** $2p^3$ 12,011 УГЛЕРОД	8 **N** $2p^3$ 14,0067 АЗОТ	9 **O** $2p^4$ 15,999$_4$ КИСЛОРОД	**F** $2p^5$ 18,99840 ФТОР	10 **Ne** $2p^6$ 20,17$_9$ НЕОН
15 **Si** $3p^3$ 28,08$_6$ КРЕМНИЙ	16 **P** $3p^3$ 30,97376 ФОСФОР	17 **S** $3p^4$ 32,06 СЕРА	**Cl** $3p^5$ 35,453 ХЛОР	18 **Ar** $3p^6$ 39,94$_8$ АРГОН
22 **V** $3d^24s^2$ 50,941$_4$ ВАНАДИЙ	23 **Cr** $3d^34s^2$ 51,996 ХРОМ	24 **Mn** $3d^54s^1$ 54,9380 МАРГАНЕЦ	25 **Fe** $3d^54s^2$ 55,84$_7$ ЖЕЛЕЗО	26 **Co** $3d^64s^2$ 58,9332 КОБАЛЬТ	27 **Ni** $3d^74s^2$ 58,7$_1$ НИКЕЛЬ	28 $3d^84s^2$
Ge 72,5$_9$ ГЕРМАНИЙ	33 **As** $4p^3$ 74,9216 МЫШЬЯК	34 **Se** $4p^4$ 78,9$_6$ СЕЛЕН	35 **Br** $4p^5$ 79,904 БРОМ	36 **Kr** $4p^6$ 83,80 КРИПТОН		
40 **Nb** $4d^25s^2$ 92,9064 НИОБИЙ	41 **Mo** $4d^45s^1$ 95,9$_4$ МОЛИБДЕН	42 **Tc** $4d^55s^1$ 98,9062 ТЕХНЕЦИЙ	43 **Ru** $4d^75s^1$ 101,0$_7$ РУТЕНИЙ	44 **Rh** $4d^85s^1$ 102,9055 РОДИЙ	45 **Pd** $4d^{10}5s^0$ 106,4 ПАЛЛАДИЙ	46
Sn $5p^3$ 118,6$_9$ ОЛОВО	51 **Sb** $5p^4$ 121,7$_5$ СУРЬМА	52 **Te** $5p^4$ 127,6$_0$ ТЕЛЛУР	53 **I** $5p^5$ 126,9045 ИОД	54 **Xe** $5p^6$ 131,30 КСЕНОН		
72 **Ta** $5d^26s^2$ 180,947$_9$ ТАНТАЛ	73 **W** $5d^36s^2$ 183,8$_5$ ВОЛЬФРАМ	74 **Re** $5d^46s^2$ 186,2 РЕНИЙ	75 **Os** $5d^66s^2$ 190,2 ОСМИЙ	76 **Ir** $5d^76s^2$ 192,2$_2$ ИРИДИЙ	77 **Pt** $5d^96s^1$ 195,0$_9$ ПЛАТИНА	78
Pb $6p^3$ 207,2 СВИНЕЦ	83 **Bi** $6p^3$ 208,9804 ВИСМУТ	84 **Po** $6p^4$ [209] ПОЛОНИЙ	85 **At** $6p^5$ [210] АСТАТ	86 **Rn** $6p^6$ [222] РАДОН		
104 $6d^27s^2$	105 $6d^37s^2$					

U	92	Атомный номер
Атомный вес **238,02**$_9$ $5f^36d^17s^2$ УРАН		Распределение электронов по уровням

Распределение электронов по застраивающимся и последующим застроенным подуровням

— s-элементы
— p-элементы
— d-элементы
— f-элементы

Атомные веса приведены по углеродной шкале (атомный вес изотопа углерода ^{12}C равен 12 точно).
Атомные веса приведены по Международной таблице 1971 года.
Точность последней значащей цифры ±1 или ±3, если она выделена мелким шрифтом.
В квадратных скобках приведены массовые числа наиболее устойчивых изотопов.
Названия и символы элементов, приведенные в круглых скобках, не являются общепринятыми.

* лантаноиды

Eu 63	**Gd** 64	**Tb** 65	**Dy** 66	**Ho** 67	**Er** 68	**Tm** 69	**Yb** 70	**Lu** 71
1,96 $4f^76s^2$ РОПИЙ	157,2$_5$ $4f^75d^16s^2$ ГАДОЛИНИЙ	158,9254 $4f^96s^2$ ТЕРБИЙ	162,5$_0$ $4f^{10}6s^2$ ДИСПРОЗИЙ	164,9304 $4f^{11}6s^2$ ГОЛЬМИЙ	167,2$_6$ $4f^{12}6s^2$ ЭРБИЙ	168,9342 $4f^{13}6s^2$ ТУЛИЙ	173,0$_4$ $4f^{14}6s^2$ ИТТЕРБИЙ	174,97 $5d^16s^2$ ЛЮТЕЦИЙ

** актиноиды

Am 95	**Cm** 96	**Bk** 97	**Cf** 98	**Es** 99	**Fm** 100	**Md** 101	**(No)** 102	**(Lr)** 103
43 $5f^77s^2$ ЕРИЦИЙ	[247] $5f^76d^17s^2$ КЮРИЙ	[247] $5f^86d^17s^2$ БЕРКЛИЙ	[251] $5f^{10}7s^2$ КАЛИФОРНИЙ	[254] $5f^{11}7s^2$ ЭЙНШТЕЙНИЙ	[257] $5f^{12}7s^2$ ФЕРМИЙ	[258] $5f^{13}7s^2$ МЕНДЕЛЕВИЙ	[255] $5f^{14}7s^2$ (НОБЕЛИЙ)	[256] $6d^17s^2$ (ЛОУРЕНСИЙ)

‹‹ BEFORE

A key step in the development of the table was the determination of the atomic weights of elements in the early 19th century.

DÖBEREINER'S TRIADS

In 1829 German chemist **Johann Döbereiner** noted **several triplets of similar elements**, where the middle element had an atomic weight equal to the average of the other two. This **introduced the concept of repetition** in the properties of elements.

JOHANN DÖBEREINER

LAW OF OCTAVES

In 1864 English chemist **John Newlands proposed the "law of octaves."** He noticed similarities between every eighth element, ordered by atomic weight, **but the pattern petered out** after the first 17 elements.

‹‹ SEE ALSO
pp.150–51 ORGANIC CHEMISTRY
pp.230–31 MENDELEEV'S TABLE

The Periodic Table

The periodic table currently contains 118 elements, of which around 90 occur naturally on Earth. The position of an element in the table is determined by its atomic number (the number of protons in the nuclei of its atoms), while the properties of each element depend on the configuration of the electrons in its atoms.

The periodic table is structured in order of increasing atomic number. It starts with hydrogen or H (atomic number 1) positioned top left, followed by helium or He (atomic number 2) at top right, then proceeds row by row through the rest of the elements in sequence by their atomic numbers. (Before 1914 the elements had been ordered according to their relative atomic masses, or weights.) Two rows of "inner transition elements" are set aside at the bottom of the table to avoid making their rows too long to fit on the page.

Since the 1920s chemists have known that the electrons within atoms are arranged in "shells," which occupy regions of space at differing distances from the nucleus (see pp.286–87). Progressing through the elements in order by atomic number, new shells are added and become sequentially filled by electrons, to produce each element's unique electron configuration (see right). This pattern forms the basis of various features of the table.

Periods in the table

The periods of the table are its rows. The first period contains just hydrogen and helium. Atoms of these elements have just one electron shell, which can hold either one electron

Total of 6 electrons in two shells (2 + 4)

Nucleus

CARBON (C) ATOM

Nucleus

Total of 35 electrons in four shells (2 + 8 + 18 + 7)

BROMINE (Br) ATOM

Electron configuration in atoms

Atoms of different elements vary in their electron configurations—that is to say, the total number of electrons they hold and the way that these are arranged in shells around the nucleus.

> " The **elements**, if arranged according to their **atomic weights**, exhibit an apparent **periodicity.**"
> DMITRI MENDELEEV, RUSSIAN CHEMIST, 1869

Group 1																		Group 18
Periods 1	1 H Hydrogen 1	**Group 2**												**Group 13**	**Group 14**	**Group 15**	**Group 16**	**Group 17**	2 He Helium 4
2	3 Li Lithium 7	4 Be Beryllium 9												5 B Boron 11	6 C Carbon 12	7 N Nitrogen 14	8 O Oxygen 16	9 Fl Fluorine 19	10 Ne Neon 20
3	11 Na Sodium 23	12 Mg Magnesium 24	Group 3	Group 4	Group 5	Group 6	Group 7	Group 8	Group 9	Group 10	Group 11	Group 12		13 Al Aluminum 27	14 Si Silicon 28	15 P Phosphorus 31	16 S Sulfur 32	17 Cl Chlorine 35	18 Ar Argon 40
4	19 K Potassium 39	20 Ca Calcium 40	21 Sc Scandium 45	22 Ti Titanium 48	23 V Vanadium 51	24 Cr Chromium 52	25 Mn Manganese 55	26 Fe Iron 56	27 Co Cobalt 59	28 Ni Nickel 59	29 Cu Copper 64	30 Zn Zinc 65	31 Ga Gallium 70	32 Ge Germanium 73	33 As Arsenic 75	34 Se Selenium 79	35 Br Bromine 80	36 Kr Krypton 84
5	37 Rb Rubidium 85	38 Sr Strontium 88	39 Y Yttrium 89	40 Zr Zirconium 91	41 Nb Niobium 93	42 Mo Molybdenum 96	43 Tc Technetium 98	44 Ru Ruthenium 101	45 Rh Rhodium 103	46 Pd Palladium 106	47 Ag Silver 108	48 Cd Cadmium 112	49 In Indium 115	50 Sn Tin 119	51 Sb Antimony 122	52 Te Tellurium 128	53 I Iodine 127	54 Xe Xenon 131
6	55 Cs Cesium 133	56 Ba Barium 137		72 Hf Hafnium 178	73 Ta Tantalum 181	74 W Tungsten 184	75 Re Rhenium 186	76 Os Osmium 190	77 Ir Iridium 192	78 Pt Platinum 195	79 Au Gold 197	80 Hg Mercury 201	81 Ti Thallium 204	82 Pb Lead 207	83 Bi Bismuth 209	84 Po Polonium 209	85 At Astatine 210	86 Rn Radon 222
7	87 Fr Francium 223	88 Ra Radium 226		104 Rf Rutherfordium 267	105 Db Dubnium 268	106 Sg Seaborgium 271	107 Bh Bohrium 270	108 Hs Hassium 277	109 Mt Meitnerium 276	110 Ds Darmstadtium 281	111 Rg Roentgenium 280	112 Cn Copernicium 285	113 Nh Nihonium 286	114 Fl Flerovium 289	115 Mc Moscovium 289	116 Lv Livermorium 293	117 Ts Tennessine 294	118 Og Oganesson 294

LANTHANOIDS

57 La Lanthanum 139	58 Ce Cerium 140	59 Pr Praseodymium 141	60 Nd Neodymium 144	61 Pm Promethium 145	62 Sm Samarium 150	63 Eu Europium 152	64 Gd Gadolinium 157	65 Tb Terbium 159	66 Dy Dysprosium 163	67 Ho Holmium 165	68 Er Erbium 167	69 Tm Thulium 169	70 Yb Ytterbium 173	71 Lu Lutetium 175
89 Ac Actinium 227	90 Th Thorium 232	91 Pa Protactinium 231	92 U Uranium 238	93 Np Neptunium 237	94 Pu Plutonium 244	95 Am Americium 243	96 Cm Curium 247	97 Bk Berkelium 247	98 Cf Californium 251	99 Es Einsteinium 252	100 Fm Fermium 257	101 Md Mendelevium 258	102 No Nobelium 259	103 Lr Lawrencium 262

ACTINOIDS

Atomic number
The number of protons in the nuclei of the element's atoms.

Chemical symbol
Used as a form of shorthand in chemical equations (in this case for aluminum).

13
Al
Aluminum
27

Relative atomic mass
The average number of nucleons (protons and neutrons) in the nuclei.

Key

The elements fall into several categories, or blocks, in which they share common characteristics. Nine color-coded categories are shown here; some correspond closely or exactly to groups (columns) of the table.

Alkali metals
A group of soft metals that are extremely reactive.

Alkaline-earth metals
A group of silver-colored, moderately reactive metals.

Transition metals
A varied group of metals, many with highly valued properties.

Inner transition elements
(Also known as lanthanoids and actinoids) Reactive metals, some rare or synthetic.

Other metals
A group of mostly rather soft metals with low melting points.

Metalloids
Elements with properties halfway between metals and nonmetals.

Other nonmetals
A collection that includes the most critical elements in living matter, notably carbon.

Noble gases
Colorless, odorless gases that have very low chemical reactivity.

Halogens
Highly reactive nonmetals that comprise two gases, a liquid, and two solids.

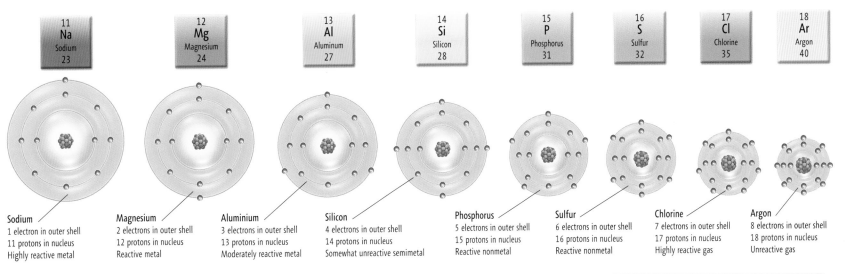

| 11
Na
Sodium
23 | 12
Mg
Magnesium
24 | 13
Al
Aluminum
27 | 14
Si
Silicon
28 | 15
P
Phosphorus
31 | 16
S
Sulfur
32 | 17
Cl
Chlorine
35 | 18
Ar
Argon
40 |

Sodium
1 electron in outer shell
11 protons in nucleus
Highly reactive metal

Magnesium
2 electrons in outer shell
12 protons in nucleus
Reactive metal

Aluminium
3 electrons in outer shell
13 protons in nucleus
Moderately reactive metal

Silicon
4 electrons in outer shell
14 protons in nucleus
Somewhat unreactive semimetal

Phosphorus
5 electrons in outer shell
15 protons in nucleus
Reactive nonmetal

Sulfur
6 electrons in outer shell
16 protons in nucleus
Reactive nonmetal

Chlorine
7 electrons in outer shell
17 protons in nucleus
Highly reactive gas

Argon
8 electrons in outer shell
18 protons in nucleus
Unreactive gas

▽ A group (column) of the table
The elements in group 1 all have one electron in their outermost shell, making most of them very similar in their properties: each of the elements shown here is a soft, highly reactive metal. Working down the group, the size of the atoms steadily increases, as with each element an extra electron shell has been added.

△ A period (row) of the table
The elements in period 3 have three electron shells. The number of electrons in the outer shell increases left to right, causing a change in properties. The atoms also get smaller, since at each step there is an extra proton in the nucleus, making it more positively charged so it pulls the negatively charged electrons in with increasing force.

DMITRI MENDELEEV

Born in Siberia, Russia, Mendeleev was Professor of Chemistry at the University of St. Petersburg. When formulating the first periodic table in 1869 (see pp.230–31), he left spaces for three undiscovered elements, whose properties he predicted. By 1888 these three elements—gallium (Ga), scandium (Sc), and germanium (Ge)—had all been discovered, with exactly the properties Mendeleev had predicted. Mendeleev also helped to develop Russia's oil and vodka industries, and introduced the metric system there. In 1955 the synthetic element mendelevium (Md) was named in his honor.

Lithium
3 electrons in two shells

| 3
Li
Lithium
7 |

Sodium
11 electrons in three shells

| 11
Na
Sodium
23 |

Potassium
19 electrons in four shells

| 19
K
Potassium
39 |

Rubidium
37 electrons in five shells

| 37
Rb
Rubidium
85 |

Cesium
55 electrons in six shells

| 55
Cs
Cesium
133 |

(hydrogen) or two (helium). The second period consists of the next eight elements, lithium (Li) to neon (Ne). Progressing through these, electrons are slotted into a second electron shell until it is full. The third period also contains eight elements and represents the sequential filling of a third shell (see above). Subsequently the makeup of each period becomes more complex, as extra electrons are slotted into new shells before previous ones are completely full. Most elements in the block called the transition metals possess an incompletely filled shell adjacent to their outermost shell.

Across each period is a progressive change in the physical and chemical properties of the elements, from reactive metallic elements on the left, through less reactive metals to reactive nonmetals, and finally on the far right, a barely reactive, or "noble," gas. The main factor causing this progressive change is the number of electrons in the outermost shells of the elements' atoms. This affects the atoms' tendency to lose or gain electrons in chemical reactions (see pp.286–87), as well as properties such as melting points and electrical conductivity (see pp.164–65). Another trend across each period is a progressive decrease in size of atoms.

Uniform groups
The groups are the columns of the periodic table, which are collections of elements that resemble each other in their physical and chemical properties because they contain the same number of electrons in their outermost shells (see left). The most uniform are groups 1, 2, 17, and 18.

Group 1 (excluding hydrogen) and group 2 contain reactive metals, which easily give up the one or two electrons in their outer shells during chemical reactions. Group 18 consists of noble gases. These have full electron shells and are nearly unreactive since they have little tendency to lose, gain, or share electrons. Group 17 is the halogens—highly reactive nonmetals that tend to gain or share electrons in reactions.

Less uniform groups
Outside these four groups there is less uniformity, the elements falling instead into similar blocks and categories. An example is the metalloids (semimetals), including silicon (Si) and germanium (Ge), which are widely used as semiconductors (see pp.164–65). In the middle of the table is the large block of transition metals, which includes gold (Au), copper (Cu), and iron (Fe). The category classified as "other nonmetals" contains many critical elements, including hydrogen—the most common element—as well as carbon (C), nitrogen (N), and oxygen (O), which along with hydrogen and a little phosphorus (P) and sulfur (S) form 99 percent of all living matter.

AFTER

Over the past 70 years a number of new elements have been synthesized.

EXTENDED PERIODIC TABLE
In 1969 an **extended periodic table** was suggested by the American scientist **Glenn Seaborg**. It was a logical extension of the principles behind the standard modern periodic table to include possible undiscovered elements up to an element with **atomic number 218**. Chemists believe it may be possible to synthesize relatively stable elements—"islands of stability" in a sea of unstable elements—far heavier than any made so far.

LATEST ELEMENTS
Elements 113, 115, 117, and 118 were formally recognized by the **International Union of Pure and Allied Chemistry** in 2016. The elements were later named as nihonium, moscovium, tennessine, and oganesson. Scientists are now searching for elements 119 and 120.

SEE ALSO »
pp.234–35 CHEMICAL REACTIONS
pp.286–87 STRUCTURE OF THE ATOM
pp.288–89 CHEMICAL BONDS

« BEFORE

The nature of chemical reactions was revealed as early chemists established that the elements combined together in fixed ways.

THE BIRTH OF MODERN CHEMISTRY
The foundation for understanding chemical reactions lies with the work of those such as the French chemist **Antoine Lavoisier**. In 1789 he surveyed the elements known at the time and established that **matter cannot be created or destroyed «** 138–39, which was important in showing that chemical reactions involve the **rearrangement** of chemical components.

PROUST'S LAW
In 1797 French chemist **Joseph Proust** proposed his **Law of Definite Proportions**, which stated that chemical elements combine **in simple ratios** to form **compounds**.

JOSEPH PROUST

« SEE ALSO
pp.52–53 ALCHEMY
pp.140–43 STATES OF MATTER

Chemical Reactions

The gradual rusting of iron and the explosive response of gunpowder show that chemical reactions can take years or be over in seconds. At the heart of any chemical reaction is a substance (or mixture of substances) that changes its character so that, by the end of the process, a completely new chemical substance is formed.

Substances change their form for many reasons. In some cases, there is a change in state (see pp.140–43) but the chemical identity stays the same, such as when ice melts into water. Chemical reactions, however, involve a complete overhaul of chemical identity. In 1803 English chemist John Dalton (see p.138) showed that when chemicals react, they do so in fixed proportions because their particles (atoms) come together in set combinations—an idea later developed into Atomic Theory (see pp.286–87).

From reactants to products
The identity of a chemical substance is determined not only by the elements it contains, but also the precise way the atoms of these elements are bonded together (see pp.288–89). Even the most subtle alterations in the arrangement of atoms can radically alter the properties of a substance, and therefore its identity. Chemical reactions can involve small shifts in position or complete rearrangements. The bonds that hold the atoms together in the original structure must be broken apart and new ones formed as the initial reactants are converted into products.

Types of chemical reaction
In the 19th century scientists identified most of the types of chemical reactions recognized today. At this time Swedish chemist Svante Arrhenius (see below) also found that reactions in solution could involve charged particles called ions (see pp.288–89), as well as atoms and molecules.

Some reactions involve combining small particles to make bigger ones—synthesis—while others break particles down—decomposition. Sometimes there is displacement: an atom, or

Endothermic reaction
This reaction is endothermic because it takes in some of the energy from the heat applied. Heating blue crystals of hydrated copper sulfate causes them to release water, forming white anhydrous copper sulfate.

> **REACTANT** A chemical that undergoes a change during the process of a chemical reaction.

> **PRODUCT** A chemical that is formed from the change during the process of a chemical reaction.

group of atoms, is ousted from its position by more reactive ones. There are also redox reactions, which involve the transfer of electrons (see pp.286–87), the negative parts of atoms. Some elements, such as oxygen, have atoms that pull electrons away from other particles. The particle that loses electrons undergoes a process known as oxidation; the particle that gains electrons undergoes reduction.

Chemical reactions can also be characterized in terms of energy changes. All substances contain chemical (potential) energy. If there is more chemical energy in the reactants than the products, then this is given off as heat during the reaction, making it exothermic. Heat-giving reactions are the basis of fuels, such as gasoline. If products contain more energy than reactants, then energy must be taken in by the reaction, making it endothermic. In 1840 Swiss-Russian chemist Germain Hess found that the amounts of energy released or taken in

SCIENTIST (1859–1927)

SVANTE AUGUST ARRHENIUS

Svante Arrhenius initially studied physics at the University of Uppsala in his native Sweden, but later turned to chemistry, where he first made important studies on electrolysis (the use of electricity to drive chemical reactions). He became Professor of Physics at the University of Stockholm, and later helped to set up the Nobel Prizes. After 1900 he made contributions to biology, astronomy, and geology. He supported a theory of panspermia (that life originated by spores carried from outer space), and devised an early version of a theory that explained the greenhouse effect.

Changing chemical bonds
During this chemical reaction of combustion, a molecule of methane reacts with two molecules of oxygen to form carbon dioxide and water; the bonds between atoms must be broken and new bonds formed.

> **"Chemistry:** that most **excellent child** of **intellect** and **art."**
> CYRIL NORMAN HINSHELWOOD, BRITISH CHEMIST (1897–1967)

Oxygen molecule
Hydrogen atom
Carbon atom
Methane molecule

BEFORE COMBUSTION

1 Molecules of oxygen and methane consist of atoms linked by strong covalent bonds (where electrons are shared). The bonds are double in oxygen and single in methane.

BONDS BREAK APART

2 Combustion causes the oxygen and methane molecules to collide and react chemically. The hydrogen atoms are torn off sequentially.

Water molecule

3 Two hydrogen atoms combine with oxygen to create water, while carbon forms double bonds with oxygen to create carbon dioxide.

Carbon dioxide molecule

AFTER COMBUSTION

Exothermic reaction
Potassium metal added to water reacts vigorously, giving off a great deal of heat, typical of an exothermic reaction. Hydrogen gas is released and a solution of the alkali potassium hydroxide results.

Types of chemical reaction

Chemical reactions can be classified according to the way in which the particles behave when their atoms are rearranged as a result of the reaction.

Hydrogen molecule made up of two atoms

Ammonia made up of three hydrogen atoms and one nitrogen atom

Nitrogen molecule with a triple bond

Synthesis

Commercial production of ammonia—known as the Haber Process (see p.240)—involves a reaction between smaller molecules of hydrogen and nitrogen.

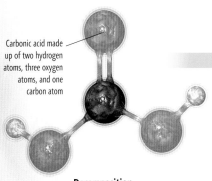

Water molecule

Carbonic acid made up of two hydrogen atoms, three oxygen atoms, and one carbon atom

Carbon dioxide made up of two oxygen atoms and one carbon atom

Decomposition

Carbonic acid is a very weak acid and spontaneously breaks down into carbon dioxide and water when warmed in water.

AFTER

by reactions was determined only by the reactants and products and not on how the reaction takes place—this had important implications in industry.

The role of electrons

The behavior of all chemical reactions depends on electron transfer, and redox reactions are explicitly defined by this process. Electrons form the cloud of negative charge around every atom, but the nature of this cloud—its size and organization—determines how an atom interacts with others. The atoms of some elements can engage by sharing electrons in covalent bonds (if they are both nonmetals), while others can only form ionic bonds. When a chemical reaction occurs, there is a redistribution of electrons as new bonds are made and old ones destroyed (see pp.288–89).

The electron configuration of an element's atom also determines how it engages with other elements. Some substances react well together, and when they do so the new bonding patterns they form are very restrictive, making the products formed very predictable. This is the case when methane and oxygen are mixed together and heated (see left).

In the 20th century the behavior of elements during chemical reactions was explained in terms of the way electrons allowed the formation of bonds.

NATURE OF THE CHEMICAL BOND

In 1939 American chemist **Linus Pauling** published a book on chemical bonding, which explained the difference between **ionic** (electron exchange) and **covalent** (electron sharing) bonding—and so established the basis of the modern understanding of chemical reactions. He was awarded the **Nobel Prize** for his work.

MOLECULAR ORBITALS

Chemical bonds involve **interaction between electrons** of atoms. In 1965 American chemists **Robert Woodford** and **Roald Hoffmann** used mathematics to describe the positions of **electrons** in orbitals around **molecules**, which helped further understanding of changes in the position of atoms during chemical reactions.

SEE ALSO ≫
pp.236–37 SPEEDING UP REACTIONS
pp.238–39 ACIDS AND BASES
pp.286–87 STRUCTURE OF THE ATOM
pp.288–89 CHEMICAL BONDS

BEFORE

Understanding of rates of chemical reactions relied on advances that encouraged chemists to think in terms of particles.

HIGH-PRESSURE GASES

In 1662 Irish polymath **Robert Boyle** quantified the way that **decreasing the volume of a gas increased its pressure**. In modern terms, he had noted that the particles of a **high-pressure gas** are brought close together.

UNDERSTANDING CHEMICAL PARTICLES

In 1803 English chemist **John Dalton** showed that when chemicals **react**, they do so in **fixed proportions** because of their particles **《 234**, heralding the arrival of **atomic theory**, which laid the foundations for **collision theory** to be developed a century later.

《 SEE ALSO
pp.100–01 THE BEHAVIOR OF GASES
pp.140–43 STATES OF MATTER
pp.234–35 CHEMICAL REACTIONS

Some chemical reactions go faster than others because the particles interact more readily with each other: for example, combustion of methane is faster than rusting of iron. The speed of a chemical reaction is determined in complex ways, and the rate of any type of reaction can vary greatly depending on prevailing conditions.

In the latter half of the 19th century chemists, such as Jacobus van't Hoff in Holland, were building theories to account for these observations, and laying the foundation of physical chemistry today.

There are various ways of measuring the speed of a chemical reaction: in some cases it is easier to measure how much reactant (chemical undergoing change) is used up in a period of time; in others, it is easier to measure the rate at which the product (chemical produced by change) is made.

Reaction processes

All chemical particles are more or less dynamic in their behavior: those in solids vibrate, while those in liquids and gases move around. For a chemical reaction to occur, particles must collide

effect makes them collide more frequently and with greater force, again increasing the rate of reaction. Light levels have a similar effect on particles that can absorb light energy. In 1889 Swedish chemist Svante Arrhenius (see p.234) proposed that there was a minimum amount of energy that a particle must have in order to react—the activation energy. When temperature or light increases, more particles reach this activation energy and so react when they collide. In 1900 Henry Le Chatelier (see below) established his law of equilibrium:

Speeding up Reactions

Although some chemical reactions are naturally more vigorous than others, all chemical reactions can be made to go faster or slower by changing conditions. It was an appreciation that chemical substances are made up of particles—atoms and molecules—that helped scientists understand how this happens.

Factors affecting reactions

Chemical particles react when they collide with each other with enough energy, and any factor that increases the chance of this happening will increase the rate of the chemical reaction.

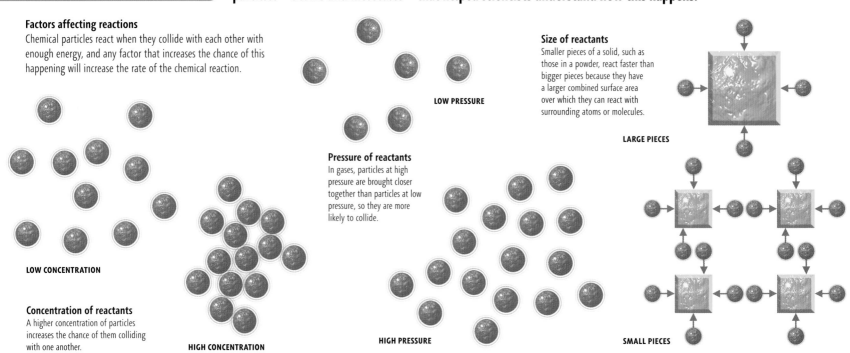

LOW PRESSURE

Size of reactants
Smaller pieces of a solid, such as those in a powder, react faster than bigger pieces because they have a larger combined surface area over which they can react with surrounding atoms or molecules.

LARGE PIECES

LOW CONCENTRATION

Concentration of reactants
A higher concentration of particles increases the chance of them colliding with one another.

HIGH CONCENTRATION

Pressure of reactants
In gases, particles at high pressure are brought closer together than particles at low pressure, so they are more likely to collide.

HIGH PRESSURE

SMALL PIECES

with one another in such a way that they can engage together in bonding (see pp.288–89). Just how many of them succeed in doing this over time determines the speed of the reaction.

If there is a high concentration of particles, they are more likely to collide, so the reaction is faster than with particles at low concentration. Gases at high pressure will give rise to faster reactions than those at low pressure, and solids with larger surface areas will react more quickly than those with smaller surface areas.

Temperature also has an effect on reactions. In 1877 Austrian physicist Ludwig Boltzmann explained that as temperature rises, particles become more energetic and move faster: the

reversible reactions are sometimes controllable; if a reversible reaction results in a decrease in pressure, then increasing the pressure of the starting materials will push the reaction forward. In the 20th century, the German chemist Max Trautz and then, independently, the British chemist William Lewis brought together ways in which particle behavior affects reactions in what became known as collision theory.

Catalysts

By the time collision theory was accepted, Latvian chemist Wilhelm Ostwald was laying the foundation of another principle affecting reaction rates. He found that the presence of a metal—platinum with rhodium— could increase the rate at which nitric acid was produced from ammonia. The metal itself remained unaffected. Half a century earlier Swedish chemist Jöns Berzelius had coined a term for this: catalysis. Catalysts often speed up reactions by bringing particles closer together, effectively lowering the activation energy in doing so.

The action of catalysts can speed up reactions that would otherwise take millions of years. Biological catalysts are known as enzymes and these are responsible for most of the chemical processes such as digestion (pp.216–17) that take place in the body. Catalysts have also had an important role in facilitating processes for industry, such as the Haber–Bosch process (see below).

Effect of acid rain on stone
Acidified by industrial pollutants in the atmosphere, rain dissolves the detail on limestone and marble statues, albeit at a slow rate.

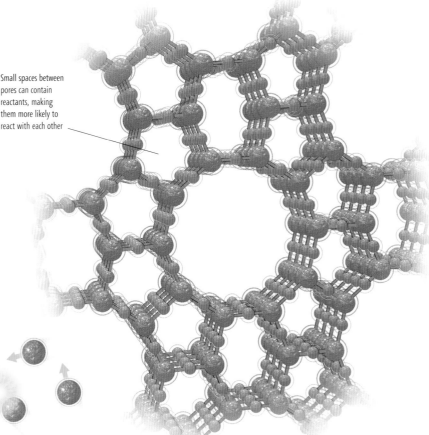

Small spaces between pores can contain reactants, making them more likely to react with each other

Zeolite catalyst
Zeolite is a mineral with a very porous structure, and is used widely as a catalyst in the petrochemical industry.

Little light energy absorbed by particles, so they are slow to react

LOW LIGHT LEVELS

Much light energy absorbed by particles, so they are quick to react

HIGH LIGHT LEVELS

Light levels
Some particles can absorb light energy, making them more reactive at high light levels than they are at low light levels.

Low-energy particles move slowly, so collide infrequently with little force

LOW TEMPERATURE

High-energy particles move quickly, so collide frequently with force

HIGH TEMPERATURE

Temperature of reactants
As temperature increases, particles move faster so collide more frequently than they do at low temperatures.

"Wherever we look, the **work** of the **chemist** has **raised** the **level** of our **civilization**."
JOHN CALVIN COOLIDGE, 30TH US PRESIDENT, 1924

AFTER »

The beginnings of atomic theory developed in the 19th century were supported by new discoveries from physics in the 1900s.

ATOMIC THEORY VINDICATED
In 1905 German theoretical physicist **Albert Einstein** demonstrated how **Brownian motion**—the visible movement of particles in gas and liquid—provided a **theoretical proof of atomic theory**, which underpinned the behavior of chemical reactions.

RISE OF CHEMICAL THERMODYNAMICS
The tendency of chemicals to **react** depends on the amount of **energy** they contain, with some chemical particles being more "energetic" than others. In 1923 American chemists **Gilbert Lewis** and **Merle Randall** determined the **energy content** of many chemical substances, which enabled scientists to calculate the energy changes that take place during reactions.

SEE ALSO »
pp.286–87 STRUCTURE OF THE ATOM
pp.288–89 CHEMICAL BONDS

IN PRACTICE

HABER–BOSCH PROCESS

The industrial production of ammonia was made possible by the process invented by German chemists Fritz Haber and Carl Bosch in 1909, using principles of equilibrium developed by French-Italian chemist Henry Le Chatelier (see opposite). Gaseous nitrogen and hydrogen are mixed together at a pressure of 200 atmospheres and at high temperatures—between 572°F (300°C) and 1,022°F (550°C). This encourages them to react with each other to make ammonia over an iron catalyst. Any nitrogen and hydrogen that does not react is recycled; this results in around 98 percent conversion. The Haber–Bosch process was an important breakthrough because it was a way of producing ammonia on an industrial scale for use in nitrogenous fertilizer for agriculture, synthetic fibres, and commercial explosives. It was also a way of harnessing elemental nitrogen, which is otherwise very unreactive (see pp.278–79).

‹‹ BEFORE

By the end of the Middle Ages scholars had succeeded in isolating some of the most important acids used today.

AQUA REGIA AND THE PHILOSOPHER'S STONE
The study of acids began when the 8th-century Persian alchemist **Geber** isolated **hydrochloric acid** by mixing common salt (sodium chloride) with **sulfuric acid**. Geber also found that **gold**, which is not affected by any other acids, would **dissolve on contact** with a mixture of hydrochloric acid and nitric acid. This mixture, which became known as **aqua regia** ("royal water"), was associated with the **philosopher's stone**, a mythical tool that alchemists believed could **turn base metals into gold**.

PURE SULFURIC ACID
The 9th-century Persian physician **Rhazes** produced **pure sulfuric acid** by heating a combination of metal sulfates. The **vapors of water** and **sulfur trioxide** reacted to form the acid in much the same way acid rain forms.

‹‹ SEE ALSO
pp.234–35 CHEMICAL REACTIONS
pp.236–37 SPEEDING UP REACTIONS

Acids and Bases

That some acids and certain bases are dangerously corrosive was known—and even exploited—by the scholars of antiquity, but it would take centuries of painstaking scientific scrutiny before the exact chemical nature of acids and bases were finally revealed by a Swedish chemist.

The sharp, stinging properties of lemon juice typify a class of natural substances that are familiar in everyday life. Organic, plant-derived acids are prominent in vinegar and sour fruit, while inorganic ones—some of which are extremely corrosive—are mainly of mineral origin.

Acids can be tempered by a class of substances called bases. However, bases can also be corrosive in high concentrations.

Properties of acids and bases
Chemically, acids and bases are characterized by what happens when they are mixed, a feature recognized by French chemist Guillaume-François Rouelle in 1754. When mixed, an acid and a base (alkali) dissolved in water

Hydrogen ion

Hydroxyl ion

WATER MOLECULE

Water splitting
Pure water is neutral because it can split (dissociate) into equal numbers of positively charged hydrogen ions (H+) and negatively charged hydroxyl ions (OH-). Acids have more hydrogen ions, and bases more hydroxyl ions.

No matter which acids or bases are mixed, the chemical reaction also forms water, and it is this that gives the biggest clue as to the true chemical identities of acids and bases.

Chemical identities
In the 19th century various chemists including the German chemist Justus von Liebig suggested correctly that hydrogen was the key element in all acids. Then in 1884 the Swedish chemist Svante Arrhenius (see p.234) provided the first precise definitions of acids and bases.

neutralize each other, forming a salt. Salts come in millions of different forms, depending on the acids and bases mixed. Common table salt (sodium chloride), for example, is formed by mixing hydrochloric acid and the base sodium hydroxide.

pH0   pH1   pH2   pH3   pH4   pH5   pH6

ACIDS

The pH scale
Acids and bases are rated on the pH scale, which ranges from pH0 (most acid) to pH14 (most basic), with pH7 being neutral. Each value produces a characteristic color in reaction to universal indicator solution.

Battery acid
Sulfuric acid, used in car batteries, is very corrosive.

Gastric acid
Hydrochloric acid made in the stomach helps digestion.

Vinegar
All of the many types of vinegar contain ethanoic (acetic) acid.

Orange juice
The citric acid in oranges and other citrus fruits is a weak acid found naturally in small amounts in all living things.

Tomato juice
The acidity of tomato juice is due to a mixture of weak acids, including citric acid, malic acid, and ascorbic acid (vitamin C).

Tea
The brown color of tea is caused by various chemicals that include very weak tannic acid.

Distilled water
The carbon dioxide dissolved in distilled water from the air makes the water weakly acidic.

"…acids and bases are substances that are capable of splitting off or taking up hydrogen ions, respectively."
JOHANNES BRØNSTED, 1923

◁ **Acid rain**
Sulfur dioxide gas from industry reacts in the atmosphere to form sulfuric acid. Acidified rain falls, damaging trees. Clean air laws in most countries introduced from the 1980s have reduced the damage.

Reducing soil acidity
Many crops grow best in soil with a pH of about 6.5. The addition of limestone to neutralize overly acidic soils can improve plant growth and encourage nutrient-releasing bacteria to thrive.

Testing for acids and bases

Litmus paper is used to determine whether a liquid is either an acid or a base. Blue paper turns red (as shown to the right) if the liquid is acidic, and red paper turns blue if the liquid is basic.

The significance of an acid's hydrogen is apparent when it forms water at neutralization. A molecule of water contains two hydrogen atoms and one oxygen atom. When an acid reacts with a base, the acid provides a positively charged hydrogen atom—a hydrogen ion (H⁺), or proton—and the base provides the remaining part of the water molecule in the form of a negatively charged hydroxyl ion: OH⁻.

According to Arrhenius, then, it was the provision of hydrogen and hydroxyl ions that defined acids and bases respectively. Stronger acids and bases release higher concentrations of their respective hydrogen and hydroxyl ions; it is the reaction of these with other materials that explains the corrosive actions of these chemicals. However, Arrhenius's definitions are applicable only for acids and bases that dissolve in

water, which not all do. Therefore in 1923 the Danish chemist Johannes Brønsted and the English chemist Martin Lowry each independently suggested that the definitions of acids and bases be changed. They each defined an acid as a chemical that donates hydrogen ions (protons), and a base as one that accepts them, so adding to Arrhenius's definition.

Measuring acidity

Following Arrhenius, scientists realized that acidity was so important in the natural world that it should be quantified. German chemist Hans Friedenthal suggested a system classifying the concentration of

ACID A chemical substance that releases hydrogen ions (protons). The stronger the acid, the higher the concentration of hydrogen ions released.

BASE A chemical substance that can accept hydrogen ions released by acids, often reacting with them to form water.

BIOCHEMIST (1868–1939)

SØREN SØRENSEN

The son of a Danish farmer, Sørensen entered the University of Copenhagen to study medicine, switching later to chemistry. He invented the pH scale while director of the Chemistry Department at the Carlsberg Laboratory in Copenhagen (a facility supported by the brewing company). He also developed an electrode to measure pH, as well as indicator solutions; both methods are used today. Sørensen won many awards and honors for his contributions to science, which also included medical research into diabetes and epilepsy.

hydrogen ions, but there was an enormous variation over the range—the most basic solutions had a hydrogen ion concentration that was around a hundred trillion times lower than the most acidic ones.

Then in 1909 a Danish chemist, Søren Sørensen, came up with a much easier, logarithmic scale that ranged from 0 (most acid) through 7 (neutral) to 14 (most basic). Each step up this simple scale represents a hydrogen ion

concentration 10 times lower than the previous one. Sørensen's scale became known as the pH scale, which stands for *pondus hydrogenii*: "potential of hydrogen." At neutral pH7 the concentration of hydrogen ions is equal to the concentration of hydroxyl ions.

Universal indicator solution is now widely used to test pH, passing from red when most acid through orange, yellow, green (neutral), and blue to purple when most basic (below).

BASES

Cow's milk While cow's milk is neutral when it is fresh, it becomes weakly acidic when bacteria in it produce lactic acid.

Seawater While mainly a solution of sodium chloride, seawater contains small quantities of other weakly basic chemicals.

Toothpaste Some of the chemicals in toothpaste are weak bases, such as sodium bicarbonate.

Antacid tablets When dissolved in water, antacid tablets produce a basic solution that raises the pH of the digestive juices in the stomach.

Ammonia Dissolved in water, ammonia forms a basic solution called ammonium hydroxide.

Household bleach The strong base sodium hypochlorite is the agent in household bleach.

Cement Made by heating limestone with clay, cement is a mix of highly basic chemicals.

Drain cleaner Most drain cleaners contain caustic soda—corrosive, concentrated sodium hydroxide.

AFTER »

Highly corrosive acids have a variety of industrial uses from removing corrosion from iron and steel to etching glass.

POLISHING SILICON CHIPS
Nitric acid, sulfuric acid, hydrochloric acid, and hydrofluoric acid are all used in their purest forms to remove silicon dioxide from the surfaces of the wafers of silicon crystal that are key components of **computer microchips**.

PICKLING AND ETCHING
Sulfuric acid, hydrofluoric acid, and hydrochloric acid are all also commonly used to remove rust and iron oxide from sheets of iron and steel—a process called pickling. **Hydrofluoric acid**, which **dissolves glass**, is also commonly used to etch or frost the surface of glass.

SEE ALSO »
pp.240–41 MASS PRODUCTION OF CHEMICALS

BEFORE

The practical applications of chemistry began with medieval Islamic scholars, who transformed alchemy into science.

ORIGINS IN ALCHEMY

In the 9th century the Islamic scholar **Geber** managed to **synthesize numerous chemical substances**, including mineral (inorganic) acids and some elements. He was able to use many of these techniques to make perfumes and drugs. A century later the alchemist **Rhazes 《 52–53** developed **methods of distillation** and also devised a way of **producing sulfuric acid**.

ALUM AND INDUSTRIAL CHEMISTRY

Alum generally refers to a variety of aluminum-based substances known since antiquity. It has been used in **numerous industrial processes**, including the production of cosmetics and as a **dye-fixer**. Alum can be extracted from clay or shale, and in Yorkshire, England, alum manufacture was **Britain's first chemical industry**.

《 SEE ALSO

pp.52–53 ALCHEMY
pp.56–57 GUNPOWDER AND FIRE WEAPONS
pp.150–51 ORGANIC CHEMISTRY

Mass Production of Chemicals

The products of chemistry have wide-ranging applications in manufacturing and design, agriculture, medicine, engineering, and warfare. Sometimes useful products can be extracted from plants or other natural sources, but the principles of chemistry have also paved the way for synthesis in the laboratory and the factory.

Early chemists developed ways of purifying key ingredients from natural mixtures, but if the extraction and purification was complicated, they sometimes tried to synthesize them. This could simply be a matter of discovering the right conditions for a reaction to happen—such as the conversion of nitrogen and hydrogen into ammonia—and once that was done synthesis could often be achieved in bulk. But in other cases, such as with the antibiotic penicillin (see pp.310–11), the chemical target was more elusive, and required several stages before the product could be isolated. Even so, a chemist who could devise a practical way of making a much-demanded product more cheaply than extracting it from natural sources could reap great financial rewards.

The color purple

Occasionally, synthesis was discovered by accident. In 1856 William Henry Perkin was an 18-year-old student at London's Royal College of Chemistry. Inspired by his mentor, August van Hofmann, he tried to produce quinine in his homemade laboratory. This was an expensive natural drug, much in demand for treating malaria, and a substance called aniline was a likely precursor. But instead of making quinine, he ended up with a purple dye. Perkin named it mauveine, and it was the first organic dye to be made artificially.

At the time dyes were costly to extract from natural products, and Perkin quickly recognized the value of his discovery. Against the encouraging background of the Industrial Revolution, he patented his dye and set up factories for its commercial production. Perkin grew rich on his lucky discovery.

Farming revolution

Perkin sold his factories at the start of the 20th century, just as the chemical dye industry was starting to expand in Germany. Here, too, a different breakthrough was taking place. Fritz Haber had succeeded in producing ammonia at relatively low temperatures from hydrogen and nitrogen using an iron catalyst. There was great demand for ammonia to make artificial fertilizers: until then, much of the nitrogen used to promote crop growth was supplied in the form of nitrates mined in places such as South America.

As populations boomed during the Industrial Revolution, demand for food spiraled. With the onset of World War I the shortage of nitrogen became a serious crisis in Germany, because only the Allies had access to South American nitrogen. Carl Bosch, an

The structure of mauveine

The molecular structure of mauveine was not known with certainty until 1994. Mauveine dye is actually a mixture of four related aromatic compounds, which differ only in the number and placement of methyl groups (groupings of three hydrogen atoms and one carbon atom).

CHEMIST (1838–1907)

WILLIAM HENRY PERKIN

Perkin was a British chemist with an entrepreneurial spirit. The family business he built on the discovery of the dye mauveine established his reputation and brought him great wealth. Perkin was one of the first chemists to make his scientific discoveries commercially viable. But he continued with his research too, discovering new dyes and successfully synthesizing perfumes. In 1906—the year of his knighthood—Perkin visited America, where he was additionally honored with the first Perkin Medal, which subsequently became the country's highest award for industrial chemistry.

◁ **Mauveine dye**
One of the earliest samples of the first synthetic dye, mauveine, paved the way for a new dye industry. Until then, the natural source of purple dye was the mucus of a marine snail from the Mediterranean region, but its extraction made it expensive.

Rings made up of six carbon atoms

Carbon atom

Nitrogen atom

Hydrogen atom

Amino group made up of two hydrogen atoms and one nitrogen

Methyl group made up of three hydrogen atoms and one carbon atom

The successful synthesis of chemical products was advanced by a greater understanding of their structures and chemical reactions.

EXTRACTING DRUGS FROM PLANTS

In the mid-20th century, American chemist Percy Julian developed techniques for **large-scale extraction of steroid medicines from plants**, making them a more affordable treatment option than steroids extracted from animals.

PERCY JULIAN

CATALYZING CHEMICAL INDUSTRY

Cracking is a process that **splits hydrocarbons** from distilled crude oil into lighter products, such as **petroleum**. In 1937 the French engineer **Eugene Houdry** devised a better way of cracking petroleum that laid the foundations for modern oil refineries.

INVENTION

DYNAMITE

Nobel's dynamite (patented 1867) made nitroglycerine safer to handle. The liquid explosive is mixed with an absorbent material, such as kieselguhr (a porous type of clay), and the mixture formed into a stick, which is then wrapped in paper. It is detonated via an electrical cable. Dynamite is still used for mining and construction, but the military use more stable explosives.

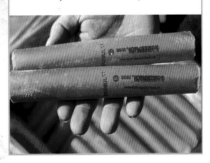

> "I intend to leave after my death a large **fund** for the **promotion** of the peace idea..."
> ALFRED NOBEL

Chemical industry during World War I
A female employee working at the Brunner Mond and Company chemical works in Northwich, Manchester, England, feeds an ash furnace in September 1918.

industrial chemist, developed Haber's method into a process that could deliver ammonia in vast amounts. Today around a third of humanity is sustained by crops grown with fertilizer made using ammonia obtained by the Haber-Bosch process.

Explosive chemistry
In World War I much of the ammonia from the Haber-Bosch process was used for making explosives. An explosive is a substance that reacts to produce a huge volume of hot gas. The reaction spreads

very quickly through the material. In gunpowder the reaction is started by a spark, but in other explosives it is initiated by a shock wave. One such explosive, the highly unstable liquid nitroglycerine, was discovered in 1847 by an Italian chemist, Ascanio Sobrero. A student of his, Alfred Nobel, made it safer to handle after a nitroglycerine explosion killed his younger brother Emil in 1864. He patented his new product as Dynamite, and made a fortune. Nobel left much of his wealth to fund the prizes that carry his name.

Explosives factory
Because of the instability of the materials used in the production of nitroglycerine, this processing plant near Cengio, Italy, used earth embankments to separate the buildings and to surround the whole complex.

The **Spread** of **Disease**

The recent coronavirus pandemic starkly demonstrates how infectious disease can devastate entire populations. Since antiquity, humans have sought to understand how disease spreads and why it affects some people but not others. Scientists call this branch of study epidemiology, and its findings have had far-reaching implications for public health.

The Broad Street Epidemic
This caricature shows people around a water pump contaminated with cholera. John Snow traced an outbreak to one such pump in Broad Street in London by plotting the occurrence of the disease in buildings in the neighborhood.

PHYSICIAN (1818–1865)

IGNAZ SEMMELWEIS

The Hungarian physician Ignaz Semmelweis successfully used one of the first hygiene regimens to reduce the spread of disease. In 1847 he was working at a maternity clinic in Vienna, where he instructed staff to wash their hands in disinfectant between performing autopsies and attending to patients. As a result patient deaths due to puerperal fever, caused by infections contracted during childbirth, dropped dramatically. Despite this, Semmelweis's theory was ridiculed and he was dismissed. He never recovered and died in a mental asylum.

As far back in history as ancient Roman times the spread of disease was reduced with the installation of the first advanced sewerage system, but scholars of the time did not understand why. This knowledge did not come until the 19th century and the advent of a new type of scientist—the epidemiologist.

It was during the 19th century that the vague notion, passed on from the ancient Greeks, that contagious diseases were caused by "noxious vapors" was finally replaced. This advance came in part through the work of scientists such as Louis Pasteur (see pp.244–45), and also the work of early epidemiologists such as William Budd, an English doctor. Budd championed the idea that disease was spread from person to person, rather than being caused by "bad air." His idea was based on observations of an outbreak of typhoid in his village in

Cholera bacterium
This color-enhanced image shows the *Vibrio* bacterium, which causes cholera. If ingested by humans, the bacterium causes diarrhea and rapid dehydration. Left untreated, cholera can kill in less than 24 hours.

Devon, in 1839, in which he noticed the disease spread quickly from sufferers to their caregivers. Today we know that contagious diseases are caused by microscopic organisms called pathogens that pass from person to person, carried in water, through the air, or transmitted during sexual contact.

Waterborne disease
William Budd continued to study the spread of typhoid and, in 1847, he made the suggestion that the disease was transmitted through infected water. He went on to make a similar case for cholera the following year. Cholera is an example of an epidemic—a disease that sweeps rapidly from place to place. In the 19th century, this intestinal disease was spreading through Europe, and the epidemic that reached London in 1854 claimed over 10,000 lives. It took a physician named John Snow to help stop it. Snow demonstrated a connection between incidence of the disease and sewage-contaminated drinking water when he traced an outbreak to a public water pump in Broad Street, London. He saw to it that the pump was put out of action and local cases plummeted. Snow's influence went on to improve public health in London and beyond, prompting many to consider him the "Father of Epidemiology."

Airborne disease
Another way in which disease spread was brought to the public's attention by the work of the German physician Robert Koch. On March 24, 1882 Koch announced that he had found the microbe responsible for tuberculosis, which had ravaging effects on the body and was responsible for nearly one-quarter of all deaths in early 19th-century England. Koch's discoveries revealed that this disease was transmitted by water droplets in the air, and could spread quickly in overcrowded slums through coughing, sneezing, and spitting. Personal hygiene became an important public concern and sufferers were quarantined.

◀◀ **BEFORE**

Before the understanding in the 19th century that contagious diseases consisted of live particles, popular belief held that disease was caused by bad gases.

ORIGINS OF THE MIASMA THEORY
The Miasma theory of **bad gases spreading disease** probably originated in Ancient Greece, where **Hippocrates ◀◀ 24–25** had already

distinguished between diseases that spread freely and those that are restricted to a particular region.

DISEASE SPREAD BY PARTICLES
In Persia, the physician **Avicenna** advocated **quarantine** in his medical encyclopedia, *The Canon of Medicine*, to reduce the spread of disease—suspecting **contamination by "bodily secretions."** However, **Ibn Khatima**, an Andalusian physician attributed the spread of **bubonic plague** in the 14th century to particles that could be carried on clothes.

PERSONAL HYGIENE
In 1546 **Girolamo Fracastoro**, an Italian physician, implied that disease-causing particles were living. On this basis he advocated a **regimen of personal hygiene**.

◀◀ SEE ALSO
pp.72–73 THE HUMAN BODY REVEALED

THE CANON OF MEDICINE

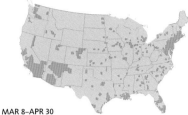

Protecting troops from influenza in 1918
In an attempt to prevent the spread of influenza among the armed forces during World War I, troops were issued with gauze masks.

▽ **Disease hot spots**
The maps here show the areas of the US most affected by the COVID-19 pandemic during the course of about four months. Identifying hot spots with localized increases in the number of patients affected helps governments understand the changing epidemiology of the disease and implement appropriate public health measures.

MAR 8–APR 30

MAY 1–31

JUN 1–JUL 15

MAR 8–JUL 15

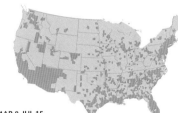

Another airborne disease, influenza, is caused by a microbe that is constantly mutating; this results in periodic epidemics, as people's natural defenses are caught unprepared. A particularly devastating strain of the virus caused a pandemic (an epidemic of continental or global scale) in 1918–19, aided by the movements of troops in World War I, and by sailors traveling on international trade routes.

New diseases

While some diseases can be contained by public hygiene measures or effective quarantine, others, such as syphilis, are spread by sexual contact. By the 20th century, many infections could be cured by drugs, such as antibiotics, (see pp.312–13) or even prevented by vaccination (see pp.248–49), but the continual emergence of new diseases has meant that research in these areas can never be entirely finished.

New strains of viruses and bacteria arise by genetic mutation, and some pathogens can even "jump species"—switching from animal to human hosts. The resulting animal-derived diseases, called zoonoses, are among the most devastating. Ebola, HIV, and coronaviruses all have zoonotic origins. COVID-19—a coronavirus strain first identified in 2019—may have originated in bats and has spread rapidly from its source to cause a global pandemic.

AFTER ››

Better understanding of the spread of disease in the 20th century led to more effective control measures.

MATHEMATICS AND SPREAD OF DISEASE
British physician **Ronald Ross** (1857–1932) applied mathematics to epidemiology. He developed **mathematical models** in order to study the **spread of malaria** in India and provided a method for calculating the rate of the spread of disease and controlling the economics of the disease (the material and work force required to track the epidemic). This approach enabled the global elimination of smallpox in 1977 258–59 ››.

EPIDEMIOLOGY OF CANCER
British epidemiologist **Richard Doll** pioneered methods that link disease to **environmental factors**. His research team was the first to confirm the link between smoking and lung cancer.

SEE ALSO ››
pp.248–49 IMMUNIZATION AND VACCINATION
pp.412–13 DISEASE CHALLENGES

BEFORE

The earliest references to bacteria and viruses were about their effects in causing disease or spoiling food.

HIDDEN LIFE

Early humans were aware of the effects of bacteria and viruses long before they could see them—they **spoiled foods** but also allowed production of **vinegar** and **cheese**. Early scholars (Arab Ibn Khatima in the 14th century and Fracastoro in the 16th century) proposed that epidemics of diseases were caused by **invisible "seedlike" particles**, but the truth of these claims was not realized until the 19th century.

≪ SEE ALSO
pp.242–43 SPREAD OF DISEASE

GRAM STAINING

Developed by Danish biologist Hans Gram in 1884, Gram staining identifies different types of bacteria: Gram-positive bacteria with thick walls stain purple and Gram-negative ones with thinner walls and an outer oily layer stain pink. Many Gram-negative bacteria are pathogenic (disease-causing) due to the presence of an outer oily layer.

Gram-positive   Gram-negative

Typical bacterium shapes

The shape of a bacterium depends on the cell-wall protein filaments inside the cell and is determined by its DNA. It affects mobility and the bacterium's ability to infect other organisms.

Rods
One of the most abundant and most studied forms of bacteria, *Escherichia coli* is a bacillus.

Spheres
This form of bacterium occurs singly, in pairs (*diplococci*), chains (*streptococci*), or in groups (*staphylococci*).

Corkscrews
A form of bacteria that has a flagellum, which helps them move in a twisting fashion.

Commas
This form of bacteria, including the cholera bacterium, is comma shaped due to the flagellum.

Despite their microscopic size, bacteria have a profound influence on the rest of life on Earth. They recycle resources and affect the atmosphere, but also cause diseases. Viruses cause disease too, by attacking living cells and sabotaging the machinery inside.

The smallest living things

A bacterium is a small cell with no nucleus—a prokaryotic organism. Bacteria come in a range of shapes, but their basic structure is quite uniform. They reproduce by binary fission—a bacterium splits into two cells, with each receiving a copy of the parent cells' genetic material (see pp.256–57). Bacteria can also join and exchange small rings of DNA called plasmids by way of conjugation—this explains the rapid spread of antibiotic resistance (see pp.412–13 and opposite).

The first person to see bacteria was probably the Dutch microbiologist Antony van Leeuwenhoek (see pp.94–95) in 1674. In 1828 German biologist Christian Ehrenberg named them after *bakterion*, Greek for "little stick." Later that century, another German, Ferdinand Cohn, observed other shapes, and classified them into rods, spheres, commas, and spirals, a system still used today (see below).

Good bacteria, bad bacteria

Bacteria are an essential part of the cycle of life because they feed on dead matter, decomposing it and recycling many of the chemical constituents. Bacteria themselves have astonishing chemical diversity and can perform many processes that are impossible in plants and animals. In the 1880s Russian microbiologist Sergei Winogradsky discovered that some bacteria get energy from minerals, much like plants get energy from the Sun (see pp.256–57). Instead of

LOUIS PASTEUR

Pasteur is one of the founding fathers of microbiology: his work on microorganisms led to pasteurization of milk and the first vaccines. Despite a modest background in working-class France, Pasteur became one of the country's greatest scientists. After his death, his remains were interred beneath the Institut Pasteur, named in his honor.

Bacteria and Viruses

A bacterium is the smallest living cell. As many as 1,000 bacteria could comfortably line up across just one period on this page and still have room to spare. Smaller still are viruses, infectious particles that can multiply only inside host cells of living bacteria, animals, or plants.

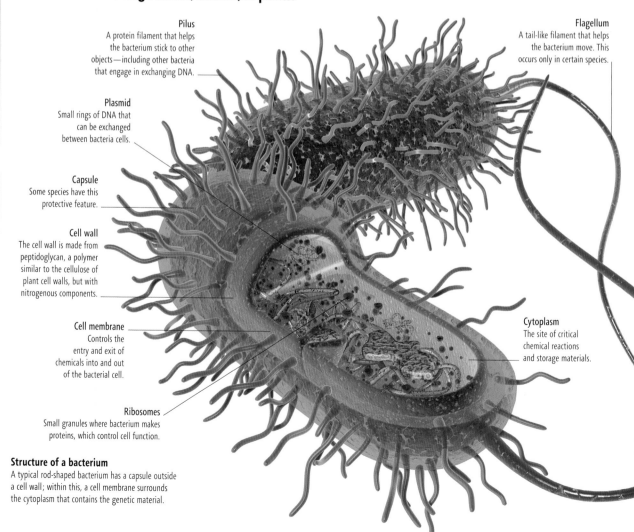

Pilus
A protein filament that helps the bacterium stick to other objects—including other bacteria that engage in exchanging DNA.

Plasmid
Small rings of DNA that can be exchanged between bacteria cells.

Capsule
Some species have this protective feature.

Cell wall
The cell wall is made from peptidoglycan, a polymer similar to the cellulose of plant cell walls, but with nitrogenous components.

Cell membrane
Controls the entry and exit of chemicals into and out of the bacterial cell.

Ribosomes
Small granules where bacterium makes proteins, which control cell function.

Flagellum
A tail-like filament that helps the bacterium move. This occurs only in certain species.

Cytoplasm
The site of critical chemical reactions and storage materials.

Structure of a bacterium
A typical rod-shaped bacterium has a capsule outside a cell wall; within this, a cell membrane surrounds the cytoplasm that contains the genetic material.

Spreading antibiotic resistance

Recent years have seen antibiotic resistance rise. It can spread by exchange of DNA between conjugating bacteria.

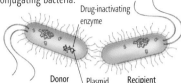

1 Duplicating plasmid Plasmids (rings of DNA) help bacteria make enzymes that enable the bacteria to resist drugs called antibiotics. Bacteria replicate these plasmids.

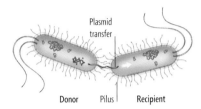

2 Plasmid transfer Once the donor and recipient cells join together, the plasmid copy is passed through a pilus to the recipient bacterium—a process known as conjugation.

3 Drug-resistant strains Recipient bacteria inherit a gene that makes new enzymes. This process produces populations of antibiotic-resistant bacteria.

7 BILLION is the number of bacteria that can live in a human mouth—approximately equivalent to the number of people living on the whole planet.

using light as an energy source, they use ammonia released from decomposers, changing it into nitrates, so making energy for themselves and nitrogen available to plants (see pp.278–79). However, some bacteria can be harmful not only to plants but also to humans and other animals, causing diseases such as tuberculosis and tetanus.

Discovery of viruses

Viruses differ from bacteria in that they are not cells. In fact, a virus is nothing more than a particle of genetic material surrounded by a protein casing (see right). In 1898 Dutch microbiologist Martinus Beijerinck first became aware of viruses while studying mosaic disease on tobacco plants. These tiny particles were finally seen by German physician Helmut Ruska and his team, in 1939, using the electron microscope that had been invented just eight years earlier.

Cycle of a virus

A virus is not made of cells, so has no cellular "machinery" to reproduce or perform metabolic processes on its own. It can only complete its cycle by invading and taking over a living host cell.

Communities of bacteria

The chemical versatility of bacteria is demonstrated by stromatolites—films of photosynthetic bacteria that bind rock particles into mounds. Examples like these, off the coast of Australia, were among the earliest life forms on Earth.

AFTER

Understanding the genetic systems of bacteria has opened up the possibility for genetic modification of microorganisms, which has important medical implications.

VACCINES FROM BACTERIA TOXINS
The harmful effects of disease-causing bacteria are caused by **poisonous substances called toxins**. Many toxins are neutralized by the body's immune system and have been used to develop **harmless forms** called **toxoids**, which are the **basis of vaccines** that protect against diseases such as tetanus and diphtheria. Tetanus and diphtheria vaccines were routinely combined with pertussis (whooping cough) vaccination from the 1940s.

SEE ALSO >>
pp.246–47 NATURAL DEFENSES
pp.248–49 IMMUNIZATION AND VACCINATION
pp.310–11 THE DISCOVERY OF PENICILLIN

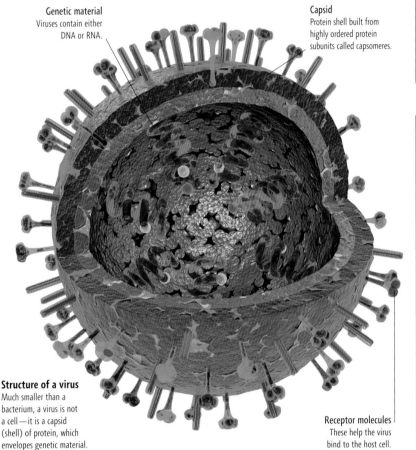

Genetic material
Viruses contain either DNA or RNA.

Capsid
Protein shell built from highly ordered protein subunits called capsomeres.

Structure of a virus
Much smaller than a bacterium, a virus is not a cell—it is a capsid (shell) of protein, which envelopes genetic material.

Receptor molecules
These help the virus bind to the host cell.

Typical virus shapes

There are many types of virus. The varying shape or make-up of viruses may assist in the replication cycle by helping penetrate a host cell or resist attack by the host defense system.

Icosahedral
Most animal viruses are roughly spherical, with at least 12 units making up the protein capsid.

Spiral-helical
Some viruses have capsid units arranged in a helix around the length of genetic material.

Complex
These viruses do not come in just one shape, but have variations to the capsid, such as a "tail."

Enveloped
These viruses coat themselves in host-cell membrane as a disguise from the host immune system.

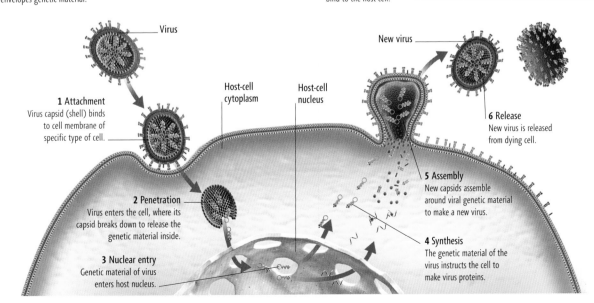

Virus

1 Attachment
Virus capsid (shell) binds to cell membrane of specific type of cell.

Host-cell cytoplasm

Host-cell nucleus

New virus

6 Release
New virus is released from dying cell.

5 Assembly
New capsids assemble around viral genetic material to make a new virus.

4 Synthesis
The genetic material of the virus instructs the cell to make virus proteins.

2 Penetration
Virus enters the cell, where its capsid breaks down to release the genetic material inside.

3 Nuclear entry
Genetic material of virus enters host nucleus.

The interior of the human body encourages the growth of many different types of microbes. Once inside, certain kinds of bacteria and viruses breed (see pp.244–45) and can cause harm. Such disease-causing organisms are known as pathogens. Their ubiquity means that invasion is inevitable, but the body puts up an impressive defense.

Barriers and fighters

Human skin forms a natural barrier to attackers: its outer layers are made of hardened, dead cells, which are lubricated by impervious oil. If this barrier becomes torn, a self-healing process automatically starts. A substance dissolved in the exposed blood, called fibrinogen—with the help of another blood component called platelets—rapidly solidifies into fibers, which are known as fibrin. These trap the precious blood cells within the body, slowing the bleeding and eventually stopping it entirely. Over the coming days, as the skin is repaired beneath the wound, the drying scab is pushed to the surface until it eventually falls away from the repaired skin.

Inside the body, a defending army of cells is mobilized to deal with invasions at the source of infection: phagocytes (pathogen-eating cells) attack microbes. Russian microbiologist Ilya Mechnikov discovered them in 1882 when he observed mobile cells in the larvae of starfishes. To put this idea to the test, he inserted small thorns into the larvae, and, the next day, he observed that the thorns were surrounded by mobile cells. However, a focused attack by phagocytes, no matter how effective, gives no long-term protection against a disease. Long-term immunity requires a much more sophisticated type of cell.

Understanding immunity

The body can distinguish its own cells from alien ones, such as bacteria, by recognizing chemical markers, known as antigens, on their surfaces. Any cell with an unfamiliar, "non-self," marker is targeted for attack by phagocytes.

> **PHAGOCYTES** White blood cells that protect the body by swallowing and digesting bacteria or harmful particles.
>
> **LYMPHOCYTES** White blood cells that can release antibodies to provide long-term immunity.

a substance dissolved in the blood, something that is undoubtedly secreted by blood cells.

Targeting specific invaders

In 1891 microbiologist Paul Ehrlich identified the protective substances in blood and called them antibodies. They are produced by white blood cells called lymphocytes, which are coated in antibody molecules with shapes that fit particular antigens. The result is that any invading microbe is likely to find itself stuck via their antigens to these lymphocytes. When this happens the lymphocytes are triggered to divide and release their surface antibodies, which circulate through the bloodstream, binding and disabling any microbes with the right-shaped antigens. This accounts for the observations of Shibasaburo and Behring that serum can enhance immunity. The system is efficient because a specific microbe will only

Phagocytes: blood cells that swallow bacteria
Around 70 percent of white blood cells are phagocytes. They contain little sacs of digestive fluid, which they use to destroy bacteria that they swallow.

« BEFORE

Immunity is gained naturally (by contracting a disease) or artificially (by vaccination) « 248–49. Immunity was appreciated from antiquity, but not fully understood.

IMMUNITY RECOGNIZED

The concept of immunity is an old one. In 430 BCE Greek historian Thucydides recorded that in Athens **people surviving the plague** were not attacked a **second time**. Al-Razi, a medieval Islamic physician, observed a similar thing with measles and smallpox.

IDENTIFYING BLOOD CELLS

White blood cells—distinguished from more abundant red blood cells by their prominent nuclei—were first seen by **Renaissance microscopists**, but their roles in defense would not be appreciated for over 200 years.

RED AND WHITE BLOOD CELLS

« SEE ALSO

pp.242–43 THE SPREAD OF DISEASE
pp.244–45 BACTERIA AND VIRUSES

Natural Defenses

The body is under constant attack from microorganisms, which can cause damage and disease. However, a natural, effective system of defenses barricades their entry, and swiftly disables any that get through. This system is so sophisticated that it remembers attackers, and makes the body immune from future invasion.

Blood clotting is fast, but, while the wound is open, bacteria can infect the exposed area. Elsewhere in the body, where the skin barrier is necessarily compromised—such as the eyes and mouth—this invasion is a constant threat, even though tears and other fluids contain natural antiseptics, which stave off some attackers.

In 1890 Japanese bacteriologist Kitasato Shibasaburo and German physiologist Emil von Behring injected the antigens of tetanus bacteria into animals, making them immune to the disease. They discovered that serum (a fluid component of blood) extracted from these animals could be injected into others to make them immune too. Shibasaburo and Behring had discovered humoral immunity: immunity that depends on

stimulate the lymphocytes that are able to release the exact type of antibody to target that particular microbe, an idea first explained in the 1950s by an Australian scientist called Frank Burnet in his Clonal Selection Theory.

Remembering the invader

Burnet also explained long-term immunity. He discovered that some of the lymphocytes stimulated in the first invasion do not release their antibodies as part of the attack, but are held back in reserve. They become memory cells, and can remain in circulation in the body for months or even years. It is the presence of memory cells for a particular pathogen that makes the body immune to the disease caused by it. A second invasion "awakens" the memory cells: they proliferate more quickly and release more antibodies. The result is that this second response staves off the damaging effects of the disease so the symptoms may not develop.

Killer cells
These specialized lymphocytes (shown here in orange) were discovered in the 1970s. They bond to diseased cells, such as those of cancerous tumors or infected with viruses, and kill them by making them autodestruct.

Immune response

The body's ability to kill harmful invading microbes involves a chain of defensive white blood cells, triggered by microbe-swallowing phagocytes, and culminating in an army of different lymphocytes.

1 An attacking microbe enters the body and is swallowed by a defensive white blood cell called a phagocyte.

Microbe

Antigen
These molecules (usually proteins) sit on the surface of the attacking microbe

Red blood cell

Phagocyte

3 Presented antigens stick to receptors on the surface of T lymphocyte cells. The cells are activated and release cytokine, a regulating chemical that stimulates other lymphocytes.

"Presented" antigen

2 Inside the phagocyte, the microbe is stripped of its antigens (surface particles), which are then presented at the phagocyte's surface and stimulate the immune response.

T lymphocyte cell
An infection-fighting type of white blood cell that helps control the immune response

Receptor
Protein on surface of T lymphocyte that recognizes a specific antigen

Antibody
A protein that binds to the antigens on microbes and destroys them

4 The activated T lymphocyte cells transfer antigens to another type of lymphocyte cell, known as a B lymphocyte cell, which divide to form either plasma cells or memory cells.

Detached antibody

Plasma cell
A blood cell that releases antibodies into blood plasma to destroy microbes

Activated T lymphocyte

Cytokine

B lymphyocyte cell
The presented antigen, and cytokine, trigger the B lymphocyte cells to divide; some form plasma cells and some memory cells.

Antigen

Antibody

Destroyed microbe

Memory cell
The presence of these long-lived cells means the body is immune to the disease caused by the microbe. If the microbe invades again they become plasma cells

5 Plasma cells formed from the B lymphocyte cells rapidly divide and release many antibodies into the blood. These antibodies bind to antigens, destroying them and the microbe. Some lymphocytes become long-lasting memory cells, giving the body immunity to the microbe.

IN PRACTICE

THE SECONDARY IMMUNE RESPONSE

Initial exposure to a pathogen (foreign particle) stimulates a primary immune response—the slow production of a moderate amount of antibodies. But a second exposure causes a bigger, faster secondary immune response—production of enough antibodies to destroy the invader without causing symptoms.

Relative concentration of antibodies in blood

Primary response to pathogen A

Secondary immune response to pathogen A

time/days   5   10   15   20   25   30   35   40   45

First exposure to pathogen A

Second exposure to pathogen A

AFTER »

Modern understanding of the immune system appreciates the specific action of lymphocytes at a molecular level, but has also shown that their workings are much more complex than formerly appreciated.

SPECIFIC ANTIBODIES

Antibodies are **types of proteins** and their **structure** was discovered in 1969 by American biologist Gerald Edelman and English biochemist Rodney Robert Porter. Some years later, in 1975, the work of German immunologist George Köhler, Argentine biochemist César Milstein, and Danish immunologist Niels Jerne exploited the fact that **specific antibodies bind to specific targets** in forensic and diagnostic procedures. For example, the color change of a pregnancy test strip relies on **specific antibodies targeting a hormone** that occurs only in the urine of a pregnant woman.

LYMPHOCYTES: DIVERSE FUNCTIONS

After the 1950s the work of French scientist Jacques Miller showed that certain types of **lymphocytes** were involved in the immune response **without releasing antibodies**. Instead, they **activate phagocytes** or prompt self-destruction of cells infected with viruses. This is significant in understanding HIV, transplant rejection, and cancer.

SEE ALSO »
pp.248–49 IMMUNIZATION AND VACCINATION
pp.412–13 DISEASE CHALLENGES

A body is immune to a disease when its defense system has been primed to fight it. It can do this because white blood cells called lymphocytes release substances called antibodies, which disable harmful agents (see pp.246–47). This process happens naturally when a body contracts a disease. Medicine can help in two ways. First, passive immunization involves injecting the antibody directly and is done when immediate treatment is needed, against for example fast-acting food poisoning, such as botulism, or to treat the bites of a venomous animal. Second, active immunization involves injecting a "trigger" to "fool" the immune system. This trigger, called an antigen, is sometimes just an innocuous molecule from the outer coat of the microbe (bacteria or virus). If this molecule is isolated

Early vaccination tools
Edward Jenner was the first doctor to vaccinate his patients against smallpox. His tools, right, consisted of small sharp blades, which were used to pierce the patient's skin so the vaccine could be applied.

and injected, the army of lymphocytes is still activated even though the disease itself does not break out. In fact, sometimes the trigger doesn't even have to come from the dangerous microbe itself.

Discovery of vaccinations
English doctor Edward Jenner's experiments famously induced immunity to smallpox, not with a smallpox trigger—but with cowpox (see pp.156–57). He called the process "vaccination" from the Latin vacca for cow, a term that has stuck ever since.

At the time disease was thought to be caused by "bad gases" or similar vague notions. It wasn't until the late 19th century when the work of French chemist Louis Pasteur and German physician Robert Koch showed that disease is caused by living particles. Pasteur began experiments to develop immunity triggers—which he called vaccines—for diseases, and the direction of his research came about by accident. In his laboratory, a neglected culture of cholera bacteria became weakened because the culture medium they were growing in was "rejected" so that they no longer produced the disease, but they still triggered immunity. It spurred a flurry of work. Pasteur successfully produced a vaccine for cholera in 1879 and for anthrax in 1881. Perhaps his most celebrated success came with his work on rabies. Pasteur and French physician Emile Roux produced a

IMMUNIZATION Any process by which the body is made immune to a particular disease.

VACCINATION The administration of a vaccine (containing an antigen) in order to trigger immunity to a particular disease.

INOCULATION The administration of a mild pathogen, in order to trigger immunity or for another purpose, such as producing a disease for study.

vaccine, but it had been tested only on animals. Pasteur risked prosecution by trying it on a boy who had been attacked by a rabid dog. The boy never developed rabies and Pasteur was fêted as a hero.

Safer vaccines
The vaccines of Jenner and Pasteur were "live" vaccines: they contained living microbes, but far milder than the ones that inflicted terrible disease. Pasteur's work led to more vaccines being made in the 20th century, and the realization that live microbes were not always

Immunization and Vaccination

There are few other areas of medical science that have had as great an impact on the health of humankind as that of vaccination: medicine creates immunity to disease where previously there was none. The first vaccination protected people against smallpox, and since then vaccines have been developed to immunize people against many more diseases, saving millions of lives.

« BEFORE

The principle of vaccination can be traced back to medieval times, but the procedure was not without risk.

EARLY PRACTICES OF VARIOLATION
Variolation was deliberate inoculation with a mild form of the smallpox virus in order to induce smallpox immunity « 156–57. This was practiced in early China and India and was attempted in Europe by the Greek physician Giacomo Pylarini in 1701. Lady Montagu, wife of the British ambassador in Constantinople in 1716, was so impressed by its effects that she had her own children variolated, and later encouraged others to do the same.

SMALLPOX SUFFERER

« SEE ALSO
pp.242–43 SPREAD OF DISEASE
pp.244–45 BACTERIA AND VIRUSES
pp.246–47 NATURAL DEFENSES

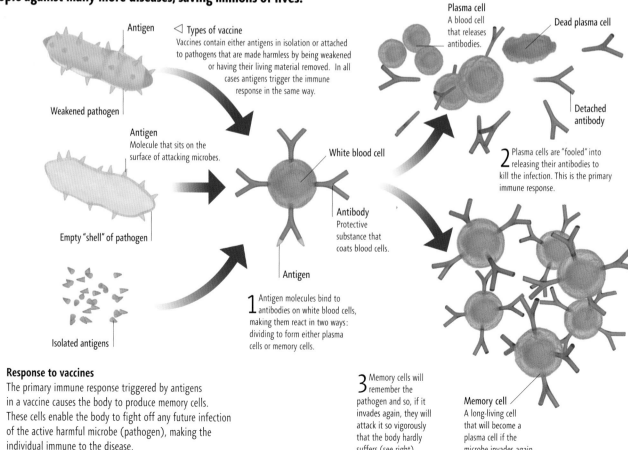

◁ **Types of vaccine**
Vaccines contain either antigens in isolation or attached to pathogens that are made harmless by being weakened or having their living material removed. In all cases antigens trigger the immune response in the same way.

Antigen

Weakened pathogen

Antigen
Molecule that sits on the surface of attacking microbes.

Empty "shell" of pathogen

Isolated antigens

White blood cell

Antibody
Protective substance that coats blood cells.

Antigen

1 Antigen molecules bind to antibodies on white blood cells, making them react in two ways: dividing to form either plasma cells or memory cells.

Plasma cell
A blood cell that releases antibodies.

Dead plasma cell

Detached antibody

2 Plasma cells are "fooled" into releasing their antibodies to kill the infection. This is the primary immune response.

3 Memory cells will remember the pathogen and so, if it invades again, they will attack it so vigorously that the body hardly suffers (see right).

Memory cell
A long-living cell that will become a plasma cell if the microbe invades again.

Response to vaccines
The primary immune response triggered by antigens in a vaccine causes the body to produce memory cells. These cells enable the body to fight off any future infection of the active harmful microbe (pathogen), making the individual immune to the disease.

Childhood vaccination
Programs for vaccinating children have been in place since the mid-20th century and have played a key role in controlling infectious diseases. Here, Native American Shosone children line up for vaccines in Wyoming in 1955.

SMALLPOX ERADICATION PROGRAM

In 1959 the World Health Organization (WHO) of the United Nations launched an intensive program to eradicate smallpox. Immunizing everyone would have been impracticable, so instead the strategy was one of ring immunization: cases were isolated by vaccination of neighbors, so containing the disease and preventing it from spreading. Smallpox was declared eradicated on December 9, 1979, and is the only disease that has been completely wiped out by humanity.

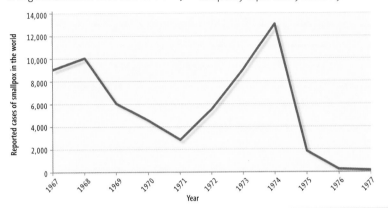

needed: that the triggers for immunity were specific molecules—the antigens—which could be isolated from harmful microbes to make vaccines even safer.

Vaccination programs

If enough people in a population are immune to a disease, it is difficult for the disease to spread. This idea, called "herd immunity," is behind government initiatives that have made vaccination compulsory—such as the 1853 English law for smallpox vaccination. This has drastically reduced the incidence of infectious disease in many parts of the world. However, it was not until the middle of the 20th century that the World Health Organization was able to oversee a vaccination program (see above) that ultimately led to the complete eradication of smallpox in 1979.

AFTER ➤➤

Organizations around the world are fighting to combat the spread of disease with the development of new vaccination programs.

NEW VACCINES

The start of the 20th century saw the development of vaccines for **diphtheria**, **pertussis** (whooping cough), **tuberculosis**, and **tetanus**. In 1952 American virologist Jonas Salk developed the first vaccine for **polio**, with an oral version appearing 10 years later. In the 1960s and 1970s, more vaccines significantly decreased infant mortality, including vaccines for **measles**, **mumps**, and **rubella**.

POLIO VACCINE

CHILDHOOD PROGRAMS

In 1900 the smallpox vaccine was the only one routinely administered to children. By the **1960s broader child vaccination schedules** had been developed to offer the most effective, economical **control of the diseases** concerned and were introduced in many countries worldwide.

mRNA VACCINES

Vaccines developed in the 1990s trigger immunity using mRNAs—copies of genes that contain information for making proteins. The mRNAs used in vaccines **encode for harmless viral proteins** that stimulate production of antibodies. In 2020, mRNA vaccines were approved for **fighting COVID-19**—less than a year after the start of the pandemic.

SEE ALSO ➤➤
pp.412–13 DISEASE CHALLENGES

1 Potentially pathogenic, naturally occurring microbe carries antigens identical to those used in the vaccine.

Antibody

Destroyed microbe

5 By binding to the antigens on the microbe, antibodies mark it for destruction or prevent it from doing harm.

Antigen

Microbe
An attacking foreign substance entering the body.

Detached antibody

4 New plasma cells can now release enough antibodies to kill the infection.

Plasma cell

2 This time, molecules of a real infectious organism bind to the memory cells produced from the vaccination.

Antibody

Memory cell

Response to infection after vaccination

After receiving the vaccine, the secondary immune response will be more effective when the body is infected naturally by the pathogen. This means that the pathogen will be destroyed before it can cause disease.

3 New plasma cells are formed from the memory cells being triggered by antigens.

« BEFORE

Early humans had no artificial light until they learned to make fire. From then on they built fires at cave mouths to provide light (and warmth) for the early part of the night.

NATURAL LIGHT

The first people did their work **during the hours of daylight**, so they had far more usable hours in summer than in winter. Hunting during the winter months would have been difficult and dangerous; many animals are most easily hunted in the evening, but pursuing them would have entailed a dark walk home, in danger from predators with **better night vision**.

OIL LAMPS

Once humans began cooking, they had access to animal fats. They soon began to make **simple tallow lamps**, which emitted a weak, smoky, yellow flame but provided light throughout the night.

EARLY EGYPTIAN OIL LAMP

LIGHT FROM CANDLES

Animal fat was also used to make **candles**. The ancient Egyptians are thought to be the first to use them; however, it was the Romans who introduced **the wick** that runs down the center. It was not until the Middle Ages that **beeswax** started to be used in the production of candles instead of animal fat.

« SEE ALSO
pp.114–15 THE NATURE OF LIGHT
pp.116–17 SPLITTING AND BENDING LIGHT
pp.164–65 ELECTRIC CURRENT

Artificial Light

Until 200 years ago the best artificial light was candlelight, and even today many millions of people still rely upon simple oil lamps. The widespread use of indoor and outdoor lighting across the developed world became possible as scientists harnessed the powers of gas and then electricity, and invented new forms for their use—such as the light bulb.

Living without any artificial light can be restricting, so the development of oil lamps and tallow candles made life much easier for early humans. The invention of gas lighting at the end of the 18th century marked a significant breakthrough in lighting technology, as did incandescent electric lamps at the end of the 19th century.

Oil and gaslight

Oil will not burn easily at low temperatures, and it can be lit with a match only if it has a wick: a piece of absorbent material—such as string or tape—that soaks up the oil. When a match is put to the wick, the heat from the flame turns a little oil into vapor, which catches fire.

Coal gas was thought to have no real use until William Murdoch used it for lighting his house in Cornwall around 1799. Two years later he lit all of the Soho Foundry in

Adjustable lamps
Still the only source of light for many people, oil lamps of some form have been used since 70,000 BCE. This 19th-century lamp has a fuel reservoir in the base, a knob for adjusting the wick, and a glass chimney to protect the flame.

JOSEPH SWAN

Born in Sunderland, northern England, in 1828, Joseph Swan earned his living by running a pharmacy in nearby Newcastle, and did his own scientific research in his spare time. He invented the dry photographic plate and several other photographic processes. Swan also produced the world's first artificial filament. He did this by dissolving blotting paper in zinc chloride solution, and then using a syringe to squirt this solution into alcohol. By carefully carbonizing this filament in the absence of air, he managed to produce the world's first successful light bulb.

Filament

Thin glass bulb

All air pumped out of bulb's interior

Supporting wires

Connecting wires

Electrical leads pass through insulated airtight base

Edison's bulb
Although Edison was not the first to create a light bulb, his model was more commercially viable because it was long-lasting and formed part of a whole lighting system.

Swan's light bulb
This lamp is similar to the first one demonstrated in February 1879. The filament is a single coil of carbonized artificial cellulose fiber, in a glass bulb from which all the air has been pumped.

"I have not **failed**, I have found **10,000 ways** that do not work."

THOMAS ALVA EDISON, PHYSICIST, c.1910

Birmingham, and gradually gasworks and gaslight spread across Britain, both in the streets and in homes. The bare gas flame was a dim flickering yellow until 1890, when it was transformed into bright white by the gas mantle—a fishnet of metal oxide placed over the gas. But by then electricity had started to become a viable alternative.

The electrical arc lamp

Humphry Davy invented the arc lamp in around 1810. Carbon rods connected to a large battery were touched together and then separated by a short distance. The electric current jumped the gap, creating a plasma that glowed brightly. This type of lamp is difficult to manage, as the electrodes continually burn away, and its extreme brightness can cause eye damage. However, arc lamps were used in street and theater lights and in lighthouses.

Inventing the light bulb

Passing an electric current through a wire makes it heat up and glow, and this can form the basis of an incandescent lamp. The problem is that when the wire reaches a dull red heat, it melts or burns. Two scientists competed to create the first successful light bulb: the American Thomas Edison (see pp.264–65) and the British chemist Joseph Swan. Swan won the race in February 1879. His lamp used a filament of pure carbon in a vacuum within a glass bulb.

Fluorescence

Fluorescent tubes work by discharging electricity through a gas at low pressure; this produces ultraviolet light, which makes the phosphorescent coating of the tube glow with a bright white light. Fluorescent lamps are about six times more efficient than incandescent lamps. Compact fluorescent lamps (CFLs) replaced incandescent light bulbs from about 1990 to 2010.

Light-emitting diodes

Light-emitting diodes (LEDs) produce light through an effect called electro-luminescence, in which electrons combine with "holes" (spaces lacking an electron) in a semiconductor material. Although their light is emitted in a narrow band of wavelengths, when combined with suitable phosphors, LEDs can produce light varying from warm white to daylight. With their long life, pleasant color, and high efficiency, LED lamps have now replaced incandescent and fluorescent lamps for most lighting.

IN PRACTICE
LIMELIGHT

In the 1820s British scientist Goldsworthy Gurney invented a blowpipe that burned hydrogen gas in oxygen, creating an extremely hot flame—it had a temperature of around 3,500° F (2,000° C). He then discovered that by heating a piece of chalk or limestone within the flame he could make an intense white light. This "limelight" was used for stage lighting in theaters—hence the expression "in the limelight"; theater spotlights are called "limes" to this day. The high temperature excites the electrons within the calcium atoms, making them emit an intense white light.

AFTER

Scientists and engineers are continually looking for new and more efficient ways of turning electricity into light, and at the same time to improve its color.

SAVING ENERGY

In 2016, researchers at the Massachusetts Institute of Technology developed a coating for the filaments of incandescent light bulbs that makes them more efficient than fluorescent lamps.

SMART LIGHTING

Lighting that turns on or off automatically is not new, but control is now available via the internet, allowing users to change the brightness or color of lighting by voice commands from a smartphone.

SEE ALSO »
pp.252–53 GENERATING ELECTRICITY
pp.364–65 LASERS AND HOLOGRAMS

Neon street signs in Hong Kong
Electrified glass tubes containing neon or other gases are often used for street signs because they are long lasting and can be made in any shape. Their color depends on the gas used; a high voltage is applied across the gas, causing it to emit light.

Glass tube

Electron

Inert gas, such as argon, neon, krypton, or xenon, at very low pressure (less than one percent of atmospheric pressure)

Phosphor coating on inside of glass tube

Contact pins to plug into socket and carry current from lighting circuit

Tungsten electrode, coated with a mixture of barium, strontium, and calcium oxides

Mercury

Mercury vapor exists as single atoms

The fluorescent lamp

The glass tube contains an inert gas, usually argon, and a little mercury. A high voltage causes electrons to move through the gas, vaporizing the mercury. Atoms of mercury excited by electrical energy give out ultraviolet light, which is then reradiated as visible light by the "phosphor" compounds that coat the inside of the tube.

Electricity from Heat

Current electricity was invented in 1799, but it remained a laboratory curiosity for much of the 19th century. The invention of practical electric lighting, however, followed by other electrical machines, created a rapidly growing demand for electricity. Most of the world's supply is now generated in large power stations.

⟨⟨ BEFORE

Transmission of electricity on an industrial scale required metals such as aluminum and copper: excellent and inexpensive conductors.

INSULATORS
All electricity would have been lost from cables underground or under the sea without **gutta percha**—a natural latex material related to rubber—whose insulating properties were discovered in 1842.

⟨⟨ SEE ALSO
pp.164–65 ELECTRIC CURRENT
pp.166–67 ELECTROMAGNETISM
pp.170–71 MICHAEL FARADAY

The Gramme dynamo
Built in 1871 by the Belgian engineer Zénobe Gramme, this dynamo was among the first to be used on a commercial scale.

Although batteries had been developed by 1800, scientists sought more constant and reliable ways to produce an electric current. A device developed by Michael Faraday in 1831 showed the way forward. Faraday's dynamo consisted of a copper disk that was spun between the poles of a magnet, making a current flow from the center of the disk toward its rim.

Direct and alternating current
The dynamo produces direct current (DC), but when the principles of alternating current (AC) were established in the late 1880s (see pp.164–65), dynamos were replaced by alternators. These use a rotating magnet inside a coil of wire or a coil of wire in a magnetic field (below) to produce AC. Motor vehicles use alternators to convert some of the energy produced by the internal combustion engine (see pp.254–55) into electrical energy.

The challenge for electricity producers is to find effective ways of generating the rotary motion that can run alternators and produce electricity. Hydroelectric plants and windmills do it directly, with spinning blades driven by water or wind. In the UK Parsons (see right) developed steam turbines in the 1880s that allowed production of cheap and plentiful electricity. This is the most common form of generation today.

Universal generation
In the early days there was much debate about whether it would be better to use DC or AC. The great inventor Thomas Edison championed DC, while the Serbian-born American Nikola Tesla favored the AC system.

In 1886 the first AC general-purpose generators were built in Massachusetts, and at the Rome-Cerchi Italian power station. Two years later Tesla introduced the idea of greatly stepping up the voltage for transmission (in order to reduce power loss), and then stepping it down again for use at lower voltages.

Power stations
Despite growth in the use of renewable sources, about 80 percent of the world's energy is still produced in large power stations. Heat is made by burning fossil fuels or through controlled nuclear fission; this heat is used to turn water into high-pressure steam, which drives steam turbines that use dozens of blades to extract kinetic (motion) energy from the steam (see right).

Transmission of electricity
Electricity can be transmitted across great distances through power lines, generally suspended from pylons. There is some energy loss on the way, since resistance in the wires causes energy to be converted to heat. This is reduced by transmitting at high voltage.

AFTER

The demand for electricity is rising; many countries are seeking clean, new, renewable ways of generating electricity.

FUEL CELLS
Fuel cells are **clean sources of energy** that generate electricity locally in small amounts.

EXPANDING USE OF ELECTRICITY
Threats of global warming 422–23 ⟩⟩ have intensified efforts to tap renewable energy sources, such as wind and wave power.

SEE ALSO ⟩⟩
pp.324–25 FISSION AND FUSION
pp.424–25 RENEWABLE ENERGY

How an alternator works
Alternators produce current either by spinning a coil of wire through a fixed magnetic field or by spinning a magnet around a fixed coil. Current is generated as the wire moves through the magnetic field.

Shaft driven by turbine
Magnetic field
North pole of magnet
South pole of magnet
Wire coil rotating counterclockwise
Carbon or metal brushes take current from rotating coil
Bulb lights up as current flows
Current flows clockwise in bulb circuit

1 Clockwise current
As one side of the coil sweeps past the north pole of the magnet a current is induced through the coil, and in this case clockwise through the bulb circuit.

Shaft
South pole of magnet
Wire coil rotating counterclockwise
North pole of magnet
Carbon or metal brushes take current from rotating coil
Bulb lights up as current flows
Current flows counterclockwise in bulb circuit

2 Counterclockwise current
Half a turn later, the other side of the coil sweeps past the north pole, and the current is again induced into the coil, but in the opposite direction, so the current runs counterclockwise in the bulb circuit.

Direction of thrust applied to the wire (thumb)
Direction of current (second finger)
Direction of magnetic field (first finger)

△ **Fleming's right-hand rule**
Hold out your right hand with thumb up, first finger at 90° to it, and second finger pointing back to you at 90°: they show directions.

Cooling tower
After use the low-pressure steam is sprayed into cool air inside cooling towers, where it condenses back to water.

Boiler
Coal is burned in the boiler to produce heat, which boils the water and carries on heating it until it turns into high-pressure steam.

Tall chimney
Fumes from the boiler escape up a chimney to be released high above ground.

◁ **Power generation and transmission**
Most electricity is generated in large power stations, often by burning coal to make high-pressure steam, which is used to drive steam turbines. The power is then "stepped up" to high voltage for transmission via power lines to the end users, or consumers.

Steam
The high-pressure steam is piped to the turbine.

Generator
The spinning turbine shaft is connected directly to the alternator, which produces alternating current at 25,000 volts (V).

Steam turbine
The steam is allowed to expand through the turbine, making the blades spin at great speed (see below).

▽ **How a steam turbine works**
Multiple-bladed fans of increasing size are mounted along a shaft. High-pressure steam hits each fan in turn, driving them around and gradually losing its energy, until the steam is reduced to atmospheric pressure.

Steam pushes through fan blades, losing pressure at each stage

Generator

Steam enters at very high pressure from boiler

Drive shaft rotates counterclockwise as steam pushes fan blades

Fan blades spin

Steam exits turbine at atmospheric pressure, ready for cooling

Step-up transformer
For long-distance transmission the transformer steps up the voltage to 110,000V or more, to minimize power loss.

25,000 VOLTS
Produced by the generator at a power station.

Domestic consumer
Household electricity reaches the home, where it is available at the press of a switch

Step-down transformer
The voltage is reduced to the level needed locally; homes might require 110V (US) or 240V (UK).

Pylons
Steel towers keep the high-voltage power lines high above ground.

The Internal Combustion Engine

The internal combustion engine replaced steam as the driving force behind industrial expansion, and put the world on wheels. These engines are at the heart of the automobile, as well as providing power for machines from lawn mowers to aircraft. The first engines appeared in the 19th century, but the technology actually had its roots in a much earlier age.

There are an estimated 1.4 billion cars on today's roads. The majority, as well as most trucks, motorcycles, trains, aircraft, boats, and generators, are powered by an invention with its origins in the 13th century—the internal combustion engine.

The principle behind the internal combustion engine is straightforward enough: take a small amount of a high-energy liquid fuel, ignite it, and use the energy released to move a piston a little bit. Do this hundreds of times a minute, and you move a piston a lot. All you have to do now is connect your piston to some wheels, and you are on the move.

Early combustion engines

The roots of combustion engines can be traced back, via the gunpowder-fueled designs of Leonardo da Vinci, Dutch physicist Christiaan Huygens, and British polymath Sir Samuel Morland, to the genius of 13th-century Arab inventor Aljazari. His 1206 *Book of Knowledge of Ingenious Mechanical Devices* included the first known description of the crankshaft, a device that is essential to the combustion engine. It would be another 600 years, however, before an actual combustion engine was created.

In 1806 Swiss inventor François Isaac de Rivaz built the first working model, powered by a mixture of hydrogen and oxygen, and he knew what he wanted to use it for; just a year later Rivaz had built a wooden automobile around his revolutionary engine. The machine, however, was not commercially successful. Mass production of internal combustion engines would not start in earnest until the mid-19th century.

Taking to the road

The turning point came with the invention of the four-stroke, petrol-driven engine by German inventor Nikolaus Otto. This was the first engine able to burn fuel efficiently in the

△ **An early production line**
Production line workers assemble motor cars at the Ford Motor Company factory in Dearborn, Michigan. Henry Ford paid his workers enough for them to afford to buy the cars they were making—a ready-made market.

‹‹ BEFORE

Some of the components and concepts associated with internal combustion engines have developed over several centuries.

ENGINES IN THEORY
Centuries before the first internal combustion engines, **Leonardo da Vinci** imagined a compressionless engine that used **gunpowder as a fuel**. He sketched his ideas in 1509 but, as far as we know, his machine was never built.

HUYGENS'S GUNPOWDER ENGINE
Having never seen Leonardo's drawings, in 1666 Dutch genius **Christiaan Huygens** tried to build an engine that **pushed a piston upward** when **gunpowder was ignited under it**, but he had no materials that could take the strain.

STEAM FOLLIES
The **very first automobiles** were steam-driven military tractors invented by French engineer **Nicolas Cugnot** in 1769. They had three wheels and reached speeds of 2½ mph (4 kph). Cugnot's machines met a premature end when one of his patrons died and the other was exiled.

‹‹ SEE ALSO
pp.168–69 THE ELECTRIC MOTOR
pp.170–71 MICHAEL FARADAY

The Otto four-stroke cycle
Powering nearly all modern cars and trucks, this process, named after its 1876 inventor Nikolaus Otto, takes place at the heart of the internal combustion engine.

Air and fuel

Air and fuel are drawn into cylinder through inlet valve

Piston

Crankshaft

1 Intake stroke—inlet valve opens and piston moves down, allowing cylinder to fill with air and fuel.

Exhaust valve remains shut

Inlet valve shuts

Air-fuel mixture is compressed

Crankshaft rotates to move piston

2 Compression stroke—piston is pushed back upward, which increases pressure in the cylinder and makes fuel burn more forcefully.

Exhaust valve opens as piston rises

Exhaust

Crankshaft continues to rotate

4 Exhaust stroke—when piston reaches the bottom, exhaust valve opens. As piston rises again, it pushes waste gases out into exhaust pipe.

Spark plug has ignited fuel

Inlet valve shut

Piston at top of its stroke

Exhaust valve shut

Crankshaft continues to rotate

3 Combustion stroke—at top of piston's stroke, a spark plug fires, igniting fuel and causing it to expand rapidly. This shoots piston downward again.

piston chamber, making it practically and commercially viable. Designed as a stationary engine, in which the action of the pistons is maintained as back-and-forth motion, it was soon adapted into an automobile engine, complete with crankshaft to produce circular motion. Finally, here was an engine that could power an automobile far more practically than the earlier steam cars, which relied on heavy boilers.

The man who achieved this feat was Karl Benz, who later founded Mercedes Benz; for this reason he is generally regarded as the inventor of the motor car. His three-wheeled Benz Patent Motorwagen, rated at less than one horsepower, was first built in 1885 and went on sale in 1888. The other major pioneer at this time was Gottlieb Daimler, who patented the prototype of the modern gas engine and made the world's first four-wheeled cars.

Diesel's influence

There is one other man who made a major contribution—Rudolf Diesel. Fascinated by the developments in

ENGINEER (1858–1913)

RUDOLF DIESEL

In 1889, French-German inventor Rudolf Diesel was a well-respected refrigerator engineer. By the end of the 1890s he had created the Diesel engine and was a millionaire. He had a keen interest in social theories as well as engineering, and conceived of his engine as an energy-efficient machine, using local fuels, for craftsmen and artisans who could not afford the steam engines used by large companies.

engine technology, Diesel set about creating his own version, which did not require a spark to ignite the fuel. His endeavors almost came to an end when, in 1894, his new "diesel engine" exploded, almost killing him. To Diesel, however, this proved that his idea worked. His first successful engine fired into life in 1897, and all diesel engines in use today are refined versions of his original concept.

Diesel engines owe their success mainly to their efficiency. Unlike gas engines, a diesel engine compresses air

and then injects fuel into the chamber. This means that, while gas engines compress at a ratio of about 10:1, diesel engines have a compression ratio as high as 25:1. The higher compression ratio leads to better fuel efficiency.

Diesel engines power nearly all heavy machinery and large vehicles such as trucks and buses, and around half of all cars sold in Europe have diesel engines. Concerns about air pollution, and new developments in battery technology have seen global sales of electric and hybrid cars rise to over one million.

The internal combustion engine radically changed not only the way we travel but our entire economic system.

THE IMPORTANCE OF OIL
Rudolf Diesel's first engines ran on peanut oil, but **fossil fuels** quickly took precedence. In 1900 the world produced about **100 million barrels of oil**. In 2020 it produced 32 **billion barrels**.

TAKING TO THE SKY
Daimler's engine was the first with a **power-to-weight ratio** high enough to enable **powered flight**, which was achieved just 18 years later.

END OF THE LINE
By the middle of the 21st century, most **internal combustion engines** will be replaced by electric motors. Governments worldwide have pledged to make the machine that has contributed so much to global warming a thing of the past.

SEE ALSO »
pp.290–91 TAKING FLIGHT
pp.338–39 ROCKET PROPULSION

OIL PUMPJACKS, CALIFORNIA

Model T Ford
Created in 1908, the Ford Model T, or Tin Lizzie, was the first affordable car the world had ever seen. Ford designed his "universal car" so that it would be easy for anyone to run and maintain.

The Nature of Sound

The Ancient Greek philosopher Aristotle was the first to note that sound traveled through the air as a vibration. After Isaac Newton identified sound as a longitudinal wave in 1687, scientists of the 18th and 19th centuries were able to investigate its characteristics, such as wavelengths and frequencies.

The principle that sound is a wave transmitted through the air from a vibrating object to the human ear was universally accepted from the 17th century onward. However, the nature of sound waves remained obscure. Most early scientists assumed that sound waves, like the vibrating strings that were often used to produce them, were transverse waves—waves whose vibrations oscillate at right angles to the direction of their travel.

Longitudinal waves

This view was abandoned when Newton described the motion of sound in *Philosophiae Naturalis Principia Mathematica* (1687) as a moving region of high and low pressure "nodes"—a longitudinal compression wave. He calculated a theoretical value of 979 ft (298 m) a second for the speed of sound. The English scientist William Derham found this to be only slightly inaccurate when he tested it a few years later.

« BEFORE

The exact nature of sound has fascinated people since the earliest times.

EARLY WAVE THEORY
The ancient Greek philosopher **Aristotle** developed a theory of sound as a wave in the 4th century BCE.

THE MATHEMATICS OF MUSIC
Galileo Galilei and **Marin Mersenne** rediscovered the **laws of vibrating strings** in the 17th century; Mersenne developed mathematical **equations** for calculating the frequency of waves produced by different types of vibration.

IN A VACUUM
Using an air pump made for him by his assistant **Robert Hooke**, 17th-century scientist **Robert Boyle** demonstrated that a bell is inaudible in a vacuum. This proved that **sound needs a medium** such as air to pass through.

BOYLE'S BELL IN A VACUUM

HIGH–FREQUENCY WAVE   LOW–FREQUENCY WAVE

Frequency and pitch
Pitch is dependent simply on the frequency of the sound, and therefore on its wavelength, because higher pitches equate to higher frequencies, which equate to shorter wavelengths.

High amplitude

Low amplitude

Loudness
The amplitude of the sound wave is equivalent to its intensity, or the pressure it exerts. But the perceived "loudness" of a sound heard by the human ear depends not only on the wave's intensity, but also on its frequency and pitch.

Measuring sound

Derham measured the speed of sound through air as 1,116 ft (348 m) a second. Today the accepted average speed of sound through air at sea level is 1,115 ft (340 m) a second, though it varies with air humidity.

Derham's method, which has often been repeated, was to send a colleague to various distant locations with orders to fire a gun at a certain time. To this end, they synchronized their watches. Watching through a telescope from a church tower, Derham was able to time the delay between the flash of the gun and the arrival of the sound waves it triggered over the various distances.

Range of hearing in different animals

The frequency, or pitch, of sound is measured in hertz (Hz). While we can typically detect sound waves with frequencies between 20Hz and 20,000Hz, bats and dolphins can hear much higher frequencies. Elephants and some other animals can hear and communicate through low-frequency "infrasound."

> **The loudest sound ever produced was the explosive eruption of the Krakatoa volcano off Indonesia in 1883. Estimated at 180 decibels (see AFTER), it was heard worldwide.**

The basic physics of sound were now understood, but there remained the problem of relating the perceived qualities of sound to the properties of the waves themselves.

In the 1630s the French philosopher priest Marin Mersenne had shown that the frequency of the sound made by a vibrating string was twice that of the sound made by a string vibrating an octave lower. This proved that the pitch of a sound was predominantly due to the frequency of its vibrations in air.

Although there was at first no way of measuring the exact frequencies of sound waves, it was possible to figure out the wavelength of the vibrations within various instruments. Organ pipes, for instance, contained either a half-wavelength or quarter-wavelength of sound depending on their design. Since the frequency of a wave is its speed divided by its wavelength, it was then relatively simple to find the frequency of the waves being produced.

Sound patterns

In the 1780s German musician and physicist Ernst Chladni developed an ingenious way of measuring sound waves directly. He used sound to vibrate a thin metal plate with a fine powder on its surface, and found that the powder collected into various elegant patterns as the vibrations interfered with one another. He then developed formulas for figuring out the properties of the sound from the patterns that formed. By observing the way that closed organ tubes filled with various gases produced notes of different pitch, Chladni was also able to calculate the speed of sound in these gases.

Wavelengths and frequencies

The physics of sound advanced rapidly from the early 19th century onward. Austrian physicist Christian Doppler predicted that the Doppler effect—the change in pitch experienced by a person as a noisy object approaches and then

Faster than the speed of sound

Ernst Mach first described how a projectile breaking the sound barrier builds up a shockwave in front of it, creating a "sonic boom." At this point water vapor condenses in the low-pressure region behind the shock, shown here around the plane's wing.

AFTER

The 20th century saw major advances in measuring and recording sound.

SOUND UNITS

The "**bel**" and "**decibel**"—units for measuring **sound intensity**—were devised by telephone engineers in the 1920s as a convenient means of comparing sound levels.

RECORDING SOUND

The ability to **record** and **reproduce sound** has also led to a deeper understanding of its other physical properties and their relation to human perception.

SEE ALSO »
pp.260–61 TELEGRAPH TO TELEPHONE
pp.266–67 CAPTURING SOUND
pp.268–69 RADIO AND RADIO WAVES

▷ **Sound harmonics**
Harmonics are the key to music. Related waves with frequencies that are whole-number multiples of a single fundamental frequency, they can be overlaid on each other for a harmonious effect.

Half a wave

Guitar string

First harmonic
An instrument's fundamental, lowest frequency (longest wavelength) is its first harmonic. For the first harmonic of a guitar string, the wavelength is twice as long as the string.

Whole wave

Second harmonic
The second harmonic is twice the frequency and half the wavelength of the first harmonic. For the second harmonic of a guitar string, the wavelength is the same length as the string.

1 1/2 waves

Third harmonic
The third harmonic has three times the frequency and one-third the wavelength of the first harmonic. For the third harmonic of a guitar string, the string is equal to 1 1/2 wavelengths.

---

passes (see opposite)—was due to a change in the frequency of the sound wave in 1842. Dutch scientist Christopher Buys Ballot proved it three years later with a trumpeter playing a calibrated note while aboard a passing train.

In 1866 the German scientist August Kundt developed an ingenious means of precisely studying sound waves: he used a sealed transparent tube in which fine powder such as talc collected at the nodes of a sound wave vibrating within it. This allowed the wavelength to be measured directly, and led to the most precise calculations of the speed of sound ever made.

Supersonic predictions

One of the most intriguing discoveries about sound came from the Austrian Ernst Mach, who correctly described the shock wave generated when objects move at supersonic (faster than sound) speeds. In 1877 Mach predicted many of the effects aircraft would encounter when they broke the so-called sound barrier in the middle of the next century.

1 Vehicle approaches
Frequency of sound waves passing observer is increased; siren sounds higher to observer.

Siren approaching

Observer

Passing siren

Observer

2 Vehicle passing
Sound waves have "normal" frequency and are regular in pitch.

Siren moving away

Observer

3 Vehicle retreats
Frequency of sound waves reaching observer is reduced; pitch sounds lower to the observer.

The Doppler effect

The most familiar version of the Doppler effect is the distortion of sound waves experienced when a speeding emergency vehicle approaches, passes, and retreats from a stationary observer.

IN PRACTICE

ACOUSTICS

For centuries architects knew that a building's internal shape affects the way that sound reverberates—its acoustics—without understanding how. Then in 1895 a talented young Harvard physicist, Wallace Clement Sabine, developed a formula for accurately calculating the sound-absorbing properties of a building, such as London's Royal Albert Hall (shown here), in terms of its volume, shape, and materials.

Across the spectrum

The electromagnetic spectrum forms a "continuum" of radiation waves with a potentially infinite range of wavelengths. All these forms of radiation have the same speed (the speed of light), and their frequency increases as their wavelength decreases. High-frequency waves carry more energy than lower-frequency ones, and are normally generated by hotter or otherwise more violent processes. Earth's atmosphere blocks many forms of radiation from space, only allowing through radio waves, visible light, the near-infrared, and some ultraviolet.

Radio waves

The longest radio waves are absorbed in Earth's upper atmosphere by the electrically charged ionosphere. Those with shorter wavelengths are absorbed by atmospheric carbon dioxide and water vapor.

Radio window

Intermediate radio waves with wavelengths between a few centimeters and about 10 meters, and with frequencies extending from six to 30,000 megahertz (MHz), pass through Earth's atmosphere and reach the surface.

Microwaves

The shortest form of radio waves, with wavelengths between 1cm and 100µm, are known as microwaves. They are widely used in communications and microwave ovens.

RADIO WAVES   MICROWAVES

| 1 Km | 100 m | 10 m | 1 m | 10 cm | 1 cm | 1 mm |

BEFORE

Long ago, people noticed that water drops could split sunlight into colors. The first mathematical study of the phenomenon was carried out by Persian scientist Kamal al-Din al-Farisi in the 13th century.

THE COLOR OF LIGHT

There was much debate about the **nature of light** in the late 1600s Europe. **Huygens** put forward a **wave theory**, which was rivaled by **Newton's particle theory**. Newton had also shown that "pure" sunlight is in fact a mix of different colors « 114–15.

AN ELECTROMAGNETIC WAVE

Thomas Young's "double slit" experiment « 114–15 reopened the debate around 1800, and pushed the consensus back toward a **wave interpretation**, setting the stage for **Maxwell's** wave-based theory of light. Scientists continued to speculate on human vision and the perception of color « 166–67.

« SEE ALSO
pp.114–15 THE NATURE OF LIGHT
pp.116–17 SPLITTING AND BENDING LIGHT
pp.166–67 ELECTROMAGNETISM

HERSCHEL SPACE OBSERVATORY
Sun shade
11.5ft (3.5m) diameter mirror
Cryostat to cool instruments

Electromagnetic Spectrum

Throughout the 19th century scientists discovered previously unknown types of radiation within the electromagnetic spectrum, beyond visible light. As they uncovered radio waves, infrared, ultraviolet, X-rays, and gamma rays, they reached a new understanding of the nature of radiation itself.

Before 1800 nobody even suspected the existence of invisible forms of "light" beyond the extremes of the visible spectrum. The discovery was an unexpected result of an experiment carried out in 1800 by William Herschel, the German-born British astronomer already famous for his discovery of the planet Uranus.

Color and heat

During his observations, Herschel noticed that differently colored filters seemed to transmit different amounts of heat, and he began to suspect that the colors might have different temperatures associated with them. As he carried out his measurements,

One wavelength   Rarefaction

Compression

One wavelength

Undisturbed position

Vibration

Longitudinal wave

A longitudinal wave is much like the waves that travel along a spring—it forms regions of compression and rarefaction (decrease in density) parallel to the direction in which it travels. Sound waves are longitudinal, and some early physicists thought that light was too.

Transverse wave

In a transverse wave, the direction of the disturbances caused by the wave is at right angles to its direction of motion—as in water waves and light. The shape of the wave determines properties such as wavelength and amplitude.

INFRARED IMAGE OF STAR BIRTH

Herschel Space Observatory

Named after the discoverer of infrared radiation, the Herschel observatory used a large mirror to study the infrared sky. It was launched into orbit in 2009 and operated until 2013.

he found that the temperatures increased noticeably at the red end of the spectrum compared with the blue, and so he decided to measure the apparently empty region where the spectrum's red light faded into invisibility. To his surprise, this turned out to be the hottest region of all. He concluded that the Sun emitted invisible, very hot rays, and named this new type of radiation "calorific rays."

Chemical rays

Herschel's discovery inspired German chemist Johann Wilhelm Ritter, who found another type of invisible radiation beyond the violet end of the spectrum. Ritter is known today for his invention of modern electroplating techniques, but it was his investigations into the way that silver salts darken

when exposed to light—crucial to the eventual development of photography (see p.262)—that proved key to the new discovery.

In 1801 Ritter investigated how the different colors of light affected silver salts. He found that they darkened more rapidly in blue light than in red. Ritter tested the salts' reaction when placed beyond the violet end of the spectrum, and found it was stronger still. Ritter's new radiation became known as "chemical rays," and their ability to stimulate chemical reactions was soon being widely used.

Energy travels in waves

Intriguing though they were, the new rays did not immediately offer further insights into the nature of light—the physics of the time really had nothing

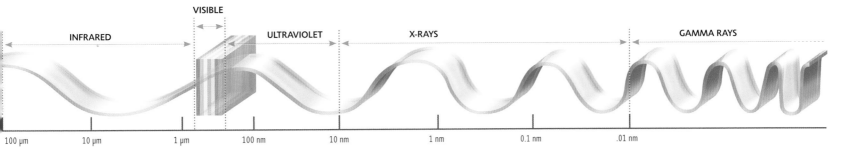

Infrared
Emitted by warm objects, infrared radiation is mostly absorbed in Earth's atmosphere by gases such as carbon dioxide and water vapor.

Optical window
Wavelengths measured in billionths of a meter pass straight through Earth's atmosphere. Animal eyesight has evolved to be sensitive to radiations in this "window".

Utraviolet
The wavelengths of ultraviolet rays are shorter than visible light. Most ultraviolet rays from the Sun are blocked by the atmospheric ozone layer.

X-rays
With wavelengths from 10 nanometers (nm) down to 0.01nm, X-rays carry large amounts of harmful energy. Fortunately Earth's atmosphere acts as a good barrier to them.

Gamma rays
Even shorter than X-rays, gamma rays have an extreme potential for penetrating materials and damaging tissues. As with X-rays, our atmosphere prevents them from reaching Earth's surface.

VISIBLE

INFRARED | ULTRAVIOLET | X-RAYS | GAMMA RAYS

100 μm   10 μm   1 μm   100 nm   10 nm   1 nm   0.1 nm   .01 nm

The ultraviolet world
Many flowers fluoresce and are far more strongly marked when seen in ultraviolet light. This evolutionary adaptation aids pollination by attracting insects that can see in the ultraviolet part of the spectrum.

to say on the possibility of radiations invisible to the human eye. The context for these discoveries would have to wait until later in the 19th century, and Scottish physicist James Clerk Maxwell's discovery of the equations linking electric and magnetic effects into a single phenomenon of electromagnetism (see pp.166–67).

(see pp.166–67)

INVENTION

INFRARED NIGHT VISION

Direct imaging of infrared light was made possible by the invention of electronic detectors that are sensitive to long wavelengths. Developed in the 1950s, they are widely used in military and rescue operations. They turn infrared information into visible light, sometimes in various colors and intensity, reflecting variations in an object's temperature. In this image the blue areas are cold, the red warmest.

Maxwell's work described light as a "transverse" wave, where particles oscillate up and down about their individual "home" positions (see opposite). He also suggested that the wave had interlinked electric and magnetic elements. However, he erroneously assumed that such a wave would require a medium through which to move, and the search for this "luminiferous (light-carrying) aether" became a major concern for late 19th-century physics.

A continuous spectrum

Maxwell's equations demonstrated how the different properties of light arose from the changing wavelengths of the electromagnetic waves. They explained why shorter-wavelength blue light was more susceptible to diffraction than the longer-wavelength red light. They also showed that the spectrum is continuous, and established that higher-energy objects and events emit waves of higher frequencies and shorter, bluer wavelengths.

Maxwell's equations made it clear that there was no physical reason why electromagnetic waves could not be many times longer or shorter than the known spectrum. In fact, the wavelengths of visible light, infrared, and ultraviolet all proved to be in the order of nanometers (billionths of a meter), but the first immediate proof of Maxwell's theories came nine years after his death, with Heinrich Hertz's demonstration of microwaves. These were radio waves with longer wavelengths than infrared (see p.268).

X-rays and gamma rays

Despite these realizations, Wilhelm Röntgen's 1895 discovery of X-rays (see p.294) still came unexpectedly, and the rays were not automatically identified as a new form of electromagnetic radiation. Nor were gamma rays (see p.298), which were discovered five years later. Within a hundred years, the rainbow spectrum

(see p.268)
(see p.294)
(see p.298)

BREAKTHROUGH

SPECTROSCOPY

As scientists improved their ability to split light into a spectrum in the 19th century— in a technique known as spectroscopy— it became possible to identify finer details. The spectrum of sunlight was found to be crossed with many dark "absorption lines" at specific colors, while light from heated laboratory samples was found to be a series of bright "emission lines" of certain colors. This led to studying the spectral "signature" of chemical elements in gaseous form. Each element has its own unique signature.

Emission lines of helium
The spectrum comprises red, yellow, cyan (two lines), and blue (three lines).

The gamma knife
Highly focused beams of gamma radiation delivered by an instrument called a gamma knife are used to treat specific types of brain tumors. The high energy carried by the radiation is used to kill cancerous cells.

of light had been opened out to reveal a huge range of new radiations, each with its own properties. The quest to put these newly discovered waves into use would be a driving force throughout the following century.

AFTER

Maxwell's theories allowed scientists to develop their understanding of light, and led to new theories and discoveries.

LIGHT AS A WAVE AND A PARTICLE
Ultraviolet fails to follow the pattern of Maxwell's equations. This ultimately led to the development of quantum physics 314–17 ❯❯.

LASER TECHNOLOGY
Scientists invented a means of producing "coherent light," with the peaks and troughs of all its waves "in step" in the 1950s. This is the key to laser technology 364–65 ❯❯.

IMPROVED IMAGING
Radio telescopes and satellite observatories gather information about the Universe from invisible radiation. The Fermi space telescope maps gamma rays to reveal high-energy sources, such as pulsars and black holes.

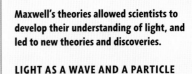

SEE ALSO ❯❯
pp.314–17 QUANTUM REVOLUTION
pp.364–65 LASERS AND HOLOGRAMS

FERMI IMAGE OF THE WHOLE SKY

BEFORE

Long-distance messaging has existed since ancient times, in the form of smoke signals, burning torches, waving flags, and mirrored light reflections.

SIGNAL FIRES
One of the earliest means of communicating over long distances was to use fire. American Indians were able to send complex signals by controlling the **smoke** from a fire, while in other countries **chains of fires** within sight of one another were used to pass messages.

NAPOLEONIC SEMAPHORE
In the 1790s Frenchman **Claude Chappe** devised a semaphore system that used **telegraph towers** with signaling arms mounted on the roof. This device spread widely across Europe.

A CLAUDE CHAPPE SIGNAL TOWER

« SEE ALSO
pp.252–53 GENERATING ELECTRICITY
pp.256–57 NATURE OF SOUND
pp.258–59 ELECTROMAGNETIC SPECTRUM

Early methods of sending signals instantaneously over long distances were known as telegraphy, from the Greek for "distant writing." There were several problems with the various optical telegraph systems developed by the early 1800s: they were slow and limited in the detail they could convey, they required a direct line of sight between the stations, and they were heavily reliant on the weather. Semaphore stations could only operate in daylight, and even signal fires and heliographs (messages sent by mirrors) were useless in fog or low cloud.

Signaling with electricity
Electricity offered an obvious means of communication, since it propagated over long distances almost instantaneously, but the problems of producing a working system based on electrostatic discharges initially proved

Women working at a telephone exchange
In 1882 London had around 35 operators; by 1910 New York employed 6,000. The operators plugged cords into a panel: one for the incoming call, and a second for the receiving party's local circuit. Long-distance calls were connected to a trunk circuit between exchanges.

laboratory curiosity for another two decades or more. Three key discoveries were needed to make the telegraph and all later wired telecommunication systems possible. The first was the

Telegraph to Telephone

Leading on from the line-of-sight telegraphs of the 1800s, the Victorian age saw a rapid development in methods of long-distance communication, making full use of the growing knowledge and use of electricity to develop more efficient telegraph and telephone devices.

Movable metal armature
Spring holding up armature so that movable and fixed contacts are separated
Electromagnet (wire wrapped around an iron core)

Armature telegraph relay
Joseph Henry's relay machine was used to telegraph messages across long distances. When the coil is energized, it attracts the armature, pulls it down, closes the contacts, and completes the circuit.

SCIENTIST AND INVENTOR (1847–1922)
ALEXANDER GRAHAM BELL

Born in Edinburgh, Scotland, Bell moved to Canada with his family at the age of 23, and later settled in Boston, Massachusetts. His family had a long tradition of teaching linguistics and elocution, and Bell too became a distinguished teacher in this field. However, he combined it with a passion for experimentation, and persuaded his father-in-law, a wealthy Boston lawyer, to back his plans for an improved telegraph. In 1874, he employed an electrical engineer, Thomas Watson, who did much to make the telephone a reality.

insurmountable. It was only after the Italian scientist Alessandro Volta discovered a means of generating steady current in 1799 that an electric telegraph became a real possibility. German experimenter Samuel von Sömmerring demonstrated a working system just a few years later in 1809, although the telegraph would remain a

130,000 POLES
held up the 14,000 miles (22,500 km) of copper wire used to make connections from New York to San Francisco.

galvanometer, invented by the Dane Hans Christian Ørsted in 1820. This made the strength of an electric current visible through the deflection of an electric needle. The second was English scientist William Sturgeon's electromagnet (invented in 1825, and considerably improved in the US by Joseph Henry in 1828), which provided a means of multiplying the magnetic force generated by a comparatively feeble current. Using these two inventions alone, the great German scientists Karl Friedrich Gauss and Wilhelm Weber were able to build the first operational telegraph system in 1833. The final piece of the jigsaw, invented by Henry in 1835, was the relay: a device capable of boosting a weak signal, so that it could travel down a much greater length of wire.

Telegraphy in action
In 1837 US inventors Samuel Morse and Alfred Vail devised the system that would eventually become the standard: the Morse Code. This entailed sending pulses of current of different lengths

("dots and dashes") down a single wire. The code was then easily transcribed into letters. A series of high-profile demonstrations secured US government backing for the system, and it was rapidly adopted, spreading widely across the continental United States. In 1866, Europe and America were linked via the first transatlantic cables.

Further innovations were rapid: the addition of a telegraph "sounder" that

An early telegraph machine, 1837
The telegraph system devised by British inventors Cooke and Wheatstone used six wires, activated in pairs to make compass needles point to letters on a dial. It was expensive to install and slow to operate.

generated sound from the current pulses made the Morse system even easier to use, since it freed the operator's eyes for transcribing the message. Teletype machines soon automated the transcription process, and by 1843 Alexander Bain had even invented a primitive machine that could transmit a picture to a remote receiver using only telegraph signals.

Birth of the telephone

The story of the telephone's early development is a complex one, confused by rival claims, but the patent for the first telephone system was awarded to Scottish-born US inventor Alexander Graham Bell in 1876. Bell was being funded to work on a system for sending multiple telegraph messages simultaneously using different signal frequencies (using a technique known as multiplexing).

In order to convert sound into an electrical signal, Bell's telephone used a thin membrane that vibrated when sound was directed toward it, moving a rod of magnetized iron in front of an electromagnet. At the other end of the wire, the signal passed into a receiver electromagnet, causing a thin metallic disk to vibrate and produce sound waves. Within a short time news of the invention had spread around the world.

AFTER »

The arrival of radio waves and wireless transmission caused another telecommunication revolution in the 20th century.

WIRELESS COMMUNICATION
Wireless broadcasting of a single signal to a large number of receivers caught on rapidly, as did **wireless telegraphy** for long-range communications such as those between ships.

2021 SMARTPHONE

CELL PHONES
Wireless telephony is now commonplace. **Mobile telephones**, which rely on land-based radio masts, and **satellite telephones** that operate via orbiting satellites, provide global communication and powerful computing in a small handset.

SEE ALSO »
pp.268–69 RADIO AND RADIO WAVES
pp.368–69 ARTIFICIAL SATELLITES
pp.378–79 THE INTERNET

Earpiece

Mouthpiece

NATIONAL TELEPHONE SERVICE

◁ **Ericsson telephone, 1890**
The Ericsson company was founded by Lars Magnus Ericsson (1846–1926), a Swedish telegraph instrument maker who began repairing Bell's early telephones.

Crank handle
Crank drives a dynamo that sends a current to the exchange or another telephone, signaling that the user wishes to call.

Bell
An incoming signal from the exchange causes a bell to ring.

Diaphragm
The diaphragm moves in and out, driven by the electromagnet, alternately increasing and lowering the pressure of the air in front of it.

Wire coils
Coils act as an electromagnet, moving the permanent magnet and causing the diaphragm to vibrate.

Sound emission
Regions of compression and rarefaction move out as sound waves that enter the listener's ear.

Cavity space
Gap between magnet and diaphragm allows air to reverberate, amplifying sound.

Electrical current
Electricity flows between wires through a carbon "button," creating a variable current depending on the resistance of the carbon and the strength of the sound waves.

Perforated screen
Screen protects the delicate diaphragm behind it.

Sound absorption
Sound waves of compressed and rarefied air produced by the speaker's voice enter the phone.

Carbon-granule button
Carbon granules held behind the diaphragm are compressed or released by the pressure of the sound waves, causing their electrical resistance to vary.

Input signals

Output signals

INSIDE A MODERN TELEPHONE

Motion study
This series of motion study photographs was taken by the British photographer Eadweard Muybridge in 1887. Muybridge made his name when he photographed horses in motion. He proved that all the animal's hooves leave the ground and are tucked under the body when galloping.

Photography

From its early beginnings as the preserve of dedicated enthusiasts, photography was to become crucial in transforming our view of the world. Geographic expeditions, wars, everyday human society, and flora and fauna could all be recorded, leading to a much greater understanding of our shared environment.

‹‹ BEFORE

While the way in which light rays react when passing through a small hole was understood more than 2,000 years ago, it was not until the 19th century that the phenomenon was translated into a way of making a permanent image.

FIRST PRINCIPLES
In the 5th century BCE the Chinese philosopher **Mozi** discovered that **light rays passing through a pinhole** into a darkened room would create an **inverted image** on the surface opposite the hole. In the 4th century BCE **Aristotle** was also aware of the principle. **Alhazen** wrote about the pinhole camera in 1021 CE in the *Book of Optics*.

LIGHT INTO ART
In the 16th century **Giovanni Battista della Porta** (1538–1615) described the use of the **camera obscura** ("darkened room") as a drawing aid. Artists including **Canaletto** and **Vermeer** are thought to have used the camera obscura to help them draw more accurately.

PRINTING THE IMAGE
In Britain in 1802, **Thomas Wedgwood** and **Sir Humphry Davy** presented a paper on the results of their work in the field of photography. They had experimented with **creating images on light-sensitive materials**, but they were unable to make the images permanent.

‹‹ SEE ALSO
pp.60–61 **ALHAZEN**

The 19th-century pioneers of photography often worked in isolation, experimenting with different ways of making light-sensitive surfaces. In Brazil, Antoine Florence covered glass plates with gum arabic (a natural adhesive taken from the acacia tree) and soot, while in England Thomas Wedgwood and Humphry Davy immersed paper and leather in silver nitrate. In France, Joseph Nicéphore Niepce tried similar methods; they all came up against the same problem: the light-sensitive surface continued to darken and the image was lost.

Fixing the image
In about 1826 Niepce exposed a pewter plate coated with bitumen of Judea, a light-sensitive form of natural tar, to daylight for several days. After rinsing the plate with a mixture of lavender oil and petroleum

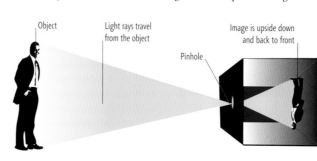

An early daguerreotype
This photograph of Le Sueur's statue of Charles I on Whitehall, London, was taken in 1839. The image has been reversed to show the scene as it existed, since daguerreotypes only produce reversed views.

oil, he found he had captured a negative image—a process he called heliography.

In 1833 Louis-Jacques-Mandé Daguerre discovered that a silver plate exposed to iodine would develop a latent image when exposed to light.

The image could be made visible by treating it with mercury fumes and then, crucially, fixed with a solution of table salt dissolved in hot water. The British chemist John Herschel also managed to fix a photographic image. His cyanotype process, using iron salts

DESIGNER (1789–1851)

LOUIS DAGUERRE

Born in Val-d'Oise, France, Daguerre was a noted designer of theater sets and dioramas, for which he used a camera obscura as a drawing aid. His experiments in fixing a photographic image came from the desire to speed up his work rather than the interest in the science of photography that drove other pioneers in the field. Daguerre took little interest in further explorations in photography and retired the year after his process was publicized.

Pinhole camera
The most basic camera is a pinhole, consisting merely of a box that is light-tight except for one small aperture through which light can pass.

Object

Light rays travel from the object

Pinhole

Image is upside down and back to front

How black-and-white film works
The film consists of a cellulose acetate base on to which the other layers are coated. The image is formed in the emulsion layer, which contains the light-sensitive silver halides, a chemical mixture that usually includes silver bromide, silver iodide, and silver chloride. Until the film is developed, the image is latent.

SCRATCH-RESISTANT COATING
EMULSION
CELLULOID
ANTI-CURL LAYER
ANTI-HALATION LAYER

Anti-halation layer stops light from scattering

How color negative film works
Silver halides in the emulsion layers are mixed with spectral sensitizers to make them respond to different colors of the spectrum. The silver halides react to light in the same way as in black-and-white film, but the image is converted to color by dye couplers in each layer.

SCRATCH-RESISTANT COATING
BLUE-SENSITIVE EMULSION LAYER
YELLOW FILTER
GREEN-SENSITIVE EMULSION LAYER
MAGENTA INTERLAYER
RED-SENSITIVE EMULSION LAYER
CELLULOID
ANTI-CURL AND -HALATION LAYER

AFTER

Digital cameras continue to develop, offering higher-quality images with each new model. Cameras enable us to find out more about the Universe.

PHOTOGRAPHING THE UNIVERSE
Cameras sent into space on probes and satellites send back **images of our galaxy and others**, extending our knowledge about stars 328–29 **»** and possible **life on other planets**.

PRESERVING THE RAIN FORESTS
Launched in February 2021, the satellite **Amazônia-1** 368–69 **»** is equipped with a wide-angle camera able to resolve 60m (195ft) objects on Earth. It will record illegal logging in the Amazon basin.

SEE ALSO »
pp.378–79 THE INTERNET
pp.396–97 SPACE PROBES AND TELESCOPES

Digital SLR (single-lens reflex)
The light comes into the lens of the camera and is reflected off the mirror and upward. It hits the top and side of the prism before passing through the viewfinder. The lower mirror flies upward on exposure to allow light to hit the sensor.

Built-in pop-up flash unit
Light is reflected around the pentaprism to the viewfinder
Built-in focusing motor
Zoom lens groups
Path of light through the camera
Circuit block for connectors
Sub-mirror for auto-focus detection
Mirror reflects light up toward the prism

1 MILLION Number of 35 mm cameras produced by Leica up to 1961. Over three decades the Leica became a favorite of many important photographers, including Henri Cartier-Bresson.

Although it could not capture as much detail as the daguerreotype, the calotype process enabled the multiple reproduction of the image—the photograph as it became universally recognized.

Eleven years later Frederick Scott Archer's wet collodion process, using glass coated with silver salts, largely took over from the calotype and daguerreotype processes, since its greater sensitivity to light much reduced exposure times. In 1871, English physician Richard Maddox introduced gelatin-coated dry plates that gave faster exposure times still.

The arrival of film
In the 1880s flexible nitrate-based plastic roll film was developed by George Eastman, along with the first mass-market cameras loaded with enough film to take 100 photographs. Once the customer had exposed the film the camera was returned to Kodak, where the film was developed and prints made. These were then sent back to the customer along with the camera, loaded

to produce a blue image, was inexpensive and simple, making it commercial and popular.

In 1840 William Fox Talbot developed the calotype, the first method to use the negative–positive process that is still in use today.

with fresh film. The first color transparency film was produced by Kodak in 1935. This was followed, in 1941, by color negative film that could be purchased from most pharmacies. Photography was no longer exclusive to professionals.

The digital era
Digital technology was first used by television in the 1950s, and later for mapping the Moon. The first professional digital stills camera was an adapted Nikon F3, released in 1991, but it was not until the 21st century that digital gained dominance over film in this area. For consumers, the digital camera was immediately successful, allowing instant gratification and the ability to distribute images across the Internet.

Development of cameras
Originally bulky pieces of equipment with a basic mechanism, cameras have become highly sophisticated and so small they can be incorporated in a mobile phone.

c.1840 Fox Talbot's camera from which he produced the first photographs using the negative–positive process.

c.1885 Handmade wooden cameras were used in daylight studios to produce fashionable portraits.

1925–36 The Leica A series introduced small, portable 35 mm film cameras that produced high-quality images.

1969 Thirteen modified Hasselblad 500E cameras were taken on the Apollo Moon landings.

1980 Nikon introduced the F3. In 1991, with a Kodak digital back, it became the first digital camera for professional use.

1984 The Nikonos V was the latest in the range of the legendary underwater cameras from Nikon. It remained in production until 2001.

2016 High-resolution digital sensors in smartphones largely replace consumer-level cameras, forcing some manufacturers out of business.

1840 | 1860 | 1880 | 1900 | 1920 | 1940 | 1960 | 1980 | 2000

c.1840 The world's first widely available daguerreotype camera (below).

1871 The gelatin dry plate was introduced by Richard Maddox.

1915 The Minnigraph, made in Germany, was the first European camera to use 35 mm cine film, producing half-frame 18 x 24 mm negatives.

1957–75 The Hasselblad 500C (left) was the first of the company's V series, which became the industry standard for professionals.

1963 Kodak released the Instamatic, an inexpensive snapshot camera. By 1970, sales topped 50 million.

1990 The Polaroid Spectra Pro offered instant results along with multiple exposures and self-timed sequential pictures.

2008 Canon's PowerShot SX200 IS joined the travel-zoom camera market, aimed at travelers wanting a lightweight digital model.

INVENTOR AND ENTREPRENEUR **Born 1847** Died 1931

Thomas Edison

"**Genius** is one percent **inspiration** and 99 percent **perspiration**."

THOMAS EDISON

Thomas Alva Edison is considered to be the greatest inventor of his time. Among his world record 1,093 patents were key inventions that ushered in the modern world. He was one of the first men to realize the true potential of electricity and harnessed it to bring electric lights, sounds, and even moving pictures into our homes.

Edison was born in Milan, a port on the Huron River in Ohio. By the age of seven, his insatiable curiosity had worn his teacher's patience so far that his mother decided that home schooling was the only option. Convinced that her son's idiosyncrasies resulted from a remarkable brain, she gave "little Al" the self-belief to change the world.

A young entrepreneur

Edison's voracious inquisitiveness was soon joined by entrepreneurship. After selling newspapers, snacks, and candy on the local railroads, at the age of 12 he started his own business selling fruit and vegetables. The newspaper he started at 14, the *Weekly Herald*, soon had 300 commuting subscribers.

One day in 1862 Edison saved the three-year old son of J. U. MacKenzie, a Michigan station master, from an oncoming boxcar. In appreciation, MacKenzie taught him Morse code and telegraphy. By the age of 16 Edison had produced his first invention, an "automatic repeater" to transmit signals between unmanned stations.

Having moved to Boston to work as a telegrapher for Western Union, Edison gave up full-time employment in 1869

to become an inventor. By June he had produced an electric vote recorder. However, his "beautifully constructed" machine met a distinctly lukewarm reception—it was simply too quick for the politicians' liking. From then on, Edison would waste no more time on inventions that no one wanted.

A succession of patents

After moving to New York, Edison patented several more improvements in telegraphy, culminating in the development of a quadruplex telegraph that could send two messages in each direction simultaneously.

By 1876 Edison was able to open a laboratory in New Jersey. Here, he and his staff would develop several different inventions at any one time. One example was a telephone transmitter that improved the clarity and volume of the voices transmitted over Alexander Graham Bell's nascent system of communications.

It was Edison's work on telephones that led to his first major invention: the phonograph. Realizing that sounds could be recorded as indentations on a rapidly moving piece of paper, he soon produced a machine that used a needle to score grooves on a tinfoil-coated cylinder. He was justifiably delighted when it replayed his own voice saying "Mary had a little lamb." Despite initial enthusiasm, however, the phonograph was destined to remain a virtual

novelty for another 10 years, by which time Edison's attention had shifted to other even more ambitious projects.

Lighting the world

On November 15, 1878, the inventor formed the Edison Electric Light Co., with the intention of producing, not just an incandescent bulb, but a lighting system that could be extended throughout an entire city. The success of Edison's system drew on other technological advances. These included the development of a longer-lasting carbonized filament by British inventor Joseph Swan (see p.250),

Thomas Edison
Photographed c.1900, Edison examines motion picture film threaded through one of his film projectors. His first projector, known as a projecting kinetoscope, was patented in 1887. The Edison Company also produced film footage for home entertainment.

Edison's 1877 phonograph
Sounds entering the central mouthpiece caused a diaphragm to vibrate, making indentations with a stylus on tinfoil wrapped around the rotating drum. A second stylus and diaphragm could play back the recording.

and a host of other crucial components: the parallel circuit; an improved dynamo; an underground conductor network; various devices to maintain constant voltage; safety fuses; and light sockets with on–off switches. By 1882 the world's first commercial lighting

Menlo Park
Edison stands in his chemical research laboratory at Menlo Park, New Jersey. The electric light bulb and many other patented inventions were developed in electrical laboratories on the same site.

system—400 lamps in total— had been installed in Pearl Street, Lower Manhattan. A year later, 513 customers were using 10,300 lamps. The electricity that powered them was provided by arguably Edison's most important invention—the first commercially viable power station. Electricity would eventually bring light to millions of homes worldwide.

Man of extraordinary vision

Edison was not the first to invent a light bulb, or the power station, but he was the man who understood how to put all the components together and demonstrate to the world the possibilities afforded by electricity. By the end of his life Edison held 1,093 patents for his inventions and innovations. He returned to the phonograph, developing the most advanced sound systems of his day, invented the alkali storage battery, and also helped to kick-start the age of motion pictures with the kinetoscope, one of the world's first projectors of moving images. "I never quit until I get what I'm after," he once said.

East River Power Plant
The world's first power station at Pearl Street, New York City, burned down in 1890. It was succeeded in 1901 by the Waterside plant on Fourteenth Street and the East River; its capacity of 120,000 watts was 10 times that of Pearl Street.

TIMELINE

- **February 11, 1847** Thomas Alva Edison is born in Milan, Ohio.
- **1859** Starts work on the Grand Trunk Railroad, selling newspapers and confectionery.
- **1863–67** Works as a telegraph operator in various cities of the Midwest.
- **1868** Moves to Boston to work for Western Union and invents an automatic vote recorder.
- **January 1869** Becomes a full-time inventor.
- **December 25, 1871** Marries one of his employees, Mary Stillwell.
- **1874** Invents the quadruplex telegraph for Western Union.
- **1876** Moves to Menlo Park, New Jersey, and constructs his first full-scale research laboratories, designed to perform electrical, chemical, and mechanical experiments.
- **December 7, 1877** Demonstrates the phonograph at the offices of Scientific American.
- **November 15, 1878** Forms Edison Electric Light Co.
- **December 31, 1879** Demonstrates the first fully functioning incandescent lighting system at the Menlo Park laboratories.

EDISON'S INCANDESCENT LIGHT BULB

- **September 1882** Opens the first permanent central power station and public lighting system, on Pearl Street. He joined forces with Bergmann & Company to market electrical equipment; the company eventually becomes General Electric.
- **1884** Mary (Stillwell) Edison dies.
- **1886** Marries Mina Miller.
- **1887** Moves into a new laboratory in West Orange, New Jersey.
- **1888** Becomes embroiled in the "war of the currents" with George Westinghouse and Nikola Tesla. His promotion of direct current (DC) systems is ultimately unsuccessful.
- **1893** Demonstrates the kinetograph and kinetoscope.
- **1896** Introduces the home phonograph.
- **1899** Unsuccessfully promotes concrete for cheap housing.

WAX MUSIC CYLINDER MADE BY EDISON

- **1900** The first modern research laboratory opens at General Electric. Begins working on a battery for use in electric cars.
- **1915** Employed to lead naval investigations into new military technologies.
- **1920s** With deteriorating health, he continues his experiments at home. Spends much effort trying to find an alternative to rubber.
- **October 18, 1931** Dies in Llewellyn Park, New Jersey. Lights are dimmed across the US.

Capturing Sound

During the second half of the 19th century inventors competed to build the first commercially viable machine that was capable of recording and reproducing sound. The recording industry that emerged was transformed during the 20th century by the invention of magnetic tape and digitally encoded sound.

In 1857 Frenchman Edouard Leon-Scott patented a device that was capable of recording – in visible form – the shape of sound waves. The "phonautograph" used a horn-like receiver to focus sound on to a membrane, which vibrated a bristle that in turn scratched patterns on a glass plate blackened with smoke. The phonautograph was used in the study of sound, but it would be another 20 years before someone successfully recorded and reproduced sound.

First sound recording

The first person to record sound was US inventor Thomas Alva Edison (see pp.264–65), who demonstrated his "phonograph" device late in 1877. Edison's system used the pressure of sound waves to create up-and-down motions in a stylus, which in turn created indentations in a tinfoil sheet wrapped around a rotating cylinder and guided by spiral grooves. After a recording had been made, the stylus was allowed to ride along the grooves again, and the resulting vibrations recreated the original sound.

Edison's phonograph was an immediate sensation, but had many limitations. Perhaps the most obvious problem was that each recording was singular, while each time the cylinder

Sound recorded as indentations in tin foil wrapped around grooved cylinder

Combined stylus and loudspeaker apparatus

Handle to rotate cylinder

Edison's first machine
Edison got the idea for sound recording from hearing the sounds made by a telegraph recording device. His first crude machine used tin foil as the medium for preserving sound waves.

was played back, the wave pattern was worn down a little bit and the sound quality deteriorated.

Early improvements

Wax cylinders, introduced in the late 1880s, were more durable than Edison's tinfoil originals, and could more easily be produced in greater numbers. Mechanical devices were used to copy a groove pattern onto several cylinders at the same time.

However, in 1887 Emile Berliner invented the "gramophone," which used flat disks instead of cylinders; these were easier to store, were more durable, and allowed for easier sound reproduction. The disks had a single spiral groove from the center to the edge, and each disk could simply be "pressed" on a metallic master to reproduce the grooves. Sound was recorded as a side-to-side movement rather than an up-and-down one. By the 1920s gramophone disks had almost completely replaced cylinder recordings.

Electronic recording

Sound recording was a mechanical process until the 1920s, when advances in valve technology led to electrical recording, in which changing current was used to drive an electromagnetic recording head. Now microphones, which transformed the oscillation of sound waves into an electrical signal, could be used in recording. Amplifier circuits removed limits on the volume of playback, allowing louder sound.

Electronic technology also opened the way for a new recording medium: magnetic tape. The principle of capturing a varying wave pattern by lining up the iron oxide particles on a wire or tape had been established in the 1890s, but the microphone made it possible to record sound waves in this format. The first commercial systems appeared in the US during the 1940s. Magnetic tape and electronic

The gramophone
Early gramophones relied on a large speaker horn to amplify the sound from a vibrating diaphragm naturally. In order to achieve sufficient volume, the stylus was held down onto the recorded track with considerable force, causing rapid deterioration of the stylus and disk.

 BEFORE

Before sound could be recorded, it was first necessary to understand its nature.

SOUND WAVES
Work done on the properties of sound from the 1600s onward had established that sound was a **wave of compression** carried through the air, which **stimulated the eardrum** by causing it to vibrate.

EARLY STORED MUSIC
Automated musical instruments that could **play back musical notes** stored on a rotating cylinder or paper punch card first appeared in Baghdad, Iraq, as early as the 9th century.

The telegraphone
The earliest magnetic recording system, pioneered by Danish engineer Valdemar Poulsen in 1899, used wire rather than a magnetic tape to record sound signals.

Breathing and Respiration

All living organisms burn fuel like engines, although the fuel they burn is sugar and fat. The energy released from this process is used for movement, growth, and a host of other functions that sustain life. Breathing and respiration are the fundamental characteristics of life that enable this energy release to take place.

B reathing is the process that brings fresh air into the lungs and allows oxygen in air to pass into the bloodstream for transport to body cells, where it is used in respiration to release energy from food.

Breathing

When the body inhales, the intercostal muscles between the ribs and the diaphragm both contract, moving the ribs up and out, and flattening the diaphragm so that air flows into the lungs. When the body exhales, these muscles relax, moving the ribs down and in, and making the diaphragm dome-shaped so that air is pushed out of the lungs.

The breathing process enables an exchange of gases within the lungs. Air enters minute air-filled sacs called alveoli, then oxygen diffuses into the blood via millions of microscopic capillaries, which permeate the lungs. Waste carbon dioxide diffuses in the opposite direction to be expelled in air from the body on exhalation.

The blood then carries oxygen to the rest of the cells in the body. In the mid-1800s German physiologist Felix Hoppe-Seyler discovered how this happens—oxygen is picked up by the blood flowing through lung capillaries by binding to hemoglobin (a red iron-containing pigment) found inside red blood cells. When blood reaches other cells in the body where oxygen levels are much lower, hemoglobin gives up the oxygen to the cells.

Aerobic and anaerobic respiration

Respiration releases heat and keeps mammals and birds warm-blooded. Respiratory energy is also used for cell activities: building protein and DNA,

Red blood cells
Disk-shaped cells have oxygen attached to their hemoglobin.

3 Respiring cells
As blood cells reach capillaries surrounded by respiring cells, oxygen bound to their hemoglobin is freed for use in aerobic respiration.

Matrix contains enzymes to initiate energy release

Cristae are the sites where most of the energy is released

Inner membrane folded into cristae

Outer membrane

△ **Mitochondrion: site of aerobic respiration**
Mitochondria are found in the cytoplasm of cells. Each capsule is bounded by a double membrane: the inner is folded into cristae to maximize the surface area for enzymes, which catalyze (speed up) the reactions of aerobic respiration; the outer is a smooth boundary.

performance, extra energy can be extracted without oxygen through breaking down molecules including sugar. Lactic acid, rather than alcohol is the by-product of this process in animals.

In 1897 German chemist Eduard Büchner found that whole cells were not needed for fermentation—extracts of yeast could ferment glucose. He had found enzymes—the chemical catalysts vital to numerous metabolic processes.

dividing cells, or pumping substances in and out of cells (see pp.214–15). After food is digested in the gut (see pp.216–17), glucose is carried in the blood to cells. Each glucose molecule is divided into molecules of pyruvic acid and transferred into organelles called mitochondria (see above). Here the pyruvic acid reacts with oxygen in aerobic respiration, releasing energy in the form of adenosine triphosphate (ATP).

Anaerobic respiration (also known as fermentation) was first recognized in microorganisms; in 1838 French physicist Cagniard de la Tour discovered that a yeast culture produces alcohol (a by-product of anaerobic respiration) when supplied with glucose. Anaerobic respiration also occurs in animals: when the demands of vigorous activity reach the limits of cardiovascular

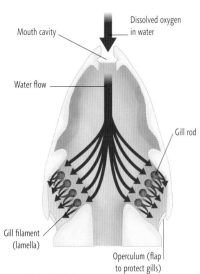

Dissolved oxygen in water

Mouth cavity

Water flow

Gill rod

Gill filament (lamella)

Operculum (flap to protect gills)

Breathing in fish
Aquatic organisms have gills instead of lungs. Water enters the mouth of a fish and flows over gill filaments, which are packed with blood capillaries to exchange oxygen and carbon dioxide between water and blood.

The Five Senses

Sensation is a fundamental characteristic of life: animals use their senses to take in information about the environment around them. Sensory cells called receptors are specialized to detect factors such as light, sound, pressure, taste, and smell.

Sense organs contain receptor cells that can detect changes in our environment and so gather information about the world around us. Different kinds of stimuli, such as light, sound, and pressure are detected by specific receptors. The strength of a stimulus determines the frequency of electrical impulses that travel along sensory nerves to the brain. Detection of stimuli is usually beneficial: the pleasurable sensation of taste is the promise of nourishment. Even pain (or fear of it) can help us avoid potentially damaging stimuli, such as burning-hot objects.

Detecting light

The eye focuses light onto the retina with its cornea and lens (see right), producing an inverted image that is later corrected by the brain. In 1856 German physician Hermann von Helmholtz proved how the eye focuses light: ciliary muscles encircling the lens control its shape, contracting to bend light from near objects or relaxing to adjust it from objects farther away. In the same decade Heinrich Müller showed that light is detected by photoreceptor cells (rods and cones) in the retina. These cells contain rhodopsin (a light-absorbing pigment), which makes light trigger electrical signals that are relayed to the brain, for processing into visual images.

Detecting chemicals

The mechanisms of our senses of taste and smell are the same: substances reach the sensory parts of the mouth and nose (whether as odor or food) to stimulate receptor cells, and the brain interprets this as flavor (see below).

❮❮ BEFORE

The ancient Greek idea that sensation was a separate process from thought promoted studies on how the eyes and ears could pick up signals of light and sound.

DETECTING THE SIGNAL

During the Renaissance many still agreed with the Roman physician **Galen** that the **eye's lens** was the **site** of **visual sensation**. In 1583 Swiss physician **Felix Platter** proved that vision was possible even after disconnecting the lens and correctly surmised that the **retina** was the part of the eye **stimulated by light.**

ANATOMY IN MINIATURE

Revealing the **structure of the ear** was a challenge to early anatomists. Andreas Vesalius **❮❮ 72–73** advised that the inner ear should be removed during dissection to allow close scrutiny; Italians **Eustachi** and **Fallopio** dissected its parts—**identifying** the **eardrum, tiny bones**, and its intricate **labyrinth of tubes** for the first time.

❮❮ SEE ALSO
pp.220–21 THE NERVOUS SYSTEM
pp.222–23 THE BRAIN

Choroid
Dark, blood-rich layer that cuts internal reflection of light in the eyeball.

Superior rectus
Small muscle that is used for moving the eyeball.

Iris
Ring of muscles that relax or contract to let more or less light through the central pupil.

Sclera
Outer opaque (white) wall of the eyeball.

Pupil
An opening in the iris, through which light passes to the lens.

Cornea
Transparent layer to let light into the eye. It is coated with the thinner protective conjunctival membrane.

Lens
Soft transparent disc that focuses light onto the retina.

Ciliary muscle
Ring of muscle that changes the shape of the lens to focus light.

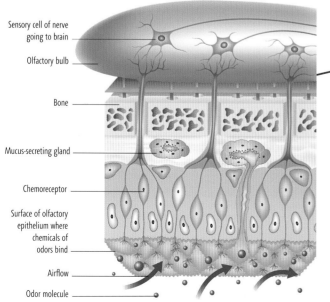

Sensory cell of nerve going to brain

Olfactory bulb

Bone

Mucus-secreting gland

Chemoreceptor

Surface of olfactory epithelium where chemicals of odors bind

Airflow

Odor molecule

NERVE PATHWAYS FOR TASTE AND SMELL

◁ **How we detect smell**
When odorous chemicals are breathed in, they dissolve in the mucus lining of the nasal cavities. Here, chemoreceptor sensory cells have surface proteins, which bind to specific types of chemicals. Once stimulated, the cells transmit nerve impulses along nerves to the brain.

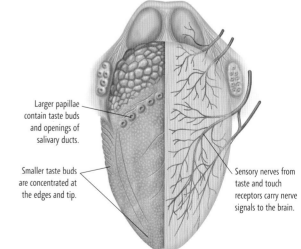

Larger papillae contain taste buds and openings of salivary ducts.

Smaller taste buds are concentrated at the edges and tip.

Sensory nerves from taste and touch receptors carry nerve signals to the brain.

△ **How we detect taste**
The tongue has a surface coating of taste buds, each of which contains around 100 chemoreceptor cells. When these are stimulated, they fire electrical impulses along sensory nerve fibers to the brain.

ARCTIC OCEAN

KENYA

Lake
Natron

TANZANIA

Lake
Eyasi

Wildebeest migration
On the African savannah, wildebeest migration ensures that the young are generally born in areas where the rains ensure plentiful grass for them to eat.

PACIFIC
OCEAN

PACIFIC
OCEAN

ATLANTIC
OCEAN

INDIAN
OCEAN

KEY
Blue wildebeest (inset)
Arctic tern
Atlantic salmon
Rorqual whales
European eel
Monarch butterfly
Barn swallow

SOUTHERN OCEAN

Migration routes
Migration is triggered by a cue, such as change of day length. Knowing the route is innate (a "fixed behavior pattern") may be modified by learning.

AFTER »»

3,500 MILES (5,600 kilometers) of a connected hive of billions of Argentine ants, stretching from Portugal to Italy, the largest supercolony ever recorded.

is associated with another. A Russian physiologist, Ivan Pavlov, discovered simple conditioning with his experiments on dogs in the 1890s (see below). Another form of conditioning was discovered 30 years later by American psychologist B.F. Skinner. In a laboratory, he found that an animal who accidentally triggered a switch to deliver food would learn from this experience, so the switch-flicking

"We animals are the most complicated things in the known universe."

RICHARD DAWKINS, BIOLOGIST, 1986

would become reinforced because of promise of food. In the wild, a bird that stumbles across food by flicking over leaves will learn to associate leaf-flicking with the promise of food, so this behavior is reinforced.

For some time, and largely due to the influence of American psychologist Edward Thorndike, most apparently intelligent animal behavior was thought to be explained by a process of trial and error. But then a German psychologist proposed something different. In his studies of captive chimpanzees, Wolfgang Köhler found that solving problems (such as stacking crates in order to reach bananas) required more insight. This kind of

behavior is the most sophisticated of all—it is behavior with purpose and guided by inspiration. Problem solving explains the superior skills involved in the most basic of biological functions, including acquiring food, avoiding predators, and raising offspring. It also encourages the use of tools and tool modification—something that reaches its pinnacle of achievement in the human species.

Jane Goodall
Regarded as the world's leading chimpanzee expert, Jane Goodall spent several decades living among chimpanzees in Tanzania. Her observations of their behaviour led to critical steps in understanding our close relatives; she showed that they have distinct personalities, can use tools, and are not vegetarian.

In the 1970s the science of ethology (animal behavior) became increasingly influenced by genetics and ecology.

EVOLUTIONARY GAMES
In 1976 English evolutionary biologist Richard Dawkins published *The Selfish Gene*, which stressed that the **evolution** of **organisms** is driven by **competition between genes**: an individual body is a "community" of working genes and the behavior of the body maximizes the reproduction of these genes.

SOCIOBIOLOGY
American biologist Edward O. Wilson was instrumental in urging that **animal behavior could evolve by natural selection**, and that this could even lead to the **evolution of complex societies**. The nonbreeding drones in a beehive, for example, can pass on their genes indirectly by helping the queen, with whom they share genetic material.

SEE ALSO »
pp.332–33 ECOLOGY AND ECOSYSTEMS
pp.334–35 CONSERVATION BIOLOGY

INNATE BEHAVIOR Pre-set by genes, this is often the predominant type of behavior seen in animals with simple nervous systems. Because it is inherited, it can evolve by the process of natural selection.

LEARNED BEHAVIOR This behavior is modified by experience so it is not stereotyped. It is most sophisticated in animals with well-developed nervous systems. Learned behavior is not inherited, but the potential to learn is.

BREAKTHROUGH

PAVLOV'S DOG

Ivan Pavlov's experiments on dogs in the 1890s—initially aimed at investigating digestion—proved to be a landmark in the understanding of animal behavior. He was the first scientist to demonstrate classical conditioning—where an animal learns to associate one stimulus (such as the sound of a bell) with another (food). Pavlov measured the degree of salivation by dogs in response to these stimuli. He found that if food was presented in association with a sound, then eventually the dogs would learn to associate the two stimuli, and salivate in response to sound alone.

Cycles in the Biosphere

Compared with Earth's total mass of nonliving matter, the biosphere has evolved as a thin veneer on its surface, comprising the living world on the land and in the oceans. In the biosphere life interacts with the atmosphere above and the soil below through the processes of respiration, photosynthesis, death, and decay.

Organisms do not live in isolation from the environment around them. Organisms affect the air when they extract oxygen and excrete carbon dioxide during respiration

Effect of combustion on carbon cycle
A forest fire, such as at Puebla, Mexico (see below), causes carbon to be released into the air. But the ash returns minerals to soil, stimulating regrowth in plants, which absorb carbon dioxide as they photosynthesize.

(see pp.270–71). And every living thing is reduced to its separate chemical components during the inexorable advance of decay after its death. Organisms are made up of the same sorts of atoms that exist in the nonliving world and these atoms shuffle from life to nonlife as chemical processes recycle them through the biosphere. Such ideas were first proposed by a Russian geochemist called Vladimir Vernadsky (see above), who popularized the idea of the biosphere in his book of the same name, first published in 1926. One element is abundant in all living bodies:

carbon. Carbon atoms can bond with other atoms to form complex organic molecules, which are the basis of life. By eating food, animals convert one complex molecule into another. But, from 450 million years ago, evolving land plants have made food from simple inorganic materials: carbon and oxygen from carbon dioxide in the air; hydrogen from water; and nitrogen from nitrates in the soil.

The carbon cycle
The carbon atoms that make up biological molecules are returned to the atmosphere as carbon dioxide whenever an organism respires. Decay involves the same process: decomposers obtain food and energy from dead material. Plants recapture this carbon when they make food—using the energy of the Sun in the process of photosynthesis (see pp. 154–55) to convert carbon dioxide and water into sugars, fats, and protein. In doing so, they effectively make the carbon cycle complete.

There is a natural balance between the carbon locked in living bodies and the carbon of the atmosphere and other nonliving reservoirs, such as chalk. But if combustion speeds the rate of carbon emission, and this is not matched by a corresponding increase in photosynthesis, then carbon dioxide

levels rise. A great deal of carbon has (become locked into organic matter—the so-called carbon sinks—over extraordinary periods of time: hundreds of years for long-lived trees, and millions for coal and oil deposits. But combustion can happen in an instant, causing rapid rises in atmospheric carbon dioxide.

The nitrogen cycle
A significant amount of organic material contains nitrogen, mostly in the form of protein and DNA. However, even though about 70 percent of the air is pure nitrogen, it is inaccessible to most living organisms in this form. Plants rely on the action of various types of bacteria and decomposers to convert nitrogen in dead matter first into ammonia, then into nitrites, and finally into nitrates. Animals obtain their nitrogen in the food they eat (see opposite). Nitrogen in the atmosphere can only be "fixed" into more usable forms naturally by nitrogen-fixing bacteria or the action of lightning.

Decomposition
The breakdown of dead organic matter in compost happens because bacteria and fungi feed on this material and digest it. In doing so, they release chemicals, such as nitrates, which plants absorb for growth.

BEFORE

The elements of carbon and nitrogen were characterized by chemists in the 1700s.

IDENTIFYING THE CYCLES
The **carbon cycle was identified** in the 18th century, culminating in 1796, when Swiss pastor **Jean Senebier** showed that inorganic carbon dioxide was taken up by plants in photosynthesis. Meanwhile, French chemist **Antoine Lavoisier** discovered that **much of the gas in air was inert—nitrogen**. By 1785 French chemist **Claude Berthollet** found that **nitrogen was a component of ammonia**, but processes of the cycle were not understood for another 100 years.

72 TRILLION TONS of carbon is locked up in the rocks of our planet.

87 percent of carbon dioxide from human sources comes from burning fossil fuel; changes in land use account for much of the rest.

‹‹ SEE ALSO
pp.146–47 DISCOVERY OF GASES
pp.154–55 HOW PLANTS WORK
pp.270–71 BREATHING AND RESPIRATION

"Living matter is the most powerful geological force."

VERNADSKY, "THE CHEMICAL STRUCTURE OF THE BIOSPHERE AND ITS SURROUNDINGS," 1965

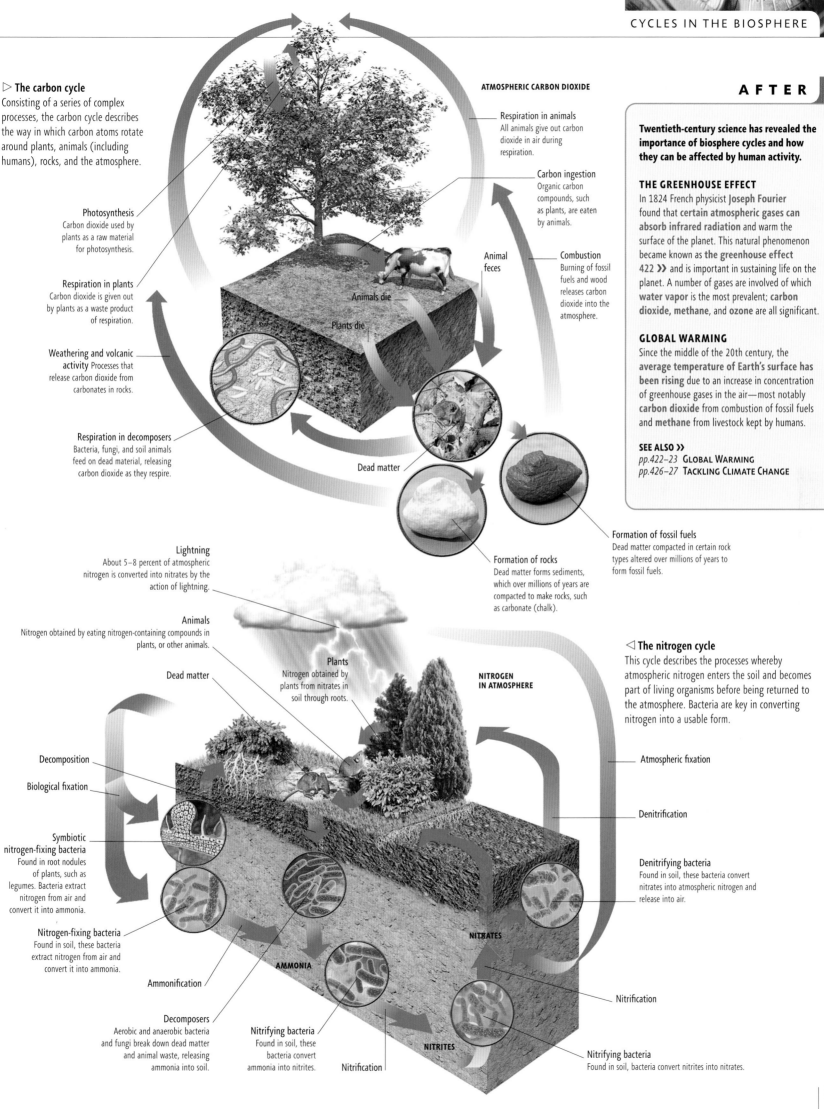

▷ The carbon cycle

Consisting of a series of complex processes, the carbon cycle describes the way in which carbon atoms rotate around plants, animals (including humans), rocks, and the atmosphere.

Photosynthesis
Carbon dioxide used by plants as a raw material for photosynthesis.

Respiration in plants
Carbon dioxide is given out by plants as a waste product of respiration.

Weathering and volcanic activity Processes that release carbon dioxide from carbonates in rocks.

Respiration in decomposers
Bacteria, fungi, and soil animals feed on dead material, releasing carbon dioxide as they respire.

ATMOSPHERIC CARBON DIOXIDE

Respiration in animals
All animals give out carbon dioxide in air during respiration.

Carbon ingestion
Organic carbon compounds, such as plants, are eaten by animals.

Animal feces

Combustion
Burning of fossil fuels and wood releases carbon dioxide into the atmosphere.

Animals die

Plants die

Dead matter

Formation of rocks
Dead matter forms sediments, which over millions of years are compacted to make rocks, such as carbonate (chalk).

Formation of fossil fuels
Dead matter compacted in certain rock types altered over millions of years to form fossil fuels.

Lightning
About 5–8 percent of atmospheric nitrogen is converted into nitrates by the action of lightning.

Animals
Nitrogen obtained by eating nitrogen-containing compounds in plants, or other animals.

Dead matter

Plants
Nitrogen obtained by plants from nitrates in soil through roots.

NITROGEN IN ATMOSPHERE

Decomposition

Biological fixation

Symbiotic nitrogen-fixing bacteria
Found in root nodules of plants, such as legumes. Bacteria extract nitrogen from air and convert it into ammonia.

Nitrogen-fixing bacteria
Found in soil, these bacteria extract nitrogen from air and convert it into ammonia.

Ammonification

AMMONIA

Decomposers
Aerobic and anaerobic bacteria and fungi break down dead matter and animal waste, releasing ammonia into soil.

Nitrifying bacteria
Found in soil, these bacteria convert ammonia into nitrites.

Nitrification

NITRITES

NITRATES

Atmospheric fixation

Denitrification

Denitrifying bacteria
Found in soil, these bacteria convert nitrates into atmospheric nitrogen and release into air.

Nitrification

Nitrifying bacteria
Found in soil, bacteria convert nitrites into nitrates.

◁ The nitrogen cycle

This cycle describes the processes whereby atmospheric nitrogen enters the soil and becomes part of living organisms before being returned to the atmosphere. Bacteria are key in converting nitrogen into a usable form.

AFTER

Twentieth-century science has revealed the importance of biosphere cycles and how they can be affected by human activity.

THE GREENHOUSE EFFECT

In 1824 French physicist **Joseph Fourier** found that **certain atmospheric gases can absorb infrared radiation** and warm the surface of the planet. This natural phenomenon became known as **the greenhouse effect 422 »** and is important in sustaining life on the planet. A number of gases are involved of which **water vapor** is the most prevalent; **carbon dioxide, methane**, and **ozone** are all significant.

GLOBAL WARMING

Since the middle of the 20th century, the **average temperature of Earth's surface has been rising** due to an increase in concentration of greenhouse gases in the air—most notably **carbon dioxide** from combustion of fossil fuels and **methane** from livestock kept by humans.

SEE ALSO »
pp.422–23 Global Warming
pp.426–27 Tackling Climate Change

4

THE ATOMIC AGE

1890–1970

Just when it seemed in 1890 that science had discovered all there was to know, the atom was broken open to reveal a dizzying array of smaller particles along with enormous amounts of energy. Two wholly new kinds of science—quantum theory and Einstein's relativity—were called into being to try to understand it all.

THE ATOMIC AGE

1890–1970

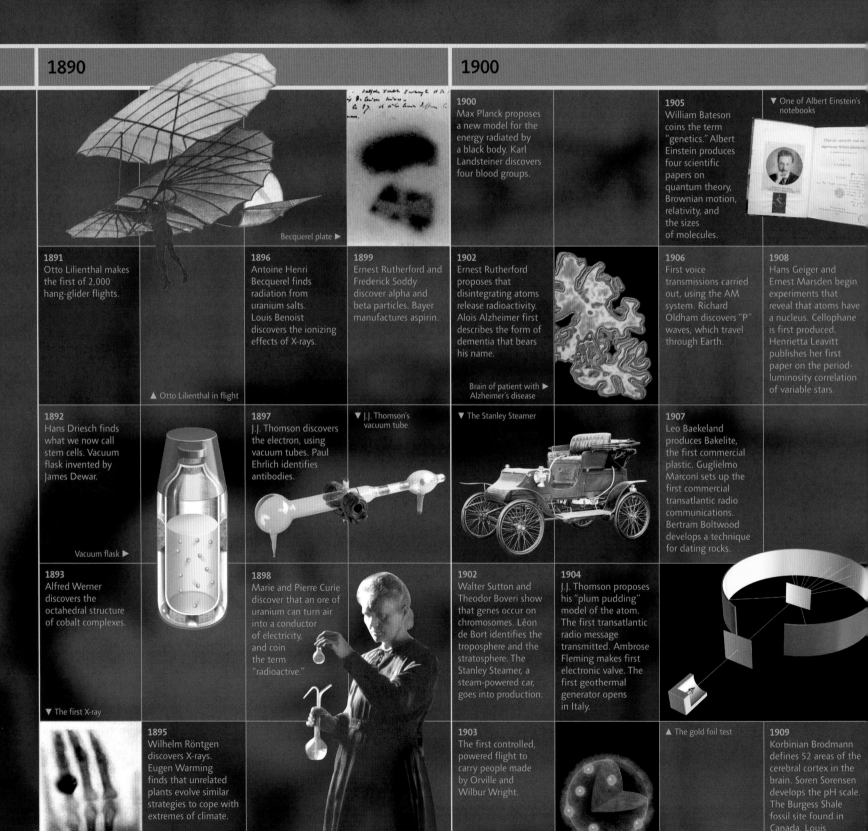

1890

1891
Otto Lilienthal makes the first of 2,000 hang-glider flights.

▲ Otto Lilienthal in flight

1892
Hans Driesch finds what we now call stem cells. Vacuum flask invented by James Dewar.

Vacuum flask ▶

1893
Alfred Werner discovers the octahedral structure of cobalt complexes.

▼ The first X-ray

1895
Wilhelm Röntgen discovers X-rays. Eugen Warming finds that unrelated plants evolve similar strategies to cope with extremes of climate.

Becquerel plate ▶

1896
Antoine Henri Becquerel finds radiation from uranium salts. Louis Benoist discovers the ionizing effects of X-rays.

1897
J.J. Thomson discovers the electron, using vacuum tubes. Paul Ehrlich identifies antibodies.

1898
Marie and Pierre Curie discover that an ore of uranium can turn air into a conductor of electricity, and coin the term "radioactive."

Marie Curie ▶

1899
Ernest Rutherford and Frederick Soddy discover alpha and beta particles. Bayer manufactures aspirin.

▼ J.J. Thomson's vacuum tube

1900

1900
Max Planck proposes a new model for the energy radiated by a black body. Karl Landsteiner discovers four blood groups.

1902
Ernest Rutherford proposes that disintegrating atoms release radioactivity. Alois Alzheimer first describes the form of dementia that bears his name.

Brain of patient with ▶ Alzheimer's disease

1902
Walter Sutton and Theodor Boveri show that genes occur on chromosomes. Léon de Bort identifies the troposphere and the stratosphere. The Stanley Steamer, a steam-powered car, goes into production.

1903
The first controlled, powered flight to carry people made by Orville and Wilbur Wright.

Plum pudding model ▶ of the atom

▼ The Stanley Steamer

1904
J.J. Thomson proposes his "plum pudding" model of the atom. The first transatlantic radio message transmitted. Ambrose Fleming makes first electronic valve. The first geothermal generator opens in Italy.

1905
William Bateson coins the term "genetics." Albert Einstein produces four scientific papers on quantum theory, Brownian motion, relativity, and the sizes of molecules.

▼ One of Albert Einstein's notebooks

1906
First voice transmissions carried out, using the AM system. Richard Oldham discovers "P" waves, which travel through Earth.

1907
Leo Baekeland produces Bakelite, the first commercial plastic. Guglielmo Marconi sets up the first commercial transatlantic radio communications. Bertram Boltwood develops a technique for dating rocks.

▲ The gold foil test

1908
Hans Geiger and Ernest Marsden begin experiments that reveal that atoms have a nucleus. Cellophane is first produced. Henrietta Leavitt publishes her first paper on the period-luminosity correlation of variable stars.

1909
Korbinian Brodmann defines 52 areas of the cerebral cortex in the brain. Soren Sorensen develops the pH scale. The Burgess Shale fossil site found in Canada. Louis Blériot flies across the English Channel.

The startling discovery that the atom is not the smallest particle spurred scientists to explore the extraordinary "quantum" rules that apply at the smallest scale. In one direction, research led in 1945 to the unleashing in the atom bomb of the extraordinary energy bound within the atomic nucleus. In another, it led in the 1950s to the revelation of the structure of DNA, the molecule inside every living cell that carries all life's instructions as a chemical code. At the other end of the scale, Einstein's theories overturned the accepted view of space and time—just as Hubble revealed that our galaxy is but one of many galaxies in a gigantic, ever-expanding universe, flung out from its origin long ago in the "Big Bang."

1910

1920

1910
Paul Ehrlich discovers that salvarsan kills syphilis bacteria.

1911
Ernest Rutherford proposes that a cloud of electrons surrounds an atomic nucleus. Thomas Hunt Morgan finds genes are carried on chromosomes.

1912
Alfred Wegener proposes the theory of continental drift. Max von Laue proves X-rays are short-wave electromagnetic radiation. Vesto Slipher measures redshifts of galaxies.

▲ Ernest Rutherford's atomic model

▼ Niels Bohr's model of the atom

1913
Arthur Holmes publishes *The Age of the Earth*, in which he estimates that Earth is around 1.6 billion years old. Niels Bohr proposes the Bohr model of the atom.

▼ Albert Einstein

1915
Albert Einstein's General Theory of Relativity published.

1916
Harlow Shapley estimates the size of the Milky Way. Lennart von Post studies pollen grains to record the distribution of plant species.

▼ Cross section of kidney

1917
Arthur Cushny finds how urine is formed in the kidneys. The Hooker telescope is completed and is the world's largest telescope until 1948.

1918
Sonar systems are used by the US and British navies. Ernest Rutherford discovers protons and hypothesizes the existence of neutrons. The Enigma cipher machine built.

◄ Enigma cipher machine

1919
Arthur Eddington finds supporting evidence for Einstein's General Theory of Relativity, by measuring stars during a solar eclipse.

▼ Solar eclipse

1920
Hans Winkler coins the term "genome." Frederick Hisaw, Harry Fevold, and Samuel Leonard find that hormones from the pituitary gland cause changes in the ovaries.

1921
Edward Mellanby discovers Vitamin D and shows that its absence causes rickets.

1923
Edwin Hubble uses the Cepheid variable stars in the Andromeda galaxy to prove that it is outside the Milky Way.

Andromeda galaxy ▶

1924
Edwin Hubble confirms that many faint "nebulae" are galaxies beyond the Milky Way. Arthur Eddington discovers the relationship between mass and luminosity in stars.

1925
The Mid-Atlantic ridge is discovered. Wolfgang Pauli proposes his exclusion principle. Cecilia Payne-Gaposchkin proposes that stars are mostly composed of hydrogen and helium.

Mid-ocean ridge ▶

1926
First successful rocket launch by Robert Goddard. Vladimir Vernadsky publishes *The Biosphere*, stating that all matter rotates in natural cycles. John Logie Baird gives the first demonstration of a working television system. Vinyl made.

1927
Werner Heisenberg publishes the uncertainty principle of quantum physics. The quartz clock invented. Farnsworth demonstrates a practical TV-picture scanning system.

Werner Heisenberg ▶

1928
Mark Lidwell and Edgar Booth develop an artificial pacemaker for the heart. Baird Television broadcasts the first transatlantic television signal.

▼ *Penicillium* mold

1929
Alexander Fleming discovers penicillin. Edwin Hubble shows that the Universe is expanding.

1930
Paul Dirac publishes *The Principles of Quantum Mechanics*. Neoprene and polyester first made.

1931
The electron microscope invented.

▼ Splitting uranium-235

1935
Arthur Tansley develops the concept of an ecosystem. Charles Richter devises the Richter magnitude scale to measure earthquakes. Nylon first made.

▲ Charles Richter

c.1940
New layers in the atmosphere are discovered—the mesosphere and the ionosphere.

The atmosphere ▶

▲ Germany's V-2 rocket

1937
Grote Reber builds the first radio telescope. The first shipborne radar systems available.

1938
Otto Hahn discovers an isotope of barium produced by the breakdown of uranium.

1941
Herman Kalckar discovers that adenosine triphosphate (ATP) is the main energy-transferring molecule inside cells. Styrofoam (expanded polystyrene) first produced.

1944
Germany's V-2 rockets, the first ballistic missiles, fall on London.

1932
Cockcroft and Walton carry out the first artificial nuclear fission. James Chadwick proves the existence of neutrons in an atom.

1934
Kodak produces the first color transparency film. Irène Joliot-Curie and Frédéric Joliot create the first artificial radioactivity.

▼ Linus Pauling

1939
Helmut Ruska first sees a virus using an electron microscope. The first jet aircraft has its maiden flight. Linus Pauling publishes *The Nature of the Chemical Bond*. The Cygnus A galaxy is discovered by its radio output.

1945
Dorothy Hodgkin discovers the structure of penicillin. The Trinity Test is the first artificial nuclear explosion.

Dorothy Hodgkin ▶

1933
Thomas Hunt Morgan wins the Nobel Prize for discovering the role of chromosomes in heredity. Fritz Zwicky infers that dark matter must exist in the Universe.

▼ Human chromosomes

1942
The first fission reactor, Chicago Pile-1, goes into operation as part of the Manhattan Project.

Manhattan Project ▶

1946
Genetic recombination demonstrated.

1948
George Gamow and Ralph Alpher predict the proportions of light elements produced by the birth of the Universe.

▼ Structure of a virus

1943
Oswald Avery proves that DNA is the carrier of genes in cells. The Colossus code-breaking machine and first electronic computer is built.

1947
The transistor is invented. Dennis Gabor develops theory of the hologram.

Security hologram ▶

"The stellar system is a **swarm of stars** isolated in space. It drifts through the Universe as a **swarm of bees** drifts through the summer air."

EDWIN HUBBLE, "THE REALM OF THE NEBULAE," 1936

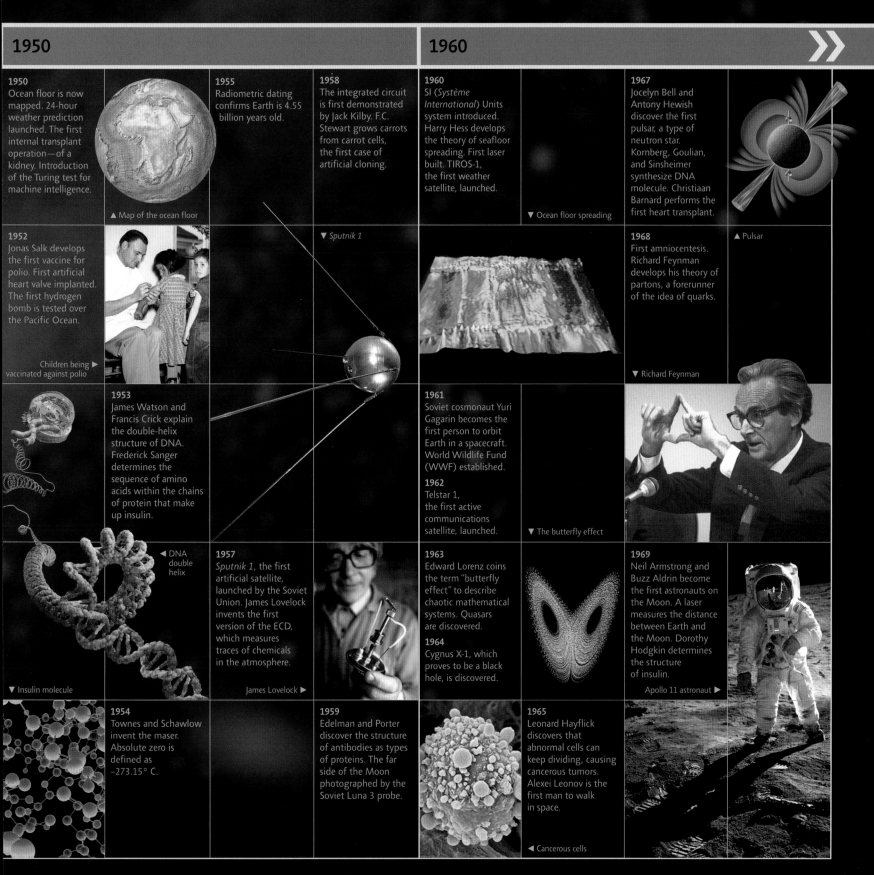

1950

1950
Ocean floor is now mapped. 24-hour weather prediction launched. The first internal transplant operation—of a kidney. Introduction of the Turing test for machine intelligence.

▲ Map of the ocean floor

1952
Jonas Salk develops the first vaccine for polio. First artificial heart valve implanted. The first hydrogen bomb is tested over the Pacific Ocean.

Children being ▶
vaccinated against polio

1953
James Watson and Francis Crick explain the double-helix structure of DNA. Frederick Sanger determines the sequence of amino acids within the chains of protein that make up insulin.

◀ DNA double helix

▼ Insulin molecule

1954
Townes and Schawlow invent the maser. Absolute zero is defined as −273.15° C.

1955
Radiometric dating confirms Earth is 4.55 billion years old.

1957
Sputnik 1, the first artificial satellite, launched by the Soviet Union. James Lovelock invents the first version of the ECD, which measures traces of chemicals in the atmosphere.

James Lovelock ▶

1958
The integrated circuit is first demonstrated by Jack Kilby. F.C. Stewart grows carrots from carrot cells, the first case of artificial cloning.

▼ *Sputnik 1*

1959
Edelman and Porter discover the structure of antibodies as types of proteins. The far side of the Moon photographed by the Soviet Luna 3 probe.

1960

1960
SI (*Système International*) Units system introduced. Harry Hess develops the theory of seafloor spreading. First laser built. TIROS-1, the first weather satellite, launched.

▼ Ocean floor spreading

1961
Soviet cosmonaut Yuri Gagarin becomes the first person to orbit Earth in a spacecraft. World Wildlife Fund (WWF) established.

1962
Telstar 1, the first active communications satellite, launched.

▼ The butterfly effect

1963
Edward Lorenz coins the term "butterfly effect" to describe chaotic mathematical systems. Quasars are discovered.

1964
Cygnus X-1, which proves to be a black hole, is discovered.

1965
Leonard Hayflick discovers that abnormal cells can keep dividing, causing cancerous tumors. Alexei Leonov is the first man to walk in space.

◀ Cancerous cells

1967
Jocelyn Bell and Antony Hewish discover the first pulsar, a type of neutron star. Kornberg, Goulian, and Sinsheimer synthesize DNA molecule. Christiaan Barnard performs the first heart transplant.

▲ Pulsar

1968
First amniocentesis. Richard Feynman develops his theory of partons, a forerunner of the idea of quarks.

▼ Richard Feynman

1969
Neil Armstrong and Buzz Aldrin become the first astronauts on the Moon. A laser measures the distance between Earth and the Moon. Dorothy Hodgkin determines the structure of insulin.

Apollo 11 astronaut ▶

285

BEFORE

Throughout the 19th century scientists worked to piece together the mysteries of the elements and their chemical reactions.

ATOMIC THEORY
John Dalton's atomic theory of matter **《** 138–39, published in the early 1800s, proved very successful in explaining both physical phenomena and chemical processes, and led to a **revolution in chemistry** in the 19th century.

THE PERIODIC TABLE
Mendeleev's periodic table **《**230–33 revealed that **elements fell neatly into groups** with similar chemical reactions and apparent patterns to their masses. **There were gaps** in the table as many elements had not yet been discovered.

《 SEE ALSO
pp.138–39 THE NATURE OF MATTER
pp.232–33 THE PERIODIC TABLE
pp.234–35 CHEMICAL REACTIONS

Static electrons
Thomson recognized the existence of negatively charged electrons within the atom, but thought that they were static.

Cloudlike body
The body of the atom was thought to be cloudlike.

Thomson's "Plum Pudding" model (1904)
In J.J. Thomson's model of the atom, negatively charged electrons were dotted randomly, like plums in a pudding, through a positively charged sphere that contained most of the atom's mass. There is no nucleus.

Nucleus

Orbiting electrons

Rutherford's model (1911)
Ernest Rutherford proposed that an atom consisted of a small, dense core of positively charged particles (the nucleus), around which negatively charged electrons orbited, like moons around a planet.

Structure of the Atom

At the beginning of the 20th century J.J. Thomson's discovery of the electron revolutionized our view that the atom was the basic unit of matter and gave rise to the first model of the atom. Over the next 30 years further insights into atomic structure led to the development of other more refined models.

New Zealand-born chemist Lord Ernest Rutherford is often known as the "father" of nuclear physics, both for his own pivotal discoveries and for his influence over later generations. The son of an immigrant farmer from Britain, he became Professor of Physics at the University of Manchester in 1907, and won the 1908 Nobel Prize for chemistry for his groundbreaking proof that radioactivity was caused by the disintegration of atoms. He won many awards, gaining a life peerage in 1931.

In 1897 British physicist Joseph John Thomson began an exhaustive series of experiments into cathode rays that would lead to a revolution in our understanding of the atom. Using a type of cathode ray tube called a Crookes Tube (see pp.292–93), he was able to test how the rays behaved in electric and magnetic fields.

As the rays passed through these fields, they bent toward a positively charged electric plate, so Thomson realized they must be negatively charged particles generated by the cathode itself (rather than electromagnetic radiation as others had proposed). He called these particles "corpuscles," but they later became known as electrons.

The particles had several other characteristics: they always appeared to have the same charge, and they had a mass less than one thousandth of a hydrogen atom. It seemed clear that the particles were coming from inside the atoms of the cathode—proving that the atoms themselves were made up of smaller parts hidden within them.

From plum pudding to nucleus
Thomson's discoveries won him the Nobel Prize for physics in 1906, but this was just the beginning of the investigation. Since it seemed impossible to extract positive charge from an atom, and yet atoms as a whole were known to be electrically neutral, Thomson developed his own theory of the atom to explain the anomaly. He suggested that the negative corpuscles (electrons) were moving around inside a uniform

Nucleus   **Gold atom**

Most alpha particles pass straight through the atom or are slightly deflected by the nucleus.

Scintillation screen
A screen coated in zinc sulfide emits light when struck by alpha particles, revealing their path through the gold foil.

DETAIL OF GOLD FOIL

A few particles hit the nucleus head-on and rebound.

Slit in lead sheet
This slit ensures the alpha particles reach the foil as a narrow beam, and filters out particles from other directions.

Alpha beam
Alpha particles are beamed on to the gold foil.

Alpha source
Alpha particles emitted by this radioactive source spread out in all directions.

Gold foil
Beaten to an extremely thin sheet, the gold foil is just a few atoms thick.

Lead block
This encases and shields the radium, which produces the rays.

The gold leaf experiment
The ingenious experiment developed by Rutherfordand his students, Geiger and Marsden, used alpha particles to probe the properties of individual atoms, revealing that they are mostly made up of empty space.

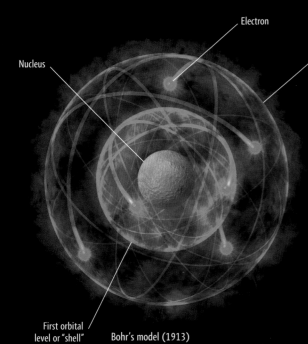

Nucleus

Electron

Second orbital level or "shell"

First orbital level or "shell"
The "shell" with the lowest energy level contains a maximum of two electrons.

Bohr's model (1913)
To explain how atoms emitted and absorbed light, Niels Bohr developed a model in which the orbits of electrons were fixed by atomic "shells" that each contained a fixed number of electrons.

Proton
The number of positively charged protons determines an element's atomic number.

Neutron
Uncharged neutrons with the same mass as a proton help to make up the atom's overall mass.

Electron
The number of negatively charged electrons balances the number of protons in a neutral atom.

Nucleus of protons and neutrons
The discovery of protons and neutrons within the nucleus further refined Bohr's model. The number of protons determines an element's chemical properties.

sphere of positive charge that accounted for most of the atom's mass. The particles were like plums in a Christmas pudding, so his model duly became known as the "plum pudding" model.

In 1909, while working under Ernest Rutherford, Hans Geiger and Ernest Marsden tested this idea. They fired alpha particles from a radium source at a thin sheet of gold leaf and measured the paths taken by the particles. The plum pudding model suggested the rays would only be slightly deflected, but in fact some were deflected by more than 90° and others rebounded from the foil. Rather than being like plum puddings, atoms actually concentrated their mass and positive charge in a small area at the center—the nucleus. In 1911 Rutherford proposed that electrons orbited the nucleus of an atom like moons around a planet.

Electrons in fixed orbits
Another of Rutherford's students, Danish physicist Niels Bohr, used the new ideas emerging from quantum mechanics to explain what kept the electrons in orbit. The Bohr model suggested that the electrons were confined in fixed orbits by atomic "shells," and that the chemical properties of an element were governed by the number of electrons in the outer shell. He also said that the amount of energy an electron had was closely linked to how far it was from the nucleus of the atom. Electrons, he suggested, could absorb or emit radiation of specific wavelengths and colors in order to move between

inner and outer shells, or orbitals. According to Bohr, when a particle carrying energy—either another atom, an electron, or even a photon— collides with the atom, an orbiting electron is given a boost of energy and jumps into an empty space within the outer higher-energy shell. If there is an empty space in the lower-energy shell below it, the electron drops back into it, losing the energy it had gained and releasing a photon, a particle of light.

Nuclear structure
There were still some problems however. The patterns in the reactions of elements (signified by their atomic numbers) did not always match the patterns in their masses. In 1913 Henry Moseley showed that atomic number was a real property that indicated the

positive electric charge in the nucleus. But if this was not always proportional to its mass, how were the two properties decided? Rutherford solved part of the problem in 1918 by identifying particles with a positive electric charge that was equal and opposite to the electron's negative charge. These particles were soon named protons, but they did not account for an atom's entire mass— the remainder seemed to lack any electric charge at all. Rutherford named these particles neutrons. It was another 14 years before the neutron's existence was finally proven, by James Chadwick, who won the Nobel Prize for his work in 1935.

AFTER »

Understanding the basic units of atomic structure led to further advances that would revolutionize the 20th century.

NUCLEAR PHYSICS
Nuclear fission 324–25 » was discovered by **Hahn, Meitner,** and **Strassmann** in 1938, leading ultimately to nuclear power stations and the fission (atomic) bomb.

PARTICLES TO WAVES
By 1926 electrons were no longer thought to act as particles but to exist as waves within electron clouds. A **mathematical model of the atom** was devised by the quantum physicist **Erwin Schrödinger**. This model

PROBABLILITY DENSITY OF AN ELECTRON IN A HYDROGEN ATOM

reimagined the electron not as a single object orbiting a nucleus, but as a probability function describing an area where the electron was likely to be found. This area—an orbital—could be represented as a cloud.

SEE ALSO »
288–89 CHEMICAL BONDS
364–65 LASERS AND HOLOGRAMS
382–83 SUBATOMIC PARTICLES

THEORETICAL PHYSICIST (1906–72)
MARIA GOEPPERT MAYER

Born and educated in Germany, Goeppert Mayer moved to the US in the 1930s with her husband, Joseph Edward Mayer. Throughout her scientific career, despite being a clear talent in theoretical physics, Goeppert Mayer had to fight to be taken seriously and even get paid for her work. She made her major contribution during an unpaid professorship at the University of Chicago, where she resolved the structure of the nucleus in her nuclear shell model. She shared the Nobel Prize in 1963 with German physicist Hans Jensen and his colleagues, who had independently made the same discovery.

BEFORE

Once the idea of matter being made of small particles became accepted, there had to be some mechanism to hold them together.

ROMAN IDEAS

Around 60 BCE the Roman poet **Titus Lucretius Carus**, who is more commonly known simply as Lucretius, wrote an **epic philosophical poem** called *De rerum natura* ("On the nature of the Universe"). In this poem, which ranged over fundamental topics such as the **mind, spirit, atoms, and the Universe,** Lucretius suggested that the **world,** including people and their minds, was **made entirely of atoms.** These **tiny spheres** were **identical in substance,** but varied in shape and size. He believed that these atoms were in **perpetual motion** at great speeds, with random "swerves" to provide humans with free will, and that they had little **"hooks" to hold them together**—"tightly packed and closely joined."

LUCRETIUS

« SEE ALSO
pp.22–23 ELEMENTS OF LIFE
pp.138–39 THE NATURE OF MATTER
pp.286–87 STRUCTURE OF THE ATOM

Chemical Bonds

Atoms, the smallest particles of elements, are held together by five major types of bond: covalent, ionic, metallic, and hydrogen bonds, and van der Waals forces. All of these bonds are formed by the movements of electrons, but the electrons are shared or transferred in various ways, according to the nature of the atoms.

Covalent bonds often form between light atoms; molecules of hydrogen, oxygen, chlorine, and water are all held together by covalent bonds. A hydrogen atom has just one electron whizzing around the nucleus, in a region called an orbital. When two hydrogen atoms combine to form a molecule—H_2—they share the electrons, so that each hydrogen nucleus has a pair of electrons in an orbital.

The electrons orbiting the nucleus of an atom are arranged in a number of "shells" (see below). Compounds are likely to be stable if the shells of their atoms are full, or complete. When atoms form covalent bonds they share electrons to achieve this stability. An atom's innermost shell is full when occupied by two electrons, so hydrogen, with a single electron, forms one bond to become stable, as in H_2. Helium, with two electrons, forms no bonds at all,

since its shell is already full. The second shell needs eight electrons to be complete; so carbon, with four electrons in its second shell, forms four covalent bonds, as in CH_4 (methane); nitrogen, with five, forms three bonds (see below); oxygen, with six, forms two, as in H_2O; and chlorine, with seven electrons, forms one bond, as in HCl (hydrogen chloride).

Ionic bonds

Mineral salts such as sodium chloride, NaCl (common salt), do not have shared electrons. Instead the atoms lose or gain electrons to complete their shells. Sodium has one electron in its outer shell. It can achieve a filled shell by losing this electron to become a positively charged sodium ion, Na^+. Meanwhile the chlorine atom can fill its outer shell by gaining one electron and becoming a negatively charged chloride ion, Cl^-. The sodium atom

gives an electron to the atom of chlorine, so that both atoms have complete shells, and the substance becomes Na^+Cl^-. Its crystals do not contain molecules, but millions of alternating ions arranged in a regular, 3-D pattern and held together by the attraction between the positive and negative charges.

Hydrogen atom in bond has slight positive charge

Electrostatic attraction forms a hydrogen bond

The oxygen atom in the bond takes more than an equal share of the pair of electrons, and so is slightly negative

Hydrogen bonds
Hydrogen bonds are weak electrical attractions between hydrogen atoms in groups such as OH (oxygen–hydrogen) or NH (nitrogen–hydrogen) in one molecule and an oxygen or nitrogen atom in another. Here they are holding together water molecules (H_2O).

Central nucleus — One complete inner shell and four extra electrons in outer shell — Two complete inner shells and one extra electron in outer shell — Each of these two oxygen atoms now has a full outer shell—four of its own electrons and four shared

CARBON ATOM   **SODIUM ATOM**

Electron shells
Atoms are stabilized when the nucleus is surrounded by a filled "shell" of electrons—two in the first shell, eight in the second, and eight in the third.

Double bonds
Double (covalent) bonds are formed between two atoms when both have two spare electrons ready to pair. This is the structure of an oxygen molecule, O_2 or O=O.

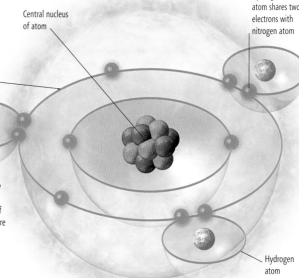

Hydrogen atom shares two electrons with nitrogen atom

Central nucleus of atom

When bonded with three hydrogen atoms, the nitrogen atom has eight electrons in its outer shell—two of its own and six shared with hydrogen atoms

Hydrogen nucleus

Covalent bonds
Covalent bonds are formed mainly between light elements sharing electrons. In this representation of the ammonia molecule (NH_3) there are covalent bonds between the larger nitrogen atom and each of the smaller hydrogen atoms.

Hydrogen atom

CHEMIST (1829–96)

AUGUST FRIEDRICH KEKULÉ

Born in Darmstadt, Germany, August Kekulé was educated in Germany, Paris, Switzerland, and London before becoming professor at the University of Bonn. He studied the valency of elements (the number of bonds their atoms can form), and in 1857 he announced that carbon has a valency of 4. Kekulé later became the primary originator of the idea of chemical structure. His most dramatic discovery was that the structure of benzene, C_6H_6, is a symmetrical ring of carbon atoms. Kekulé allegedly "saw" this arrangement after having a dream of ouroboros—a snake biting its own tail—which is a symbol common to many ancient cultures.

Metallic bonds and hydrogen bonds

Atoms in solid metals sit in a regular, 3-D pattern, and share the loose outer electrons from all the atoms. It is like a grid of positive metal ions bathed in a sea of loose electrons that can move freely through the grid (see below). Metals are good conductors, since these free electrons can travel easily through the metal carrying electrical current or heat.

Covalent compounds with OH and NH groups often have hydrogen bonds between their molecules. Oxygen is more "electronegative" than hydrogen; it takes more than its fair share of the electrons that form a bond between them, giving the oxygen atom a slight negative charge and the hydrogen a slight positive charge. The opposing charges result in attractions between neighboring molecules, especially in water, where a hydrogen atom on one molecule forms a hydrogen bond with the oxygen atom on another molecule.

Van der Waals forces

Substances made of lone atoms, such as neon and argon, or small molecules, such as nitrogen and oxygen, tend to be gases with low boiling points. The only forces between the lone atoms or the separate molecules are van der Waals forces. These forces are named after the Dutch scientist Johannes Diderik van der Waals, who, in 1873, was one of the first to postulate their existence. Van der Waals forces are much weaker than any other type of bond, and are rather like a slight stickiness between the particles.

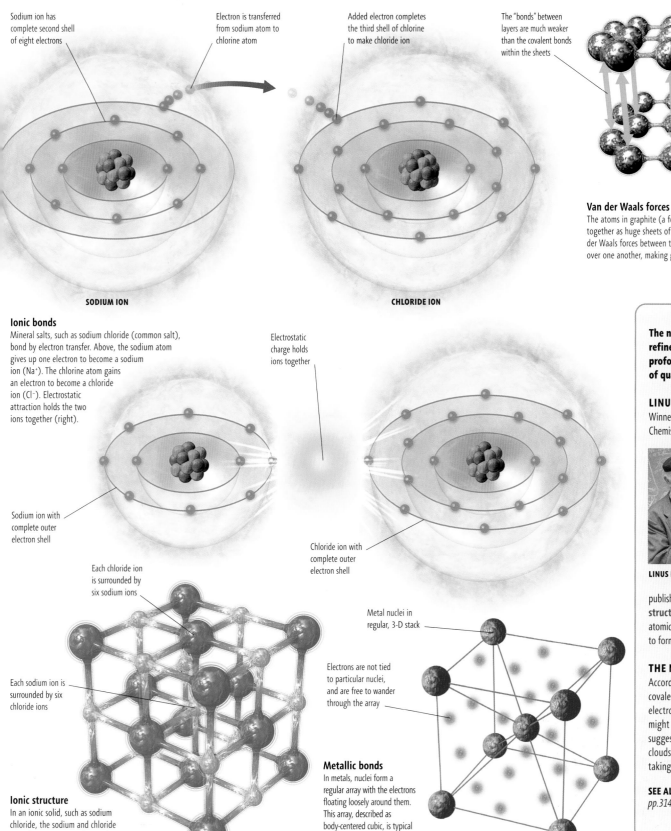

Sodium ion has complete second shell of eight electrons

Electron is transferred from sodium atom to chlorine atom

Added electron completes the third shell of chlorine to make chloride ion

The "bonds" between layers are much weaker than the covalent bonds within the sheets

Bonds between carbon atoms

SODIUM ION

CHLORIDE ION

Van der Waals forces
The atoms in graphite (a form of pure carbon) are bonded together as huge sheets of hexagons. There are only weak van der Waals forces between the sheets, so they can slide easily over one another, making graphite a good lubricant.

Ionic bonds
Mineral salts, such as sodium chloride (common salt), bond by electron transfer. Above, the sodium atom gives up one electron to become a sodium ion (Na$^+$). The chlorine atom gains an electron to become a chloride ion (Cl$^-$). Electrostatic attraction holds the two ions together (right).

Electrostatic charge holds ions together

Sodium ion with complete outer electron shell

Chloride ion with complete outer electron shell

Each chloride ion is surrounded by six sodium ions

Each sodium ion is surrounded by six chloride ions

Metal nuclei in regular, 3-D stack

Electrons are not tied to particular nuclei, and are free to wander through the array

Ionic structure
In an ionic solid, such as sodium chloride, the sodium and chloride ions alternate in a regular, 3-D array.

Metallic bonds
In metals, nuclei form a regular array with the electrons floating loosely around them. This array, described as body-centered cubic, is typical of iron and chromium.

AFTER

The nature of chemical bonds has been refined during the last 150 years, and was profoundly changed by the development of quantum mechanics.

LINUS PAULING
Winner of Nobel Prizes for both Peace and Chemistry, **Linus Carl Pauling** was an American

scientist and peace activist. He was an early exponent of the ideas of **quantum mechanics in chemistry**, and his book *The Nature of the Chemical Bond* (1939) is one of the most influential chemistry books ever

LINUS PAULING

published. He set out five rules for finding the **structure of ionic crystals**, and described how atomic orbitals in **covalent compounds** combine to form molecular orbitals.

THE NATURE OF THE CHEMICAL BOND
According to Pauling, when two atoms form a covalent bond, they **share electrons**, but those electrons are **not fixed between them**, as might be implied by the short line often used to suggest a bond in scientific diagrams. Instead the clouds of electrons around the nuclei **hybridize** taking on more favorable positions for bonding.

SEE ALSO »
pp.314–17 QUANTUM REVOLUTION

BEFORE

Early attempts at flight were in hot-air balloons and gliders. Often optimistic, they were in some cases involuntary.

EARLY GLIDER

When he became a Buddhist, the brutal 6th-century CE Chinese emperor **Kao Yang** "released" all prisoners by making them jump from a high tower with makeshift bamboo wings. One of them, Yuan Huang-Thou, managed to **glide around 1½ miles (2.4 km)**, and land. The emperor later had him starved to death.

FLYING MACHINE

Leonardo da Vinci was an artist with a technological mind and a vivid imagination. Among his notebooks were several superb sketches of **flying machines**.

ONE OF LEONARDO'S FLYING MACHINES

HOT-AIR BALLOONS

The first hot-air balloon was built by French paper manufacturers and brothers **Jacques-Etienne** and **Joseph-Michel Montgolfier**, and demonstrated in June 1783. The second flight, in September 1783, carried a sheep, a duck, and a chicken. The third, in November, carried a doctor and an army officer 6 miles (9 km) over Paris.

MONTGOLFIER BALLOON TAKE OFF

« SEE ALSO
pp.254–55 THE INTERNAL COMBUSTION ENGINE

Taking Flight

People have always longed to fly like birds. The first heavier-than-air machines limped off the ground during the 19th century, but by the end of the 20th century military aircraft were essential to every major power, and cheap flights carried millions of people to vacations destinations all over the world.

In order to get off the ground, a flying machine must have more lift than weight. It can achieve this either by being less dense than the surrounding air, or by having the power to push suitably shaped wings through the air at high speed. Balloons and airships are lighter than air; helicopters pull themselves up with huge rotors; other aircraft rely on the lift from their wings.

Pioneers of aviation

Inspired by the Montgolfier Brothers (SEE BEFORE), in the 1790s George Cayley investigated aerodynamics, watching crows, and using a whirling-arm machine to study lift and drag at his home in Yorkshire, England. In 1853 he achieved a manned flight with a glider, volunteering his coachman to be the world's first test pilot. He said that he would be able to power his aircraft "if he could get a hundred horsepower into a pint pot." This was achieved with the internal combustion engine (see pp.254–55), invented some 30 years after his death in 1857.

Powered flight

Inventors experimented with powered flight in the 19th century. In 1848 English industrialist John Stringfellow tested a steam-powered model aircraft, which covered a distance of a few metres before crashing. Longer test flights were performed with steam-powered aircraft near Washington D.C. in 1896 by Samuel Langley.

However, the first controlled, powered, man-carrying flight was

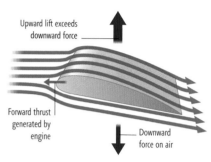

How lift works

When an angled wing is driven forward, it forces air downward, pushing the wing upward and generating most of the lift. Further lift comes from the Bernoulli principle (see p.485)—air pressure reduces as its flow rate increases over the curved upper surface of the wing.

achieved on December 17, 1903, by Orville and Wilbur Wright, owners of a bicycle shop in Dayton, Ohio. They chose to take their "Wright Flyer" or "Flyer 1" to Kitty Hawk, North Carolina, because they were informed by the US Weather Bureau that Kitty Hawk had some of the steadiest winds in the country, and they wanted a reliable headwind to assist takeoff.

They had experimented extensively with gliders and box kites, finding out how to achieve both lift and control. Their successful aircraft was essentially a box kite, fitted with a gas engine.

Once the Wright Brothers had shown the way, others followed. In 1909 Louis Blériot flew across the English

Wright Brothers' test flight

One of a series of test flights by Orville Wright at the US Army base Fort Myer, Virginia, in 1908. On September 17, a propeller broke, causing a serious crash in which Orville was injured and his passenger killed.

AVIATION PIONEER (1848–96)

OTTO LILIENTHAL

A dedicated champion of flight, Otto Lilienthal, from Anklam in Pomerania (now in Germany), made a series of hang gliders, and starting in 1891 used them for more than 2,000 flights, taking off from the top of an artificial hill he built near Berlin.

Given a reasonable headwind, he could hold the machine stationary, and shout instructions to a photographer on the ground below.

Unfortunately he made one flight too many; on August 9, 1896 he crashed and broke his back. The next day, before he died, he said "Small sacrifices must be made."

Aircraft development

Getting off the ground was difficult in the 19th century, but once the internal combustion engine arrived in the 1880s, technological advances came fast.

1891 The "winged Prussian" Otto Lilienthal starts building hang gliders and makes 2,000 flights in them, jumping from a hilltop.

1852 Henri Giffard flies the three horse-power airship Aerial Steamer in Paris at 7 mph (11 kph).

1936 The Supermarine Spitfire becomes a legend in World War II.

1939 The first jet aircraft, the Heinkel HE 178, is produced in Germany.

1970 The first jumbo jet, the Boeing 747, a wide-body airliner with twin aisles, is able to carry more than 500 passengers.

2015 The Airbus 350, designed to reduce fuel consumption and noise, enters service.

1850	1875	1900	1925	1950	1975	2000

1853 George Cayley's coachman becomes the world's first test pilot; Cayley's glider carries him some 200 yd (180 m) across Brompton Vale, UK.

1903 The Wright Brothers achieve the world's first controlled powered flight at Kitty Hawk, on the Atlantic coast, US.

1928 The first commercial aircraft the Ford Trimotor is able to carry just 17 passengers.

1949 The first commercial jet De Havilland Comet is hailed as a triumph, but is withdrawn after two calamitous crashes caused by failure of cabin pressure.

1976 Concorde is the first supersonic passenger aircraft.

2003 Concorde is taken out of service after 28 years of carrying VIPs.

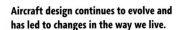

> "To invent an **airplane** is nothing. To **build one** is something. But **to fly** is **everything**."

OTTO LILIENTHAL (1848–96)

Channel, a distance of around 25 miles (40 km), and in 1927 Charles Lindberg flew the 3,000 miles (5,000 km) across the Atlantic Ocean.

Such record-breaking flights continue to this day; in 1986 Dick Rutan and Jeana Yeager flew around the world in a single flight, a record-breaking 25,000 miles (40,000 km) without refueling. Meanwhile, human-powered flight was launched by Gossamer Condor, which won the Kremer Prize by flying a 1 mile (1.6 km) figure-of-eight course on August 23, 1977. Two years later the human-powered Gossamer Albatross crossed the English Channel.

The jet engine

In 1928 air cadet Frank Whittle wrote an essay, "Future developments in aircraft design," predicting that to fly faster, aircraft would have to fly much higher where the air is thin and drag is reduced. To do that, he said, would require a different sort of engine—a jet engine. No one paid much attention, and when he went ahead and built a prototype the UK government refused to back it; so the technology passed to the US, and German engineers also ran with the idea. The first jet aircraft, the German Heinkel HE 178, flew in 1939, and the Messerschmitt 262 saw limited active service late in World War II.

6 MILLION The total number of parts in a Boeing 747–400 airliner.

565 MPH (910 KPH) The usual cruising speed of a 747–400.

AFTER »

Aircraft design continues to evolve and has led to changes in the way we live.

NASA'S HELIOS

In 1999 NASA launched an unmanned, **solar-powered aircraft**, Helios, with a wingspan of 247 ft (75.3 m). In 2001 Helios achieved a **world altitude record** for propeller-driven aircraft of 96,863 ft (29,524 m). Designed as a vehicle for atmospheric research and as a communications platform, she was brought down by turbulence over the Pacific Ocean in June 2003.

GROWTH IN TRAVEL AND TOURISM

Since the **Wright Brothers** made their first unsteady hops in 1903 the travel industry has taken to the air with enthusiasm. In 2019 scheduled airlines offered more than 3 billion seats on 38.9 million flights and on average carried more than **4.2 billion passengers every day**.

AFFORDABLE FLIGHTS LEAD TO CROWDED AIRPORTS

SEE ALSO »
pp.338–39 ROCKET PROPULSION

Rotating compressor blades draw air in

High-speed stream of exhaust gases

Fuel sprayed into compressed air burns continuously

Turbine and compressor driven around by hot gases

How a jet engine works

Air is pulled in at the front by a fan, mixed with fuel and ignited, and blasts out behind through a nozzle. The force pushing the exhaust gases out of the back is matched by an equal force pushing the aircraft forward.

BEFORE

Scientists had been working with vacuums for nearly 200 years before the invention of the vacuum tube.

FIRST VACUUMS
In 1643 **Evangelista Torricelli** showed that vacuums **‹‹** 96–97 might exist. The space above the **column of mercury** in his barometer seemed completely empty (in fact, it is filled with very low-pressure mercury vapour).

VACUUM PUMPS
In 1650 Otto von Guericke produced the first **vacuum pump**, which he used in a dramatic

demonstration in his home town of Magdeburg **‹‹** 96–97. He fitted two hemispheres together and pumped out most of the air from inside. The hemispheres held together so strongly that **they could not be separated**, even by two teams of horses pulling in opposite directions.

‹‹ SEE ALSO
pp.96–97 DISCOVERY OF THE VACUUM
pp.98–99 ROBERT BOYLE
pp.286–87 STRUCTURE OF THE ATOM

MAGDEBURG HEMISPHERES

Vacuum Tubes

During the 19th century experiments with electricity in glass vacuum tubes led to the discovery of cathode rays, electrons, and X-rays. The tubes themselves enabled the development of sound amplification and many broadcasting systems, such as radio and television. They were also essential in building the early computers.

In 1838 English scientist Michael Faraday fitted metal electrodes inside either end of a sealed glass tube from which he had removed much of the air. The air pressure inside the tube was around a hundredth that of the air outside. He connected the electrodes to a high-voltage electricity supply and noticed a strange glow in the tube: electricity was flowing between the electrodes and "exciting" the low-pressure air to produce light.

Nearly 20 years later, in 1855, German scientific-instrument maker Heinrich Geissler made a vacuum pump that could achieve much lower pressures, so the tubes were "evacuated" (nearly completely emptied). He was also a talented glassblower, and managed to find a way to seal platinum wires inside these intricately designed Geissler Tubes.

Cathode rays
One of Geissler's colleagues, German physicist Julius Plücker, experimented with the tubes. He found that the glow inside a tube vanished at one ten-thousandth atmospheric pressure because there was virtually no air to excite. But electricity was still flowing across the vacuum between the two electrodes. Plücker also noticed that the glass at one end was glowing.

It was always the end furthest from the cathode (the electrode connected to the negative terminal) that glowed. Something was emanating from the cathode and travelling the entire length of the tube to hit the glass opposite. In 1876 German physicist Eugen Goldstein named this "cathode rays."

Objects placed between the two electrodes cast shadows, suggesting that cathode rays travel in straight lines. The fact that the rays came from the cathode and travelled towards the positive electrode (the anode) suggested that they were carrying negative electric charge.

Cathode

Negative terminal

Shadow cast by the cross on the back of the tube

Crookes Tube
In William Crookes' Maltese cross experiment of 1879, the cross in the path of the cathode rays cast a crisp shadow on the phosphor screen. The experiment proved that cathode rays travel in straight lines.

Positive terminal (anode)

Maltese cross mask placed in the path of the rays

Calibrations to show deflection

Vacuum tube

Metal plates deflect electrons electrostatically

Electromagnetic coils deflect moving electrons magnetically

Anodes

Cathode

Discovery of the electron
J. J. Thomson used this vacuum tube to investigate the nature of cathode rays. Some people suspected that they were streams of particles, but Thomson worked out the electric charge and mass of the particles, by measuring to what extent magnetic and electric fields deflected them.

Early vacuum tubes
In 1857 Geissler, a skilled glassblower, made fancy-shaped tubes with electrodes at each end, and pumped out all the air.

CATHODE RAY TUBE (CRT) TELEVISION TUBES

Television sets in the 20th century used the cathode ray tube (CRT), invented in 1897 by German scientist Karl Ferdinand Braun. At the back of the tube a "gun" fires high-speed electrons. The front of the tube forms a screen, the inside of which is coated with phosphors—metal compounds that glow when the electron beam hits. The beam varies in strength, and is scanned from side to side by the electromagnets around the tube. This builds up a picture on the screen, line by line. A color CRT has three beams and three different-colored phosphors.

Electron beams fired from three guns

Vacuum allows electrons to move freely

Inside of screen coated with phosphors

Vacuum tube valves in computing
Colossus was the name given to a series of computers used in Britain to decipher German codes in World War II. They made use of vacuum tubes to carry out logical operations, and are considered to be the first programmable digital computers ever built.

WILLIAM CROOKES

As well as inventing the vacuum tube that bears his name, physicist and chemist William Crookes made two important discoveries. In 1861 he discovered the element thallium, and in 1895 he identified the first known sample of helium. Crookes also carried out scientific studies of spiritualism, in which he believed. A fellow of the Royal Society, he was knighted in 1897.

Discovery of the electron

In 1875 English scientist William Crookes managed to make tubes with even better vacuums, and his tube—the Crookes Tube—became the standard apparatus for investigating cathode rays. Crookes found that the rays could make a paddle wheel turn when it was placed between the electrodes. This suggested that the rays were streams of particles.

In 1897 English physicist Joseph John (J. J.) Thomson used a vacuum tube to investigate the particles. He measured the ratio of their charge to their mass, and made the assumption that their charge was the same—but of opposite sign—to the charge on ionized hydrogen (now called a proton). From this, he calculated the mass of the particles, and found that they were tiny—less than a thousandth the mass of even the lightest atoms. Thomson had discovered the electron.

Another key discovery that was made using vacuum tubes was the existence of X-rays (see pp.294–95). Cathode rays sometimes cause the emission of these penetrating, invisible rays.

Rapidly developing technology

It was found that the cathode rays could be deflected by magnetic fields. This, combined with the fact they could make a glow on the end of the tube, led to the development of the cathode ray tube (see above), and ultimately both television and radar.

In 1904 the world's first electronic component—Ambrose Fleming's "diode valve"—used the fact that electrons travel in only one direction in vacuum tubes. In 1907 American engineer Lee de Forest added a metal grid between the two electrodes. By supplying a small electrical signal to the grid he could control the large electric current passing through the valve. His device, the Audion, was the world's first amplifier. It was used first in radio, and later in television and the film industry. Subsequent configurations of valves eventually led to the consumer electronics industry.

Valves can also be used as electronic switches: when there is no charge on the grid, current flows across the valve (the device is "on"). By supplying a negative charge, the electron flow stops (turning it "off"). Early computers, used thousands of vacuum tube valves as switches to carry out operations and calculations.

AFTER »

Vacuum tubes are still used to produce X-rays. But their role in electronic circuits and televisions has been superseded.

THE TRANSISTOR
Transistors 366–67 » were invented in 1947. They do the same job as valves—amplifying and switching—but they are smaller, cheaper, more reliable, and use much less power.

FLAT-SCREEN TELEVISION
Televisions with cathode ray tubes have been largely replaced by flat screens. Two main types are in use. LED TVs use liquid crystal displays with filters of three colors, backlit by white LEDs. OLED TVs use a three-color matrix of organic LEDs.

SEE ALSO »
pp.294–95 THE DISCOVERY OF X-RAYS
pp.366–67 MICROCHIP TECHNOLOGY

Vacuum tube development
From the introduction of Geissler Tubes to the decline of the use of valves in electronic equipment, the history of vacuum tubes spans almost 100 years.

1897 Braun Tube is the forerunner of the oscilloscope, and television and radar tubes.

1878 Crookes Tube contains a higher vacuum than Geissler's; produces first true cathode rays.

1907 de Forest invents the Audion: the forerunner of the transistor, which amplifies signals and acts as an electronic switch.

1913 Coolidge's improved "thermionic" X-ray tube uses a heated cathode.

1931 DuMont develops the first practical cathode ray TV tube.

1948 Semiconductor diodes and transistors begin to replace valves in electronic circuits.

1850	1875	1900	1925	1950

1855 Geissler produces a glow at the cathode of his evacuated tubes.

1895 Röntgen discovers X-rays using a Crookes Tube.

1897 J.J. Thomson discovers the electron using magnetic and electric fields to deflect cathode rays.

1904 Fleming's valve is the first electronic component: it is crucial in the development of radio.

1920 RCA begin commercial production of thermionic valves.

1927 Farnsworth demonstrates a practical TV-picture scanning system.

The **Discovery** of **X-rays**

The discovery of X-rays in 1895 came as a complete surprise, encouraging a fresh wave of investigations into the nature of electromagnetic radiation, instantly paving the way for a variety of new life-saving technologies, and leading to novel ways of exploring the structure of materials.

No one suspected X-rays existed before German physicist Wilhelm Röntgen stumbled across them in his laboratory, but their discovery was to revolutionize imaging techniques across the sciences.

Unblockable light

In 1895 Röntgen was investigating the properties of various types of vacuum tube, and the mysterious "cathode rays" that they produced. He erected a cardboard screen to block out the effects of visible light from the tube, so as to isolate the effects of the cathode rays. But as he began experimenting, he noticed that a nearby fluorescent screen had begun to glow, despite the heavy cardboard. It could not be due to light, nor to the cathode rays (since they emerged from the tube in just one direction); it had to be something new.

Putting X-rays to work

Röntgen threw himself into the study of his new rays, naming them "X" rays after the mathematical term for something unknown. Two months later he published the first account of them, sparking a sensation.

One of Röntgen's first steps had been to test the power of the rays to penetrate different objects. He found that they were blocked by metal and other dense materials, but not by paper

PHYSICIST (1845-1923)

WILHELM CONRAD RÖNTGEN

Wilhelm Röntgen was born in Lennep, Germany, in 1845. Unexceptional at school, he had a natural aptitude for making mechanical devices, but after being unfairly expelled from technical school he gained a place to study physics at the University of Utrecht. Here he studied under the theoretical physicists Rudolf Clausius and August Kundt, who were huge influences on his work. Röntgen initially focused on the conductivity of crystals, but he was best known for the discovery of X-rays, winning the Nobel Prize in 1901.

Early Crookes tubes

In the 1850s British scientist Sir William Crookes experimented with passing electrical discharges through gases at very low pressures in glass tubes, which came to be called Crookes tubes. These Crookes tubes show discharges through air at different pressures.

or human skin. He realized that the rays affected photographic plates, devised a simple X-ray camera, and used it to make an image of the bones in his wife's hand. This new technique for looking at the inside of the human body had obvious medical applications, and X-rays were soon being used for the diagnosis of a huge range of problems. In 1896, the French physicist Louis Benoist discovered the ionizing effect of X-rays, finding that they left air electrically charged as they passed through. As early as 1900, this ability to irradiate matter was adapted to treat skin cancers.

Recognizing the dangers

At around the same time, physicians and experimenters began to realize that overexposure to X-rays could cause problems, damaging skin cells, leaving radiation burns, and even triggering cancers. From the 1900s onward X-rays began to be used with more caution and in smaller doses when creating diagnostic images of the body. However, it was not until the second half of the 20th century that the full risks of X-rays were

The first X-ray

One of the most famous images in photographic history, this X-ray of the hand of Röntgen's wife, Anna, clearly shows her bones and wedding ring. The image was taken in December 1895, and immediately revealed the medical implications: fractured bones were being X-rayed by January 1896.

≪ BEFORE

The 17th and 18th centuries saw breakthroughs in the understanding of light and radiation.

TYPES OF LIGHT

Isaac Newton proved in 1666 that sunlight was made up of different colors. In 1800 **William Herschel** discovered infrared radiation, and in 1801 **Johann Ritter** discovered ultraviolet radiation.

ELECTROMAGNETIC WAVES

James Clerk Maxwell described light as an electromagnetic wave in the 1850s, and it became clear that the wavelengths of infrared radiation were longer than visible light, and those of ultraviolet radiation were shorter.

EARLY VACUUM TUBES

Sir William Crookes proved cathode rays existed by making early versions of the **cathode ray tube**. The discovery of X-rays would not have been possible without them.

≪ SEE ALSO
pp.114–15 THE NATURE OF LIGHT

△ Early chest X-ray
Wilhelm Röntgen prepares a patient for an X-ray in 1896. X-rays were quickly adopted by doctors as a diagnostic tool, and the first radiological journal —*Archives of Clinical Skiagraphy*—was documenting progress in print by April 1896.

recognized and rigorous guidelines were introduced to minimize exposure.

There were still uncertainties about the true nature of X-rays. These were addressed in 1912, when Max von Laue of Munich University designed a series of experiments in which X-rays were passed through the regular three-dimensional structure of a crystal. The observation that X-rays were diffracted by the atoms in the crystal showed that they had the properties of very short-

wavelength radiation. His experiments also showed that the structure of the crystal itself affected the pattern of the diffracted X-rays, so offering a potential means of probing the arrangement of materials on the atomic scale.

Continuing developments
In the 1940s the X-rays' extremely short wavelengths were used in microscopes to produce images of detail far smaller than that detectable with visible light. The rays' ability to penetrate bodies has continued to be developed in medicine through new forms of computer tomography, and in engineering to look deep into materials for indications of strain or degradation before there are any outward signs.

X-rays allowed scientists to delve deeper into structures ranging from the human body to black holes in space.

THE STRUCTURE OF DNA
X-ray crystallography offered clues to the double-helix **structure of DNA**, which in turn gave rise to the modern revolution in **genetics** and **biotechnology**.

MEDICAL DIAGNOSES
X-rays are still used today in **medical imaging** techniques, such as computerized tomography (CT) scans.

X-RAY ASTRONOMY
Orbiting **telescopes** are used to study the X-ray radiation that is produced by some of the most **energetic events** and objects in the Universe, including **black holes**.

AIRPORT SECURITY
X-ray machines are used to **scan baggage** at airports, revealing the contents on a monitored computer screen.

SEE ALSO »
pp.296–97 MARIE CURIE
pp.396–97 SPACEPROBES AND TELESCOPES

LUGGAGE SCANNED AT AIRPORT

▷ **Medical X-ray machine**
The penetrating power of X-rays makes them well suited for studying the human body. Body tissues absorb X-rays to varying degrees, depending on their density, and so appear as different shades—from black through to white—on radiographic film.

2 Free electrons
The electrons are emitted from the heated filament and accelerated into a metal target—the anode—producing high-energy photons.

1 Electron production
A heated filament in the cathode ray tube generates high-energy electrons.

3 Stream of photons
The photons pass out of the cathode ray tube, channeled through a lead cylinder.

4 Low-energy filters
The photons pass through filters that block the low-energy rays, so that only the high-energy X-rays go through toward the patient. They are focused onto one particular part of the patient's body.

5 Body imaging
Dense tissues, such as bone, absorb X-rays almost completely and appear white on radiographic film; soft tissues absorb X-rays less well and appear gray.

Perforated filter
The filter screens out scattered X-rays that would blur the image.

X-ray plate
The plate is coated with silver halide emulsion, which blackens when exposed to X-rays.

Screen glow
A fluorescent screen emits light when struck by X-rays.

6 Unabsorbed X-rays
The areas where X-rays pass through the body will show as black on the X-ray.

Image of bone

PHYSICIST AND CHEMIST **Born 1867** Died 1934

Marie Curie

"**Nothing** in life is to be **feared**, it is **only** to be **understood**."

MARIE CURIE

A world-famous scientist and remarkable role model for women, Marie Curie was the first woman to win the Nobel Prize and the first person to be awarded two Nobel Prizes in different sciences. She was a pioneering scientist who introduced the world to the marvels of radioactivity and worked selflessly to promote the use of radium to alleviate human suffering. By a sad irony it was her groundbreaking work, and the exposure to dangerous radiation that went with it, that ultimately caused her death.

Marie Curie was born Maria Sklodowska in 1867 in Warsaw, at that time part of the Russian Empire. Her father Wladyslaw was a teacher of physics and mathematics, whose pro-Polish

sentiments lost him his well-paid job as a school principal; her mother ran a girls' boarding school, but resigned from her position after Maria's birth. Life was hard for the Sklodowska family, and the eldest daughter, Zosia, died of typhus when Maria was eight. Two years later her mother, who had been suffering from tuberculosis, also died.

Maria and her siblings gained their passion for education from their parents. Maria's brother, Josef, went to medical school in Warsaw, but women were not admitted to the Russian-run university. Maria and her elder sister, Bronya, agreed that Maria would work as a governess to help pay for Bronya's medical studies in Paris. In return, once Bronya was qualified, she invited Maria to Paris so that she could continue her studies, and in 1891 Maria enrolled at the College de Sorbonne. It was on arrival in France that Maria became a naturalized French citizen and registered her name as Marie.

Radioactive notebooks
Marie and Pierre Curie meticulously recorded their observations in notebooks. Marie Curie's notebooks, now held in the National Library, Paris, remain radioactive to this day and carry a health warning.

Nobel Prize awards
This picture of Pierre and Marie Curie working together in their laboratory appeared on the cover of *Le Petit Parisien* in 1901. Two years later they won the Nobel Prize for physics.

$$= \frac{mc^2}{\sqrt{1 - \dfrac{q^2}{c^2}}}$$

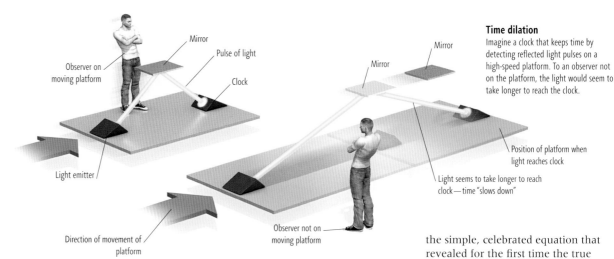

Einstein's first theory of relativity, proposed in 1905, is known as the special theory. It is built on two assumptions. The first, set out by Galileo within the 17th century, is the principle of relativity—that physical laws should be the same for all observers. The second, predicted by Maxwell's equations (see p.259), was that the speed of light is a constant, regardless of the motions of the light source and the observer.

Einstein began by considering how these rules would affect two observers in different "inertial reference frames" (environments free of acceleration or deceleration). He realized that if these

Observer on moving platform
Mirror
Pulse of light
Clock
Light emitter
Direction of movement of platform

Mirror
Mirror
Observer not on moving platform
Position of platform when light reaches clock
Light seems to take longer to reach clock—time "slows down"

Time dilation
Imagine a clock that keeps time by detecting reflected light pulses on a high-speed platform. To an observer not on the platform, the light would seem to take longer to reach the clock.

Theories of Relativity

Einstein's special and general theories of relativity resolved many unexplained features of the Universe, and predicted others. Based on simple assumptions, they showed that space and time are intimately linked dimensions that can be stretched, warped, and even traded off against one another.

observers were to experience gravity, mechanics, and the speed of light in the same way, then many phenomena would, of necessity, seem to behave differently when one observer looked at the other reference frame. Events that seemed to occur simultaneously for one observer, for example, might look to the other as if they were separated by a substantial period of time. This effect, called the "relativity of simultaneity", shows that something very strange is happening—when two observers are in motion relative to each

other, each will say time is running differently for the other (see above).

In everyday experience the effect is negligible, but the greater the relative speed of objects, the more obvious it is. If an observer on Earth could view a rocket moving at a substantial fraction of the speed of light relative to him (a relativistic speed), then time on board the rocket would seem to have slowed down, or dilated. The rocket would also seem to have contracted in a way that matches the Lorentz-Fitzgerald contraction (see p.300). But those on board the rocket would say it was Earth that was slowed and distorted.

There are several ways to interpret these effects, but all of them rely on an important realization – that space and time are intrinsically linked in a

"continuum", and our perception of objects in this continuum depends on relative motion. Einstein's former teacher, Hermann Minkowski, did much to explain the properties of this four-dimensional "space-time."

Mass-energy equivalence

One result of Einstein's theories was a rewriting of the laws of momentum and energy. While it is theoretically possible for a rocket to continue accelerating forever, it will never actually be able to move faster than the speed of light. In normal movement, an object's kinetic energy (energy of motion) is defined as $\frac{1}{2}mv^2$, where m is mass and v is velocity—but if the rocket's velocity has an unbreakable limit in c, the speed of light, then where does the additional energy from acceleration "go" at relativistic speeds? Einstein realized that if energy was to be conserved, it must in fact be used to increase the object's mass, m. Energy and mass, then, are interchangeable, and Einstein's calculations led him to

the simple, celebrated equation that revealed for the first time the true nature of their relationship: $E = mc^2$.

Generalizing relativity

In the decade following publication of the special theory, Einstein continued to work on the consequences of the theory, extending the principles he had already established to look at "non-inertial" situations—those involving acceleration and deceleration.

To do this, Einstein focused on the nature of gravity. He showed that a gravitational field is physically equivalent to a frame of reference that is under constant acceleration. Since beams of light will seem to bend in rapid acceleration, he predicted that they would also bend in strong gravitational fields. He found that strong gravity would cause distortions of time, mass, and appearance—large objects warp space-time around them.

Einstein used the theory to solve a long-running astronomical mystery. The elliptical orbit of Mercury, the closest planet to the sun, "precesses" or rotates around the sun at a faster rate than it should according to classical Newtonian physics. Einstein's theory explained it perfectly as a result of Mercury's passage through the distorted space-time around the sun. Despite this success, however, it was not until 1919 that Arthur Eddington's solar eclipse expedition to Principe, off West Africa, demonstrated the accuracy of the theory: he showed that starlight really did bend in the sun's gravity.

« BEFORE

Although the physics of the early 1900s was effective when applied to make predictions and advance technology, it left some fundamental problems to be solved.

THE NATURE OF GRAVITY
Newtonian mechanics was good for describing everyday interactions between objects, while his theory of gravity seemed to give accurate predictions for the behavior of larger bodies such as planets and stars.

THE SPEED OF LIGHT
The discovery that the speed of light is constant and independent of relative motion posed awkward questions about how objects might behave in extreme circumstances and at speeds approaching the speed of light.

« SEE ALSO
pp.108–09 GRAVITATIONAL FORCE
pp.258–59 ELECTROMAGNETIC SPECTRUM
pp.300–01 EINSTEIN AND RELATIVITY

The constancy of the speed of light
Light from a moving source travels at the same speed as light from a stationary source—so the speed of light from a car's headlights does not increase when the vehicle accelerates, or decrease when it slows down.

> " Put your hand on a hot stove for a minute, and it seems like an hour. Sit with a pretty girl for an hour, and it seems like a minute. That's relativity."
>
> ALBERT EINSTEIN (1879–1955)

True position of star

Apparent position of star to observers on Earth, who assume light has traveled in a straight line

Sealed box undergoing uniform acceleration upward

Beam of light bends downward

Sealed box in a uniform gravitational field, caused by planet's gravity

Light source

Beam of light bends downward through gravitational field

Massive planet creates strong gravitational field

Thought experiment with light
The equivalence of acceleration and gravity is the cornerstone of general relativity. This experiment shows that, just as extreme acceleration should deflect the path of light beams, so should extreme gravitational fields.

AFTER

Since their publication, Einstein's theories have played a vital part in areas such as nuclear fission and fusion, and the evidence supporting them continues to mount.

TIME DILATION
This effect has been proved by **flying atomic clocks** at high speed around the world, and comparing the time that passes for them with that registered by similar devices on the ground.

GRAVITATIONAL LENS
The **Hubble Space Telescope** has used "gravitational lensing," caused as light bends around galaxies, to make images of very distant objects in the Universe.

GRAVITATIONAL WAVES
Einstein's theory of general relativity implied that extreme astronomical events, such as black hole collisions, **cause ripples through space-time (gravitational waves) 402–403 ≫**. These were detected by scientists a century after they were predicted.

SEE ALSO ≫
pp.304–05 **ALBERT EINSTEIN**
pp.398–99 **DARK UNIVERSE**
pp.406–407 **STRING THEORY**

Sun

Distortion of space-time caused by sun's mass deflects light from distant star

Space-time around sun is warped by the sun's mass, creating a so-called gravitational well

Two-dimensional rubber sheet represents four-dimensional space—dents in sheet represent distortions of space-time

Telescope on Earth

The proof of Einstein's theory
Arthur Eddington's measurements of stars near the sun during the solar eclipse of 1919 showed that their positions were distorted in accordance with Einstein's theory—a result of the sun's mass deflecting their light.

Distorted space-time
Demonstrations of relativity often reduce the four dimensions of space-time to a two-dimensional "sheet" to visualize how large masses distort space-time. In this case, a star such as the sun creates a "dent" in space. The path of light from distant stars and galaxies is also deflected, causing them to change their apparent positions.

PHYSICIST **Born 1879** Died 1955

Albert Einstein

"The **whole of science** is nothing more than a **refinement** of everyday thinking."

ALBERT EINSTEIN, "PHYSICS AND REALITY," 1936

German-born physicist Albert Einstein was at the heart of two great revolutions in 20th-century physics. Among his important contributions were his special and general theories of relativity and the discovery of the quantum nature of light. In later years his reputation and personality transcended scientific limits, transforming him into a political figure and popular icon.

Early schooling

Born and raised in Germany, Einstein showed remarkable perception from an early age. He read widely and had mastered Euclidean geometry by the age of 12. However, he rebelled against the unimaginative educational system of his time. Fate intervened when his father's business failed in 1894 and the family moved to Italy in search of work. Einstein eventually completed his schooling in Switzerland and graduated from Zürich Polytechnic in 1900. To make a living, he

initially offered private lessons in mathematics and physics, before a friend helped him secure a job at the Swiss patent office in Bern. In 1902, he become a technical expert, third class—a relatively lowly position.

Scientific papers published

In his spare time Einstein was working on the series of scientific papers that would make his reputation. His working methods, often based on intuitive "thought experiments," led to a series of remarkable breakthroughs. Four scientific papers, published in the leading journal *Annalen der Physik* during 1905, brought Einstein to the world's notice in a single year. The first paper—on the photoelectric effect—was to become a keystone of quantum theory. The second investigated the issue of Brownian motion.

◁ **Theories of relativity**
Einstein's theories were first published in detailed papers in German-language scientific journals, but he later wrote several book-length explanations for a more general readership.

▷ **Bern apartment**
While he worked at the Bern patent office, Einstein lived with his wife Mileva and their first son in the second-floor apartment of Kramgasse 49. Today the apartment is maintained by the Albert Einstein Society, with further exhibits on the third floor.

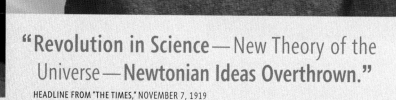

"**Revolution in Science**—New Theory of the Universe—**Newtonian Ideas Overthrown.**"
HEADLINE FROM "THE TIMES," NOVEMBER 7, 1919

1 Individual electrons striking the screen fall roughly where one would expect from classical physics.

2 As more electrons pass through the apparatus, the interference pattern starts to emerge.

3 Over time a complex pattern of light and dark bands, formed by constructive and destructive interference of the waves, builds up on the screen.

Quantum dual-slit experiment

When electrons or other particles are passed through a dual-slit apparatus similar to that of Thomas Young, classical physics would lead us to expect two narrow bands of "illumination." But because these particles have wavelike properties, they will actually interfere and produce a banded effect.

Particle source

Particle

Pair of narrow parallel slits

Fringe pattern on detector screen

wavelengths emitted when certain substances were heated in a laboratory. By modeling the structure of the atom, with electrons orbiting in discrete "shells" whose distance from the nucleus was related to their energy, Bohr was able to explain the "emission spectra" of atoms in terms of photons given off as electrons jumped between different orbits (see pp.286–87).

By the late 1910s most physicists were starting to accept the quantum model of light. One other advantage was the fact that it did away with the need for a

> **" Anyone who is not shocked by quantum theory has not understood it."**
>
> NEILS BOHR, 1958

supposed light-transmitting "ether" in the vacuum of space, therefore meshing neatly with that other great theory of the 20th century— relativity (see pp.300–01). However, Bohr's quantum model of the atom lacked a theoretical foundation, and so far could only accurately predict the emissions from hydrogen, the simplest atom. »

feeble source of ultraviolet light would cause a current to flow. If light was a continuous wave, then the effect of either source should be the same.

Einstein's solution was to take Planck's theory one step further, and suggest that light itself was "quantized"—broken down into discrete packets, or "quanta," each with its own wavelike properties of frequency, wavelength, and energy—what we now call photons. The release of an electron from the charged surface, Einstein said, required the absorption of a photon with the right wavelength

and energy, so it didn't matter how many infrared photons bombarded the surface—since the photons did not have a short enough wavelength and high enough energy, no charge would be emitted.

From Bohr to de Broglie

It took some time for the full implications of this explanation to sink in. One of the first people to grasp that it was saying something fundamental about both the nature of light and the nature of atoms was Danish physicist Niels Bohr, who by 1913 had used it to crack another longstanding problem of light and atoms. This was the precise and unique nature of the light

INVENTION

SCANNING TUNNELING MICROSCOPE

The scanning tunneling microscope is an ingenious means of studying the surface of materials at an atomic level. It relies on the principle of quantum tunneling (see left), in which the wavefunction of electrons allows them to overcome an apparently insuperable barrier. In this image, a scanner with an extremely sharp probe is brought close to iron atoms (in blue on the image below) on a

copper surface. A voltage applied between the probe and the material causes electrons to flow through the gap. Measurements of the current flow as the scanner is drawn across the surface allow the surface to be mapped out to a precision of less than 0.1 nanometers. The probe can even be used to manipulate atoms on the material.

The classical world

In classical physics, a particle cannot escape from within a "well" without gaining enough energy to get over the walls and out of the other side. This should make phenomena such as radioactive alpha decay impossible.

Quantum tunneling

But the uncertainty principle (see over) makes it possible for the particle to "borrow" energy from its surroundings for a brief instant, allowing it to "tunnel" through the barrier and out the other side.

THEORETICAL PHYSICIST (1901-76)

WERNER HEISENBERG

Physicist Werner Heisenberg was born in Würzburg, Germany, and studied at the universities of Munich and Göttingen. Aware of Heisenberg's fascination with atomic physics, his tutor arranged for him to meet Niels Bohr in 1922, marking the beginning of a fruitful relationship. He played a crucial role in developing "matrix mechanics", used to calculate electron energy levels, and helped to develop the Copenhagen interpretation (see opposite). In World War II he worked on the German nuclear project, and afterwards became the first director of the Max Planck Institute for Physics.

>> But this was only the beginning of the story. The next chapter opened in 1924, when a young French physicist called Louis Victor de Broglie submitted a PhD thesis in which he made a highly daring suggestion: if light could have the properties of both a wave and a particle, then might matter have the properties of both a particle and a wave?

De Broglie backed up his proposal that "wave-particle duality" might extend beyond light to matter itself with a theoretical means of calculating the "wavelength" of a particle. Such a wavelength turned out to be inversely proportional to the mass of the particle in question, and becomes vanishingly small for any object above the atomic scale. But on the subatomic level the wavelength was measurable, and by 1927 scientists at two separate laboratories had shown that electrons diffracted and interfered with one another in the same way as photons of light.

Exclusion and entanglement

Meanwhile, two young Austrian physicists added their ideas. The first was Wolfgang Pauli, who put Bohr's model of the hydrogen atom orbitals on a firmer footing with his "exclusion principle" of 1925. Reasoning that the overall "quantum state" of a particle was described by certain properties, each with a fixed number of possible discrete values, Pauli declared that it was impossible for two particles in the same system to occupy the same quantum state at the same time.

In order to explain the pattern of electron shells that was apparent from the periodic table (see pp.230–31), Pauli realized that the electrons must be defined by four quantum numbers,

Superposition
The cat exists in a "superposition" of two different states, both alive and dead. As long as the box is not opened, there is a 50 percent chance the cat is still alive.

Schrödinger's cat
Schrödinger used this "thought experiment" to point out a problem in the Copenhagen interpretation: while the box remained unopened, the cat would hover in an indeterminate state between life and death.

1 Particles are independent
In this example, a laser beam is fired at two unentangled particles.

2 Particles become entangled
Action of the laser beam forces the particles into an entangled state.

3 Particles operate as one
The particles can be separated by any distance, but remain entangled. Manipulating one particle affects the other instantaneously.

4 The bond is broken
However, a manipulation that reveals the state of the particles causes entanglement to break down.

Laser beam

5 All is revealed
Once the link between the particles is broken, the state of each of them becomes evident once again.

Quantum entanglement
Quantum entanglement happens when a pair of particles interact. When separated, even across a galaxy, the particles stay "entangled" so if a measurement is performed on one particle, the other will be instantly affected. The effect was called "spooky action at a distance" by Albert Einstein.

which added a further property to the three already recognized. This property came to be known as "spin." It became clear that spin values separated two distinct families of fundamental particle: "fermions," with spins that have half-integer values, and "bosons," with spins that are whole-number values or zero. All matter particles are fermions, and obey the exclusion principle, while bosons, which do not obey this principle, are "force-carrying"

3 Fatal hammer blow
Once the electrical circuit has been completed, the hammer will be released and the bottle of poison will be smashed open, killing the cat.

Hammer

1 Radioactive decay
There is a probability the radioactive material (below) will release a decaying atom over the course of an hour. But it may not.

2 Completing the circuit
The experiment is set up so that, if a nucleus decays, the Geiger counter will detect the radiation and the electrical circuit will be completed.

Geiger counter

Radioactive material

Bottle of poison

particles, such as photons. Since Pauli's four quantum numbers defined the orbital shell and energy level in which an electron moved, they provided a neat theoretical explanation of how electrons distribute themselves around the nucleus of an atom, the transitions they make between energy levels, and therefore their emission spectra and even chemical reactions. The exclusion principle therefore underpins all of modern chemistry, and has other significant consequences—most notably the phenomenon of "entanglement," by which information in one place can seemingly be transferred to another instantaneously.

> "The **more success** the quantum theory has, the **sillier** it looks."
>
> ALBERT EINSTEIN TO HEINRICH ZANGER, 1912

principle was a huge advance in the understanding of subatomic particles, other events would soon alter the way in which these particles were themselves considered.

The uncertainty principle

In 1926 Austrian physicist Erwin Schrödinger took de Broglie's theories of wave-particle duality a step further by devising a means of calculating a particle's "wave function"—the distribution of its energy in space. Schrödinger argued that "duality" was not really an issue, and that the wave alone represented the reality.

Electron cloud
Thanks to quantum physics, modern chemists know that the atomic structure of elements such as helium is in fact composed of clouds of probability rather than particles in distinct orbits.

This approach sparked fierce debates about the meaning of quantum theory, but Schrödinger's equation also allowed one of his strongest opponents, Werner Heisenberg, to develop another important idea —the "uncertainty principle."

Heisenberg realized that the wave function meant a particle could not be "localized" to a point in space and at the same time have a well-defined wavelength; a wave that is concentrated at a specific point cannot reveal its wavelength, and a wave with a measurable wavelength cannot be exactly located in space. In particle terms, the more accurately one measures a particle's position, the less accurately one can know what its momentum is.

Copenhagen and beyond

The interpretation of the uncertainty principle and the wave function remains controversial, with deep implications that are as much philosophical as they are scientific. Perhaps the most influential approach to the problem is the "Cophenhagen interpretation," developed in Denmark by Heisenberg, Niels Bohr, and others in the late 1920s. They proposed that the act of measurement and interaction with a "classical" (i.e. large-scale) system caused the wave function of a particle to collapse, and that the very nature of the measurement influenced the way in which the wave function collapsed.

This explains, for example, why it is possible to devise an experiment that measures either the wavelike or the particle-like properties of an electron or photon, but not both at the same time. Taken to extremes, it implies that no subatomic event is decided on until it is observed—an idea that led Schrödinger to invent the "Schrödinger's cat" thought experiment (see left).

However, the Copenhagen interpretation is far from being the only way to resolve the uncertainty principle. Other ideas include the "ensemble interpretation," which states that quantum physics only really

applies to large samples of particles; the "objective collapse" theory, in which the wave function collapses regardless of whether it is measured or not; and the "Bohn interpretation," in which particles always have discrete positions, but these are guided by the wave function.

However, perhaps the best-known alternative is the "many worlds" interpretation, in which the different possible results of the collapse all come true in mutually unobservable "histories"—effectively an infinite number of universes representing every possible quantum outcome.

AFTER »»

Although many of the findings of quantum theory seem extraordinary, they have had some important implications for the modern world.

ATOMIC ENERGY

Understanding of the **atom's deep structure** and the quantum principles behind **radioactive decay** has led to the development of **nuclear fission 322–23 »** for both bombs and power stations. In the future, this same knowledge may open the way for clean and unlimited energy from **nuclear fusion**.

NUCLEAR POWER STATION

UNDERSTANDING LASERS

Quantum physics also underlies **laser technology 364–65 »**, in which cascades of intense, synchronized, or "coherent" **photons** are generated in **tightly focused beams**. Lasers are used for a variety of medical, scientific, and manufacturing applications.

QUANTUM COMPUTING

In the future, it is hoped that quantum concepts such as **entanglement** of quantum bits (**qubits**) and **superposition** of multiple states may be used to solve **complex problems** rapidly, beyond the ability of traditional computers, heralding a new generation of **superfast computers**. So far quantum computing **experiments** have been limited to manipulating small amounts of data, but the potential is **vast**.

SEE ALSO »
pp.382–83 SUBATOMIC PARTICLES
pp.404–405 GRAND UNIFIED THEORY
pp.406–407 STRING THEORY

A universe of galaxies
An estimated 10,000 galaxies have been caught
on camera in this Hubble Space Telescope image.
The patch of sky viewed is just one-tenth the
diameter of the full Moon and appears largely
empty in ground-based images.

The Expanding Universe

At the start of the 20th century, astronomers believed the Universe was a finite size and that it contained only one galaxy, the Milky Way. By 1930 the American astronomer Edwin Hubble had shown that the Universe contains many galaxies all moving away from us.

For decades astronomers had observed a number of fuzzy-looking objects they termed "nebulae," which were believed to be part of the Milky Way. Hubble observed these in the 1920s using the 100-inch Hooker telescope, on Mount Wilson, California, then the largest telescope in the world.

Hubble established distances to the nebulae by observing Cepheid stars (variable giant yellow stars) within them. It had been shown that the regular variation in brightness of these stars could be used to determine distance. His calculations revealed that most nebulae were outside the Milky Way, which led him to suggest these were independent galaxies and that the Universe was millions of times bigger than previously thought.

Hubble and other astronomers, notably American Vesto Slipher, had also been looking at dark lines in the light spectra of these galaxies. Slipher had found the lines were displaced toward the red end of the spectrum, a displacement known as "red shift." This implied that galaxies were speeding away from us and that the Universe was expanding.

In 1929 Hubble noted a relationship between red shift and distance; the more distant the galaxy, the faster it sis receding. This is now known as the Hubble-Lemaître Law. Hubble's initial value for the expansion rate, now known as the Hubble Constant, was around 300 miles (500 km)/sec per 3.26 million light years. This was soon proved wrong since it indicated the Universe was younger than Earth. The figure is now believed to be around 43.4 miles (70km)/second per 3.26 million light-years.

Hubble's findings proved what some earlier astronomers had theorized about. He opened the way for astronomers to speculate on how expansion began and, in time, to develop the Big Bang theory of the origin of the Universe.

Red shift

Dark lines within an object's light spectrum shift depending on whether the object is moving away from or toward us. When a galaxy is moving away the wavelengths are stretched out (red shift). If it is moving closer the wavelengths are squashed up (blue shift).

Wavefront of light from galaxy

Direction of travel

Galaxy recedes from observer 1 and approaches observer 2

Wavefront squashed up

Wavefront stretched out

OBSERVER 1

OBSERVER 2

Red-shifted spectrum line

Blue-shifted spectrum line

> **"The history of astronomy is a history of receding horizons."**
>
> EDWIN HUBBLE, AMERICAN ASTRONOMER, 1889–1953

« BEFORE

Cosmology—the study of the origin and structure of the universe—did not exist as a scientific subject before the 20th century.

CREATION BELIEFS

Traditionally, **different cultures** and religions had their own views of the creation of the universe. Many believe that it was created by a god.

THEORIES OF COSMOLOGY

Before 1920 people thought the universe consisted of one huge galaxy and that the solar system lay near its center. Astronomers used **photography and improved telescopes**

to prove both of these facts wrong during the early decades of the 20th century.

« SEE ALSO
pp.74–75 THE SUN-CENTERED UNIVERSE
pp.318–19 THE EXPANDING UNIVERSE

"GOD OVERLOOKING HIS EARTH-CENTERED UNIVERSE,"FROM THE LUTHER BIBLE, 1534

The Big Bang

Scientists questioned the origin of the universe for the first time in the 20th century. Could it really have existed in one form, and forever—or was there a starting point, and a process of continual change? The evidence pointed more and more toward the Big Bang, and a universe 13.8 billion years old.

Until around 100 years ago the universe was thought not to change over time. But during the 20th century scientists collected evidence proving that the universe is not static, but is in a constant state of expansion and change. Its expansive state is linked to an explosive event known as the Big Bang.

Origin of the Big Bang theory

The seeds for the new view of the universe were planted in 1916 when Albert Einstein published his general theory of relativity, from which the Dutch astronomer Willem de Sitter developed an imaginary expanding universe. The idea became a reality in 1929, when Edwin Hubble (see p.341) showed the universe is expanding, which means that it was once smaller and denser. It was, however, Belgian

Georges Lemâitre's suggestion, in 1931, that the universe's origin was a primeval "cosmic egg" that exploded, creating an expanding universe, which served as the first Big Bang model.

The evidence mounts

In the 1940s Fred Hoyle, together with Hermann Bondi Bondi and Thomas Gold, put forward an alternative view of the universe. The "Steady State" theory proposed that the universe appears the same in every location and at all times. It has no beginning or end, and matter is continuously created.

In 1948 George Gamow outlined how the relative proportions of the hydrogen and helium in today's universe could be produced in a Big Bang Universe. The case for the Big Bang strengthened in the mid-1960s, when Martin Ryle showed that the distant, older, radio galaxies

were more numerous and densely packed than those nearby. This disproved a vital characteristic of the hypothetical Steady State Universe, in which density would always stay the same.

Direct evidence for the Big Bang also came around that time. George Gamow had predicted that a remnant of the universe's initial radiation would remain: the microwave background that would have permeated all space, before starting to cool. This was detected in 1964 by Arno Penzias and Robert Woodrow Wilson, confirming that the universe was intensely hot in the past.

The start of the universe

We now think that the Big Bang occurred 13.8 billion years ago, starting our universe. It produced all energy, matter, and space, and marked the recognized start of time. At the outset

▽ Origin of the Universe
The Big Bang produced all the energy, matter, and space in the Universe.

DIAMETER	3×10^{-27} ft/10^{-26} m	33 ft/10 m	10^5 m (62 miles/100 km)
TEMPERATURE	10^{27}K (1800 trillion trillion ° F / 1000 trillion trillion ° C)	10^{27}K (1800 trillion trillion ° F/1000 trillion trillion ° C)	10^{22}K (18 billion trillion ° F/10 billion trillion ° C)

Inflation era
The universe expands from being billions of times smaller than a proton (a subatomic particle) to approximately the size of a football field within a tiny fraction of a second.

Particle soup
The universe is a "soup" of fundamental particles and antiparticles, which are continually forming and then reverting back to energy.

Separation of forces
Initially, the universe had one superforce. As the universe cools, forces separate. From now on the forces of nature (such as gravity) and physical laws are as we experience them today.

TIME	A hundred-billionth of a yoctosecond (10^{-35} seconds)	A hundred-millionth of a yoctosecond: 10^{-32} seconds	1 yoctosecond: 10^{-24} seconds

The Big Bang

After its initial inflation, the universe settles into a steadier

Early particles include quarks, gluons, and bosons

Photon particle – a basic unit of light

FRED HOYLE

Sir Fred Hoyle was one of the greatest scientific thinkers of the 20th century. As the main exponent of the Steady State theory, this Englishman single-handedly stimulated the worldwide development of cosmological research in the 1950s and 1960s. His demonstration of the ways in which heavier elements are made in stars was equally significant. He coined the phrase "Big Bang" while making a radio broadcast in early 1950.

300,000 YEARS AFTER THE BIG BANG (BEFORE STARS)

Hydrogen 76% | Helium 24% | Trace of lithium

TODAY, AFTER MANY CYCLES OF STAR BIRTH AND DEATH

Hydrogen 74% | Helium 23%

Oxygen 1%
Carbon 0.5%
Neon 0.5%
Iron 0.1%
Nitrogen 0.1%
Traces of other elements

Composition of the Universe
The early Universe consisted of hydrogen and helium, with a trace of lithium. Today it still consists mainly of hydrogen and helium, along with other chemical elements that were created in the stars.

the universe was dense, hot, and contained pure energy. It rapidly ballooned in size, before settling to a steadier rate of expansion. It consisted of tiny particles of energy, which became tiny particles of matter within seconds. Within minutes the universe had turned almost entirely into the nuclei of hydrogen and helium atoms.

The first stars
Over the next 300,000 years or so the universe expanded and cooled. When the temperature had dropped enough, the first atoms formed and the universe became a transparent mix of hydrogen and helium, in a 3:1 ratio. This matter was unevenly distributed and, over millions of years, density irregularities

formed into clumps, giving birth to the first stars and galaxies, possibly as early as 200 million years after the Big Bang.

The first stars were vast, and made almost entirely of hydrogen and helium. Nuclear reactions deep within their cores produced new chemical elements such as carbon, oxygen, and silicon which were dispersed into space after the stars' death. The universe's chemical mix slowly changes as successive stars produce elements in this way, or as they die explosively.

The Big Bang theory continues to be refined. Yet, the biggest question of all—"what came before?"—remains completely unanswered.

EXPLORING THE COSMIC BACKGROUND
In 1992 the Cosmic Background Explorer satellite (COBE) recorded tiny variations in the intensity of the **cosmic microwave background**, from which present-day galaxies and clusters grew.

HADRON COLLIDER
Physicists are using the world's **largest scientific instrument**, the Large Hadron Collider 404–05 >>, to study **particle collisions** that would have occurred immediately after the Big Bang.

DARK ENERGY
The expansion of the universe is accelerating because of dark energy, which makes up about 68 percent of the total energy of the universe.

SEE ALSO >>
pp.324–25 FISSION AND FUSION
pp.400–01 DARK UNIVERSE

---

60 billion miles/100 billion km
10^{13}K (18 trillion ° F/10 trillion ° C)

1,000 light-years
10^8K (180 million ° F/100 million ° C)

100 million light-years
3,000K (4,900° F/2,700° C)

First protons and neutrons
The universe has cooled enough for particles to combine, producing the first protons and neutrons. During the next three minutes these combine to form atomic nuclei of hydrogen and helium.

Opaque era
For the next 300,000 years or so, matter particles are in a continual state of interaction with photon particles. The movement of the photons is restricted and the universe appears "foggy."

Matter era
Protons and atomic nuclei begin to capture electrons and form the first atoms. The photons now travel freely through the universe as radiation, making the universe transparent.

1 microsecond:
10^{-6} seconds (1 millionth of a second)

200 seconds

300,000 years

The First Atom Bomb

On July 16, 1945, a huge fireball rose into the skies above the Alamogordo Test Range in the New Mexico desert. The "Trinity Test" was the culmination of decades of research on atomic structure and the nature of radioactivity. With energy equivalent to 16,000 tons of TNT, this was the first artificial nuclear explosion—herald of a new technology that would transform the world, for good or ill.

As early as 1919 physicist Ernest Rutherford had shown that it was possible to trigger "transmutations" from one element to another artificially, by bombarding a material with alpha particles. It was clear that "splitting the atom" (see pp.324–25) could release enormous energy, but scientists had no idea how this might be done in practice on a large scale.

Then in 1932 came confirmation of the existence of the neutron, followed by the realization that bombardment with high-energy neutrons provided a way to trigger radioactive decay and release further neutrons in the process. In 1933 Hungarian-born physicist Leo Szilard suggested that, in the right conditions, a densely packed "critical mass" of radioactive material could undergo a "chain reaction," with neutrons from one decay event triggering others and potentially releasing a huge amount of energy.

Certain heavy elements were known as potential sources of radioactive emissions, and in the mid 1930s experiments by the Joliot-Curies and Enrico Fermi successfully created artificial radioactivity by bombarding stable heavy elements with alpha particles. In 1938 German physicist Otto Hahn discovered an isotope of barium produced by the breakdown of uranium. His former student Lise Meitner and her nephew Otto Frisch showed how this could be the result of "nuclear fission"—the breaking apart of heavy isotopes to form much lighter ones. This fission of heavy elements clearly offered the potential for Szilard's chain reaction.

On the brink of World War II eminent physicists urged US President Franklin D. Roosevelt to obtain this technology before Germany did. The resulting Manhattan Project brought together many of the finest minds in 20th-century physics—under the scientific direction of American physicist J. Robert Oppenheimer—to figure out the practicalities of nuclear reactors and weaponry, and the means of refining and concentrating "fissile" material. The test detonation was watched in secrecy by fewer than 300 people, but people all around the world were to see images of the "mushroom cloud" over Hiroshima, Japan, just three weeks later.

> "I am become **death**, the **destroyer of worlds**."
>
> J. ROBERT OPPENHEIMER

BEFORE

The exploitation of nuclear energy was the culmination of several breakthroughs in the understanding of atoms.

ATOMIC STRUCTURE
In the early 1900s **Ernest Rutherford** and his team began to explore the **internal structure of atoms**. Frederick Soddy identified the existence of **isotopes ‹‹ 299** and noted that heavier isotopes tended to be radioactive.

NEUTRON BREAKTHROUGH
James Chadwick's 1932 discovery of **neutrons ‹‹ 287** provided the means to **manufacture radioactive elements**.

‹‹ SEE ALSO
pp.286–87 STRUCTURE OF THE ATOM
pp.298–99 RADIATION AND RADIOACTIVITY
pp.322–23 THE FIRST ATOM BOMB

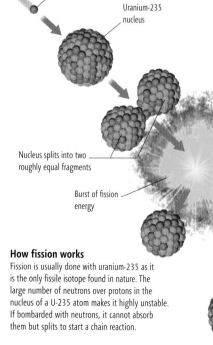

Lise Meitner
Austrian physicist Lise Meitner worked with chemist Otto Hahn, and together they discovered the stable isotope of protactinium and collaborated on nuclear fission. After Meitner had fled to Sweden as anti-Semitism rose in Germany, her role in the discovery was downplayed, and Hahn alone received the Nobel Prize for it.

Neutron

Uranium-235 nucleus

Nucleus splits into two roughly equal fragments

Burst of fission energy

Lighter products (e.g. barium and krypton nuclei)

Shower of excess neutrons

Further U-235 nuclei in chain reaction

How fission works
Fission is usually done with uranium-235 as it is the only fissile isotope found in nature. The large number of neutrons over protons in the nucleus of a U-235 atom makes it highly unstable. If bombarded with neutrons, it cannot absorb them but splits to start a chain reaction.

Fission and Fusion

There are two possible ways to extract energy from atomic nuclei—either by fission (splitting apart heavy nuclei) or by fusion (joining together light nuclei). There are about 440 fission reactors in operation today, and fusion holds the promise of unlimited clean energy if it can be harnessed in a controlled way.

The principles of nuclear fission were figured out through the 1930s and put to the test during World War II. The bombs that were detonated over Hiroshima and Nagasaki relied on an uncontrolled fission "chain reaction"— a phenomenon first identified by Hungarian-born scientist Leo Szilard as early as 1933. The principle is simple. Once a lump of fissile material reaches a critical mass or critical density, neutrons escaping from one nuclear decay trigger another, releasing further neutrons, and so on. Much of the material fissions almost instantly, with an enormous burst of energy.

Fission versus fusion

Fission releases energy because the mass of the fragmented products from each decay is slightly less than the mass of the original nucleus—the lost mass is converted directly into energy in accordance with Einstein's equation $E = mc^2$ (see p.300). As physicists learned more about the structure of particular atomic nuclei, it became clear that this release of energy occurred only in the fission of elements heavier than iron. Lighter nuclei, in contrast, absorb energy when they are split, but the process of joining them together—nuclear fusion— has the potential to release larger amounts of energy than fission. The Sun and other stars release energy through fusion, but on Earth it is far harder to achieve than fission, because nuclei can

only be forced together at tremendous temperatures and pressures. However, after World War II scientists striving to develop more powerful weapons for the encroaching Cold War realized they had already built a means of producing such conditions—the fission bomb. The hydrogen bomb, first tested over the Pacific Ocean in the 1950s, relies in essence on channeling the

explosive force, radiation, and neutrons from a small fission "trigger" into a secondary charge of fusion fuel in order to create a reaction.

Putting fission to work

The first man-made fission reactor, "Chicago Pile-1," began operation as part of the Manhattan Project in December 1942. By the mid-1950s nuclear power plants, able to extract energy from controlled reactions and convert it into electricity, were running in both the US and the Soviet Union.

Nuclear fusion test facility
Attempts at fusion usually involve a doughnut-shaped "torus," such as this one at the KSTAR facility in South Korea, which compresses and heats fusion fuel until conditions mimic those in the core of the Sun.

Regulated nuclear fission involves a core containing pellets of enriched radioactive fuel, usually moderated by an intervening material (such as water or graphite) that slows the movement of the "fast neutrons" that are released, making them more likely to interact with other atoms and sustain the chain reaction. Control rods made from neutron-absorbing materials are inserted or removed to manage the flow of neutrons and the rate of the chain reaction. Power plants are typically designed so that the reaction will be automatically halted in the event of a catastrophic failure. They usually extract energy by allowing the

SEA POWER

Almost as soon as they became feasible, nuclear reactors were adopted by both the US and Soviet navies. Building compact reactors was a major challenge, but their efficiency and low fuel use made them ideal power sources both above the waves and below during the tense years of the Cold War. They radically reduced the need for refueling and allowed ships and nuclear-armed submarines to stay at sea for months at a time. Perhaps the most impressive of all nuclear surface ships is the USS *Theodore Roosevelt*: commissioned in 1986, the vessel is powered by two nuclear reactors.

reaction to heat water in the reactor pile, and transferring that heat to spin the blades of steam turbines.

Fusion power

While fission has proved effective in generating energy, attempts to harness fusion have been less successful. The usual method is to heat and compress the fusion fuel using electrical and magnetic fields, but producing the right conditions for controlled fusion tests modern science and engineering to the limit. So far none of the prototype reactors has been able to produce more energy than it consumes. Translating the process to an industrial scale presents further challenges. However, the potential for fusion power, which can generate energy from common elements with none of fission's harmful waste products, is ample for the foreseeable future.

The Sun
The core of the Sun reaches temperatures of 15 million °C and densities of 165 tons per cubic meter—enough to trigger nuclear fusion.

Deuterium nucleus:
1 proton, 1 neutron

Fusion releases energy

Excess neutron

Tritium nucleus:
1 proton,
2 neutrons

Excess energy

Helium nucleus:
2 protons,
2 neutrons

How fusion works
Fusion is easier to achieve with heavier isotopes of hydrogen such as deuterium and tritium. When forced together, these merge to form helium nuclei, releasing a burst of energy in the process.

AFTER

Nuclear power is likely to be with us for years to come because it is a viable alternative to fossil fuels and its use can help reduce carbon emissions.

NUCLEAR DANGERS
The waste from nuclear fission is hazardous to transport and store. Disasters at nuclear power stations release dangerous radioactivity into the atmosphere. An explosion at the reactor in **Chernobyl, Ukraine, in 1986** directly claimed 56 lives, and its long-term effects may be far more profound. The cost of the accident in **Fukushima, Japan, in 2011,** in which three reactor cores

melted after being hit by a tsunami, is still being assessed.

FUSION PROGRESS
The world's **largest fusion reactor**, called ITER, is being built in southern France by an international collaboration of scientists. The goal of the project is to demonstrate for the first time the feasibility of fusion by releasing more energy than was initially used to heat a plasma (an electrically charged gas). This would be a step toward building commercial fusion reactors.

ITER EXPERIMENTAL FUSION FACILITY

PHYSICIST **Born 1918** Died 1988

Richard Feynman

"I think I can safely say that **nobody** understands **quantum mechanics**."

RICHARD FEYNMAN, FROM "IN CHARACTER OF PHYSICAL LAW," 1964

The American physicist Richard Feynman combined an astonishing aptitude for solving impenetrable problems in physics with being a brilliant teacher and communicator. He developed quantum electrodynamics — what he called "the jewel of physics." This was the theory of how electrically charged particles (such as electrons) interact with photons (particles of light). He also made significant contributions to other areas of physics, including superfluidity and the development of the quark model of particle physics.

Early life and career

Feynman was born into a Jewish family in New York City in 1918. From an early age he showed a strong interest in scientific matters. In his school years he enjoyed tinkering with radios, clocks, and any other gadgets he could lay his hands on. He obtained a degree in physics from MIT (the Massachusetts Institute of Technology). While completing a Ph.D. in the same subject at Princeton, Feynman was approached in 1941 to work on the Manhattan Project—the secret development of a weapon based on nuclear fission (an "atomic bomb") at Los Alamos, New Mexico. He initially refused, but then agreed to participate, worried by thoughts of what might happen should Nazi Germany first develop such a weapon. Much of his time on the project was involved in supervising computational work, although he also developed the formula for calculating the yield of a fission bomb.

Throughout his time at Los Alamos, Feynman gained respect from many of the top minds in physics because of his openness to new ideas and willingness to speak his mind, argue with distinguished scientists such as Niels Bohr, and challenge ingrained ideas.

The Manhattan Project

Assigned to the top-secret Manhattan Project, Feynman conducted bomb research at the remote National Laboratory site at Los Alamos, New Mexico. He and the many other personnel involved were accommodated in the site's sprawling trailer park.

Richard P. Feynman

The postwar years

After World War II, in the fall of 1945, Feynman was appointed professor of theoretical physics at Cornell University. Soon he was to return

Richard Feynman
This 1959 photograph of Feynman, in characteristically animated mood, was taken during one of his lectures at the California Institute of Technology (Caltech), Pasadena.

FEYNMAN DIAGRAMS

The Feynman diagram is a pictorial method that he invented to describe the possible interactions between subatomic particles (see pp.382–83). Feynman diagrams are widely used today by particle physicists.

The diagrams represent fermions (matter particles and their antiparticles, such as electrons and positrons), as straight lines. Bosons ("messenger" particles), such as photons, are depicted by means of wavy lines. In each diagram, time is represented along one axis and space along another axis. Each point of intersection indicates an interaction between two particles, but is also a type of shorthand notation for a series of (often highly complicated) equations. The diagrams represent particle interactions in a similar way that curves on ordinary graphs can represent mathematical equations.

One application of the diagrams has been to calculate the probabilities of various particle interactions happening in particle accelerators such as the Large Hadron Collider, built near Geneva, Switzerland.

Examples of Feynman diagrams
The top diagram represents electromagnetic repulsion: two electrons (e-) approach, exchange a photon, and then move apart. In the lower diagram, an electron (e-) and positron (e+) annihilate to produce a photon, which turns into a muon (μ-) and antimuon (μ+).

to developing ideas in quantum electrodynamics, or QED. This research continued after he became professor of theoretical physics at the California Institute of Technology (Caltech) in 1950.

Mid-life researches

QED is the most precise and successful theory in physics and is important to the deep understanding of phenomena involving electromagnetic radiation and electrons. This includes all of chemistry but also such questions as why the sky is blue, why diamonds are hard, and how the retina in the eye works. While at Caltech, Feynman also worked to explain superfluidity—a phenomenon in which liquid helium, when cooled close to absolute zero, is able to flow without viscosity. In 1965 Feynman was awarded (jointly with

Shuttle Commission public hearing
Feynman uses his hands to illustrate a point about the functioning of O-ring seals during the public hearing into the causes of the Space Shuttle *Challenger* disaster, in which the seven crew members died.

Julian Schwinger and Sin-Itiro Tomonaga) a Nobel Prize for his work on quantum electrodynamics.

In the late 1960s Feynman developed a model of the subatomic world called the theory of "partons," which was a precursor of the modern theory of quarks. Meanwhile, he gained fame for brilliant and entertaining lectures.

Later years

In the 1970s and early 1980s Feynman helped introduce nanotechnology (devices that operate at the molecular scale), and pioneered the development of quantum computing, in which the switches, or bits, in a computer exist in more than one state simultaneously.

He died on February 15, 1988, in Los Angeles as a result of abdominal cancer. His last words are said to have been either "I'd hate to die twice. It's so boring" or "This dying is boring."

Microchip from a quantum computer
Digital computers work by manipulating bits that are limited to one of two states: 0 and 1, or off and on. In quantum computers off and on occur simultaneously, making their processing speed millions of times faster.

May 11, 1918 Richard Feynman is born in New York City. In childhood he is fascinated with science and puzzle solving.

1939 Graduates in physics from MIT (Massachusetts Institute of Technology).

1942 Completes PhD at Princeton University in New Jersey. Marries Arline Greenbaum, even though she has been diagnosed as having tuberculosis and his family oppose the marriage.

1942–45 Joins the Manhattan Project (to develop the first nuclear bomb). As a junior physicist, Feynman is not central to the work.

June 16, 1945 Suffers a personal tragedy as his wife Arline dies as a result of her illness.

1945–50 Appointed professor of theoretical physics at Cornell University (New York State), where he returns to developing the theory of quantum electrodynamics (QED).

1950 Becomes professor of theoretical physics at the California Institute of Technology (Caltech), in Pasadena, where in addition to continuing his work on QED, and inventing Feynman diagrams, he helps further develop the theory of superfluidity.

FEYNMAN LECTURES

1952 Is briefly married to Mary Louise Bell.

1960 Marries Gweneth Howarth. (They have a son in 1962 and adopt a daughter in 1968.)

1961–63 Delivers a series of lectures that become *The Feynman Lectures on Physics*, a textbook first published in 1964.

1965 Awarded (jointly with Julian Schwinger and Sin-Itiro Tomonaga) the Nobel Prize in Physics for his theory of quantum electrodynamics.

1968 Develops the theory of partons— hypothetical hard particles inside the protons and neutrons in the center of atoms. His work in this area contributes significantly to the modern understanding of quarks.

1970s Gives many public lectures, such as an address on the topic of Cargo Cult Science.

1979 Has successful surgery for stomach cancer.

1982 Conceives quantum computing, in which the effects of quantum mechanics are harnessed in the working of computers.

1985 Publishes *QED: The Strange Theory of Light and Matter.*

1986 Serves on a panel that investigates the causes of the explosion of the Space Shuttle

CHALLENGER DISASTER

Challenger. His written account of the experience reveals a lack of harmony between NASA's engineers and executives.

February 15, 1988 Dies in Los Angeles as a result of cancer.

« BEFORE

In premodern times astronomers viewed the stars simply as a backdrop to the motions of the Sun, Moon, and planets.

STAR BIRTH

Chinese astronomers were the first to record the appearance of what seemed to be a new star in the night sky. In the **Astrological Annals of the Houhanshu of 185 CE**, they noted that the new object was bright, but did not move, thus ruling out the presence of a comet. Modern astronomers believe that their predecessors may have witnessed a **stellar explosion**—the oldest supernova on record.

SUN'S ENERGY

The question of how the Sun generated its heat and stayed **luminous** for so long was a physical and intellectual problem. In 1854 the German physician and physicist **Hermann von Helmholtz** calculated that the Sun generated heat by shrinking 246 ft (75 m) per year.

STARLIGHT

In 1666 **Isaac Newton split light** into a spectrum of colors « 110–11. In 1863 **William Huggins** studied the spectra of stars and identified their **chemical elements**, proving that stellar atmospheres consist of the same elements found on Earth.

« SEE ALSO

pp.26–29 THE FIRST ASTRONOMERS
pp.38–39 ANCIENT IDEAS OF THE WORLD
pp.182–83 THE SOLAR SYSTEM

Life cycle of a solar-mass star
All stars form from vast cold and dense interstellar clouds. The images below show the life cycle of a Sun-like star that has planets (not all do). The mass of the star will determine the length of its life cycle, with lower mass stars lasting longer.

The Life Cycle of Stars

From the late 18th century astronomers began to question the nature of the Sun and the stars. By the early 1900s they had established that there is a range of star types, including red giants and white dwarfs. But these types are actually stages in a star's life cycle—each star moves from one type to another as it ages.

The groundwork for our present understanding of the nature and evolution of stars started over a century ago, when astronomers in North America and Europe independently classified stars according to their spectra. They split the light of an individual star into its spectrum (its rainbow band of colors) and analyzed this to establish a range of stellar properties such as temperature and luminosity.

Giants and dwarfs

The US Harvard *Annals* of 1897 included Antonia Maury's classification of stars, and this caught the attention of Danish astronomer Ejnar Hertzsprung. From 1905 he used the data to establish a relationship between stellar properties. His work was independently corroborated by the American astronomer Henry Norris Russell in 1910.

Both astronomers showed that when the surface temperature of stars is plotted against their luminosity on a graph, well-defined groupings appear. They found that stars divide into two main groups: dwarfs and giants. In fact, about 90 percent of stars are dwarfs—they make a band that sweeps diagonally across the graph, known as "the main sequence." The remainder form other groups: red giants, supergiants, and white dwarfs. The

Star-forming nebula
New stars are forming within this portion of the Eagle Nebula, which is illuminated by nearby, hot young stars. The Eagle Nebula is an immense starbirth cloud in one of the spiral arms of the Milky Way galaxy.

graph these astronomers used—now generally referred to as the H–R (Hertzsprung-Russell) diagram (see p.439)—is the world's most important astronomical diagram.

Energy and luminosity

The H–R diagram revolutionized the study of stellar evolution, and it is still a fundamental tool in our understanding of the stars, their lives, ages, and composition. It allowed astronomers to theorize on stellar evolution, but they still lacked an understanding of the stars' internal structure and mechanism. They knew a star's energy was produced according to Einstein's equation ($E = mc^2$), but they did not know how.

Two English astronomers provided new information in the 1920s. Cecilia Payne (later Payne-Gaposchkin) showed that stars are made primarily of hydrogen and helium, although her work was not seen as important at the time. But Arthur Eddington's suggestion—that a star's energy could come from nuclear fusion reactions

1 A cloud of (mainly) hydrogen begins to collapse owing to a trigger, such as a collision with another cloud. A pressure wave pushes the cloud in; it becomes unstable and fragments.

2 A fragment of cloud collapses as its gravity pulls material into the core, where the density, pressure, and temperature build up. It is now a protostar.

3 Infalling gas picks up speed, creating a rotating disk of stellar material from which two jets of gas are expelled. The protostar's center becomes dense and hot enough for fusion.

4 The protostar's outward pressure and its inpulling gravity balance each other out. The protostar has stabilized and is now a main sequence star.

5 Gas and dust unused in the star-making process surround the star. They cool and condense to form tiny pieces that join together to form planets.

Earth's local star
The Sun is a main sequence star. It has been shining steadily for around 5 billion years and will do so for about 5 billion more. Like most other stars, approximately three-quarters of it is hydrogen.

deep within the star itself, where hydrogen converts to helium—had an immediate impact.

More important still was Eddington's formalization of the relationship between the mass (amount of

material) of a main sequence star and its luminosity (energy output). By knowing a star's luminosity it is possible to calculate its mass—it is this property, more than any other, that dictates the pattern and length of a star's life. The more massive the star, the shorter its life. Eddington's discovery was the key to a true understanding of the life cycle of stars. »

Cecilia Payne-Gaposchkin
In 1925, Cecilia Payne-Gaposchkin became the first woman to earn a PhD in astronomy from Harvard University. Her bold thesis made the revolutionary claim that the Sun was largely composed of hydrogen and helium. The theory only became accepted once others made similar discoveries.

ASTRONOMER (1873–1967)

EJNAR HERTZSPRUNG

Born in Denmark, Hertzsprung trained as a chemical engineer but turned his interest in astronomy into a full-time career at the age of 29. He spent ten years at Potsdam Observatory in Germany, followed by 25 at Leiden Observatory in the Netherlands—becoming its director—before retiring to Denmark, where he worked into his nineties. His name lives on in the Hertzsprung-Russell diagram (see p.437), but he also greatly contributed to the study of binary stars, Cepheid variable stars, and open star clusters.

6 The main sequence star is stable and shines steadily. As it continues to convert hydrogen into helium in this longest stage, it gradually becomes larger and more luminous.

7 When the core has been converted to helium, it collapses: the core gets hotter and the outer layers are pushed out. The star becomes a red giant, up to 100 times larger than before.

8 The core is so hot that nuclear reactions convert helium to heavier elements. As the core becomes hotter still, the star's surface is blown off and starts to form a planetary nebula.

9 The dying star is surrounded by a planetary nebula—a glowing shell of gas and dust that slowly disperses. The remains shrink to form a white dwarf.

10 All that is left is a white dwarf. The nuclear fuel has been spent and this Earth-sized star fades until eventually it will be a cold, dead black dwarf star.

« During the 1950s a host of scientists, including Fred Hoyle, worked on the nuclear reactions inside stars. They established how energy is produced; explained how increasingly heavy elements are formed in the reactions; and outlined star formation and the stellar life cycle.

The stellar life cycle

Stars are vast balls of spinning, hot, luminous, mainly hydrogen gas, which start life in clouds of gas and dust. Their light and heat is produced by nuclear reactions, and mass lost in the

100
BILLION °C is the temperature inside a dying star just before it explodes and becomes a supernova. Only some stars do this; so supernovae are rare.

process is converted into energy. Stars like the Sun, and those up to about eight times the Sun's mass, eventually become red giants; then they evolve into planetary nebulae, and finally Earth-sized white dwarfs. White dwarfs are extremely common. Over 97 percent of all stars in the Milky Way will eventually become white dwarfs.

Higher-mass stars

Stars made of more than about eight times the Sun's mass die in a more spectacular fashion. The star suddenly explodes and temporarily becomes a more brilliant star: a supernova. The gas and dust blasted away forms into a supernova remnant. Supernovae are rare; each galaxy has one supernova every 200 years or so. The original star's core remains after the explosion and its fate depends on its mass.

The black hole Cygnus X-1
The largest black dot on this image is a blue supergiant star. It has a less massive but invisible companion: a black hole that is slowly pulling material from the supergiant.

In 1932 Sir James Chadwick discovered the neutron (the subatomic particle). One year later, astronomers Wilhelm Baade and Fritz Zwicky suggested that the core remains of a supernova would be an extremely small, dense star made completely of neutrons: a neutron star. These city-sized stars are the smallest,

densest stars we can detect. In 1967 Jocelyn Bell and Antony Hewish discovered the first pulsar, which turned out to be a rotating neutron star.

In 1939 Robert Oppenheimer and George Volkoff showed there is an upper limit for a stable neutron star of about three times the Sun's mass—beyond this the star continues to collapse and forms a black hole. These

A nebula of gas and dust surrounds the central pulsar

A wind of particles from the pulsar ploughs into the surrounding nebula, creates a shock wave, and forms the inner ring

The rapidly rotating pulsar

Material hit by one of the two beams of radiation that flow away from the star

Supernova core
The Crab Nebula pulsar, discovered in 1968, is the core of a supernova. It rotates around 30 times per second, surrounded by stellar material thrown off in the explosion. This image is the central detail of the Crab Nebula (see right).

Supernova remnant
The Crab Nebula is the remains of a supernova that was seen from Earth in 1054. Strands of glowing gas were pushed off by the supernova explosion. The remains of the original star (pulsar) is seen as a white dot in the center.

SUBRAHMANYAN CHANDRASEKHAR

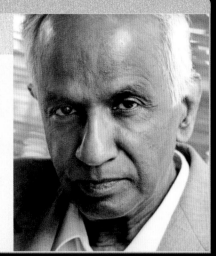

Born in Lahore (then in India), the third of ten children, Chandrasekhar spent most of his working life at the University of Chicago. In 1983 he won the Nobel Prize for Physics for his work on stellar evolution and structure. The "Chandrasekhar Limit" states that the upper limit of a white dwarf is 1.4 times the mass of the Sun. This is important when looking at stars that initially have a mass of more than about eight times that of the Sun. If the mass of the star's core at the end of its life exceeds this limit, the core will produce a neutron star or a black hole.

are so densely compacted that even light cannot escape them. Their existence was first suggested by John Michell in 1783, but largely forgotten. In 1939 Oppenheimer proposed that a black hole could be produced in a supernova explosion, but it remained a theoretical concept until Cygnus X-1 was identified as one in the early 1970s. Even after that, it was several more years before most astronomers accepted Cygnus X-1 as a black hole.

White dwarf
A dim white dwarf (circled) is outshone by the other stars in the cluster NGC 6397. It is the core remains of a star that was originally up to eight times the mass of the Sun.

Stellar recycling

Higher-mass stars produce elements such as carbon, oxygen, silicon, and progressively heavier elements in nuclear reactions, either before or during supernova explosions. Discarded material moves off into space, eventually joining with other interstellar material to form the clouds that give birth to new stars. Supernovae are vitally important in seeding the Universe with elements heavier than iron—elements that are essential for terrestrial planet production, and for the generation and evolution of life forms.

Stellar cycle
Material discarded by previous generations of stars becomes part of a cloud that in turn gives birth to a new generation of stars.

Nuclear reactions in star produce heavier elements

Young star shines brightly

Mature star sheds material

Gas and dust shed by stars joins with interstellar material

Star-forming nebula

Fragments of cloud condense to form stars

"Stars are born, **live**—often for billions of years—and **die**…sometimes in a…**spectacular manner**."

CARL SAGAN, ASTRONOMER, 1978

Rotational axis

Direction of rotation

Rotating star

Beam of radiation faces away from Earth

PULSAR OFF

Rotational axis

Direction of rotation

Beam aligned with Earth

PULSAR ON

Beam swings away from Earth

Direction of rotation

Rotational axis

PULSAR OFF

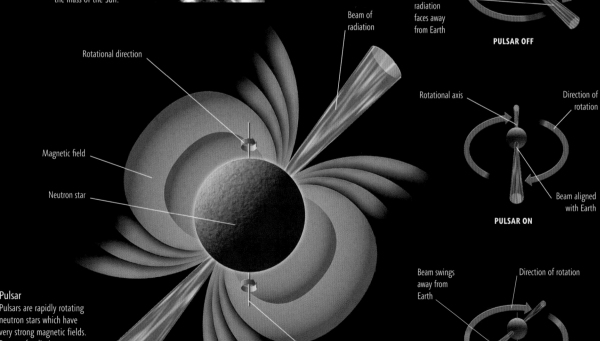

Beam of radiation

Rotational direction

Magnetic field

Neutron star

Pulsar
Pulsars are rapidly rotating neutron stars which have very strong magnetic fields. Beams of radiation are emitted from their magnetic-polar regions, and we can detect these if Earth is within the path of a beam.

Rotational axis

AFTER

Astronomers continue to learn about the stellar world as space-age advances transform the studies of stellar objects.

BROWN DWARFS
The existence of **brown dwarfs** (and their name) was suggested by **Jill Tarter** in 1975; the first was found in 1988. Their mass and temperature are too low to sustain fusion of hydrogen.

SUPERNOVA EXPLOSION
In 1987 the supernova SN1987A exploded in the nearby **Large Magellanic Cloud galaxy**. Astronomers were able to test their predictions of the theory of supernova explosions.

SPACE TELESCOPES
The 8ft (6.5m) mirror of the **James Webb Telescope**, scheduled to launch in late 2021, will give astronomers views of the very first stars.

FAR FUTURE
In the far future stars will have no usable **hydrogen and helium** left for nuclear reactions, so the stellar universe will be very much fainter than it is now.

SEE ALSO »
pp.340–41 GALAXIES, CLUSTERS, AND SUPERCLUSTERS
pp.394–95 INSIDE THE SOLAR SYSTEM

BEFORE

Much ecological thinking has its origins in studies of natural history, which flourished particularly during the age of exploration.

EARLIEST ECOLOGICAL IDEAS
The medieval Arabic scholar **al-Jahiz** wrote the *Book of Animals*—an encyclopedia of **zoology** that included numerous ideas that presaged later biological theories, including **natural selection** and the concept of the **food chain**.

EXPLORERS AND NATURALISTS
The first naturalists to make **accurate observations** of the living world laid the foundations of **field ecology**: Englishman **Joseph Banks** famously **recorded many new species** on explorer Captain James Cook's first voyage across the Pacific (1768–71). In England, **Gilbert White** is widely considered to be one of the first ecologists because of his **observations of the animals and plants** at his home in Selborne.

« SEE ALSO
pp.200–01 HOW EVOLUTION WORKS
pp.202–03 CHARLES DARWIN
pp.278–79 CYCLES IN THE BIOSPHERE

Ecology and Ecosystems

Millions of different species of organisms live on Earth. Each individual species is the product of a long evolutionary process, which has been profoundly influenced both by the conditions and by the resources of each and every species' surroundings.

In the conclusion to his famous book *On the Origin of Species* in 1859, English naturalist Charles Darwin (see pp. 202–03) contemplated: "an entangled bank, clothed with many plants... birds... insects... and worms... these elaborately constructed forms, so different from each other, and dependent on each other in so complex a manner, have all been produced by laws acting around us."

Darwin's description foreshadowed the rise of ecology: a new type of science that would seek to explain the scene that he saw on the "entangled bank." Ecology is the study of the distribution and abundance of organisms, and of their interactions both with their environment and with each other. It examines each individual organism in all levels of biological hierarchy: population, community, ecosystem, and biosphere.

The term ecosystem encompasses not only living organisms but also nonliving physical features of an environment, all of which function together as an independent unit.

Effect of environment on organisms
A living organism affects, and is affected by, the world around it. Environmental conditions, such as temperature, determine whether an organism can live and breed. In 1895 Danish botanist Eugen Warming proved this by showing that unrelated plants coexisting in the same regions had evolved strategies to cope with temperature extremes.

Struggle for survival, as outlined by Charles Darwin, also plays a key role in shaping natural groups of organisms. Living things use up resources from the environment (food, oxygen, water, and salts). If the number of individuals increases, competition for resources becomes greater, which means that each individual gets a smaller share and the weaker produce fewer offspring.

Interactions between organisms
A population is a group of individuals of the same kind of organism, united by an underlying genetic bond: all belong to the same species. A habitat ordinarily harbors numerous coexisting species. The genetic difference between species is reflected by their different capabilities and the roles that they play. In order for species to coexist, each must occupy a subtly different role—its niche—or competition for resources would make

Rain forest pyramid of biomass
Chemical energy flows through a food web, from photosynthetic plants (producers) to animals (consumers). Each organism in this web gives off heat energy—and loses chemical waste—meaning that the amount of energy available to successively higher levels of the food chain diminishes. Consequently, there is always a lower combined weight of top-level predators than organisms at lower levels.

Jaguar — TERTIARY CONSUMERS: TOP PREDATORS

Motmot bird Tarantula Coati — SECONDARY CONSUMERS: CARNIVORES AND OMNIVORES

Tapir Butterfly Mouse Agouti — PRIMARY CONSUMERS: HERBIVORES

Elephant ear Aroid vine Acacia tree Coccoloba tree Passion flower vine Strangler fig — PRODUCERS: PHOTOSYNTHETIC PLANTS

BIOLOGIST (1907–1964)

RACHEL CARSON

American marine biologist Rachel Carson shook the world with her book *Silent Spring* (1962), in which she linked the decline in bird populations with the widespread use of the insecticide DDT. The title of her book is based on the prospect of springtime without birdsong. Carson's research showed how the use of chemicals in agriculture and industry polluted streams and soils and, travelling through the food chain, damaged animal populations. Her work is credited with helping to advance the environmental movement, and her campaigning led to the banning of DDT in the USA.

Mutualism
Clown fish live among the tentacles of sea anemones and are resistant to their sting. Both species benefit from such a relationship: the clown fish is protected from predators and the anemone eats leftover food brought in by the clown fish.

Parasitism
Rafflesia is the world's largest single flower, capable of reaching a diameter of 3 ft 3 in (1 m). It is a parasite that grows on tropical vines. The relationship is beneficial only to the parasite, which has no leaves and so cannot make its own food via photosynthesis.

one drive the other to extinction. Russian scientist Georgii Gause proved this competitive exclusion theory in 1932 when, in experimental cultures of microorganisms, one species consistently drove another to extinction.

Interactions between species were defined in the mid-1900s by two brothers, American ecologists Eugene and Howard Odum, who classified species based on whether interrelationships were beneficial or harmful to individuals. Thus, some interactions benefit one organism and harm another, such as predation (the predator hunts and feeds on the prey) or parasitism (the parasite takes from the host). Competition is harmful to both; mutualism is beneficial to both (see above).

Ecosystems: cycles and energy flow

In 1935 English botanist Arthur Tansley (see left) developed the concept of an ecosystem—how organisms of a community interact with the nonliving world within a unified system. In 1926 Russian geochemist Vladimir Vernadsky articulated the idea that all matter is rotated as part of natural cycles (see pp.278–79), meaning that all organisms add waste to their surroundings and most take oxygen from it. At the same time the ecosystem is fuelled by energy from the Sun. Developing this idea, Tansley, along with American botanist Frederic Clements, proposed that there was a continual flow of energy through a food chain, from the power of the Sun to the release of heat by every living thing. Their theory was later elaborated into the idea of the food web (see opposite), where energy does not follow a simple linear chain, but passes through a complex network of interactions.

If an ecosystem is destroyed, for instance, by fire (see below), volcanic eruption, or landslide, there follows a process of succession. Early stages of this process usually give rise to fast-growing, well-dispersed species, but as succession continues, more competitive species may replace them.

AFTER »

As ecology became a quantitative discipline, scientists developed mathematical methods to help explain this complex subject.

CONTROLLING POPULATIONS

In the 1920s two mathematicians, American **Alfred Lotka** and Italian **Vito Volterra**, devised an equation to describe the **relationship between the population sizes of predators and prey**. This equation helped scientists to predict how **populations change over time** and how different species can interact. Scientists use such equations to develop **strategies to control** crop pests or parasites that cause disease.

CONCEPTS OF PRODUCTIVITY

The productivity of a plant or animal is a measure of how much **living material is built up during growth**, in relation to the amount of **material the organism takes from the environment**. Productivity assessments are important in helping agriculturalists balance the cost of maintaining crops or livestock, and the potential returns in terms of food production.

SEE ALSO »
pp.334–35 CONSERVATION BIOLOGY

> **"Nature** undisturbed seems like a **symphony** whose harmonies arise from variation and change.**"**
>
> DANIEL B. BOTKIN, AMERICAN ECOLOGIST AND WRITER, "DISCORDANT HARMONIES", 1990

IN PRACTICE

RELATIONSHIP BETWEEN PREDATOR AND PREY

Interactions between predators and prey result in oscillations in population sizes of both species. As more prey are born, there is more food for predators, so more predator offspring survive; the increase in numbers of predators then reduces the number of prey. The result is that the pattern of predator numbers is slightly out of phase with the pattern for prey.

Succession after destruction
Habitat destroyed in the Shoshone Fire of 1988 in the Yellowstone National Park, Wyoming, showed recovery about 10 years later. Combustion enriches the soil and encourages growth of plants, such as this fireweed (right).

NATURALIST (1838–1914)

JOHN MUIR

Born in Scotland, Muir moved with his family to Wisconsin when he was just 11 years old. He studied science and medicine at university, but his passionate interest in nature drove him to explore his adopted country, and to study its plants, animals, and landscapes. He funded his own travels and studies, and wrote many books and articles promoting nature conservation. Muir argued powerfully that the government should preserve forest habitats, and that Yosemite Valley should be established as a national park. In 1892, he set up the Sierra Club, one of the first organizations dedicated to conserving natural areas; the club remains active to this day, with a reported 2.4 million members.

BEFORE

Concern for degradation of the environment has ancient origins, but concerted efforts to recognize and address many issues only began in the 20th century.

ORIGINS OF ENVIRONMENTALISM

Many **Islamic scholars** during the **Middle Ages** voiced concerns about the environment. Prominent among them was the philosopher **Al-Kindi**, who wrote, during the **9th century** CE, about the **contaminating effects** of pollutants on **water supplies, air,** and **soil**.

THE INDUSTRIAL AGE

In **19th century Britain**, declines in water and air quality caused by industry prompted the formation of environmental groups, such as the **Commons Preservation Society**.

« SEE ALSO
pp.332–33 Ecology and Ecosystems

Conservation Biology

Biological diversity is a precious resource that humans have long taken for granted. The extinction of species and the impoverishment of natural habitats resulting from our activity has brought together scientists from various disciplines—such as ecology and genetics—to understand how best to safeguard life on Earth.

On September 7, 1936—in Hobart Zoo, Tasmania—Benjamin, the world's last remaining thylacine, died. Two decades before, a similar fate had befallen the passenger pigeon—once one of North America's most common birds—when the species was eliminated by overhunting. As the century progressed, ever more people began to see that wildlife was not limitless. Species were endangered by hunting or habitat loss, and in vulnerable ecosystems, such as islands, they were being wiped out by introduced alien animals—often rats and cats. In 1948 an organization, later known as the International Union for the Conservation of Nature (IUCN), was set up to examine how resources could be used sustainably—and how living diversity could be protected, and 1961 saw the founding of the World Wildlife Fund (WWF), now Worldwide Fund for

The last of a species
Benjamin, the last thylacine, died as a result of being inadvertently locked out of his sleeping quarters on a freezing night. Thylacines were wolflike marsupials, relentlessly hunted to extinction by humans.

the area of habitat available but with its degree of isolation. In their study of life on islands, they found that smaller, more remote islands were more impoverished. This island concept applies to real islands, surrounded by

> "In conservation, the **motto** should always be '**never say die.**'"
>
> GERALD DURRELL, CONSERVATIONIST, "THE AYE-AYE AND I: A RESCUE MISSION IN MADAGASCAR," 1992

Nature). Even as the conservation movement gained momentum, noted academics including American biologist Thomas Lovejoy and British ecologist Norman Myers were saying that species loss was an environmental crisis.

Fragmented habitat

Habitat loss remains the biggest single threat to wildlife, but ecologists also recognize that habitat fragmentation has a profound impact on diversity. In 1967 American biologists Robert MacArthur and Edward O. Wilson published a paper that showed that species numbers declined not just with

water, and virtual islands—such as an island of woodland surrounded by an ocean of intensive farmland.

Large predatory animals, which need plenty of space to hunt, are especially vulnerable to extinction. Not only are there fewer individuals that can reproduce, but small breeding groups are prone to expressing genetic weaknesses that lower their fitness. In 1980 two geneticists—the Australian Ian Franklin and the American Michael Soulé—studied the "critical" size for a viable population and came up with the "50/500" rule: 50 was the minimum number of individuals

required in a population to avoid inbreeding; 500 was the number needed for long-term success.

Preserving diversity

In 1985 American botanist Walter Rosen coined the term biodiversity for wildlife's all-encompassing variety, and the 1992 UN Earth Summit in Rio de Janeiro defined its three components: species; genetic (i.e. within species); and ecosystem. Scientists became focused on quantifying biodiversity to prioritize areas for conservation, spawning new efforts to document and map Earth's living landscapes. Meanwhile, new assessments were being made of the scale of the biodiversity crisis, with some scientists fearing that human impact on the environment had increased "background" species extinction rates by a factor of 100 to 1,000. The magnitude of this rate suggests that humans are currently causing a mass extinction event comparable to the one that eliminated dinosaurs 65 million years ago.

CATEGORIES OF THREAT

Scientists at WWF (Worldwide Fund for Nature) and IUCN (International Union for the Conservation of Nature) have gathered information on the population ecology of many species, allowing them to be ranked in terms of conservation priority, from "least concern" to "critically endangered." Listed below is a selection of species in each of the recognized categories of threat. The rankings have been important in guiding the priorities of these—and other—organizations.

Least concern
Kirk's Dik-dik (*Madoqua kirkii*)
House Sparrow (*Passer domesticus*)

Near threatened
European Otter (*Lutra lutra*)
Rock Partridge (*Alectoris graeca*)

Vulnerable
Hawaiian Goose (*Branta sandvicensis*)
Polar Bear (*Ursus maritimus*)

Endangered
Mountain Gorilla (*Gorilla beringei beringei*)
Pink Pigeon (*Nesoenas mayeri*)

Critically endangered
Javan Rhinoceros (*Rhinoceros sondaicus*)
Spix's Macaw (*Cyanopsitta spixii*)

Extinct in the wild
Père David's Deer (*Elaphurus davidianus*)
Scimitar-horned Oryx (*Oryx dammah*)

Extinct
Golden Toad (*Incilius periglenes*)
Thylacine (*Thylacinus cynocephalus*)

Logging practices
When tropical forests are razed to harvest timber and make room for agriculture, biodiversity is reduced. However, selectively felling trees in forests is sustainable, and in some cases can increase biodiversity.

Conservation efforts can be focused on one species, or group of species, or on the positive management and restoration of habitat.

EXTINCTION

GOLDEN TOAD

The golden toad, which once inhabited a small patch of **cloud forest** in Costa Rica, was last recorded in 1989. Its loss may have been due to climate change, disease, or to the particular susceptibility of amphibians to **air pollution**. Its sudden extinction has drawn attention to the loss of amphibian diversity worldwide.

CONSERVING FORESTS

The **Queen's Commonwealth Canopy**—an initiative set up in 2015—aims to establish a network of protected forests and currently involves all 54 countries of the Commonwealth.

SEE ALSO >>
pp.390–91 **CLONING AND STEM CELLS**
pp.422–23 **GLOBAL WARMING**

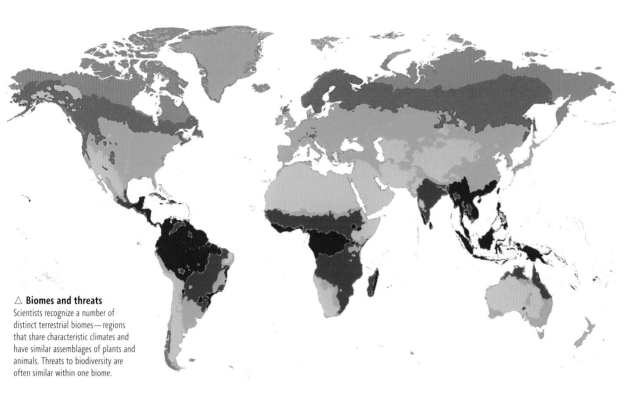

△ **Biomes and threats**
Scientists recognize a number of distinct terrestrial biomes—regions that share characteristic climates and have similar assemblages of plants and animals. Threats to biodiversity are often similar within one biome.

Tropical forest
Half of the world's animal and plant species are found here. These forests are particularly threatened by clearance for cattle grazing, plantations, and mining.

Temperate forest
These forests of broad-leaved deciduous trees occur in temperate regions, where they compete with human land use, and are affected by pollution and forestry.

Boreal forest
Conifers predominate in these cold forests. Exploitation of tar sands, and the disruption of migration routes by development are major concerns.

Polar and tundra
The slow- and low- growing vegetation in these polar regions is very vulnerable to disturbance and to the effects of climate change.

Ice
The ice caps of the North and South Poles, and snow-capped mountain peaks are threatened by airborne pollutants, fossil fuel extraction, and climate change.

Mediterranean
With its ideal climate this biome, which is characterized by its woodlands and aromatic shrublands, is threatened primarily by human development.

Desert
Desert communities have long been affected by overgrazing, which can permanently affect biodiversity; they are particularly vulnerable to climate change.

Savanna
This grassland, with its small or sparse trees, is subject to overgrazing and has been harmed by the introduction of exotic species.

Temperate grassland
These grasslands of North America and Eurasia, have been widely converted to agricultural land since the development of the steel plough.

Cultivated land
The only biome created by humans would have previously formed part of another biome, such as grassland or savanna.

WILDLIFE CORRIDORS

Nature reserves are wild—or often managed—areas designed to maintain a certain habitat or group of species. They are effectively islands of habitat. Biologists discovered that more species could be better protected if small reserves were connected by wildlife "corridors" through which animals could pass, to form larger, more viable, breeding groups.

« BEFORE

Natural rubber has been used since ancient times, but it was not until the 19th century that it became a truly useful material.

VULCANIZATION OF RUBBER

The **rubber tree** is native to South and Central America, where the ancient Mesoamericans used it for various purposes, including making balls for games, as far back as 1600 BCE. Ordinary natural rubber is **prone to snapping**, **stretches too much**, and is **sticky**. In 1839 American inventor **Charles Goodyear vulcanized rubber** with heat and sulfur, making it **tougher, harder wearing**, less elastic, and less sticky. Huge rubber plantations were created in Africa and Asia to meet demand.

CHARLES GOODYEAR

The Age of Plastics

Until the early 20th century natural materials were used in construction, engineering, and the home. But as more and more ways of making inexpensive synthetic polymers emerged the situation changed radically, so that by the end of the century we were living in what has been called the Age of Plastics.

People have used natural polymers for millennia—a polymer is a chemical compound made up of long-chain molecules (see below). Natural polymers occur in plants and wood. As chemists began to understand natural polymers, the idea of trying to mimic them emerged. This led to the production of many useful synthetic materials, including plastics, which have had great impact in many areas.

The first synthetic polymer

In the 1830s it was found that treating starch or wood fibers with nitric acid could nitrate the cellulose, and in 1856 English inventor Alexander Parkes treated these nitrocellulose materials with solvent. Thus he invented the first synthetic polymer, named Parkesine, which was an early form of celluloid: a class of materials that resemble animal horn, which are waterproof, but also elastic. Parkes intended the material to be used as waterproofing for fabrics, but commercial success eluded him.

Practical plastics

In the 1860s American inventor John Wesley Hyatt was looking for a replacement for ivory in billiard balls, and experimented with nitrocellulose and camphor to make celluloid in 1872. While highly flammable and easy to decompose, this material was also easy to mold and shape, making it the first thermoplastic—polymer that is liquid when heated, and solid when cooled (see below left). Celluloid was the material of choice for film in the early years of photography (see pp.262–63) and cinema, in spite of its notorious flammability.

By 1909 Belgian chemist Leo Baekeland had accidentally discovered that mixing phenol and formaldehyde with powdered wood produced a resinlike material he called Bakelite, which was the first thermosetting plastic (polymer that is soft until set solid by heat, see below left). Bakelite became the design material of the Art Deco movement owing to its durability and ability to be

▷ **Structure of starch, a natural polymer**
Starch, a natural polymer, is made up of lots of glucose molecules strung together in chains. Scientists made different long-chain molecules when manufacturing synthetic polymers.

Chemical bond

Carbon atom

Oxygen atom

Hydrogen atom

Long chains of polymers of high molecular weight

THERMOPLASTIC POLYMER CHAINS

Cross-links between long chains of polymers form when heated

THERMOSETTING PLASTIC POLYMER CHAINS

Thermoplastics
Polymers that are liquid when heated and solid when cooled, thermoplastics are often used for pipes since they can be blow-molded to make hollow tubes with no joints, and can withstand high gas and liquid pressures.

Thermosetting plastics
Thermosetting polymers can be molded and then irreversibly set by heating into solid objects, such as this skateboard wheel, which is hard-wearing yet has some of the useful properties of rubber.

Plastics: an industrial revolution
From the mid-19th century chemists experimented with natural substances to produce semisynthetic plastics. Wholly synthetic materials soon followed, culminating in rapid progress in polymer production in the 20th century.

1872 Henry Wyatt invents celluloid, the first thermoplastic.

1887 American clergyman Hannibal Goodwin unsuccessfully files a patent claim for celluloid film. Two years later Eastman Kodak succeeds with its own claim.

1909 Belgian chemist Leo Baekeland unveils Bakelite at the February meeting of the New York branch of the American Chemical Society.

1926 Researcher Waldo Semon invents vinyl—flexible, plasticized PVC, the first commercially useful form of PVC—while working for the B.F. Goodrich Company in Ohio.

1935 DuPont patents nylon. Three years later a nylon toothbrush is launched. Stockings follow in 1940.

1860	1870	1880	1890	1900	1910	1920	1930	19

1862 Alexander Parkes unveils Parkesine, the first synthetic plastic—made by treating cellulose with nitric acid—at the London International Exhibition.

1872 German chemist Eugen Baumann makes the first polyvinyl chloride (PVC), but not until the 1920s is the material made flexible enough to be commercially useful.

1894 English chemist Charles Cross and his colleagues Edward Bevan and Clayton Beadle patent viscose rayon, a semisynthetic silklike fiber made from cellulose.

1912 Swiss chemist Jacques Brandenberger patents cellophane, a semisynthetic polymer still used in packaging.

1933 Eric Fawcett and Reginald Gibson of ICI in Northwich, England find a practical way to synthesize polythene.

'Well, it was kind of an **accident**, because **plastic** is not what I meant to **invent...**"

LEO BAEKELAND, BELGIAN CHEMIST, ON INVENTING BAKELITE

▷ **Plastics in the home**
Bakelite telephones, such as this 1930s model, became a domestic icon of the early 20th century. The sculpted, curving lines of the instrument were made possible by the use of plastic.

shaped. With the addition of colorful dyes, it lent itself to many practical uses, from light switches to lamps, and telephone handsets to pipe stems.

Synthetic clothing fibers

The Age of Plastics truly began in the DuPont chemical company's laboratory, run by American chemist Wallace Carothers (see below). In 1930 his team made an artificial rubber, dubbed neoprene, and also discovered the

Bulletproof Kevlar
Kevlar is a synthetic fiber that was developed by Polish-American chemist Stephanie Kwolek during the 1960s. Its incredible strength means that it is used to make bulletproof and stabproof vests.

starting point for the synthesis of an artificial clothing fiber. In 1934 Carothers' team succeeded at this when they developed nylon, which was the perfect substitute for expensive, fragile silk. Nylon became symbolic of wartime strife in the US with the rush to acquire it in World War II, when new supplies of silk could not be imported from Asia.

Modern plastics

Once chemists began making polymers, more and more followed. In 1938 American chemist Roy Plunkett accidentally discovered the "nonstick" material polytetrafluoroethylene (PTFE), or Teflon, widely used to coat metal cooking pans, and also used to make highly airtight and inert seals for uranium enrichment equipment during the building of the atomic bomb in World War II. In 1965 Polish-American chemist Stephanie Kwolek from DuPont invented a lightweight but bulletproof polymer, Kevlar, a modified

form of nylon also used in stabproof vests. A porous form of PTFE emerged later as the water-repellent, but breathable, clothing material Gore-Tex,

4 TRILLION	The number of disposable

plastic bags made globally each year. This presents a huge disposal problem, which only reusable, recyclable bags may solve.

bringing the history of polymers full circle to Parkes's dream of plastic waterproofing.

The price of polymers

Our ever-growing worldwide demand for polymers has come at a price. They are made from raw materials that come from oil—a finite resource—and, not being biodegradable, they pose a huge waste problem. Plastic buildup in the seas is now proving difficult to prevent.

CHEMIST (1896–1937)

WALLACE CAROTHERS

First of four children born to a teacher father, American Wallace Carothers worked as an instructor at Harvard before joining the DuPont chemical company's fledgling research laboratory, where he led a team in the production of many important materials, including synthetic rubber (neoprene) and nylon. His life was blighted by depression and alcoholism—he reputedly carried cyanide tablets with him. He married Helen Sweetman in 1936, but took his own life after the death of a favorite sister in 1937, so never saw his daughter nor the plastics revolution his pioneering research spawned.

IN PRACTICE

RECYCLING PLASTICS

Domestic plastic waste once went to landfill sites like the one here. Now it is increasingly being recycled. But first it must be sorted. This is mainly done by hand, but automatic sorting using X-ray fluorescence, electrostatics, flotation, and infrared and near-infrared spectroscopy is increasingly being used.

AFTER ≫

As oil supplies worldwide dwindle, the long-term future of plastics may lie in mining landfill sites or in bioplastics.

MINING PLASTICS

Experiments in the US have shown it is possible to **mine landfill sites** and **recycle used plastics** as "raw materials" for new plastics. °A major problem, though, is the buildup of methane gas from decomposing organic waste buried with everything else at old landfill sites.

BIOPLASTICS

Biodegradable and compostable bioplastics are derived from renewable natural biomass sources such as corn starch and vegetable oil. They can be used for nondisposable as well as disposable products.

1937 The Dow Chemical Company introduces polystyrene —developed in Germany by BASF—to the US.	1954 Italian chemist Giulio Natta makes polypropylene, a tough plastic widely used in many products.
1950	**1960**
1941 Kinetic Chemicals Inc, a joint venture between DuPont and General Motors, patents Teflon—PTFE.	1966 Stephanie Kwolek of DuPont patents Kevlar, first used in the 1970s in tires.

BEFORE

The rocket is a simple piece of technology for propelling a payload, from fireworks and explosives to satellites and astronauts.

ROCKET FIREWORKS

The first rockets were Chinese fireworks « 56–57, used for celebrations and in warfare, to set fires and spread panic, from around the 13th century.

EARLY CHINESE ROCKETS

NEWTON'S LAWS

In the late 1600s, Newton's laws of motion clarified the principle of action and reaction « 104–05 that

underpins a rocket's operation. Explosive gases igniting at the base of a rocket are forced out in one direction, and this action creates a reaction—an equal and opposite force—that pushes the rocket in

the opposite direction. When this thrust exceeds the weight of the rocket, the rocket begins to rise. Rockets produce thrust even in a vacuum, so can propel a craft in space.

CONGREVE'S ROCKETS

Metal-cased **gunpowder-propelled rockets** designed by Englishman William Congreve were used against wooden ships in the early 19th century.

« SEE ALSO
pp.290–91 Taking Flight

Rocket Propulsion

The exploration of space depends on a form of propulsion that works beyond our oxygen-rich atmosphere. Rockets that were developed from instruments of destruction in World War II launched the first satellites into space in the 1950s, and continue to provide the power for today's space programs.

Konstantin Tsiolkovskii
The Russian pioneer of rocket dynamics was inspired to investigate the science of rocketry after reading fictional stories about space travel written by Jules Verne.

Although people had fantasized for centuries about journeying into space, the origins of modern rocket propulsion can be traced to Konstantin Tsiolkovskii (1857–1935), a Russian schoolteacher. He realized that Newton's third law (see pp.104–05) would enable a rocket to generate thrust even in the vacuum of space beyond Earth's atmosphere, provided it carried both fuel and an "oxidizer" (a chemical with which the fuel could react).

Tsiolkovskii made a series of breakthroughs, including the idea of building a rocket from not one but several "stages," each with its own fuel tanks and engines. When a stage has used up its fuel, the whole stage falls away and the next stage fires its engines. A staged rocket minimizes the amount of useless weight that has to be carried all the way to space. Tsiolkovskii made models, carried out

calculations, and developed many of the techniques that are still used in rocketry today, but never attempted to launch a rocket himself.

Rocket development

The two pioneers of practical rocketry were American Robert Goddard and German Hermann Oberth. Goddard built the first liquid-fueled rocket, powered by burning gasoline with liquid oxygen. He launched it successfully in 1926 and went on to conduct a series of increasingly ambitious flights. Oberth produced a number of widely read books and gathered around him a group of ambitious young engineers, who soon began making their own rockets. After experimenting with solid fuel, the early

PHYSICIST (1882–1945)

ROBERT GODDARD

Massachusetts-born physicist Robert Goddard became fascinated with space travel after reading H. G. Wells' science-fiction novel *The War of the Worlds* at the age of 16. He did much of his work on the development of liquid-fueled rockets while teaching at Clark University, and journalists delighted in mocking what they saw as his outlandish ideas. This attitude changed after Goddard's first successful rocket launch in 1926, and increased publicity and finance from farsighted backers—such as aviation pioneer Charles Lindbergh—allowed him to conduct increasingly ambitious experiments up to his death in 1945.

Atlas V launch
First launched in 2002, the Atlas V rocket has since had more than 80 successful launches. It is a medium-sized rocket used to carry payloads weighing up to 41,570 lb (18,850 kg) to low-Earth orbit, or smaller payloads farther from Earth.

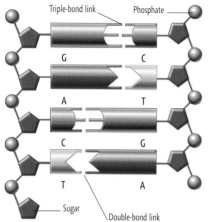

Base pairing
The bases of nucleotides of DNA can pair up only in two configurations because their sizes and shapes fit. Each pair has one large base and one small, adhered by either three bonds—guanine (G) with cytosine (C)—or two bonds— A (adenine) with T (thymine).

The **Genetic Code**

The characteristics of organisms are established by differences in their chemical make-up. One of these chemicals, DNA, carries the instructions for building another important substance—protein. It is the remarkable variety of proteins that accounts for the great diversity of life processes.

<< **B E F O R E**

The idea of genes as particles of inheritance originated with Gregor Mendel in 1866, but that the genes were composed of DNA was not widely appreciated until the 1900s.

CONFLICTING IDEAS ABOUT HEREDITY
In the early 1800s French zoologist Jean-Baptiste Lamarck proposed a theory that characteristics acquired during the life of an organism could be passed on. In 1859 Darwin's evolutionary theory << 198–99 corrected the principle of this view. But it was not until the 20th century that Gregor Mendel's Laws of Inheritance << 306–307 were combined with studies of DNA and protein production to prove that acquired characteristics could not be inherited.

DARWIN'S FINCHES

<< SEE ALSO
pp.308–09 CHROMOSOMES AND INHERITANCE

Every living organism is made up of a complex substance called protein. It is central to the structural fabric of cells and, in the form of enzymes, it controls the chemical reactions of life.

By the late 19th century many scientists already thought that the cell nucleus held the secret of inheritance. A phosphoric-acid-rich substance had been identified in cell nuclei in 1869. In 1937 the British physicist William Astbury produced X-ray diffraction patterns that showed this substance to be deoxyribonucleic acid (DNA), which had a regular structure, quite unlike the tangled form of proteins.

By 1953 James Watson and Francis Crick had confirmed that this structure was a double helix, which enabled the fundamental process known as replication to occur (see pp.346–47). Scientists knew that protein production was influenced by DNA, but they did not know how and so began an exciting new line of enquiry.

Determining characteristics
DNA provides the instructions for the formation of protein (see opposite). As an organism grows, its cells divide in such a way that every cell has a copy of the DNA instructions for making protein; although cells also use these instructions selectively. In 1902 English physician Archibald Garrod anticipated a link between genetic material and protein. He recognized that some inherited diseases upset the balance of chemical reactions: some proteins were not made properly because the instructions were faulty. He called them "inborn errors of metabolism."

In 1941 American geneticists George Beadle and Edward Tatum found they could induce similar errors in bread mold. Their work confirmed that proteins—specifically enzymes (proteins that speed up reactions)—were affected. Beadle and Tatum identified that sections of DNA—the gene—provided the codes to make enzymes. The gene concept was later modified to include all proteins.

Every organism produces thousands of proteins, each encoded by thousands of different genes packaged in every cell of the body. DNA's ability to copy itself and replicate means that the genes in all cells of the same body are identical, but vary from one body to another. Likewise the protein encoded differs, so their characteristics vary too.

Cracking the code
DNA and proteins are long-chain molecules (polymers) made up of repeated building blocks. By 1929 Phoebus Levene, a Russian–American biochemist, discovered that DNA was made up of units called nucleotides, which consist of one sugar, one phosphate, and one base. The base can be one of four types: adenine, cytosine, thymine, or guanine. Proteins, on the other hand, have 20 different types of building blocks called amino acids.

In the aftermath of the double-helix breakthrough biologists began to look at how the nucleotide sequence in

Adenine–thymine link

Guanine–cytosine link

Complementary base pairs
Pairs of nucleotides are arranged in a specific order that is replicated before cell division. The sequences of three bases (the gene code or triplet codon) determine the manufacture of specific proteins.

Cell
All plant and animal cells have a nucleus.

Nucleus
This is the cell control center and contains the DNA in the form of chromosomes.

Cytoplasm
The part of a cell in which protein synthesis takes place.

Chromosome
Tight package of DNA that develops at cell division, formed to make DNA copies easier to manipulate.

Chromatin thread
Consists of DNA coiled around proteins called histones.

Packaging of DNA
Lengths of DNA double helix are organized into threads that solidify to form chromosomes. These unravel between cell divisions at certain points, enabling the DNA molecules to expose their information for making protein and to allow replication.

Histone protein
This acts as a scaffold for the huge quantity of DNA to coil around.

Sugar-phosphate "backbones"
A double helix of DNA consists of two chains of nucleotides. The "backbones" are made up of units of sugar and phosphates either side of a core of paired bases.

Molecule of mRNA assembled from nucleotides

Triplet codon on RNA strand

DNA strands separate

Free nucleotides link to form the mRNA

Gene "template" is copied onto the mRNA

1 The first stage in making proteins is known as "transcription". Part of the DNA double helix unwinds to expose a length of one strand called a gene. A molecule of mRNA is assembled along the other strand, building a complementary "copy" of the gene sequence or gene code.

Amino acid brought to the ribosome by transfer RNA (tRNA) molecule

Ribosome straddles two base triplets to accommodate two amino acids

Amino acid specific to codon bonds to tRNA

tRNA leaves the ribosome to be used again

Chain of amino acids develops

More amino acids added till code fully translated

2 The second stage of protein production is known as "translation". A ribosome passes along the mRNA strand and assists the reaction that joins amino acids together. The mRNA provides the information and the tRNA brings amino acids to build up the chain. Once complete, the chain folds into a protein.

Protein folds up into a shape determined by its sequence of amino acids

△ **Building proteins**
Protein synthesis occurs within all cells and has two distinct stages. Transcription takes place inside the nucleus and translation outside it.

AFTER

With the understanding of the chemical basis of inheritance came the controversial possibility of manipulating genetic material.

MAPPING GENES
A genome is a complete **catalogue of the genes** of a cell or organism. The first draft of the **human genome 414–15 ≫**, published in 2001, provides a greater understanding of **genetic diseases**.

DNA-CUTTING ENZYMES
Scientists developed tools to isolate genes for further study. Bacterial DNA exists free in the cytoplasm. Bacteria contain chemicals called restriction enzymes, which the cells use to **"cut"** DNA and **exchange genetic material**; a process that can be replicated in the laboratory.

SEE ALSO ≫
pp.386–87 GENE TECHNOLOGY
pp.414–15 THE HUMAN GENOME

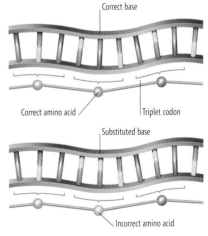

Barbara McClintock
An American geneticist and botanist, Barbara McClintock studied the genetics of corn. She challenged many assumptions about what genes are capable of and was awarded a Nobel Prize in 1983 for the discovery of jumping genes, which can change their position on a chromosome.

DNA determined the amino-acid chain of a protein. In 1961 Francis Crick and South African biologist Sydney Brenner confirmed earlier research and demonstrated that a set of three nucleotide bases (or triplet) encodes one amino acid. Crick and Brenner found that a virus stopped making protein if they inserted (or deleted) one or two nucleotides from its DNA; but if three nucleotides were inserted, the virus could still make protein.

In the same year, American geneticist Marshall Nirenberg and German biochemist J. Heinrich Matthaei cracked the first part of the code. They made a chain of a single nucleotide, which produced a protein composed of identical amino acids. Over the next ten years all 20 amino acids were decoded.

Making proteins
Protein is made on granules called ribosomes in a cell's cytoplasm. So somehow, instructions from the genes must "move out" of the nucleus. Spanish biochemist Severo Ochoa focused on a substance related to DNA called ribonucleic acid (RNA). First identified in 1909, RNA is found in both the nucleus and the cytoplasm of cells, and transpired to be the "messenger" molecule (mRNA). The final link came in 1964 when American biochemist Robert Holley cracked the structure of smaller transfer RNA (tRNA) molecules that brought the amino-acid building blocks to the ribosome for assembly into proteins. By understanding protein synthesis, molecular biologists had finally unveiled the blueprint of life itself.

Correct base

Correct amino acid

Triplet codon

Substituted base

Incorrect amino acid

△ **Gene mutation**
DNA replicates when cells divide, and normally makes an exact copy of itself. Errors in the replication process can result in alterations in the base-pair sequence of a gene, so that a different amino acid is substituted, producing a different variety of protein.

Chaos Theory

Complex systems, such as the weather, seem unpredictable, even though they follow physical laws. However, even simple systems can exhibit chaotic behavior. Chaos theory can now explain these phenomena, with surprising and often beautiful results.

For centuries scientists explained seemingly chaotic phenomena, such as weather and turbulence in fluids, as complex but ultimately predictable according to the laws that Newton and his contemporaries had proposed. This situation changed in the latter half of the 20th century, when a theory of "deterministic chaos," now better known simply as chaos theory, emerged to explain the underlying patterns of behavior.

Unpredictability

The beginnings of chaos theory lay in the discoveries of Jacques Hadamard and Henri Poincaré, who were both working in the field of differential equations and dynamical systems (systems of variables that interact and change with time). Poincaré noticed that even tiny changes in the initial conditions of a seemingly simple system often resulted in large—and unpredictable—changes in outcome; in other words, his results exhibited chaotic behavior. Although he published his findings as early as 1908, they were largely ignored by the scientific community, who were more concerned with the hot topics of the day: relativity (see pp.302–03) and quantum theory (see pp.314–17).

It was not until the 1960s that chaotic behavior entered mainstream science, largely due to the development of electronic computers. The discovery that sparked this interest was made by Edward Lorenz, who worked at the Massachusetts Institute of Technology in the notoriously unpredictable field of meteorology. Using a series of equations, he ran models of atmospheric conditions through his computer. Following one such run, he set up a second run using his figures for the initial conditions, but this time rounded down from six

MATHEMATICIAN (1854–1912)

HENRI POINCARÉ

Mathematical physicist Henri Poincaré was born in Nancy, France. In 1879 he earned his doctorate from the University of Paris, for his thesis on the properties of functions defined by differential equations. In applying these ideas to celestial mechanics, he identified some of the concepts behind chaotic systems. Poincaré made important contributions to various mathematical fields, notably algebraic topology. He also wrote extensively on science for the public.

« BEFORE

Before chaos theory, scientists thought every aspect of the Universe operated according to a "deterministic" pattern—if you knew the initial conditions, you could predict how the system would behave.

Pierre-Simon Laplace, that the **Universe** operates in an **entirely predictable way**, like clockwork.

CLOCKWORK UNIVERSE

In 1686 **Isaac Newton** published his groundbreaking *Philosophiae Naturalis Principia Mathematica* ("Mathematical Principles of Natural Philosophy"), in which he proposed that **gravity was a universal force**, acting on every kind of matter from apples to planets. Newton's discoveries gave rise to the idea, developed by

THE CLOCKWORK MODEL OF PLANETARY ORBITS

The smoke suddenly breaks into turbulent, disordered motion

The smoke initially rises in a predictable and regular flow

Unpredictable behavior
One example of a chaotic system is smoke rising from a point (such as an extinguished candle). The smoke leaves the source with only minute differences in speed, direction, and other conditions. Beyond a certain point, the increase in these differences suddenly causes the flow to break down into turbulent eddies.

Moving Continents

The development of plate tectonics—a synthesis of the earlier "continental drift" hypothesis with new discoveries, such as seafloor spreading—was the most important geological breakthrough of the 20th century.

At the beginning of the 20th century, several topics in Earth science lacked a satisfactory explanation. One of these was how mountains form. Existing hypotheses, such as the idea that Earth contracts as it cools, leaving "wrinkles" on its surface, were unconvincing. Another anomaly to explain was how rocks on the tops of mountains could once have existed on the ocean floor. Nor could anyone come up with a good reason for why there are so many similarities between parts of continents that are separated by wide oceans—in their fossils, sequence of rock strata, and so on—or why deposits of coal, formed from tropical plants, exist in the frozen wastes of Antarctica. Finally, it was not clear why there were such heavy concentrations of earthquakes and volcanoes in some parts of the world.

In 1912 a novel idea, called continental drift, which proposed that continents had once been joined but had since split apart, seemed to provide a solution to some of the questions, but was rejected for lack of an explanatory mechanism. However, in subsequent decades, a series of discoveries in various areas of geology led to a new theory, called plate tectonics. This proposed that Earth's rigid outer shell is composed of several pieces, called plates, that are slowly shifted around as a result of large-scale processes occurring deep within the planet. This new theory explained not only how continents can separate or join together; it also accounted for inconsistencies in the fossil record and, through consideration of what happens at the plate boundaries, many other phenomena—such as how mountains form, and why earthquakes occur where they do. With its ability to explain so much, plate tectonics has revolutionized the way we look at Earth and the processes that occur at its surface.

Plate motion

Plate tectonics proposes that Earth's surface is composed of separate plates, which are moved around by forces such as gravity and convection in Earth's interior. Interactions between the plate boundaries account for such phenomena as the distribution of volcanoes.

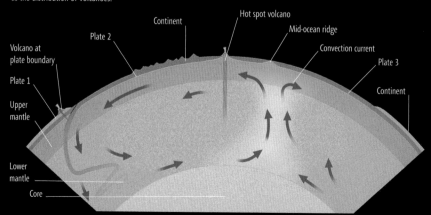

Volcano at plate boundary
Plate 1
Upper mantle
Lower mantle
Core
Plate 2
Continent
Hot spot volcano
Mid-ocean ridge
Convection current
Plate 3
Continent

"The **crust** of the **earth** must be a **shell floating** on a **fluid** interior."

BENJAMIN FRANKLIN, 1782

BEFORE

The idea that the continents were arranged differently in the past goes back centuries.

COMPLEMENTARY COASTLINES
Flemish cartographer **Abraham Ortelius**, in 1596, and English philosopher **Sir Francis Bacon** **《** 69, in 1620, noticed that the **shapes of continents** on either side of the Atlantic match, **as if they had once been joined.** Ortelius thought the Americas had been "torn away from Europe and Africa… by earthquakes and floods."

THEORY OF LAND BRIDGES
In 1885 Austrian geologist **Edward Suess** noted **similarities between plant fossils** in India, Africa, and South America, and suggested that all three may once have been **linked by wide land bridges**, forming a supercontinent, which he called Gondwanaland. He thought that the oceans had flooded these land bridges, thereby separating the continents.

《 SEE ALSO
pp.184–85 HOW ROCKS FORM
pp.186–87 THE FOSSIL RECORD
pp.354–55 MOVING CONTINENTS

Plate Tectonics

Continental drift was a revolutionary new idea proposed in 1912. Originally it was dismissed, but by the 1960s geologists had combined many new discoveries, mainly about the ocean floor, into the groundbreaking theory of plate tectonics, which explained how continents really could move relative to each other.

In 1912 German scientist Alfred Wegener proposed the new concept of continental drift, a forerunner of plate-tectonic theory later that century.

Wegener's theory
Wegener suggested that hundreds of millions of years ago today's continents were joined in a single large continent, which had since broken up, with the land masses "drifting" apart. To back up this theory, he pointed to the coastlines of the Americas, which correspond in shape to those of Africa and Europe. Similar fossils were also found in these continents, as well as similar sequences of rock layers. The Appalachians of eastern North America, for example, matched the Scottish Highlands,

magnetic poles were in the past. By the 1950s studies such as these were bringing up anomalies that could only be explained if continents had moved—otherwise the magnetic poles would have had to have been in two places at once. In this way, interest in the idea of continental drift was rekindled.

Theory of seafloor spreading
Building on an earlier idea by English geologist Arthur Holmes—that convection currents in Earth's mantle might carry large slabs of lithosphere (crust and top layer of mantle) along at the surface—in 1960 American geologist Harry Hess developed the theory of seafloor spreading. He proposed that new, relatively thin and

Earth's crust. These subduction regions are found at the edges of some oceans, where deep trenches form in the ocean

7 The number of large major tectonic plates now recognized

8 The number of smaller minor tectonic plates

floor and frequent earthquakes occur. In the 1960s symmetrical patterns of magnetic stripes in rocks were found either side of mid-ocean ridges, confirming Hess's theory (see below).

Putting the jigsaw together
The seafloor spreading theory quickly led, in the 1960s, to the idea that Earth's

Continent separation
The continental-drift hypothesis, now known to be correct, proposed that the continents were joined in the past in a supercontinent called Pangaea, and have since moved apart.

Supercontinent of Pangaea

Eurasia beginning to split from North America

200 MILLION YEARS AGO

India is in the middle of Indian Ocean

Africa has separated from the other southern continents

75 MILLION YEARS AGO

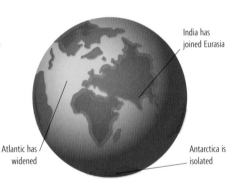

India has joined Eurasia

Atlantic has widened

Antarctica is isolated

PRESENT DAY

METEOROLOGIST (1880–1930)

ALFRED WEGENER

Born in Berlin, Germany, Alfred Wegener trained as an astronomer before becoming a meteorologist. He began collecting evidence for his continental drift theory in 1911, after reading about similarities between fossils on either side of the Atlantic. Sadly, he died on an expedition, decades before his theory was accepted as part of plate tectonics.

suggesting they had once been part of one mountain range. Wegener could not, however, explain how such large masses could move apart, so his idea was not developed.

Mid-ocean ridges and magnetism
Over the next 40 years, there were two key breakthroughs. The first involved mapping the ocean floors. In 1925 a German expedition discovered the mid-Atlantic ridge, a range of mountains running down the middle of the Atlantic seafloor. Later, in the 1950s, American oceanographers Maurice Ewing and Bruce Heezen discovered similar ridges extending into the other oceans. It was soon realized that these ridges contain narrow rift valleys, where new seafloor is created.

The second discovery concerned traces of Earth's magnetic field that are locked into certain rocks when they form. Sequences of rocks sometimes record changes in this magnetism, and by measuring rocks in different continents, it is possible to track where Earth's

dense oceanic lithosphere is continually created at mid-ocean ridges, then moves away from them. Where thicker and less dense continents are attached to this lithosphere, they move too. The creation of new oceanic lithosphere is balanced by activity at "subduction zones," where it is destroyed by descending toward the mantle under

surface consists of a jigsaw of "plates," which shift in relation to each other, driven by new plate creation at mid-ocean ridges. A new analysis of the degree of geometric fit between the coastlines of Africa and South America found that they matched even more closely than had been previously thought. In the late 1960s researchers constructed and tested

Red and yellow indicate magnetization of one polarity

Blue and green indicate magnetization of opposite polarity

Center of mid-ocean ridge (runs vertically)

Transform fault that has offset sections of ridge

Spreading of ocean floor
Reversals in Earth's magnetic field over millions of years, preserved in the rocks on either side of a mid-ocean ridge (an underwater mountain range), are visible in this color-coded image. The patterns are symmetrical on either side of the ridge—an observation that helped confirm the theory of seafloor spreading in the 1960s.

Place Alaska, US
Date March 27, 1964
Moment magnitude 9.2
Number of fatalities 131

Place San Francisco, US
Date April 18, 1906
Moment magnitude 7.8 (estimated)
Number of fatalities 3,000+

Place Mexico City, Mexico
Date September 19, 1985
Moment magnitude 8.0
Number of fatalities 5,000–45,000

Place Coast of Peru
Date May 31, 1970
Moment magnitude 7.9
Number of fatalities 70,000

Place Haiti
Date January 12, 2010
Moment magnitude 7.0
Number of fatalities 300,000

Place Izmit, Turkey
Date August 17, 1999
Moment magnitude 7.4
Number of fatalities 17,000–40,000

Place Kashmir, Pakistan
Date October 8, 2005
Moment magnitude 7.6
Number of fatalities 80,000 +

Place Tangshan, China
Date July 28, 1976
Moment magnitude 8.2
Number of fatalities 240,000–250,000

Place Off Sendai, Japan
Date March 11, 2011
Moment magnitude 9
Number of fatalities 21,000

Place Kanto, Japan
Date September 1, 1923
Moment magnitude 7.9
Number of fatalities 100,000–142,000

Place Valdivia, Chile
Date May 22, 1960
Moment magnitude 9.5
Number of fatalities 2,000–6,000
Most powerful earthquake recorded

Place Off Sumatra, Indonesia
Date December 26, 2004
Moment magnitude 9.1
Number of fatalities 225,000+
Deadliest tsunami ever recorded

Place Sichuan, China
Date May 12, 2008
Moment magnitude 7.9
Number of fatalities 69,000+

Earthquake risk
This map shows global earthquake risk, with dark red and orange indicating high risk, and green and white low risk. In addition the map shows the locations of 12 major earthquakes that have occurred since 1900, significant because of the number of human fatalities, the economic damage caused, or their sheer size.

Tectonics: the causes of earthquakes
In the 1960s and 1970s the plate tectonics theory helped clarify the underlying causes of earthquakes considerably. The largest earthquakes tend to occur at convergent or transform plate boundaries, in areas where the plates cannot move past each other smoothly. Some, however, occur

CHARLES RICHTER

Born in Ohio, Charles Richter spent most of his life in southern California, a region that suffers frequent and occasionally devastating earthquakes. In the 1930s, with German-born seismologist Beno Gutenberg, Richter developed the earthquake rating scale that bears his name. It is based on the strength of the shock waves recorded by seismometers, taking into account the distance of the instruments from the epicenter of an earthquake.

32,000 TIMES
more energy is released by an earthquake of magnitude 8 than by one of magnitude 5, or by one of magnitude 7 compared with an earthquake of magnitude 4.

along faults that are located a little distance from, and often parallel to, plate boundaries, rather than being in the same place as them. These faults have usually developed over time, due to stresses in the crust caused by plate movements. Many shallow earthquakes (with a focus less than 230 feet/ 70 meters below Earth's surface) also occur along mid-ocean ridges, which are linked to rifting and the ocean floor spreading from the ridges (see pp.358–59). Mapping earthquake locations has helped geologists work out where the plate boundaries are.

Effects of an earthquake
The effects of an earthquake derive from the violent shaking caused by the shock waves. Buildings not designed to withstand shaking are likely to collapse. Some types of soil can liquefy during a severe earthquake and lose all strength, so that buildings previously supported by them sink into the ground. Landslides are a common secondary effect when buildings collapse, as are fires, and electrical and gas supplies are damaged. If the focus of a quake is beneath the ocean,

it can cause a tsunami (see below), as chunks of ocean floor are displaced along with huge volumes of seawater.

Earthquake prediction
Seismologists can predict where, but very rarely when, quakes are likely to occur. Indicators include ground uplift within the vicinity of known faults, which can indicate strain building up in the rocks. Another technique has been to look for "seismic gaps"—earthquake zones that have had little seismic activity for years, and where strain energy may be building up.

The birth of a tsunami
Earthquakes below the seafloor are the most common cause of tsunamis. The tsunami waves increase in size as they encounter shallow water near the shore. They can grow from 15 ft (5 m) in height when traveling across the open ocean to up to 100 ft (30 m) or more in height when they reach the coast.

AFTER

Much research over the last 50 years has gone into earthquake prediction and control.

CONTROL MEASURES
Since the 1960s scientists have known that **injecting high-pressure fluids** deep into the ground can lubricate faults and so trigger several small earthquakes before enough strain energy has built up to cause a large one. **Benefits versus risks** are still, however, being evaluated.

BUILDING PROTECTION
Knowledge of the effects of earthquakes on buildings, and the use of tremor-resistant design features, has meant that **skyscrapers expected to resist earthquakes** as large as moment magnitude 8.5 have been built in vulnerable cities, such as Los Angeles and Mexico City.

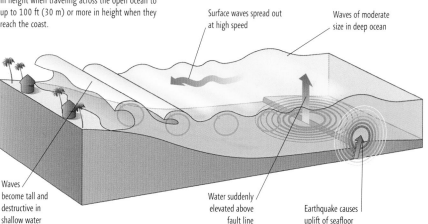
Surface waves spread out at high speed. Waves of moderate size in deep ocean. Waves become tall and destructive in shallow water. Water suddenly elevated above fault line. Earthquake causes uplift of seafloor.

Agriculture is the practice of growing plants and crops and rearing animals as livestock to satisfy human requirements. Humans derive most of their needs from a limited number of plant and animal species—descendants of wild ancestors that have been domesticated since antiquity. Closely managed, the yield of crops and livestock can be maximized by providing them with good growing conditions while at the same time controlling pests and diseases. An understanding of ecological principles has found ways of protecting the land, without undermining the ability of agriculture to provide for populations.

Domestication and breeding

Plant and animal species became domesticated when humans began raising them for their own use. The earliest selective breeding of useful varieties was purposeful, but occurred

Agriculture

Humans rely upon plants and animals as sources of food as well as for materials for clothing, construction, and even medicines. Science would show how crops and livestock could be managed to sustain the demand of growing populations and—ultimately—how to minimize damaging influences on the environment.

BEFORE

Agriculture, much like medicine, was driven by the needs of civilization. Population growth over time has led to the the need to increase food production.

ORIGINS OF CROPS

Domestication of crops can be traced back to the birth of civilization in Africa and the Middle East. **Wheat** crops originated in the "Fertile Crescent" of the rivers Euphrates and Tigris, while **rice** and **corn** were staples in the Far East and the Americas. Intensive cultivation probably began in Mesopotamia, and Egypt around 5000 BCE.

BEGINNINGS OF AGRICULTURAL SCIENCE

Mechanized farming methods and **selective breeding** increased agricultural productivity. In 18th-century England, **Jethro Tull** pioneered many farming techniques, including the design for the **seed drill** that mechanized seed sowing.

JETHRO TULL'S SEED DRILL

long before knowledge of genetics gave it a scientific basis. Scientists became aware that selective breeding was possible because of genetic variation, but the subsequent inbreeding reduced variation. Although domesticated crops can produce superior foods, they are inherently vulnerable as a result. In the 1840s a disease called potato blight devastated potato production in Ireland, causing widespread famine in a population dependent upon the crop. In 1908 American botanist George Shull showed that hybrids—plants produced by crossing different species or varieties – were more vigorous, and he used them to improve the quality of maize. By the 1920s geneticists had found a way of measuring the degree to which variation was due to genes. Selective plant and animal breeding works only when traits are genetically determined, so the understanding of heritability enables such traits to be identified with greater confidence.

Improving crop yield

The Industrial Revolution of the 19th century was made possible at least partly by improved agricultural practices. This in turn brought mechanization to agriculture with the invention of machines like the tractor.

Meanwhile, a better understanding of soil nutrients and pest and disease control helped increase yields still further. Different crop species use nutrients in different ways. A farmer can ensure that the nutrient content of soil is managed effectively by rotating his crops (see right). This also prevents build-up of pests or diseases specific to any one type of plant.

Good agricultural practice
This aerial view of fields in Burgundy, France, in early summer shows a regimen that allows cultivation of a range of crops, including potatoes (foreground), winter wheat, winter barley, oilseed rape (from the cabbage family), and corn, as well as grass for livestock.

Adding fertilizer to the soil enhances plant growth. Traditional fertilizer, such as animal manure, releases nutrients in the soil slowly as it decomposes. It also improves water retention and opens the texture of heavy clay soils, encouraging root growth and nutrient-releasing bacteria. However, inorganic (artificial) fertilizer—in the form of pure nitrate or phosphate—works faster. In the 1800s phosphate-producing industries became possible with advances in chemistry. At the same time guano (bird dung)—collected from the South Pacific—proved to be a powerful fertilizer, and Chile supplied much of the world's agricultural nitrate. In 1908 Fritz Haber first filed his patent on the synthesis of ammonia from its elements. Haber and Carl Bosch went on to develop an economic way to make nitrogenous fertilizer. The invention—known as the Haber-Bosch process (see pp.240–41)—produced ammonia for conversion to

Legumes
Peas and beans replenish nitrogen. Their roots contain nitrogen-fixing bacteria that incorporate nitrogen from the air into the soil.

First cereal crop
Wheat is grown in first year and uses nitrogen from the soil. It follows a legume crop.

Second cereal crop
Barley is often grown in third year since it uses less nitrogen than wheat.

Root vegetables
Potatoes or turnips are grown in second year, and use potassium.

Crop rotation
This is an ancient regimen popularized in England in the 18th century by Charles Townshend. Crops are "cycled" around four fields over four years for effective management of soil nutrients. Nitrogen-"hungry" cereals are alternated with less demanding crops or those that replenish nutrients. Variation also prevents build-up of pests of diseases that affect plant families. Livestock farmers add an extra three years grazing into the cycle.

Intensive olive farming
Spain produces more than 40 percent of the world's olive oil. Intensive close planting of olive trees in the south—one of the hottest, driest parts of Europe—is possible only through an ambitious government-funded irrigation system.

Developments in agriculture lie in making use of new capabilities of crop plants—either those developed by genetic modification or through new discoveries.

NEW CROPS
The **winged bean** is a plant native to New Guinea, which exemplifies the potential of **wild species** to produce **new crops**. Flour can be made from its seeds, the pods are rich in vitamins, and

PSOPHOCARPUS TETRAGONOLOBUS, OR WINGED BEAN

its roots are high in protein. Other species include the salt-tolerant Salicornia and Tamarind Tree.

GENETIC MODIFICATION
Selective breeding and more advanced techniques can **alter plants**. It is possible to manipulate the genetic material in tissues to give plants characteristics such as **pest resistance**.

SEE ALSO »
pp.386–87 GENE TECHNOLOGY
pp.390–91 CLONING AND STEM CELLS

nitrate and revolutionized productivity. Other areas of plant science were also finding new ways of improving yields. Cultivation in greenhouses enabled year-round supply of some crops. In 1938 American agriculturalists Dennis Hoagland and Daniel Arnon

> **ORGANIC FARMING** A type of agriculture that excludes artificial fertilizers and pesticides, as well as growth enhancers and genetically modified organisms.

developed an idea from William Gericke that allowed plants to be grown without soil. The technique —called hydroponics—would go on to make crop production possible for farmers with poor-quality land.

In 1939 chemistry offered agriculture something other than fertilizer. German chemist Paul Müller found that a chemical called DDT was a very effective insecticide. Until then pest control had used natural plant-based products. DDT was man-made—and proved much more effective. It was later found to be harmful to the environment and its use was banned.

Sustainable development
The age of the pesticide was not universally welcomed. British botanist Albert Howard (1873–1947) rejected polluting pesticides and synthetic fertilizers after seeing traditional agricultural practices in India in the 1920s. He went on to pioneer chemical-free agricultural methods that became known as organic farming. A landmark book of 1962, *Silent Spring*, by American biologist Rachel Carson, warned of the polluting effects of pesticides and did much to change the public view. Today environmental attitudes have shifted as a result of Howard's and Carson's efforts. Pesticides are used, but they target specific pests and are degradable. In addition, there is a move to combine their use with biological pest control.

Hydroponic lettuces
Modern hydroponics allows crops such as lettuces to be cultivated on a very large scale in any climate. Plants are grown in solutions of nutrient that permeate a soilless inert medium, such as rock wool, perlite, or gravel.

BREAKTHROUGH
BIOLOGICAL PEST CONTROL

Pest control by natural enemies of pests— their predators or parasites—prevents chemical pollution and a buildup of pesticide resistance. However, extensive research is needed to ensure that the introduced enemy does not become a pest. Biological control, for example, using predatory mites to kill spider mites (below) requires that some pest damage is tolerated. A compromise is often reached whereby biological pest control is used in combination with traditional techniques.

Plant breeding
Rice is the second most abundant cereal (grass) crop—after corn. As with many crops, varieties and hybrids have been bred from their wild ancestor (far left) to create plants with a bigger seed grain, or heavier crop (near left).

AGRICULTURAL SCIENTIST (1864–1943)

GEORGE WASHINGTON CARVER

Despite battling years of institutional racism, George Washington Carver became one of America's leading agricultural experts. To restore soils that had been destroyed by repeated cotton farming, he developed important techniques, including crop rotation and alternative cash crops. As his fame grew, he worked to publicize these approaches. He encouraged farmers to grow nitrogen-rich crops such as peanuts and soybeans to help restore soil. He also came up with derivative products that could be made from these crops to make them commercially popular.

Light show
Coordinated displays of different colored laser beams creating spectacular patterns in the air are often used to entertain audiences at rock concerts and other events.

Lasers and Holograms

The word "laser" stands for Light Amplification by Stimulated Emission of Radiation. It refers to the production of a narrow beam of light from a substance whose atoms have been "excited" artificially. Lasers have many uses, including delicate surgery, accurate measuring, and the creation of holograms.

1 Photons from a flash tube are absorbed by atoms in the ruby rod, raising the atoms' electrons to a higher energy level.

2 When the electrons spontaneously drop back to their normal energy level, the atoms emit photons in random directions.

3 If an emitted photon hits another atom, the second atom emits a photon of the same energy and direction of travel.

4 The number of photons rapidly rises, until a stream of photons emerges from the partially reflective mirror as a laser beam.

BEFORE

Laser technology stemmed from research during the early years of the 20th century into how light and matter interact.

EINSTEIN'S CONTRIBUTION

In 1917 **Albert Einstein ‹‹ 304–305** realized that a photon hitting an already excited atom might cause the emission of another photon with the same energy (wavelength) as the first.

THE MASER

In 1954 Charles Townes and Arthur Schawlow invented the **maser (Microwave Amplification by Stimulated Emission of Radiation)**, a predecessor of the laser. Masers are used today in devices such as atomic clocks.

‹‹ SEE ALSO
pp.286–87 **STRUCTURE OF THE ATOM**
pp.314–17 **QUANTUM REVOLUTION**

Theodore Maiman, an American physicist, built the first laser in 1960. When researching what to use as a lasing medium—the substance that is artificially excited in order to produce a laser beam—Maiman chose a ruby crystal, so his device was called a ruby laser. Since then, many different materials have been used as lasing media. The first laser to use a gas (a helium–neon mixture) was also devised in 1960, and the carbon dioxide laser came soon after. Diode lasers, developed in the 1970s, were the first to produce a continuous beam rather than pulses of light. These lasers use semiconductor material as the medium, and they make up the majority of lasers in use today.

Laser beams vary in intensity, but all share certain properties. All the photons (packets of radiation) in a laser beam have the same wavelength, so laser light is monochromatic—it has just one color. The photons are also coherent, or in "phase," which means that all their wave peaks and troughs line up. Laser beams are usually highly directional and do not diffuse (spread out), so they can be focused on specific targets and projected over great distances.

Lasers in everyday life

Industry employs lasers to cut through steel and other materials. Surgeons use lasers of much lower power in eye surgeries and cosmetic procedures. A laser beam cuts tissues more precisely than a scalpel, and the heat from the beam cauterizes severed blood vessels instantly, preventing bleeding. Lasers are also used to clear out deposits clogging arteries.

Lasers are ideal for making precise measurements, in areas ranging from surveying to astronomy. In 1969 a laser measured the distance to the Moon to within centimeters (see p.173). Low-cost diode lasers are used for retrieving data from optical disk, such as DVDs, and in barcode readers, photocopiers, and printers. Other applications of

How a ruby laser works
When energy is fed into a ruby rod, the ruby atoms start to release photons of a specific wavelength. When these photons strike neighboring atoms, even more photons are emitted, causing a chain reaction. The photons bounce between two mirrors until the light is so intense that it emerges as a laser beam.

Patricia Bath
An ophthalmologist and laser scientist, Patricia Bath made huge advances in the prevention of blindness. In 1986, she invented a new surgical technique that used lasers to remove cataracts.

lasers include scientific research, fiber-optic communication, guiding missiles, and holography.

Holography

Pioneered by Hungarian engineer Dennis Gabor, a hologram is an image formed by the interference patterns created by the interaction of two laser light beams—one scattered from the subject of the hologram (usually a 3-D object), the other a "reference" beam. These patterns are recorded in a transparent film or plate. To view a hologram, the plate or film has to be illuminated either by a laser or by white light. The interference patterns diffract (scatter) the light waves in such a way that, when they reach a viewer's eyes, an image of the subject appears. Laser-viewed holograms give a monochrome image, but white light holograms can produce images in all colors. Holograms

are often used as a security feature on paper currency, credit cards, and product packaging. They are also used for optical data storage, retrieval, and processing, and in instrument displays for aircraft pilots.

AFTER

Research scientists are continuously finding practical and theoretical new applications for both lasers and holograms.

GAMMA-RAY LASERS

In 2007 scientists fleetingly created a **new form of matter called positronium**. A combination of electrons and positrons, it is a **potential energy source for ultra-powerful gamma-ray lasers**, which might one day be used to **kick-start nuclear fusion reactors** 324–25 **››**.

MOVIES BY LASER

Many cinemas are now equipped with projectors that, instead of wasting energy by passing white light through color filters, **create color directly** with red, green, and blue lasers.

MILITARY LASERS

In 2014 the US Navy demonstrated an operational ship-based military laser in the Persian Gulf. With a maximum power of 30 kilowatts, the laser can destroy incoming missiles, drones, and small boats, or can be used at reduced power to disable sensors and guidance systems on larger vessels.

SEE ALSO ››
pp.410–11 **MODERN SURGICAL PROCEDURES**

ELECTRICAL ENGINEER (1900–79)
DENNIS GABOR

In 1933 the Hungarian engineer Dennis Gabor fled from Nazi Germany, where he was working in electronics. After settling in Britain he joined an engineering company. In 1947, realizing that information about the 3-D shape of an object could be stored by comparing light from the object with a "reference" beam, he developed the idea of holography. Gabor had to wait until 1962, and the invention of lasers, to see his idea become reality. In 1971 he received the Nobel Prize for Physics for his invention of holography. In later life, Gabor wrote a book on social theory and the future of humanity.

« BEFORE

Before microchips, electronic devices were bulky and had to be assembled piece by piece.

CIRCUIT BOARDS
Formerly made from separate components soldered together by hand, circuits are now built on boards with **printed wiring. Robots insert and solder microchips**, each containing many components.

EARLY ELECTRONIC COMPUTERS
The first programmable electronic computer, **Colossus, was built in 1943** as part of the British code-breaking effort in World War II. It was soon followed by the US **ENIAC** (Electronic Numerical Integrator And Computer), used primarily to carry out **ballistics calculations**.

ENIAC COMPUTER

Microchips on circuit board
The circuit boards of modern computers often contain several microchips, each with its own data processing function. A multicore microchip contains two or more microprocessors on the same microchip to increase processing power.

Try to imagine a world without microchips. The most obvious difference is that there would be no cell phones or personal computers. Without microchips, computers would be just too large and expensive. It is not just personal computers, however, that would be impracticable; an increasing number of everyday devices, from cars to toasters, use built-in computers and microchip technology. But what are microchips, and how has society come to rely on them?

Circuits in miniature
A microchip is a tiny electronic circuit in which all the components are housed on a single chip. Today most microchips, or integrated circuits, are made from silicon, but the first working chip, demonstrated by engineer Jack Kilby at Texas Instruments in 1958, was made from germanium (a chemical element with many similar properties to silicon).

Kilby's idea was to make all components from the same material so that they did not have to be wired together, thereby dramatically reducing the time and cost involved in manufacturing electronic devices. Kilby's chip allowed the electronics industry to realize the true potential of transistors—the semiconducting devices used to switch or amplify electrical signals in a circuit. Previously, engineers had painstakingly soldered hundreds of components to build complex circuits. But manufacturing them all from a single semiconducting material meant it was possible to cram thousands, and eventually millions, of transistors onto an area the size of a

Early transistor
Transistors are the building blocks of most electronic devices, acting as switches or amplifiers. Transistors from the 1950s (right) were more than 0.4in (1cm) in length; modern ones are so small that hundreds of thousands can fit on a microchip the size of a grain of sand.

Microchip Technology

Often said to be the most important invention of the 20th century, the microchip has helped put computer technology into the palms of our hands. But keeping up with the rapid pace of progress means squeezing billions of components on to every chip.

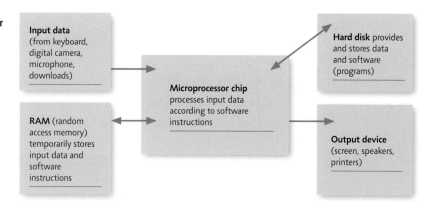

grain of rice. This paved the way for more compact and reliable electronics, including powerful computers, and in 2008 the electronics company IBM was able to announce that it had succeeded in packing more than two billion transistors onto a single chip.

Uses of microchips

Integrated circuits have become ubiquitous in modern technology, finding their way into everything from cell phones and wristwatches to kitchen appliances such as dishwashers and refrigerators. Chips are also used in credit cards to store encrypted information, which helps protect users from fraud.

The proliferation of the microchip over the last half-century is due to the development of photolithography, a mass-production process that uses light to make chips from silicon wafers. Production processes have now advanced to a stage where it is possible to make hundreds of microchips from a standard 12 in (30 cm) silicon wafer.

Microchips and computers

In the 1980s the economies of "micro" production dramatically reduced the price of silicon chips, and this helped usher in the era of personal computing. At the heart of every modern PC is a microprocessor chip, which is responsible for most of the machine's core information-processing functions (see right).

The breakneck speed at which microchip technology has advanced over the years has driven the construction of ever more powerful computers. This, in turn, has driven

How a microprocessor chip works

The microprocessor chip is the workhorse of modern computers. Once booted up, it receives input data, retrieves and processes information and instructions from the hard disk, carries out the software instructions, and passes on the processed data to output devices.

Input data
(from keyboard, digital camera, microphone, downloads)

RAM (random access memory) temporarily stores input data and software instructions

Microprocessor chip processes input data according to software instructions

Hard disk provides and stores data and software (programs)

Output device (screen, speakers, printers)

Chip under the skin
Microchips can be used to keep track of pets. Implanted under the skin (in the neck of a cat in the X-ray above), a chip contains details of the pet owner's name and address, which can be read by a scanner if a wandering pet is found.

the demand for ever smaller and more advanced microchips.

In 1965, Intel cofounder Gordon Moore described this trend in terms of the number of transistors on a microchip. Moore's Law, as it is now called, states that this number is likely to double every two years. But pushing up the number of transistors is becoming increasingly difficult. Having too many transistors on a single chip generates a lot of heat, which can prevent the tiny circuits from working properly. To move beyond these limits, scientists are developing entirely new devices called quantum computers, which rely on the physical state of individual subatomic particles—rather than miniaturized circuits based on transistors—to store and process data. Some nanotechnologists are also attempting to create wafers from potentially high-performance electronic materials such as graphene sheets—which are one-atom-thick sheets of carbon.

▷ Microchip miniaturization

The small size of microchips, and other components, makes it possible to pack immense computing power into relatively compact personal computers, such as the modern laptop shown in this X-ray.

Battery

Port for external device (such as mouse)

Hard disk

Motherboard (main circuit board)

Central processing unit (CPU, or microprocessor)

Cooling fan

AFTER »

INTEL'S FIRST MICROPROCESSOR CHIP

The microprocessor, or central processing unit (CPU), does most of the computation in a modern computer. The first ever commercially produced microprocessor chip, Intel's 4004, was introduced in 1971. By today's standards the 4004 was relatively simple, with just 2,300 transistors (compared with more than two billion on modern chips). Despite the ongoing trend, for miniaturization, the 4004 chip remains one of the smallest microprocessor designs that ever reached commercial production, measuring just over $^1/_8$ in (3 mm) by $^1/_6$ in (4 mm).

New materials and new types of microchips could replace conventional silicon microchips in the future.

PRINTABLE ELECTRONICS

New semiconducting materials, such as carbon nanotubes, could make it possible to **print electronics** in the same way that documents are printed using inkjet printers. Going one step further than microchips, these **electronic circuits** would be bendy, stretchy, and paper thin, allowing them to be **incorporated into clothing**, food packaging, and ultra-thin display screens.

QUANTUM CHIPS

In 2006, scientists at the University of Michigan created the world's first **quantum computer**, in which the spin of an electron determined the value of a quantum bit, or qubit (the quantum equivalent of a bit). In 2013, NASA and Google collaborated on a quantum computer that they claimed to be **100 million times faster** than a conventional machine—but only for a very specific type of calculation .

SEE ALSO »
pp.378–79 THE INTERNET
pp.380–81 ARTIFICIAL INTELLIGENCE AND ROBOTICS

BEFORE

The development of artificial satellites depended on several scientific and technological advances—in particular, an understanding of orbits and the development of rockets.

UNDERSTANDING ORBITS

German astronomer Johannes Kepler in 1608 **«** 76–79 described mathematically the **behavior of objects in orbit**. But it was English scientist Isaac Newton who worked out that the force that keeps an object in orbit is gravity. Any orbit is simply an **elliptical path** on which, at every point, an object's tendency to continue in a straight line is precisely balanced by the inward force of gravity.

PROJECTILES

In his 1687 book *Principia Mathematica* **«** 111, Newton proposed a thought experiment in which a **projectile shot sideways** from a mountain with enough speed would go into orbit. Satellites in **low-Earth orbit** need to move at a velocity of more than 17,000 mph (25,000 kph).

DEVELOPING ROCKETS

Launching any object into orbit requires a means of propulsion that will work beyond Earth's atmosphere. This technology only became a reality with the development of **liquid-fueled rockets** from the 1920s onward, and especially during World War II.

« SEE ALSO
pp.38–39 ANCIENT IDEAS OF THE WORLD
pp.136–37 NAVIGATING THE OCEANS
pp.338–39 ROCKET PROPULSION

Artificial Satellites

Earth has one natural satellite—the moon—and thousands of artificial ones, which have been launched into orbit. Artificial satellites have a huge range of applications in telecommunications, astronomy, weather forecasting, navigation, and scientific study.

The only significant force on an object above Earth's atmosphere is gravity. If that object is launched above the atmosphere with sufficient "sideways" speed, its fall will match the curvature of Earth's surface. Under these conditions, the object is in orbit.

First steps

On August 2, 1955, the US announced that it intended to launch the world's first artificial satellite to learn more about the near-Earth environment. A similar promise from the Soviet Union, issued four days later, was generally dismissed by Western commentators, but it was the Soviets who won this first battle in the "space race" with the launch of Sputnik 1 on October 4, 1957. The satellite broadcast regular signals that could be detected by radio receivers on the ground for 21 days. It remained in orbit for a further 71 days until the orbit decayed and Sputnik 1 burned up in the atmosphere.

Sending signals

The earliest satellites had scientific goals, but space engineers soon began to investigate other uses. One of the most obvious was communications. A satellite can act as a relay station, dramatically extending the range of radio and television broadcasts and telephone networks, which is normally limited by the curvature of Earth's surface. This restriction is overcome by relaying signals from transmitters in orbit.

Space junk
Debris, from tiny paint flakes to derelict satellites, orbits Earth as a legacy of our exploration of space. There are thought to be nearly a million pieces of space junk larger than a marble, including 34,000 that are more than 4 in (10 cm) across.

Notable early communications satellites included Telstar, which was launched in 1962 and sent live television pictures and telephone calls across the Atlantic, and Syncom 3 (1964) the first satellite that remained at the same point in the sky when seen from Earth. This property, called geosynchrony, is achieved by putting a satellite into an orbit directly above the equator at an altitude of 22,236 miles (35,786 km), where the orbital period is 24 hours—the same time it takes Earth to rotate.

> # 4.8 MILES PER SECOND
> The speed of a satellite in low-Earth orbit.

Geosynchronous, or geostationary, satellites are particularly suitable for communications because a transmitter or receiver on the ground, once locked onto the position of the satellite, will be able to send signals to a predictable fixed point in the sky. Geostationary satellites are still used for television broadcasts, as well as for relaying news feeds from journalists in far-flung places. However, they are rarely used in international telephony or for Internet links, because the radio waves' journey—to the satellite and back again to a ground station—introduces a quarter-of-a-second delay. Most Internet traffic today is carried via an extensive network of undersea cables that spans the globe, and by satellites in much lower orbits around Earth. Satellites in low-Earth

orbit (less than 1,240 miles, or 2,000 km, up) do not maintain a fixed position relative to the Earth's surface, so many are needed to achieve global coverage. Signals are intelligently routed from one satellite to another to achieve links from one part of the world to another. Similar constellations of satellites provide navigation services that can enable a suitably equipped device on the ground to work out its exact position.

Radio antenna 8–10 ft (2.4–2.9 m) long

The first satellite
Sputnik 1 completed 1,440 96-minute orbits of Earth, traveling at 18,000 mph (29,000 kph) on an elliptical path that ranged in altitude between 133 and 583 miles (215 and 939 km). The capsule deployed its long radio antennae only after separating from the upper stage of its R-7 rocket.

Highly elliptical orbit used for high-latitude communications satellites

Polar orbit used for Earth-observing satellites—rotating around Earth brings most areas into view

Sun-synchronous orbit keeps the same angle relative to the Sun, so the ground below is always lit in the same way

Geostationary, or Clarke, orbit circles Earth once a day and remains above the same point on equator

Low-Earth orbit is used by manned spacecraft and space stations

Satellite orbits
Depending on the function satellites have to fulfill, they can be launched into a wide variety of different orbits, ranging from a circular low-Earth orbit to much higher geostationary orbits (not to scale).

Earth observing satellite

Launched in 1999, NASA's Terra satellite is equipped with five sensors that are gathering data about Earth's climate and the nature of interactions between atmosphere, land, and oceans.

ASTER Provides high-resolution images to create detailed maps of surface temperature, reflectance, and elevation.

MISR Views Earth from multiple angles to reveal differences in cloud and land cover.

MOPITT Monitors the lower atmosphere, focusing on the density and distribution of pollutants.

CERES Readings from two identical instruments help scientists calculate the amounts of radiation absorbed and emitted by Earth.

MODIS This instrument tracks a range of Earth's vital signs, such as cloud cover and water vapor.

Clouds and Earth's Radiant Energy System instruments (CERES)

Advanced Spaceborne Thermal Emission and Reflection Radiometer (ASTER)

Multi-angle Imaging SpectroRadiometer (MISR)

Solar panels provide energy

Moderate-resolution Imaging Spectroradiometer (MODIS)

Measurement of Pollution in the Troposphere instrument (MOPITT)

Radio antenna

AFTER

The cost of building and launching satellites is high, but they provide services that are used by billions of people worldwide.

SATELLITE NAVIGATION

A constellation of satellites broadcasts signals carrying their exact position and the precise time on their **atomic clocks**. A receiver, such as that in a smartphone or GPS unit, works out the distance between itself and the satellites by **comparing the timings** of the signals they send. It can then work out its own position. At least four satellites in "line of sight" are required for a fix.

GPS SOFTWARE ON A SMARTPHONE

COLLISION RISK

With ever more satellites and space debris in orbit, **collisions are increasingly likely**. In 2009, a spent Russian satellite collided with a US **Iridium** communications satellite over Siberia.

SATELLITE INTERNET

Several governments and private corporations are in the process of launching constellations of **satellites to provide faster internet connections**. A constellation may consist of more than 10,000 interconnected satellites.

SEE ALSO >>
pp.372–73 HUMAN SPACEFLIGHT
pp.378–79 THE INTERNET
pp.396–97 SPACEPROBES AND TELESCOPES
pp.400–01 DARK UNIVERSE

Storm monitor

This image, captured by the MODIS instrument on board the Terra satellite, shows Typhoon Nangka off the coast of Japan in July 2015. Satellites like Terra help track destructive storms, providing early warnings of their approach.

Eyes in the sky

Satellites also provide unique opportunities for many kinds of military, (see box, right), astronomical, and scientific observation.

From their vantage point above Earth, orbiting telescopes can detect frequencies of infrared, ultraviolet, and X-ray radiation that are largely blocked by the atmosphere. Furthermore, telescopes that detect visible light can produce clearer images, unhindered by Earth's turbulent, cloudy, and polluted atmosphere. NASA's Hubble Space Telescope has been in low-Earth orbit since 1990, and is due to be succeeded by the James Webb Telescope in 2021. Many satellites gaze down on Earth rather than into space. Weather satellites produce pictures for bulletins and gather data for meteorological research and prediction. Some weather satellites are in geostationary orbits, so they can keep a gaze on a fixed area of the globe below, but most are in lower orbits, where they can resolve finer detail. Earth scientists have also benefitted enormously from satellite technology. So-called remote sensing satellites can monitor a host of different variables from their unique vantage point in space, including temperature, wind speed, sea level, ice cover in the poles and changes in land use, such as deforestation and agriculture.

SPY SATELLITES

Reconnaissance satellites are fitted with high-resolution imaging systems, which collect data on military installations or other sensitive areas on the ground. Early spy satellites carried a film camera and would automatically eject the exposed film in a canister, which was later recovered from the ground. Since the 1970s, spy satellites have used digital technology, and simply transmit their data to ground stations. Cameras aboard modern reconnaissance satellites can resolve objects on the ground the size of a tennis ball.

Microsatellites

The heaviest unmanned satellites in orbit weigh up to 11 tons and cost a small fortune to design, build, launch, and operate. The future of space exploitation lies with microsatellites such as UWE, developed at the University of Würzburg, Germany. It weighs around 2.2 lbs (1 kg); machines like this are set to open up space for a greater range of applications in the decades to come.

At 02.56 GMT on July 21, 1969, Neil Armstrong jumped lightly from a ladder attached to the side of the spiderlike lunar module, and put a human footprint on the Moon for the very first time. With up to 400,000 people working on the Apollo program, the Moon landing was a result of the US government's decision to put a man on the Moon ahead of the Soviet Union.

The Apollo missions were born in the aftermath of Yuri Gagarin's historic first orbit of Earth in April 1961. President John F. Kennedy, concerned about being repeatedly embarrassed by the Soviets in the "Space Race," demanded a realistic space goal that might allow the US to overtake their Soviet rivals. At the time, the Soviets' technological lead was so great that the prospect of a lunar landing was the only target that NASA (National Aeronautics and Space Administration) officials believed they could achieve first.

Enormous resources were thrown at a project that stamped its mark on the country throughout the 1960s. While astronauts practiced space maneuvers such as rendezvous docking, and spacewalking, a series of robot probes traveled to the Moon —crash landing, conducting orbital surveys, and ultimately making "soft" landings in order to learn as much about Earth's satellite as possible.

At the same time engineers were finalizing a flight plan called "lunar orbit rendezvous." This would involve a number of docking and undocking procedures in the Moon's orbit, which would reduce the weight of the spacecraft. But the project still required the most powerful rocket ever built, and Wernher von Braun, the man behind the German World War II V-2 rocket, would make the three-stage Saturn V his masterpiece.

After years of planning, Apollo 11 launched from Florida on July 16, 1969. From orbit around Earth, a brief burn of the upper-stage rocket engine put it on course for the Moon. The streamlined Command and Service Module (CSM) then broke free of the upper stage, turned around, and extracted the ungainly Lunar Module (LM) from its resting place within the rocket shroud. As the linked modules swung behind the Moon, a brief burst of the CSM's own engines put them into the Moon's orbit. Neil Armstrong and Buzz Aldrin boarded the LM for the journey to the surface, while Michael Collins stayed on the orbiting CSM.

As the LM drifted toward its landing site, the retrorockets slowed its descent and it touched down on the lunar surface, so realizing perhaps the greatest scientific and engineering achievement of modern times.

One giant leap for mankind
Apollo 11 astronaut Buzz Aldrin walks on the Moon. Aldrin and mission commander Neil Armstrong spent about two and a quarter hours on the Moon's surface collecting samples of lunar materials, taking photographs of their environs, and carrying out scientific experiments.

Human computers
To help launch the first American into space, a group of Black women were hired at NASA, alongside other female "computers," to run calculations by hand and plot the orbital trajectories of spacecraft. Mary Jackson (shown left), Katherine Johnson, and Dorothy Vaughan were the first Black managers at the Langley Memorial Aeronautical Laboratory in Virginia. Their contribution, critical for many of NASA's early spaceflights, is recalled in the movie *Hidden Figures*.

« BEFORE

Some of the most important developments in spaceflight evolved from research and advances in military aviation and rocketry.

PRESSURE SUITS
The development of jet engines during World War II meant aircraft could reach extremely **high altitudes**; special suits were needed to protect pilots from the **low pressures** they encountered. Early spacesuits were developed from these pressure suits.

ROCKET-POWERED FLIGHT
The first plane propelled by a liquid-fueled rocket was the German **Heinkel HE 176**, which took to the air in June 1939.

« SEE ALSO
pp.338–39 ROCKET PROPULSION
pp.244–45 ARTIFICIAL SATELLITES
pp.246–47 NATURAL DEFENSES

Timeline of humans in space
Many of the greatest advances in human spaceflight occurred within a single decade—the 1960s.

April 12, 1961 The Vostok Descent Module (left) is used by Yuri Gagarin to become the first person to orbit Earth and descend.

June 16, 1963 Valentina Tereshkova becomes the first female cosmonaut.

December 24, 1968 The crew of Apollo 8 become the first people to orbit the Moon.

September 29, 1977 Launch of Salyut 6, the first in a new wave of "modular" Soviet space stations.

April 12, 1981 The Space Shuttle *Columbia* launches a new era of reusable space vehicles.

1960   1970   1980

May 5, 1961 Alan B. Shepard Jr makes the first US sub-orbital spaceflight.

18 March 1965 Alexei Leonov leaves his Voskhod 2 spacecraft and makes the first spacewalk.

July 20, 1969 Apollo 11 astronauts Neil Armstrong and Buzz Aldrin make the first manned landing on the Moon.

January 28, 1986 The Space Shuttle *Challenger* explodes on launch and is destroyed.

1992 Russia and America announce their intention to work together on joint Shuttle–Mir missions.

Human Spaceflight

Keeping people alive in space—and bringing them safely back to Earth—were much greater challenges than launching satellites. From the cramped capsules that carried the first space travellers, spacecraft have evolved into bigger, safer, and more comfortable places for sustained work and exploration in space.

COSMONAUT (1932–68)
YURI GAGARIN

Russia's first cosmonaut was born to peasant parents on a collective farm near Smolensk—a humble background that may have influenced his selection as the pilot of Vostok 1. He studied engineering at college, then trained as a fighter pilot and was selected for cosmonaut training in 1960. He became a national hero following his 108-minute, single-orbit flight, and subsequently toured many countries to promote the Soviet space program. He died in a training accident while flying a jet from Chkalovsky Air Base.

Fit for space
Long-term spaceflight aboard the ISS causes bones to lose minerals and muscles to waste away. Astronauts have to do two hours of vigorous exercise daily to counter these effects.

The era of human spaceflight began on April 12, 1961, with the Soviet Union's announcement that cosmonaut Yuri Gagarin was in orbit around Earth. The US caught up the following year, when John Glenn became the first US astronaut to circle the Earth.

Launch and reentry
Putting people in space and ensuring their safe return has always been a huge engineering challenge. Launching a capsule capable of carrying crew, even as far as low-Earth orbit, requires a powerful rocket. On the return journey, the speeding spacecraft must be insulated from the enormous heat generated by friction with the atmosphere and then slowed using parachutes. NASA opted for splashdown at sea, while Soviet spacecraft came down over land, with a final blast from braking retrorockets just before reaching the ground.

Throughout the 1960s, the US competed with the Soviet Union for dominance in space technology and exploration, and especially in the race to put humans on the Moon—a contest that was won in 1969 by NASA's Apollo 11 mission (see pp.370–71). After this, the superpowers took

different approaches. Beginning in the 1970s, the Soviet Union developed a series of space stations that could remain in orbit for several years. Crews were ferried to and from them by Soyuz spacecraft, and automated ferry craft made regular deliveries of supplies from Earth.

NASA operated its space station, Skylab, in 1973–1974, but then concentrated on developing a large, reusable space transport vehicle—the Space Shuttle. This system consisted of a winged, rocket-propelled orbiter fitted with a large external fuel tank and a pair of external solid rocket boosters for lift-off. The orbiter could carry up to seven astronauts for two weeks or more, while its long cargo bay could ferry satellites into orbit, retrieve them for repairs, or house Spacelab—a European-built pressurized laboratory module. At the end of a mission, a brief engine burn slowed the orbiter and began its return to Earth. It landed without power, like a glider, on a runway.

Cooperation in space
In the 1990s, NASA worked with the Russian, European, Canadian, and Japanese space agencies to build and equip the International Space Station, which has been crewed continuously since 2000. Crews were transported by Space Shuttles and Soyuz craft. After the Shuttle was retired in 2011, all crews traveled in Russian Soyuz spacecraft until 2020, when both crew and cargo versions of the SpaceX Dragon 2, launched by a partially reusable Falcon 9 rocket, entered service under contract to NASA. It was

Orbiting laboratory
Construction and maintenance of the International Space Station has so far involved around 200 spacewalks, totaling more than 1,200 hours of extravehicular activity (EVA). Here, astronaut David Wolf instals a camera on a truss of the station in 2002. US astronauts practice EVAs in simulated low gravity conditions, in a huge swimming pool near NASA's Johnson Space Center in Houston.

November 20, 1998 Construction of the International Space Station (ISS) begins.

April 2001 The first space tourist, Dennis Tito, visits the International Space Station.

October 15, 2003 China launches its first astronaut, Yang Liwei , into orbit in the Shenzhou 5 space mission.

February 6, 2020 After 289 days on the ISS, Christina Koch sets record for longest spaceflight by a woman.

2000

2010

2020

June 21, 2004 Mike Melvill makes the first privately funded manned spaceflight on SpaceShipOne.

September 27, 2008 Zhai Zhigang makes first spacewalk by a Chinese astronaut.

September 29, 2011 China launches first space station, Tiangong-1

September 12, 2015 Gennady Padalka sets record for total time spent in space by one person at 878 days.

May 30, 2020 SpaceX Dragon 2 makes first commercial crewed flight to Earth orbit.

Recent years have seen more nations and companies engaging in human spaceflight.

PRIVATE SPACEFLIGHT
Virgin Galactic is aiming to offer **commercial, suborbital spaceflights** in the early 2020s. Axiom Space is planning its **first fully private flight** to the International Space Station in **2022** or later. It will use the SpaceX Crew Dragon capsule on a Falcon 9 rocket to carry the first crew without a professional astronaut.

PLANNED MISSIONS
NASA is leading an international program called Artemis to put a small space station in lunar orbit from where astronauts will travel to a **base near the moon's south pole**. The first landing, including a female astronaut, is planned for **2024**. It will use the Orion crew module and the heavy-lift Space Launch system. India is developing a crewed orbital craft called Gaganyaan and has plans for its own space station. Several nations have ambitions **to send humans to Mars**. NASA is furthest ahead with its plans, but a mission is unlikely before the 2030s.

SEE ALSO »
pp.394–95 **INSIDE THE SOLAR SYSTEM**
pp.396–97 **SPACE PROBES AND TELESCOPES**

"**Exploration** is not a choice, really; it's an **imperative**."

MICHAEL COLLINS, ASTRONAUT ON GEMINI AND APOLLO MISSIONS (1930–)

the first time a private company had launched a crewed orbital spacecraft.

Chinese space stations
In 2003 China, which had launched its first unmanned craft in 1970, became active in human spaceflight with the launch of its Shenzhou 5 spacecraft. The craft is similar to an enlarged Soyuz craft.

Following this success, China launched its first space station, Tiangong-1, in 2011. It was visited by Chinese astronauts in 2012 and 2013. A second space station, Tiangong-2, operated between 2016 and 2019. Both burned up in the atmosphere at the end of their missions, and in 2021 construction work began on a more ambitious successor, Tianhe-1.

The helmet has an antiglare visor

The backpack can sustain its wearer for at least seven hours of EVA

Helmet lights provide illumination during EVA

The suit is self-sustaining, and does not need to be connected to the spacecraft by an umbilical tether

Liquid coolant circulates through a high-tech undergarment to stop the cosmonaut from overheating

The Orlan suit weighs 265lb (120kg)

The outer layer is puncture-resistant to protect from micrometeoroids

ISS space suit
Cosmonauts on the International Space Station use the Russian-designed Orlan ("Sea Eagle") space suit when engaged in extravehicular activity (EVA). The suit's backpack contains a computer which monitors and controls the suit's multiple systems.

IN PRACTICE

REENTERING THE ATMOSPHERE

Reentering the atmosphere at speed heats a spacecraft exterior to around 1,600°C (2,900°F)—hot enough to melt steel. One solution is to fit the leading face of a reentering spacecraft with an "ablative" heat shield that burns away, protecting the hull and interior. NASA's next crewed spacecraft, Orion, which is scheduled for its first unmanned flight in 2018, will have the biggest ablative shield ever built (right).

5

THE INFORMATION AGE

1970 ONWARD

By the 1970s science and technology were in the ascendancy as astronauts landed on the Moon, and fertility experts produced the first test-tube baby. Meanwhile, advances in electronics and computing enabled almost instant global communication—and people started to talk about the Information Age.

THE INFORMATION AGE

1970 ONWARD

1970

1970
First vaccine for rubella. String theory is born. Restriction enzymes, products of bacterial cells, discovered. The Soviet Venera 7 probe lands on Venus.

▼ Intel 4004, Intel's first microprocessor chip

1975
The term "global warming" coined by Wallace Broecker, a geochemist at Columbia University.

1971
Intel introduces its first microprocessor chip (Intel 4004). Mariner 9 sends back pictures of the surface of Mars.

Surface of Mars ▶

1976
Richard Dawkins publishes *The Selfish Gene*, stressing that evolution is driven by competition between genes. Concorde is the first supersonic passenger aircraft. The down quark discovered.

1977
Human growth hormone produced. Rings discovered around Uranus. Carl Woese proposes that there are three domains of life: two bacteria and one eukaryote. Voyager 1 launched.

1972
The base pairs of a viral gene identified. The first remote sensing satellite (Landsat) launched. Moon rock is dated at 4.5 billion years old. Jacob Bekenstein suggests black holes have an entropy.

▼ Base pairs of a gene

1973
Z and W bosons (subatomic particles) produced.

▼ Concorde

1974
First close-up of the surface of Mercury. Sheldon Glashow and Howard Georgi propose unifying the strong and electroweak interactions into a Grand Unified Theory.

▼ Close-up of Mercury

1978
Louise Brown, the first baby conceived by *in vitro* fertilization (IVF), is born. Human insulin is produced in the laboratory using recombinant DNA technology.

▲ IVF procedure

1979
Photos taken by the Voyager 1 space probe reveal Jupiter's rings and the planet's innermost moon, Metis.

1980

▲ Wind farm

1980
The first wind farm built in the US.

1981
The Space Shuttle first launched. First vaccine for Hepatitis B. The Scanning Tunneling Microscope (STM) developed.

Space Shuttle launch ▶

1982
Richard Feynman conceives quantum computing.

1983
HIV identified as the cause of AIDS. The weak gauge boson discovered. The Infrared Astronomical Observatory launched.

1984
String theory developed further. Alec Jeffreys devises a genetic fingerprinting method. Kary Mullis invents the polymerase chain reaction.

▼ Genetic fingerprinting

1985
AZT, the first licensed HIV drug, launched. Buckminsterfullerene, a form of carbon, discovered.

Buckminsterfullerene ▶

1986
Giotto photographs Halley's comet.

1987
NASA measures the movement of tectonic plates. A supernova is seen to explode in the galaxy the Large Magellanic Cloud.

Exploding supernova ▶

1988
Under James Watson, scientists begin to catalog the locations of genes on chromosomes for different species.

1989
The first sheet of superclusters of galaxies, the Great Wall, is found.

▲ Superclusters

BREAKTHROUGH

NOBEL PRIZE IN CHEMISTRY

In 1964 Dorothy Hodgkin was awarded the Nobel Prize in Chemistry "for her determinations by X-ray techniques of the structures of important biochemical substances." She was only the third woman chemist to be so honored, after Marie Curie and Irène Joliot-Curie.

At Oxford University Hodgkin had pioneered the technique of using X-ray crystallography to determine the three-dimensional structure of biomolecules. A form of the same technique was employed by Rosalind Franklin at King's College, London, to produce the X-ray diffraction pictures of DNA that Watson and Crick at Cambridge University used to reveal the structure of DNA (see p.346).

Hodgkin took part in meetings in 1946 that led to the foundation of the International Union of Crystallography. X-ray crystallography is still widely used by scientists requiring knowledge of a molecule's structure in order to gain a clear understanding of its function.

largest molecule to be described using X-ray techniques. In recognition of her work, in 1947 Hodgkin was elected to the Royal Society in London.

Work on vitamin B_{12}

In the academic world, paleontologists tend to be shown fossils all the time, while ornithologists are often brought preserved birds for investigation. In Hodgkin's case, scientific friends would regularly bring crystals that might interest her. Chemist E. Lester Smith was one such friend who gave Hodgkin some cherry-red crystals of vitamin B_{12}. Characteristically, Hodgkin became determined to fathom out their structure and, using an array of complex computer models, she unraveled the answer.

She won a string of awards, most notably the Nobel Prize in Chemistry in 1964—remarkable given the sexism of the scientific community at the time. A year later she was made a member of the Order of Merit. This

institution of only 24 members had been founded by Edward VII in 1902 and is considered Britain's highest honor in science, the arts, and public life. Before Hodgkin, the only woman to have achieved this accolade was the pioneering nurse and writer Florence Nightingale. However, by Hodgkin's

a small sample of crystalline insulin by Robert Robinson. Analyzing this complex protein had been extremely difficult, requiring advanced X-ray diffraction techniques and high-speed computing. Hodgkin was careful to give due credit to the international team of young researchers who had

> "She radiated **love**: for **chemistry** … her **crystals,** and her college … Her love was combined with a **brilliant mind** and an iron will to succeed … "
>
> M. F. PERUTZ, AT HODGKIN'S MEMORIAL SERVICE, 1994

own admission, her greatest achievement did not come until 1969, when long years of patient work were finally rewarded by her determination of the structure of insulin; this was some 35 years after she was first given

helped her achieve her goal. When she retired from Somerville College in 1977, however, Hodgkin decided to turn her insight to a completely different field.

Advocate for peace

Enjoying the respect of the scientific community, Hodgkin wanted to exert her influence to help secure peace in international relations. She had long been a member of Science for Peace, a group dedicated to ending war, and in 1953 had been denied a visa to the US because the group had communist members. In 1976, though, these restrictions had been lifted and she was able to travel the world freely as a passionate advocate for peace.

Crystals never ceased to fascinate Hodgkin, however, and she continued her tours of American universities, discussing proteins, insulin, and the history of crystallography—right up until her death in England in 1994.

Distinguished company

Photographed in 1985, Hodgkin stands next to Stephen Hawking, Lucasian Professor of Mathematics. Hodgkin, like Hawking, was a member of the prestigious Royal Society.

WALLPAPER BASED ON STRUCTURE OF INSULIN

BEFORE

Gene technology became a practical possibility once the chemical basis of genetics was understood.

PROTEIN SYNTHESIS

As the result of numerous collaborative experiments in the 1950s, biologists demonstrated that the **inherited material was DNA**, and that it exerted its influence on the characteristics of organisms by instructing them to make **specific combinations of proteins**. The potential use of **genetic engineering** lay in the promise of transferring genes between organisms to **change these proteins**.

GENE EXCHANGE IN BACTERIA

Joshua Lederberg's discovery that bacteria could **exchange genes** explained observations that bacteria changed their characteristics when mixed with others. They can transfer genes by passing small rings of DNA called **plasmids** between them, which is known as **conjugation**.

BACTERIA UNDERGOING CONJUGATON

« SEE ALSO
pp.244–45 BACTERIA AND VIRUSES
pp.348–49 THE GENETIC CODE

BIOLOGIST (1925–2008)

JOSHUA LEDERBERG

Joshua Lederberg was an American biologist who was distinguished by his contributions in many areas of science. In 1958 he won the Nobel Prize for Medicine after discovering that bacteria exchanged genes in plasmids. He led the Genetics Department at Stanford University, California, and was a key scientific advisor to the US government, as well as being involved in NASA's space exploration—researching the potential extraterrestrial spread of microbes.

Gene Technology

The realization that the characteristics of living organisms were determined by sections of molecules called genes opened up the possibility of altering them in a way that went far beyond the breeding programs of plants and animals. Any organism could be endowed with features that would make it more useful.

In the 1950s scientists established that inherited characteristics are determined by sections of DNA in cells called genes (pp.308–309). Genetic engineering involves artificially transferring genes from one organism to another—usually a completely different species—so that characteristics become altered in the resulting genetically modified organism or GMO. The theory became a reality in 1970 with the discovery of restriction enzymes. Bacteria produce these enzymes to cut up the DNA of invading viruses, but unlike conventional digestive enzymes they cut the DNA at precise positions. This ability was to prove critical in their use in genetic engineering.

The benefits of genetic engineering

From the 1970s scientists became increasingly aware of the potential benefits of genetic engineering. With the knowledge that genes determined proteins came the realization that an organism's protein production could be altered by transplanting genes.

Back in 1955 English biochemist Frederick Sanger had been the first to decipher the complete amino acid sequence of a protein (that of insulin). Animals produce insulin to control blood-sugar levels; if insulin cannot be made, a disease called diabetes results. Diabetes can be controlled with insulin treatment. At that time, the only source of insulin was from the pancreases of cattle or pigs and, while purification methods were improving, there was a risk that animal version could be contaminated with viruses. Genetic engineering offered the possibility of producing insulin from bacteria, which reduced this risk; furthermore, such engineering paved the way for modifying "higher" organisms, such as plants, to improve food production.

The first GM bacteria

In 1973 American biochemists Herbert Boyer and Stanley Cohen started to use restriction enzymes to cut threads of DNA. Because these enzymes cut bacterial DNA at specific points, they could be used to cut out those precise fragments of DNA that contained genes—a process called gene splicing. Boyer and Cohen also knew that

bacteria routinely exchanged plasmids, so they used restriction enzymes to cut the rings open and another enzyme—ligase—to bond the gene fragments into the loops. In this way plasmids could be used as carriers, or vectors, for the inserted genes. The "new" genes would be transferred from one bacterium to another and replicated whenever the modified bacteria reproduced.

Using this technique they produced the human hormone somatostatin in 1977 and then followed that with human insulin in 1978. Boyer established a company for the commercial production of these and other proteins; in 1982 "man-made" insulin was the first genetically engineered product approved by the US Food and Drug Administration.

Further developments

The sequencing work of Sanger and others paved the way for scientists to assemble genes from their component building blocks, removing the complex task of isolating the proteins they code

70 PERCENT the proportion of therapeutic insulin currently used worldwide that has been produced by genetic engineering.

Luminescent fish
The first genetically modified pet to be made available to the public, this tiny fish—known as a Glofish—was produced by inserting genes for bioluminescence from jellyfish into zebra fish.

for first. Most therapeutic insulin and the blood-clotting factor used to treat the disorder hemophilia is made this way.

Restriction enzymes proved useful in other areas. In 1984 English geneticist Alec Jeffreys realized that the number and size of the DNA fragments they produce could characterize a person's genetic identity. DNA profiling is now used in forensic science and to establish family relationships.

USES FOR GENE TECHNOLOGY	
SUBJECT	POSSIBLE USES
Plants	Improving nutritional content, for example, golden rice contains a gene that produces a precursor of vitamin A.
	Lengthening shelf life by suppressing genes responsible for the production of polyphenol oxidase, which causes fruit to become brown.
	Adding pesticide and/or herbicide resistance to crops, such as soy, so that farmers can use fewer chemicals without reducing the yield.
	Ensuring good harvest by adding genes for pest and/or disease resistance, for example, virus-resistant papaya and insecticide-producing corn.
	Improving hardiness, such as adding ice resistance to crops normally grown in warmer climates to enable cultivation in colder areas.
Animals	Improving nutritional content of food, for example, growth-enhanced tilapia fish or transgenic chickens engineered for their high-protein eggs.
	Developing experimental models, such as mice engineered to develop cancer or mice with missing genes, to demonstrate gene function.
	Altering aesthetics, for example, zebra fish modified to create bioluminescent Glofish (see above).
Chemical products created by genetic engineering	Proteins for treating conditions such as diabetes (insulin), hemophilia (blood-clotting factor), and hepatitis B (interferon).
	Products that arise from protein activity, such as aspartame used as low-calorie sweetener in food.
Humans	Gene therapy (see pp.416–17) that targets body cells and overrides or replaces a faulty gene with a functioning one is increasingly being approved to tackle a wide range of genetic disorders, including leukemia and other cancers.
	Gene editing using a technology called CRISPR-Cas9 (see pp.416–17) that corrects faulty genes is being trialed to treat diseases with a genetic basis.

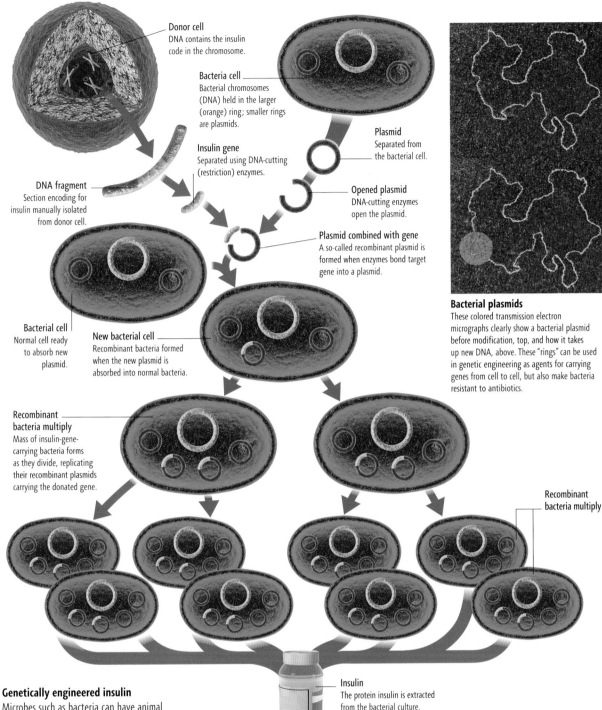

Donor cell
DNA contains the insulin code in the chromosome.

Bacteria cell
Bacterial chromosomes (DNA) held in the larger (orange) ring; smaller rings are plasmids.

Insulin gene
Separated using DNA-cutting (restriction) enzymes.

Plasmid
Separated from the bacterial cell.

DNA fragment
Section encoding for insulin manually isolated from donor cell.

Opened plasmid
DNA-cutting enzymes open the plasmid.

Plasmid combined with gene
A so-called recombinant plasmid is formed when enzymes bond target gene into a plasmid.

Bacterial cell
Normal cell ready to absorb new plasmid.

New bacterial cell
Recombinant bacteria formed when the new plasmid is absorbed into normal bacteria.

Recombinant bacteria multiply
Mass of insulin-gene-carrying bacteria forms as they divide, replicating their recombinant plasmids carrying the donated gene.

Recombinant bacteria multiply

Insulin
The protein insulin is extracted from the bacterial culture.

Bacterial plasmids
These colored transmission electron micrographs clearly show a bacterial plasmid before modification, top, and how it takes up new DNA, above. These "rings" can be used in genetic engineering as agents for carrying genes from cell to cell, but also make bacteria resistant to antibiotics.

AFTER »

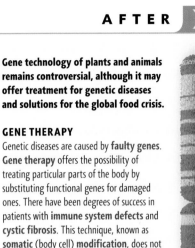

Gene technology of plants and animals remains controversial, although it may offer treatment for genetic diseases and solutions for the global food crisis.

GENE THERAPY
Genetic diseases are caused by **faulty genes**. **Gene therapy** offers the possibility of treating particular parts of the body by substituting functional genes for damaged ones. There have been degrees of success in patients with **immune system defects** and **cystic fibrosis**. This technique, known as **somatic** (body cell) **modification**, does not affect reproductive cells so cannot be passed on. **Genetic modification** alters organisms so that they can pass on changes.

GM CROPS
Crops such as rice and maize have been modified and grown worldwide. Debate exists over the safety of GM crops, but supporters point out that improved crop yields are needed to avert the possibility of a global food crisis.

SEE ALSO »
pp.390–91 **HOW CLONING WORKS**
pp.414–15 **THE HUMAN GENOME**

CHROMOSOME 7 —SITE OF CYSTIC FIBROSIS GENE

Genetically engineered insulin
Microbes such as bacteria can have animal or plant genes inserted into them so that they can be modified to produce a useful protein. The donated gene that instructs the manufacture of the protein is first isolated and then bonded into a plasmid ring—a molecule that is naturally absorbed by another bacteria.

The future of GM
Gene technology has evolved to make more ambitious modifications possible. For example, naturally occurring plasmids in some bacteria can be modified to suppress certain genes and to carry other useful genes into plant cells instead. Modification of animals necessitates manipulation of embryos. GM animals, such as insects and mice, have mainly been used to further understanding of gene function, for example, genes that cause cancer.

A key area of gene technology concerns the production of crops that are genetically modified to resist pests or herbicides or to change qualities. Such GM products first became commercially available in the 1990s, the first being the FlavrSavr tomato. Most GM crops are grown in North America but they are increasingly sown in developing countries, such as Brazil and India.

Not everyone is convinced about the safety of such GM crops and research is ongoing to assess the implications for human health. Some health and environmental organizations advocate an embargo on such foods until their safety has been undeniably proven.

Genetically modified crops
Soy crops have been genetically modified to make them resistant to herbicides, allowing farmers to exercise better control over weeds that grow alongside the soy.

BREAKTHROUGH

DNA PROFILING

This technique, also known as "genetic fingerprinting," compares samples of DNA from different people. Restriction enzymes "digest" each DNA sample, generating a pattern of bands that looks like a bar code; the bands correspond to clusters of DNA fragments. Although 99.9 percent of DNA sequences are the same in every individual, each person does have a unique pattern. Similarities usually show family relationships.

GENETIC FINGERPRINT OF A FAMILY (TOP FOUR STRIPS)

In Vitro Fertilization (IVF)

On July 25, 1978, at Oldham General Hospital, near Manchester in England, Lesley Brown gave birth to a healthy baby girl. This was the result of a groundbreaking technique that had been developed to help women with blocked or missing fallopian tubes conceive. The birth of Louise Brown was a particular cause for celebration: she was the world's first baby who had been successfully conceived by IVF.

Lesley Brown had blocked fallopian tubes, which meant that sperm could not reach her eggs for conception—but in the late 1970s scientists developed a procedure that offered her hope. In vitro (in glass) fertilization involved a woman receiving fertility medicines in order to stimulate ovulation, retrieving eggs from the woman's ovaries, and introducing sperm and egg together in a laboratory Petri dish. Scientists provided the right conditions to encourage fertilization and monitored the first few days of embryological development. The tiny embryo—no more than a ball of cells—is then implanted back into the woman's uterus for a natural pregnancy.

For nearly 20 years before Louise Brown's birth, the physiologist Robert Edwards had been perfecting the medium needed for fertilization at the University of Cambridge, England. Edwards had managed to nurture human eggs to the critical stage where they were receptive, and in 1969 he witnessed the first human in vitro fertilization: in his petri dish he saw clear evidence that sperm had successfully penetrated an egg.

Edwards developed the clinical side of his work in collaboration with a surgeon, Patrick Steptoe, who used a probe to retrieve eggs from the ovaries of volunteering patients. Together they solved other technical problems: they used drugs to control the menstrual cycle of patients so that they knew when eggs were ready for extraction, and they perfected a method of returning embryos to the uterus. During much of their work Edwards and Steptoe endured condemnation from critics who called them "dehumanizing." However, the critics were silenced when Louise Brown was born and mother and baby's condition was described as excellent.

Since the birth of Louise Brown the techniques pioneered by Edwards and Steptoe have been adopted and refined by hospitals around the world. Drugs based on natural hormones have been developed to stimulate egg production and fertilization itself can be assisted using ICSI (intracytoplasmic sperm injection), a method where a single sperm is injected directly into the egg.

The moment of fertilization
This color-enhanced image was produced by a scanning electron microscope and shows numerous sperm cells clustered around an egg. As soon as one penetrates the coat around the egg will harden, barring entry to the others.

"I don't feel any **more special than anyone else.**"

LOUISE BROWN, FIRST "TEST TUBE" CONCEPTION

BEFORE

Cloning technology arose from breakthroughs in understanding the development and genetics of embryos.

DISCOVERY OF MERISTEMS

German botanist Karl Nageli recognized parts of a plant that were **continually active** in cell division, and called them meristems in 1858.

He described how these meristems were found in the growing parts of a plant, such as the shoot and roots tips.

MERISTEM (RED) IN SHOOT

PIONEER OF EMBRYO CLONING

In 1885 German biologist Hans Dreisch separated the cells of **sea urchin embryos** and discovered that the isolated cells grew into separate embryos, showing that each cell carried the complete set of **genetic instructions**.

SEE ALSO
pp.348–49 THE GENETIC CODE

Cloned forest
One of the world's largest living organisms is an entire forest made from clones of a single aspen tree. Known as Pando, (Latin for "I spread" it is located within Fishlake National Forest in Utah.

Cloning and Stem Cells

Clones are genetically identical individuals. They occur as a natural form of reproduction across the living world. However, some organisms—and some cells—clone much more easily than others. By studying biological development, scientists have sought to understand the reasons for these differences—and devise technologies that could not only clone complex animals, but also offer new ways of treating disease.

Sexual reproduction mixes genes from two parents. Clones—a term coined by British biologist J.B.S. Haldane in 1963—are produced asexually without the involvement of fertilization. They are made, for instance, when single-celled organisms split in two, or plants grow side shoots and runners that can sprout and detach. Plant breeders have long recognized the advantage of propagating plants in this way: desirable inherited characteristics, such as big fruit or high seed yield, get

Dolly the sheep
The first mammal clone to be produced by somatic nuclear transfer—Dolly—was born on July 5, 1996. Of 277 udder cells used, 29 grew into embryos, but only one embryo developed into a fully formed lamb.

passed on, unadulterated, into cloned offspring. Cloning can be quick and productive, hundreds of descendants coming from a single parent.

Plants clone so easily because they have meristems—parts, such as the shoot and root tips, that contain undifferentiated, continually dividing cells. But in 1958 British botanist F.C. Steward went further and cloned carrot plants from nonmeristematic cells, showing that even differentiated cells have the genetic potential to form an entire organism. By adding hormones under sterile conditions, scientists can make tiny pieces of plant sprout roots and leaves. Animal tissues can be cultured too, but, even with hormones, they refuse to differentiate to form an entire body—a problem that makes cloning animals far more complicated.

The problem of differentiation

To make a complete animal, cells must be "pluripotent"—able to differentiate into all the tissues that comprise a body. This is a characteristic of fertilized eggs

and the cells in an early-stage embryo, but is absent in most mature cells, except a class known as stem cells—a term first used by Russian biologist Alexander Maksimov in 1908.

Early indications that adult cells could be pluripotent came from the work of American cancer researcher Leroy Stevens in 1953. He saw that some kinds of tumors contained mixtures of differentiated tissue—including bone, gut, and blood—suggesting that the cancerous cells were pluripotent.

Cloning animals

Meanwhile, scientists were attempting to make animal clones using early stage embryos. In 1952 American biologists Robert Briggs and Thomas King transplanted a nucleus from an early embryo of a frog into an unfertilized frog egg, from which the nucleus had been removed. The egg developed into a clone of the donor animal. Briggs and Kind tried using nuclei from later stage embryos but failed to make a viable clone, and concluded that nuclei in

Genetic parent "donor" of skin cells

Nucleus

Nucleus removed from skin cell

Skin cell containing genetic material of donor sheep

Donor egg combined with nucleus from skin cell of genetic parent using electric stimulus

Egg-donor sheep

Nucleus

Egg from donor sheep

Donor egg with nucleus removed

Artificial "zygote" created and grown in laboratory

Cell division of zygote creates embryo, genetically identical to skin-cell nucleus

Lamb, a clone of her genetic parent

Embryo implanted in a foster sheep

Somatic nuclear transfer

Dolly the sheep was genetically identical to the donor sheep (top left in the diagram). The nucleus of the donor's skin cell was inserted into an egg taken from a different ewe (centre) that had had its nucleus removed. Whereas an egg has only one set of chromosomes, the skin-cell nucleus has a double set of chromosomes—just like a fertilized egg. Replication of the skin cell's DNA produced an embryo that is genetically identical.

Nucleus removal
In the cloning process, a sheep egg has its nucleus removed with a hollow glass needle before the nucleus from an udder cell is injected into the egg.

BREAKTHROUGH

STEM CELLS

Stems cells differ from normal cells in that they can differentiate into specialized cell types, such as muscles or nerves. Stem cells from an early embryo (below) can differentiate into almost any type of tissue. Stem cells also exist in the mature adult body, but these cells can differentiate into a much narrower range of tissue types.

EARLY STAGE EMBRYO

"Clones are simply identical twins."

RICHARD DAWKINS, BBC INTERVIEW, JANUARY 31, 1999.

The outer cells of the embryo become compacted, enclosing an inner cell mass

About four days after fertilization, a human embryo is made up of tens of cells

Embryonic stem cells

Cells taken from a human, or other mammalian embryo, at the right stage are pluripotent (have the potential to develop into almost any cell type). Given the correct nutrients they can be grown in a laboratory. These embryonic stem cells can be induced to differentiate into a range of specialized cells, which can be used in research or in disease treatment.

The inner cells are harvested at this stage

The cells are cultured in a laboratory

The cells are capable of differentiating into a variety of tissues

MUSCLE CELL

EPITHELIAL OR BODY-TISSUE CELL

NERVE CELL

FAT CELL

Cloning technology has the potential to have groundbreaking applications in medicine and conservation.

TREATING DISEASE

Stem cell therapy has the potential to offer life-changing treatment for patients with damaged and diseased nervous systems. In 2010, **human embryonic stem cells** were used to heal spinal injury, and in 2012, they helped restore retinal function in patients with degenerative eye disorders. By 2020, breakthroughs in stem-cell technology were helping to grow nerve cells in the lab, paving the way for developing cures for neurodegenerative disorders, such as Parkinson's disease.

BACK FROM THE DEAD

Cloning techniques can potentially be used to build up numbers of individuals of **endangered species** and one day even to recover species that have become extinct. In 2009 the **Pyrenean Ibex** became the first extinct animal to be cloned in this way, although the clone produced did not survive. In 2013, in Australia, the gastric-brooding frog, which had become extinct in the mid-1980s, was successfully cloned from embryos using **somatic nucleic transfer**. Researchers were optimistic that this species—and possibly others—could be brought back from extinction by this or similar technologies.

PYRENEAN IBEX

SEE ALSO »
pp.414–15 THE HUMAN GENOME

developing embryos cells slowly lost their pluripotency—their potential to give rise to a complete animal. However, in 1958, British biologist John Gurdon succeeded in cloning frogs from differentiated gut cells. In the following decades, clones of mammals (which have much smaller eggs than frogs) were made by nuclear transfer from embryonic cells, and then—in a landmark experiment—from differentiated cells.

In 1996 Ian Wilmut and Keith Campbell of the Roslin Institute in Scotland demonstrated how the nucleus from a body (somatic) cell could be "reprogrammed" by injecting it into a denucleated egg cell. They used the nuclei of a sheep's udder cells, and fused nuclei and eggs with an electric shock. Of 277 udder cell nuclei, 29 grew into embryos—and one survived into adulthood. This animal, called Dolly, became the first animal clone produced by this so-called somatic nuclear transfer.

Therapeutic cloning

Embryonic cells can differentiate into almost any kind of tissue, so using such stem cells opens up the possibility of treating disease by producing tissues for transplantation. In 1998 American biologists James Thomson and John Gearhart succeeded not only in isolating human embryonic stem cells—but also culturing them in the laboratory. Similar techniques have been developed to culture stem cells taken from mature adult tissues; and in 2006, Japanese biologist Shinya Yamanaka found a way of reprogramming adult cells so that they functioned like embryonic stem cells.

This discovery addresses many of the ethical issues that have been raised around the use of cells from human embryos for research and development of therapies. By 2014 teams of researchers in America, Japan, and South Korea were demonstrating that all sorts of adult cells could be reprogrammed in this way— by-passing the need to use embryos.

The possibilities of stem cell therapy are just beginning to be explored. One day soon, heart cells grown in a laboratory may be used to rebuild healthy tissue in patients suffering from heart disease, and patients with type I diabetes may receive pancreatic cells to replace insulin-producing cells that have been lost or destroyed.

Nanotechnologies

Conceived in the 1950s, nanotechnology is engineering on a scale invisible to the naked eye. By manipulating individual molecules and atoms, nanoengineers are able to make materials with unique properties that promise enormous developments across every industry.

In December 1959, at the annual meeting of the American Physical Society, physicist Richard Feynman (see pp.326–27) laid down two challenges: the first was to take the words from the page of a book and write them on an area 25,000 times smaller, and the second was to produce a rotating electric motor smaller than a cube with sides of 0.4mm. He offered a reward of $1,000 for each challenge. British engineer William McLellan claimed the prize for the second challenge the following year, but the first was not met for more than 25 years. Tom Newman of Stanford University in the US obliged in 1986 by reducing the first page of *A Tale of Two Cities* to nanoproportions using an electron beam machine.

Small but powerful

In nature almost everything happens at the nanoscale—that is 100 nanometers and below, one nanometer (nm) being one millionth of a millimeter. But until the Scanning Tunneling Microscope (STM) was invented in the 1980s this miniature world was invisible to our eyes. Now we have microscopes capable of tracking the intimate details of our cells' inner workings. As a result, scientists in every field, from medicine to materials science, have been inspired to start working on much smaller scales.

However, particles begin to act very strangely once taken down to the nanoscale. Otherwise innocuous materials, such as metal powders and even flour, start exploding. Strange subatomic effects come into play when particles are this small, changing the electrical behavior and reactivity of the materials. It is this sort of unpredictable behavior that has led some to question the safety of using nanoparticles. Nevertheless, there are many problems for which nanotechnology might be the only solution—for instance, in keeping up with the current trend for ever-more powerful computers or delivering drugs to diseased cells in the human body.

Using nanoparticles

The power of nanotechnology has long been within our grasp: we just have not realized it. As early as the 1920s car tires were treated with a material

Colloidal gold
Gold nanoparticles, also known as colloidal gold, have many uses, ranging from staining glass to treating arthritis. Scientists also think they could be useful in detecting cancer cells.

called carbon black to cut down on wear and tear; no one had realized that the carbon particles being used were nano sized. For centuries, gold nanoparticles have featured in production of red and blue stained glass—another odd property of nanoparticles is that they do not necessarily appear the same color as their larger counterparts.

Since Newman met Feynman's challenge, nanotechnology has grown into a science in its own right. In 1985 buckminsterfullerene, a form

« BEFORE

Nanotechnology could not exist until scientists could see individual atoms and molecules and had advanced knowledge of how matter behaves at the nanoscale.

SCANNING TUNNELING MICROSCOPE

Not until the development of the **Scanning Tunneling Microsope (STM)** by Swiss researchers **Gerd Binnig** and **Heinrich Rohrer** in 1981 were scientists able to see things at a **single atom level**.

CLUSTER PHYSICS

The investigation of properties acquired by materials as they increase in size from just a few atoms is known as **cluster physics**. The science arose—in the early 1980s—because it was realized that the **behavior of materials alters with mass** and can be **unpredictable** at very small size, when **subatomic quantum effects** become more prevalent.

« SEE ALSO
pp.326–27 **RICHARD FEYNMAN**

How small is small?
Small objects can be seen with an ordinary microscope, but the nanoworld is so small it must be magnified many tens of thousands of times to be visible.

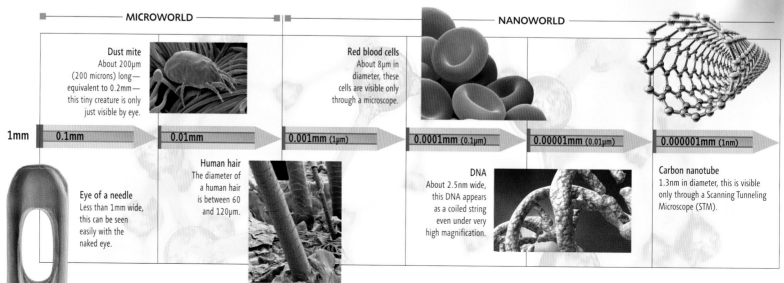

MICROWORLD — **NANOWORLD**

Dust mite
About 200µm (200 microns) long—equivalent to 0.2mm—this tiny creature is only just visible by eye.

Red blood cells
About 8µm in diameter, these cells are visible only through a microscope.

1mm | 0.1mm | 0.01mm | 0.001mm (1µm) | 0.0001mm (0.1µm) | 0.00001mm (0.01µm) | 0.000001mm (1nm)

Eye of a needle
Less than 1mm wide, this can be seen easily with the naked eye.

Human hair
The diameter of a human hair is between 60 and 120µm.

DNA
About 2.5nm wide, this DNA appears as a coiled string even under very high magnification.

Carbon nanotube
1.3nm in diameter, this is visible only through a Scanning Tunneling Microscope (STM).

Carbon buckyball
Now recognized as a form of the element carbon, buckminsterfullerene has 60 atoms arranged in a regular arrangement of pentagons (five sides) and hexagons (six sides). It occurs naturally in candle soot, but is now produced commercially by other means.

CARBON NANOTUBE

In 2005 bikes with frames made of carbon nanotubes were used in the Tour de France. Apart from making the bikes very strong and rigid, the use of nanotubes made them extremely light. Each bike frame weighed less than 2.2lb (1kg). This technology is now being used to produce other sports equipment, such as tennis rackets and golf clubs.

of carbon, was discovered. Famed for its hollow, soccer-ball shaped structures, the so-called buckyballs were named after American architect Richard Buckminster Fuller, whose famous geodesic domes had a similar structure. Buckyballs are made up of 60 carbon atoms and have remarkable properties. For example, they are hard to break apart even at high temperatures and are difficult to compress. So far, buckyballs themselves have few practical applications, but their discovery soon led scientists to a related—and just as intriguing—molecule, the carbon nanotube, arguably one of the best known and most useful products of the nanotechnology revolution. Molecular

Stem cells on nanowire
This Scanning Electron Micrograph (SEM) shows mouse embryonic stem cells growing on vertical silicon nanowires. These nanowires relay electrical signals to and between the stem cells growing on them.

models reveal a structure similar to a roll of chicken wire. It is extraordinary then to find that these tiny tubes are about 100 times stronger than steel. They are also extremely light and can be conducting or semiconducting — attractive properties for electronic engineers who want to create cheap solar panels and touch screens.

In biology, tiny fluorescent crystals called quantum dots are already used to track the movements of cells and molecules in minute detail. In 2005 scientists showed some crystals could latch on to viruses that cause respiratory infections in children, raising hopes of developing new diagnostic tests. The crystals were able to recognize and "stick" to structures in the virus's outer coating. Quantum dots could also revolutionize the computer industry, being used to increase data storage and speed up data retrieval.

Technology for everything

The potential of nanotechnology is just beginning to be tapped commercially. Nanoparticles are used in dirt-repellent paints, sun creams, and sports equipment, and nanocomposites have been used to make light, scratch-resistant bumpers for motor cars. Nanofilters made from alumina fiber are used to purify medical serums, and nanomagnets have been inserted into blood cells to provide detailed images of blood vessels using Magnetic Resonance Imaging (MRI). The Chinese coal company Shenhua is testing nanoscale catalysts that could transform coal into gasoline and other liquid fuels.

Nanomachines, including motors made from DNA (genetic material) and nanoscale computers that can perform simple calculations, could form the basis for more sophisticated nanoscale devices. Engineers are now working on self-propelled nanorobots that may one day be able to perform operations without the need for major surgery.

NANOTECHNOLOGY APPLICATIONS

Few people realize that nanotechnology is already being used in almost every industry. From energy to electronics, manufacturing to medicine, nanotechnology is proving its worth and providing solutions to problems once thought impossible to solve.

Consumer goods
Clothing and other textiles; food; home appliances; optics; sportswear and equipment; cosmetics

Construction
Paint; surface treatments and protection

Manufacturing
Aerospace; vehicle manufacturing; print and packaging; refineries

Medicine and health
Drug delivery; diagnostics; tissue engineering; skin care

Chemistry and the environment
Filtration; catalysis (speeding up of chemical reactions); chemical sensors; water quality

Electronics
Memory storage; computing

Energy
Fuels; solar cells; batteries

" It is a **staggeringly small world** that is below. "

RICHARD P. FEYNMAN, AMERICAN PHYSICIST, 1959

AFTER

Nanotechnology is still very much work in progress, and some of the most promising nanoparticles have yet to find their niche.

UNLIMITED POTENTIAL

In the future, nanotechnology could yield super-strong **bone implants**, ultra-high speed **computer chips**, and artificial, touch-sensitive "skins" for **humanoid robots**. Nanodevices circulating in the bloodstream could be used to monitor human health from within the body. Superstrong but light materials could **revolutionize the construction industry** while others may provide an economic and efficient means of harnessing energy.

« BEFORE

Early astronomers studied the positions and movements of the planets and stars. Later scientists explored their characteristics.

EARLY TELESCOPIC OBSERVATIONS
From 1609 onward, **astronomers** began to study the **planets** with **telescopes** « 84–85. In 1659 **Christiaan Huygens** recorded a dark marking on Mars and two of Jupiter's cloud belts.

ASTRONOMICAL PHOTOGRAPHY
From the mid-19th century **photography** was used to **record astronomical data**.

SPACE MISSIONS
Mariner 2 became the first spacecraft to visit a planet when it flew by **Venus** in December 1962.

« SEE ALSO
pp.74–75 THE SUN-CENTERED UNIVERSE
pp.370–71 MOON LANDING

Inside the Solar System

At the beginning of the space age in the 1960s, astronomers had identified all of the major and many of the minor solar system bodies, and had established their basic properties. Since then unexpected discoveries have been made using Earth- and space-based telescopes, and spacecraft sent to explore them.

The scale of the solar system was estimated accurately for the first time more than 200 years ago. Since then astronomers have been refining measurements, such as the size of the solar system, and the size, mass, and spin of the planets. They have also established other characteristics of the planets, such as their surface temperatures, chemical compositions, and internal structures.

A closer view
Orbiting satellites allow us to monitor the entire earth, identifying large-scale trends in its oceans, land, and atmosphere. Since the 1960s, all eight major planets, and the dwarf planets Ceres and Pluto, have been explored by spacecraft. The missions to Uranus, Neptune, and Pluto were flybys, but all other planets have been monitored by orbiters, some delivering probes or rovers to explore surfaces and atmospheres. Orbiters have carried an array of different instruments, in addition to cameras, to investigate the properties of the surfaces and interiors of planets and their moons.

The rocky planets
Mariner 4 returned the first close-up of any planetary surface, from Mars in 1965. This small planet, with deep canyons and extinct volcanoes higher than any others in the Solar System, has had the most spacecraft visits—well over 20 successful missions by 2021, including six autonomous rovers. NASA's Sojourner, the first rover to operate on the planet, landed in 1997. Whether life has ever existed on Mars has intrigued people for centuries and has been a driver behind scientific exploration.

Planetary properties
The composition of the surface and internal structure of the bodies in our Solar System. These have been studied through space probes and telescopes. Diagrams are not to scale.

Venus Earth Moon Mars
Mercury Jupiter
Saturn Uranus Neptune

The Sun
Sun — Photosphere — Convective zone — Radiative zone — Core
The Sun generates energy in its highly dense core, which is then released through its surface.

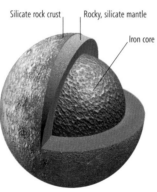

Mercury
Silicate rock crust — Rocky, silicate mantle — Iron core
Mercury is rich in metals compared to the other rocky planets; its high density indicates a huge iron core.

Venus
Silicate crust — Rocky mantle — Molten outer core — Solid iron-nickel inner core
Venus's similar size and density to Earth suggest it has a comparable internal structure.

Earth
Silicate-rock mantle — Rocky crust — Molten outer core — Solid iron-nickel inner core
Earth's surface crust is broken into plates that move against and away from each other.

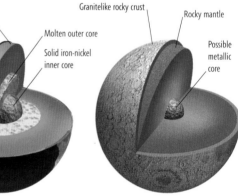

The Moon
Granitelike rocky crust — Rocky mantle — Possible metallic core
Moon's density is similar to that of the Earth's mantle—from which it may mainly have originally formed.

Cratered Mercury
This view of Mercury's cratered terrain was captured by the Messenger spacecraft in October 2008. Most of the craters were formed when the planet was bombarded by asteroids more than 3.5 billion years ago.

Volcanic Venus
The bright volcano in the foreground is Sapas Mons, one of Venus's largest shield volcanoes. On the horizon is Maat Mons, Venus's tallest volcano. The pair grew as successive lava flows accumulated.

The far side of the Moon
Spacecraft have provided images of the far side of the Moon—the side never seen from Earth. This view of the Daedalus Crater was taken during the Apollo 11 manned mission in July 1969.

AFTER »

KUIPER BELT OBJECTS

With improved technology, astronomers have detected ever-fainter objects in the Solar System, such as the small icy bodies in the Kuiper Belt beyond Neptune. Pluto, the largest of these, was discovered in 1930, but the next, Albion, was not found until 1992. In 2019, the New Horizons craft flew past Arrokoth, one of around 2,000 such objects now known. About 22 miles (36 km) long, it consists of two bodies that collided gently and fused together. It is the most distant object ever visited by a spacecraft.

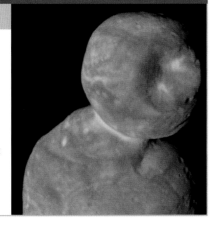

Although Mars has icy polar caps, dry river beds, and ancient floodplains, no evidence of life has been discovered.

The Soviet Venera 7 craft's soft landing on Venus in 1970 was the first on any planet. The dense, carbon-dioxide-rich atmosphere makes Venus the hottest planet due to greenhouse effect. It is also shrouded in opaque clouds of sulfuric acid. Its landscape, shaped by volcanism, was revealed in detail by radar on the NASA Magellan orbiter in the 1990s.

Mercury, the smallest planet and nearest to the Sun, was first imaged in 1974 by Mariner 10, revealing a cratered world reminiscent of the Moon. Its whole surface was mapped from orbit by Messenger between 2011 and 2015.

Giant planets

Scientists have found out more about the four giant planets from seven spacecraft missions since 1973. It is now known, for example, that Jupiter has a turbulent atmosphere and huge thunderstorms. More about the complex nature of Saturn's rings has also been revealed.

In 1977 astronomers discovered rings around Uranus. Spacecraft later found rings around Jupiter and Neptune too, and tens of relatively small moons orbiting the four giant planets. Jupiter's moon Io has around 400 active volcanoes, while images of Europa's cracked icy surface provide evidence for an ocean of liquid water below it.

Knowledge of the Solar System continues to grow rapidly, mainly through the continued acquisition of new data by spacecraft missions.

ONGOING EXPLORATION

In 2021, NASA's **Perseverance** became the latest Mars rover. The Chinese rover **Tianwen-1** was also due to land in 2021. **BepiColombo**, a European/Japanese craft launched in 2018, is due to enter orbit around Mercury in 2025. India is developing missions to Venus and Mars to be launched in 2024. A NASA-led mission to explore **Jupiter's moon Europa** is set for a 2024 launch.

SAMPLE RETURN

At least **14oz (400g) of material** from asteroid **Bennu,** collected in 2020 by NASA's OSIRIS-Rex, is due to be returned to Earth in 2023. NASA and ESA continue work on a joint plan to return a sample from Mars to Earth in 2031.

SEE ALSO »

pp.396–97 SPACE PROBES AND TELESCOPES
pp.398–99 EXOPLANETS
pp.432–33 REFERENCE: ASTRONOMY

" This is **the time** when humans have begun to sail the **sea of space**. "

CARL SAGAN, ASTRONOMER, 1980

Mars
Mars's small size and distance from the Sun mean it cooled faster than Earth and probably has a solid core.

Jupiter
Toward its center, Jupiter has rocky material mixed in with hydrogen and helium, but it has no distinct core.

Saturn
Like the other giants, Saturn's material changes state as the pressure, density, and temperature increase toward its center.

Uranus
Uranus is made mainly of water, methane, and ammonia in the form of hot liquid surrounded by a gaseous layer.

Neptune
Neptune is similar in size and structure to Uranus. Methane in the atmosphere of both planets provides their color.

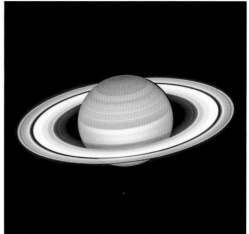

Surface of Mars
Spacecraft images supply evidence of Mars's watery past. Gale crater, seen here in a view from the Curiosity rover, has deposits that show it was once a lake.

The structure of Jupiter
The Juno spacecraft began studying the planet's structure, composition and magnetic field from orbit in 2016. Data from Juno revealed that Jupiter does not have a distinct, small solid core as once thought.

The ringed world of Saturn
All four giants have rings, but Saturn's are the only ones obvious from Earth. Its seven rings consist of ringlets, which are made of individual pieces and lumps of dirty water-ice that orbit the planet.

« BEFORE

As the limitations of earthbound astronomy became more apparent, pioneering experiments paved the way for studying the Universe from above the earth.

LIMITED VIEW
The capability of **Earth-based telescopes**, even the largest, was **limited by the atmosphere**, which blocks many kinds of radiation and distorts images with its turbulent motion.

16.7 FT (5.1 M) HALE TELESCOPE 1948

BALLOONS AND ROCKETS
Cosmic rays bombarding Earth were discovered in 1912 during a **balloon flight** by Austrian physicist Victor Hess. The first **rocket flights** carrying X-ray and ultraviolet detectors above the atmosphere took place in the 1940s.

« SEE ALSO
pp.368–69 ARTIFICIAL SATELLITES

IN PRACTICE
GRAVITATIONAL SLINGSHOTS

Exploring the solar system in a reasonable timeframe presents some difficult challenges, but since the 1970s NASA engineers have used an ingenious technique to shorten journey times without the need for additional fuel. A "gravitational slingshot" involves directing the spacecraft toward a planet, swinging around it, and leaving in the same direction as the planet is traveling, picking up significant speed in the process. The technique was first used to adjust the orbit of the Mariner 10 probe, which flew past Venus on its way to Mercury in 1973, but its most famous application was for the Voyager spacecraft, which took advantage of a rare planetary alignment to tour the gas giants in the 1970s and 1980s (see right).

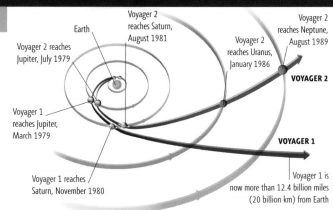

Voyager 2 reaches Saturn, August 1981
Earth
Voyager 2 reaches Jupiter, July 1979
Voyager 2 reaches Uranus, January 1986
Voyager 2 reaches Neptune, August 1989
VOYAGER 2
Voyager 1 reaches Jupiter, March 1979
VOYAGER 1
Voyager 1 reaches Saturn, November 1980
Voyager 1 is now more than 12.4 billion miles (20 billion km) from Earth

Cassini spacecraft
The Cassini Orbiter went into orbit around Saturn in 2004. It carried instruments to study the planet, its environment, and satellites, and also deployed Huygens, a small lander that sent back images from Saturn's mysterious moon Titan.

Low-gain antenna (1 of 2)

36 ft (11 m) magnetometer boom

Radar bay

Fields and particles pallet

Remote sensing pallet

Huygens Titan probe

Radioisotope thermoelectric generator (1 of 3)

Engine (1of 2)

Space Probes and Telescopes

The space age has shown us astounding phenomena, both within our solar system and beyond. Space probes have journeyed to the Moon, all the major planets, and several smaller worlds, while orbiting telescopes have brought stunning clarity to our deep views of the Universe.

It is little wonder that the Moon was the first object to be targeted by what became known as a space probe. Lying on average just under 250,000 miles (400,000 km) away, it is on our cosmic doorstep and modifications to some of the earliest satellite launchers brought it within reach. Although early US and Soviet missions ran far from smoothly, a landmark came on October 7, 1959, when Luna 3 passed behind the Moon and sent back pictures of its unseen far side. Later moon missions surveyed its entire surface—literally testing the ground in preparation for the manned Apollo landings (see pp.370–71)—but

it is only since the 1990s that spacecraft have returned to the Moon to study its surface in detail.

A series of Soviet failures in the 1960s allowed NASA to make the first flybys of both Venus and Mars. However, the Soviet Union's Venera program put it ahead in the exploration of Venus, with 13 successful missions between 1967 and 1983. The robust capsules withstood atmospheric entry and the expected high temperature and pressure. In 1967, Venera 4 returned the first analysis of a planetary atmosphere, and in 1970, Venera 7 survived a

bumpy landing to confirm a surface temperature of around 860°F (460°C) and an atmospheric pressure 100 times that on Earth.

The first flybys to Mars revealed the heavily cratered southern region, disappointing those who had hoped to find a more hospitable world. It was only in 1971 that Mariner 9 revealed the towering volcanoes and riverlike channels of the Northern

Early probe
At 20 in (51 cm) tall, NASA's tiny, conical Pioneer 4 probe, launched in 1959, had just enough room to carry an image sensor and a radiation detector.

Hemisphere. The two Viking missions of 1976, with orbiters and the first landers, were a great success. Further missions, including orbiting surveyors and robot rovers, have enabled scientists to study the surface of Mars in detail.

The giant planets are separated by far greater distances than the inner four, making mission planning and spacecraft operation even more challenging. Uranus and Neptune have been visited only by Voyager 2 (see above).

In 2019, New Horizons broke new ground in long-distance spacecraft operations. It was accurately targeted

◁ Chandra Space Telescope

The Chandra X-ray telescope is one of a series of NASA satellites designed to observe the sky at various wavelengths. Because X-rays will pass straight through conventional mirrors, it uses a series of curving metal surfaces to bring the rays to a focus on its detectors.

△ Combined image

This composite image of X-rays from Chandra and optical light data from Hubble Space Telescopes shows the largest region of star formation in a nearby galaxy, M33.

A number of ambitious new probes and observatories are being planned for the next decade, and some are already en route to their destinations.

EXOPLANETS

Since 2006, four space telescopes have been launched to search for and study planets outside the Solar System: **COROT**, **Kepler**, **TESS**, and **CHEOPS**, in addition to dozens of Earth-based instruments.

BEYOND HUBBLE

The Hubble Space Telescope's **successor**, the 21 ft (6.5 m) infrared **James Webb Space Telescope** is set for launch in 2021 after several delays. It will orbit the Sun, keeping at a distance of 1 million miles (1.5 million km) from Earth.

PROBING THE SUN

Study of the Sun from space has been ongoing since the 1970s. **NASA's Parker Solar Probe** will be the first to fly through the Sun's atmosphere, in 2025. **ESA's Solar Orbiter** started a seven-year mission in 2020.

MISSON TO MOONS

ESA's **JUICE spacecraft**, to be launched in 2022, will explore **Jupiter's icy moons** and may become the first spacecraft to orbit the moon of another planet (Ganymede), in 2032.

SEE ALSO »
pp.400–401 **DARK UNIVERSE**

X-RAY IMAGE (CENTRAL REGION)

ULTRAVIOLET

VISIBLE LIGHT

MID-INFRARED

△ Same galaxy, different wavelengths

Different types of telescope show different features, as in these images of galaxy M31. They range from those using X-rays, which detect very hot objects, to infrared telescopes, which pick up faint radiation from objects too cool to shine in visible light.

to pass 2,200 miles (3,500 km) from Arrokoth in the Kuiper Belt after a journey of more than 4 billion miles (6 billion km).

Space telescopes

Telescopes in orbit around Earth have revealed details of some of the most distant objects in the Universe. Space-based astronomy allows observation at any time of day or night, and it is not affected by weather or atmospheric turbulence. Detailed images using X-rays, infrared, and gamma rays are only possible above Earth's atmosphere. Such images reveal objects undetectable with ground-based telescopes. The first

Spitzer telescope

Launched in 2003, this space telescope monitored infrared radiation until its liquid helium supply ran out in 2009. Its three detectors showed stars and planets forming.

such telescopes included the Orbiting Astronomical Observatory (OAO-2), an ultraviolet instrument launched in 1968, and IRAS, the InfraRed Astronomical Observatory, launched in 1983. The Hubble Space Telescope was launched in 1990 and has sent back countless images (see pp.318–19), leading to important observations about the Universe. Since the 1990s NASA has launched much larger and more ambitious "Great Observatories," including the Compton Gamma-Ray Observatory (1991), the Chandra X-ray Telescope (1999), and the Spitzer Infrared Space Telescope (2003).

Sunshield incorporating solar panel

Mirror measuring 34 in (85 cm)

Insulating pipes filled with liquid helium

Star trackers and pointing systems

Communications antennae

"**Space** isn't remote at all. It's only an **hour's drive** away if your car could go **straight upwards**."

SIR FRED HOYLE, BRITISH ASTRONOMER AND COSMOLOGIST (1979)

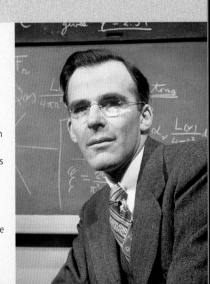

ASTROPHYSICIST (1914–97)

LYMAN SPITZER

One of the first scientists to recognize the potential of space-based astronomy was US astrophysicist Lyman Spitzer. Born in Ohio, Spitzer was head of his department at Princeton University by the age of 33. He made a name for himself through pioneering studies of the interstellar medium—the gas and dust that lies between the stars—and later of plasma physics and nuclear fusion. He first described the benefits of a space-based optical telescope in 1946, and in 1965 NASA asked him to help plan what became the Hubble Space Telescope. With his deft lobbying of politicians and scientists, he ensured that the project became a reality.

Exoplanets

Since the 1990s, the idea of discovering planets around stars other than the sun has transformed from an intriguing possibility to a flourishing science that has revealed thousands of new worlds, known as exoplanets.

The sun is not the only parent star with a planetary system. Extra-solar planets—exoplanets—have been found around other stars as well. Detecting planets around other stars presents huge technical challenges. Planets shine only by reflection, so their light is easily overwhelmed by that of the stars they orbit, making these smaller and dimmer objects very difficult to spot by direct imaging. Early hopes that the gravitational pull of exoplanets might cause a perceptible wobble in the tracks of their parent stars moving across the sky were frustrated by the limitations of the technology available at the time. Some astronomers began to ask whether, despite earlier assumptions, our own Solar System was something of a rarity. However, in the 1980s infrared astronomy revealed the presence of clouds of warm gas and dust around some of the sky's youngest stars—potential new planetary systems waiting to be born.

Young planetary system
The young star Fomalhaut is encircled by a disc of debris that has given rise to new planets. The gravitational pull of planets is the only explanation for the structure of the narrow outer ring, although none has been confirmed yet.

« BEFORE

Ever since astronomers realized that stars were other suns, they have been intrigued by the possibility that the stars have their own planetary systems.

NEBULAR HYPOTHESIS
In 1775, German philosopher Immanuel Kant proposed a **theory of star formation** in which stars and their planetary systems formed from the collapse of nebulae.

FRIEDRICH BESSELL

MOVEMENTS OF THE STARS
In 1844, Kant's countryman and astronomer **Friedrich Bessell** observed wobbles in the motion of the stars Sirius and Procyon—these were later traced to the gravitational influence of orbiting, planet-sized white dwarf stars. In 1963, Dutch astronomer **Peter van de Kamp** reported a wobble in the motion of Barnard's Star caused by a proposed planet, although other astronomers cast doubts on his claims.

The first exoplanets
The first few exoplanets were uncovered in 1992 by observing slight variations in the signals from a stellar remnant called a pulsar (see p.330) but the limited precision of measurements, through even the largest Earth-based telescopes, continued to frustrate attempts to measure the anticipated wobbles of sunlike stars.

In 1995, however, Swiss astronomers Michel Mayor and Didier Queloz finally made a breakthrough using a different approach. Instead of attempting to measure transverse (side-to-side) movement of stars across our line of sight, they used an advanced spectrometer known as ELODIE to measure Doppler shifts (variations in the color of light from the host star, see below) caused by the star's radial motions toward and away from Earth.

Doppler shifts had long been used to map the motion of stars relative to the sun, and to measure binary stars in orbit around each other, but in order to be detectable they had to be in the order of miles per second (far larger than the influence of even the largest planet). By spreading out the spectrum to an unprecedented degree, the sensitive detector ELODIE could measure shifts of just a few metres per second—enough to spot the tell-tale wobbles of a faint star called 51 Pegasi, which revealed an orbiting exoplanet some 50 light-years from Earth.

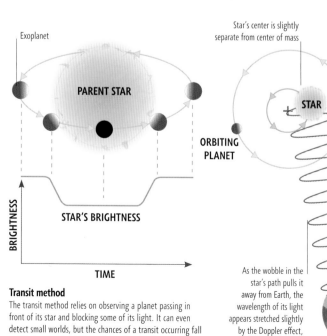

Transit method
The transit method relies on observing a planet passing in front of its star and blocking some of its light. It can even detect small worlds, but the chances of a transit occurring fall rapidly for planets orbiting at large distances from the star.

Radial velocity method
A planet's gravitational pull causes its star to wobble. This motion is easiest to detect for a high-mass planet orbiting close to its star. The ease of detection also depends on how the planet's orbit is aligned to our own point of view.

As the wobble in the star's path pulls it away from Earth, the wavelength of its light appears stretched slightly by the Doppler effect, creating a red shift.

As the star's wobble moves it toward Earth, its light appears compressed and blue shifted.

The telescope was protected by a sunshade, in order to prevent sunlight from entering the photometer as it follows an independent orbit around the sun.

Kepler space telescope
From 2009 to 2013, NASA's Kepler continuously monitored the brightness of a field of 150,000 stars in the constellation Cygnus, looking for transit dips. As its pointing system grew less reliable, engineers continued its mission until 2018 by studying a range of fields for periods of about three months each.

AFTER »

More than 4,000 exoplanets have been discovered. New missions and techniques are now being directed at revealing more about the physical characteristics of these distant worlds.

TESS
NASA's **TESS (Transiting Exoplanet Survey Satellite)** was launched in 2018. The satellite aims to identify transiting planets around bright stars across the entire sky. In 2020, TESS discovered an **Earth-sized planet TOI 700d** in its star's habitable zone.

CHEOPS MISSION
In 2019, ESA's **CHEOPS (CHaracterising ExOplanets Satellites)** was launched. It is designed to study the intensity of light from nearby stars that are already known to have orbiting exoplanets, in order to measure the transits with an even higher precision. The aim is also to observe the transits around stars with exoplanets whose radial velocity is known.

3.1 ft (0.95 m) mirror is linked to a sensitive 95-megapixel camera detector

51 Pegasi b
The first exoplanet to be discovered around a sunlike star, 51 Pegasi b (shown here in an artist's impression) is a "hot Jupiter." It circles its star in just 4.23 days and has about half of Jupiter's mass. In 2015, the planet was officially named Dimidium, meaning "half" in Latin, in reference to its mass.

Age of discovery

Even as painstaking measurements of radial velocity in the 1990s began to reveal more exoplanets around sunlike stars, plans for a more efficient detection technique were also advancing. Made practical through improvements in space telescopes, the transit method takes advantage of the fact that a small proportion of other planetary systems will be aligned in such a way that one or more planets will repeatedly pass, or transit, in front of their star as they orbit each other. Each transit will cause a brief dip in the star's brightness as seen from Earth.

Early ground-based observations, and some from space telescopes, proved to be successful, but the first large-scale application of the transit method came with NASA's Kepler mission— a spacecraft designed to "stare" continuously at stars in a crowded patch of the Milky Way and measure dips in their brightness. Kepler ultimately detected more than 2,600 exoplanets, including some Earth-sized worlds around sunlike stars.

Diverse systems

The biggest surprise in all these discoveries has been the sheer variety of exoplanets. Until the 1990s, most astronomers had assumed that other planetary systems would take the same basic form as our own—with rocky worlds close in and gaseous ones further out. Instead, they discovered an enormous variety of different systems, some containing so-called hot Jupiters—gas giants orbiting very close to their stars in a matter of days or even hours.

300 MILLION The estimated number of Earth-like planets in the Milky Way in orbits likely to make them habitable, according to Kepler data.

Both the radial velocity and transit methods are geared toward detecting hot Jupiters, but they reveal different characteristics. The radial velocity method can indicate a star's mass, while the transit method can reveal its size. Rare occasions in which both kinds of data can be combined with other ingenious techniques—such as measuring the effects of starlight passing through a planet's atmosphere—have begun to reveal intriguing details of rocky super-Earths, lightweight but small gas dwarfs, superdense chthonian planets (the stripped-down cores of hot Jupiters), and even potential ocean worlds, which might possibly provide ecosystems for plants and animals.

As detection methods continue to improve and become more sensitive, the known number of Earth-like exoplanets at distances from their stars that place them in the habitable zone (see below) is steadily increasing. These results provide intriguing new targets in the search for life elsewhere in the universe.

IN PRACTICE

THE SEARCH FOR LIFE

Since the first exoplanets were discovered, astronomers have been trying to identify planets likely to support life. These are likely to be rocky planets that orbit stable stars in a region called the habitable (or Goldilocks) zone where conditions permit the presence of liquid water. Other factors likely to favour life are a stable atmosphere that traps heat and a magnetic field that blocks harmful radiation from space. As well as looking for suitable exoplanets, astronomers use telescopes that scan the radio spectrum for signals that might have originated from intelligent life.

THE ALLEN RADIO TELESCOPE ARRAY, US

ASTROPHYSICIST (1942–)

MICHEL MAYOR

Swiss astrophysicist Michel Mayor has spent most of his career at the University of Geneva, developing spectrometers to make increasingly accurate measurements of radial motion. Working with André Baranne and Didier Queloz, he developed the ELODIE instrument, which entered service in 1994 and soon revealed the first exoplanets around sunlike stars. In 2019, he and Queloz were joint recipients of the Nobel Prize in Physics.

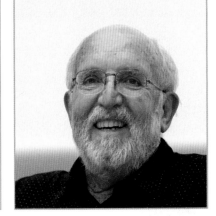

<< **BEFORE**

The Universe was previously thought to contain no more than we can see, or at least detect directly in other ways—such as by collecting X-rays or infrared radiation.

UNIVERSE MADE BY GOD

One fundamental belief for many past astronomers was that the Universe was made by God. This being true, they also supposed that God would not make unnecessary things. Therefore the idea that God made most of the Universe out of materials we cannot "see" was regarded as highly unlikely.

EINSTEIN'S THEORIES

Albert Einstein introduced his **general theory of relativity** << 300–01 in 1915, when it was believed the Universe was static. He realized that **a completely static Universe would collapse** under the influence of gravity and become a single huge central mass, so he introduced a constant to his equation: a repulsion that essentially kept the galaxies apart. Dark energy is a similar repulsion.

<< **SEE ALSO**
pp.318–19 The Expanding Universe
pp.396–97 Space probes and Telescopes

Neta Bahcall
Israeli cosmologist Neta Bahcall's work focuses on combining observational data to determine the structure of the Universe. Her foremost contribution is the determination of the mass density of the Universe.

Dark Universe

In recent decades astronomers have known that the Universe consists of much more than the familiar stars and galaxies. Recent findings suggest that in fact 95 percent of the ever-expanding Universe is made up of two completely mysterious constituents: dark matter and dark energy.

Ordinary matter is what atoms, planets, stars, and galaxies are made of. It is known as baryonic matter by physicists, and it makes up just less than five percent of the Universe. The next largest constituent of the Universe, making up 26.8 percent of it, is also a form of matter, known as dark matter, because it emits no light and we don't know what it consists of. The remaining 68.3 percent is not matter, but dark energy, whose nature is unknown.

The idea that the Universe contains unaccountable material first arose in 1933. Fritz Zwicky, a Swiss astronomer who had been studying the Coma Cluster of galaxies, concluded that each galaxy was many times more massive than inferred by their stellar content.

Evidence for dark matter

During the 1960s and 1970s, radio astronomers found that hydrogen gas orbiting in the outer parts of some spiral galaxies traveled as fast as the gas close to the center. This was inconsistent with most of the mass residing in the bright centers of these galaxies. It implied that galaxies have halos of invisible matter. In 1970, Vera Rubin showed that the same was true of glowing gas clouds in the Andromeda Galaxy and by 1985, after observing many galaxies, proved it to be a general phenomenon.

One theory is that dark matter may consist of a type of elementary particle known as a Weakly Interacting Massive Particle, or WIMP. WIMPs can pass through normal matter and the hope is that every now and then a WIMP will collide with the nucleus of an atom and physicists will be able to detect the recoil. Normal cosmic rays produce similar effects, so experiments in search of WIMPs are performed in mines at least $^2/_3$ mile (one kilometer) below Earth's surface. No WIMPS have yet been found.

In the 1990s, astronomers accepted that the Universe was expanding at an

Searching for dark matter
The world's most powerful telescopes have all at some time been part of the search for dark matter. Surveys by the Sloan Foundation Telescope, New Mexico, have also helped prove the existence of dark energy.

Dark matter distribution
A computer simulation created by an international group of astronomers in 2005 shows the large-scale distribution of dark matter within the Universe. The pattern echoes that produced by the large-scale distribution of the light from galaxies. Yellow denotes the densest matter, then red, blue, and black.

ASTRONOMER (1928–)

VERA RUBIN

Born in Philadelphia, PA, Vera Rubin was fascinated by the stars as a young girl. She became a research astronomer, and her discoveries have altered our understanding of the Universe. Rubin's work on the orbital motion of stars around the center of galaxies convinced astronomers that dark matter exists. She has received many prestigious awards, including the National Medal of Science—America's highest scientific award—in 1993.

GRAVITATIONAL LENSING

The presence of matter in space can cause light rays to be deflected. This effect is known as gravitational lensing. The light from galaxies lying directly beyond a galaxy cluster (pp.340–41), for instance, is "bent" by the cluster. The cluster works like a giant lens, deflecting the light. From our position on Earth, the light paths from a more distant galaxy appear to arrive from different directions, because we are receiving multiple distorted images of the distant galaxy. Astronomers can use the effect of gravitational lensing to determine the mass of a galaxy cluster, figure out where the mass is distributed within the cluster, and assess the amount of dark matter it contains. Gravitational lensing has also been used to confirm the theory that the expansion of the Universe is accelerating (pp.318–19).

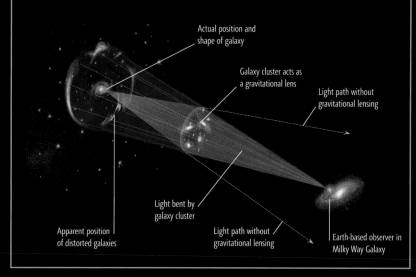

Actual position and shape of galaxy

Galaxy cluster acts as a gravitational lens

Light path without gravitational lensing

Light bent by galaxy cluster

Apparent position of distorted galaxies

Light path without gravitational lensing

Earth-based observer in Milky Way Galaxy

Gravitational lensing in action
Thin arclike features are visible in this colored image of galaxy cluster Abell 2218. They are the result of gravitational lensing and are distorted images of galaxies that are some 5–10 times farther away than the cluster.

Dark matter must have influenced the formation of the large-scale structure of the Universe, with its superclusters and voids.

FUTURE OF THE UNIVERSE
The presence of dark energy **will affect the future of our universe**. If its influence continues in the same way, the acceleration of the **expansion of the Universe will continue**. Should its influence increase, the Universe could disintegrate in a "big rip." If dark energy's repulsion turns to attraction, it will contract in a "big crunch."

ONGOING ENQUIRY
Although **no dark matter particles have been detected yet**, scientists are searching for them using increasingly sensitive underground detectors. They are also looking into space for evidence of the by-products of collisions between dark matter particles, and are attempting to **create dark matter particles** in particle accelerators such as the Large Hadron Collider.

RADICAL ALTERNATIVE
One suggestion to explain the **accelerating expansion of the Universe** is that time itself is slowing down, and our time dimension slowly turning into a new space dimension. This would explain why the far-distant, ancient stars appear as though they are accelerating.

> "The **Universe** is made mostly of **dark matter** and **dark energy**, and we don't know what either of them is."

SAUL PERLMUTTER, ASTROPHYSICIST, 1999

ever-slower rate, due to the influence of gravity. But observations of supernovae in distant galaxies contradicted this: they were found to be fainter, and hence farther away, than would be the case if the expansion of the Universe were slowing down. This indicates that the expansion of the Universe is actually speeding up.

Dark energy's influence
More recent observations have corroborated this theory. It seems the gravitational attraction which would slow the expansion rate is being

overcome by a repulsion; a sort of "anti-gravity." Detailed observations of the supernovae indicate that the Universe was slowing down until about 5–6 billion years ago, but then the deceleration somehow switched to acceleration. It is thought that the density of matter in the young Universe may initially have been high enough for gravity to dominate, but as matter thinned out, the repulsive influence took over. This influence is attributed to an unknown energy that is distributed relatively uniformly in

Galaxies in the Coma Cluster
The Coma Cluster consists of more than 3,000 galaxies. While studying the movements of these galaxies in 1933, Fritz Zwicky found that the cluster contains much more mass than its visible galaxies suggest. The unexplained mass was

Merging black holes
An artist's impression shows two black holes spiraling
together. The signal strength and time delay in
GW150914's detection at the two LIGO sites allowed its
rough distance and direction to be calculated, placing it
1.4 billion light-years away in the far southern sky.

Gravitational Waves

The first experimental detection of gravitational waves in 2015 came almost exactly a century after Albert Einstein had written the equations that predicted their existence (see pp.304–305). It not only confirmed one of the last untested predictions of general relativity but also opened a new window onto the Universe, potentially allowing future astronomers to study events during the Big Bang itself.

Gravitational waves are ripples in spacetime—the four-dimensional fabric that makes up the Universe. In around 1916, while studying the implications of his field equations of general relativity, Albert Einstein proposed that large, nonsymmetrical masses undergoing acceleration would propagate waves of spacetime distortion. Traveling at the speed of light, the waves would manifest themselves as a "flexing" in the dimensions of space, slightly reducing and then expanding measured distances.

Einstein's suggestion withstood the skeptical investigation of other physicists, and breakthroughs in astronomy during the mid-20th century (notably the discovery of binary pulsars—pairs of superdense neutron stars that spiral toward each other) provided a mechanism through which such waves might actually be produced. However, it was only in the 1990s that construction of large-scale detectors based on the principle of laser interferometry became practical (see below). Initial observations carried out in the early 2000s at the Virgo project in Italy and the two Laser Interferometer Gravitational-Wave Observatory (LIGO) sites, in Washington and Louisiana, proved unproductive. But on September 14, 2015, during engineering tests following an extensive upgrade, LIGO made the first successful detection of a passing gravitational wave. The signal, designated GW150914, took the form of a "chirp" lasting less than one-fifth of a second, with eight cycles increasing rapidly in frequency—a result of two distant black holes, each with around 30 times the mass of the Sun, spiraling closer together before merging completely.

Several more events, involving both black hole and neutron star mergers, have been detected by both LIGO and the upgraded Virgo instrument, marking the beginning of a new era of astronomy. Future detectors with greater sensitivity will be able to pick up waves from weaker sources and across greater distances, perhaps even including those from the first 400,000 years after the Big Bang—an era otherwise hidden from view by the opacity of the early Universe.

Laser interferometry
Gravitational wave detectors use a laser beam split into two components, which are reflected along perpendicular tunnels that are several miles long. The beams are recombined to form an interference pattern, which changes if the length of the tunnels is affected by a passing gravitational wave.

"The Universe has spoken and we have understood."

DAVID BLAIR, MEMBER OF THE LIGO COLLABORATION TEAM, 2016

BEFORE

The idea of forces of nature, or fundamental forces, goes back at least to the ancient Greeks. For most of history, people have thought there were two or three such forces.

THREE FORCES OF NATURE
In the 16th century the recognized fundamental forces were **magnetism**, **electricity**, and **gravity**. The English scientist **William Gilbert** « 80–81 studied magnetism and electricity, believing they were separate phenomena.

ELECTROMAGNETISM
In 1819 the Danish physicist **Hans Christian Ørsted** « 166 discovered that **electric currents caused magnetic fields**. He noticed that a compass needle deflected from magnetic north when the electric current from a battery he was using was switched on and off. In the 1820s the French physicist **André-Marie Ampère** showed that wires carrying electric currents could attract or repel each other.

FARADAY AND MAXWELL
English scientist **Michael Faraday** « 170–71 explored how electricity and magnetism were connected. In the 1860s, Scottish physicist James Clerk Maxwell developed equations that united magnetism, electricity, and light.

« SEE ALSO
pp.314–17 QUANTUM REVOLUTION
pp.320–21 THE BIG BANG
pp.382–83 SUBATOMIC PARTICLES

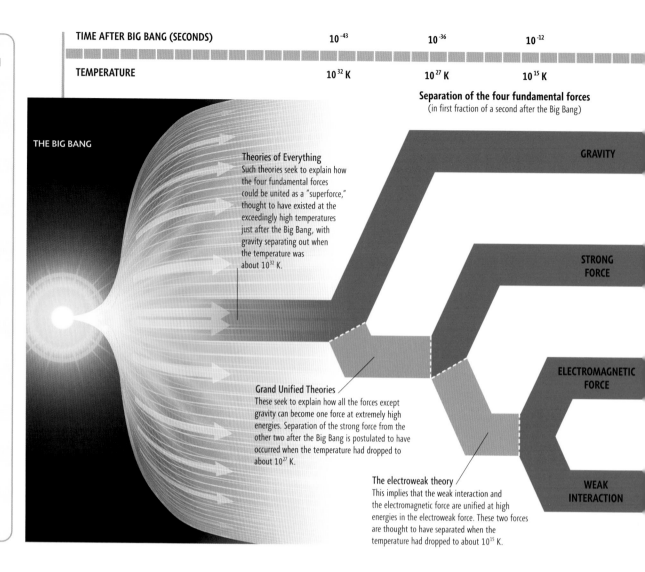

TIME AFTER BIG BANG (SECONDS)	10^{-43}	10^{-36}	10^{-12}
TEMPERATURE	10^{32} K	10^{27} K	10^{15} K

Separation of the four fundamental forces
(in first fraction of a second after the Big Bang)

THE BIG BANG

GRAVITY

STRONG FORCE

ELECTROMAGNETIC FORCE

WEAK INTERACTION

Theories of Everything
Such theories seek to explain how the four fundamental forces could be united as a "superforce," thought to have existed at the exceedingly high temperatures just after the Big Bang, with gravity separating out when the temperature was about 10^{32} K.

Grand Unified Theories
These seek to explain how all the forces except gravity can become one force at extremely high energies. Separation of the strong force from the other two after the Big Bang is postulated to have occurred when the temperature had dropped to about 10^{27} K.

The electroweak theory
This implies that the weak interaction and the electromagnetic force are unified at high energies in the electroweak force. These two forces are thought to have separated when the temperature had dropped to about 10^{15} K.

Grand Unified Theory

Grand Unified Theory (GUT) seeks to unite three of the four fundamental forces of nature: electromagnetism, the strong nuclear force, and the weak nuclear interaction. It is not to be confused with the ongoing search for a "Theory of Everything" (see pp.406–407), which seeks to include the fourth fundamental force, gravity.

A round the end of the 19th century it seemed to scientists that the electromagnetic (EM) force, which causes unlike charges to attract and like charges to repel, together with gravity, could be used to explain everything in physics.

The existence of atomic nuclei was discovered in 1908; but, by the late 1920s, some puzzling facts had emerged about these nuclei. For one thing, they contained only positively charged particles (protons) and uncharged ones (neutrons), and it was hard to explain why the protons did not fly apart, repulsed by the EM force. Another question that required explanation was why neutrons sometimes suddenly turn into protons with the emission of electrons, producing the phenomenon known as radioactive beta decay.

Consequently, in the 1930s, two new forces were proposed: the strong force, which holds protons and neutrons together in the nucleus, and the weak interaction, which (among other effects) causes radioactive beta decay. Hence, four fundamental forces, or interactions, were now recognized.

Role of messenger particles
At this time, little was known about how the four forces actually work. But over subsequent decades, a model was developed in which the forces are transmitted by different "messenger" particles called gauge bosons. By the middle of the 20th century quantum mechanics had established that particles can appear ephemerally in the subatomic world, in accordance with Heisenberg's Uncertainty Principle (see pp.316–17), and this is what happens with the messenger bosons. They are "virtual" particles that appear, take part in an interaction, and then disappear. The distance over which they can act depends on how much energy has to be "borrowed" from the vacuum of the Universe to create them. The more energy used, the shorter a particle's life

THE FOUR FUNDAMENTAL FORCES OF NATURE

The four fundamental forces differ in their messenger particles, the elementary particles that they affect, their relative strengths (calculated here at a range of 10^{-18}m), and the range over which they operate.

FORCE OR INTERACTION	ELEMENTARY PARTICLES AFFECTED	MESSENGER PARTICLES	RELATIVE STRENGTH	RANGE IN METERS
GRAVITY	All particles with mass	Gravitons (hypothetical)	10^{-41}	Infinite
STRONG	Quarks, gluons	Gluons	25	10^{-15}
ELECTRO-MAGNETIC	All electrically charged particles	Photons	1	Infinite
WEAK	Quarks, leptons	W and Z bosons	0.8	10^{-18}

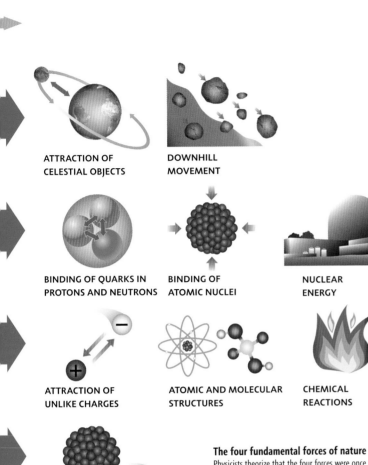

ATTRACTION OF CELESTIAL OBJECTS

DOWNHILL MOVEMENT

BINDING OF QUARKS IN PROTONS AND NEUTRONS

BINDING OF ATOMIC NUCLEI

NUCLEAR ENERGY

ATTRACTION OF UNLIKE CHARGES

ATOMIC AND MOLECULAR STRUCTURES

CHEMICAL REACTIONS

MAGNETISM AND ELECTRICITY

LIGHT AND HEAT RADIATION

BETA RADIOACTIVE DECAY

The four fundamental forces of nature
Physicists theorize that the four forces were once unified in a single force that split in the first fraction of a second after the Big Bang. After splitting, the four forces exerted different effects, producing separate phenomena such as the examples shown above.

THEORETICAL PHYSICIST (1926–96)

ABDUS SALAM

Born in Pakistan, Abdus Salam studied mathematics and physics at the University of Cambridge, and became a physics professor at Imperial College, London. There he worked on establishing the electroweak theory, for which he was awarded a share in the 1979 Nobel Prize for physics. Abdus Salam was the first Muslim Nobel Laureate in science. He contributed greatly to scientific endeavor in his home country, Pakistan, and set up a center in Trieste, Italy, to promote science in the developing world.

span and distance range. In the case of the EM force, the virtual particles are photons (tiny packets of electromagnetic radiation). The way photons act in transmitting the EM force became the basis of the successful theory of Quantum Electrodynamics (QED), developed by Richard Feynman (see pp.326–27) and others.

The electroweak theory
Physicists suspected that the weak interaction might also be mediated by messenger particles, although it was established that these would have to possess relatively high mass (which would also account for the short range of the force). During the 1960s it was realized that at very high energies these messenger particles could appear as real, long-lived particles that would act in the same way as photons. Under these conditions, the EM force and the weak interaction would become a single unified force. This was the basis on which Abdus Salam (see above) and the US physicists Sheldon Glashow and Steven Weinberg independently developed the electroweak theory, which explained not only how the weak interaction works, but also how

it could be unified with the EM force. The electroweak theory postulated the existence of messenger particles (called W+, W-, and Z bosons) for the weak interaction, with different messengers being exchanged in different instances of the interaction. It also predicted the

Hadron Collider
One use of the Large Hadron Collider (LHC), at CERN in Switzerland, is to investigate Grand Unification Theories. The collider has magnets to keep particles on the correct paths; here, a magnet is being moved into place.

Higgs boson—a particle that gives other particles mass. The existence of W and Z bosons was confirmed in 1983, and a particle with the characteristics matching those of the Higgs boson was detected in 2012 and in 2015.

Strong force and grand unification
Originally the strong force was thought to exist between nucleons (protons and neutrons), but following the discovery that these particles consist of sub-units called quarks (see pp.382–83), the strong force came to be seen as one that primarily binds quarks together within nucleons, while a "leakage" of the force binds together the nucleons.

But could the strong force be united with the electroweak force? Because of

the success of the electroweak theory and QED, physicists looked to develop a similar theory that could explain how the strong force operates. This led in the 1970s to the theory of quantum chromodynamics (QCD)—the proposal that quarks, too, interact by means of messenger bosons called gluons. QCD is based on the idea that every quark can exist in any of three so-called colors, with the quarks inside nucleons constantly changing "color" as they interact with the gluons to produce the strong force.

Regarding unification of the strong force (as defined by QCD) with the electroweak force, various GUTs have been proposed, but none is entirely satisfactory. Any such theory would need to postulate the existence of very high-mass bosons (dubbed X-bosons) that could interconvert quarks and leptons (see pp.382–83). A problem in testing any GUT is that producing these bosons requires recreating the conditions in the first fraction of a second after the Big Bang. Scientists at the Large Hadron Collider have come closest to achieving this by colliding protons or atomic nuclei at nearly the speed of light. The discovery of a particle that is almost certainly the Higgs boson confirmed that this approach can work.

AFTER

Beyond a Grand Unified Theory is the goal of unifying all four fundamental forces in an ultimate "Theory of Everything."

THINKING IN STRINGS
At present, **gravity** is most exactly described by Einstein's general theory of relativity, while the explanation of the other three forces is based on the theory of **quantum mechanics**, and these two theories appear in some respects to contradict each other. One of the best known attempts at unifying all four forces is **string theory 406–407 ≫**. In string theory, **elementary particles** are treated as **infinitesimally thin, stringlike objects** rather

than dimensionless points. But string theory is very much in its infancy.

LOOPS OF SPACE
An alternative "theory of everything" candidate, **loop quantum gravity** does not try to unify the four fundamental forces. It models gravity not as a force but as a property of spacetime, as general relativity does. Loop quantum gravity proposes that spacetime is **granular**, with the smallest units being tiny, interlinked loops of gravitational fields.

SEE ALSO ≫
pp.406–07 **STRING THEORY**

BEFORE

The ideas that led to the emergence of string theory started to appear in the 1960s, as scientists looked for a way to incorporate gravity with electromagnetism and the strong and weak nuclear forces.

WHEN FOUR BECAME THREE
The first step toward **unifying** the four forces of nature (gravity, electromagnetism, and strong and weak nuclear forces) came in the 1960s. Physicists found a mathematical package, now known as the **electroweak theory**, describes the **electromagnetic force** and the **weak nuclear force** in one set of equations.

POINTING THE WAY
In 1979 **Abdus Salam, Sheldon Glashow**, and **Steven Weinberg** were awarded the **Nobel Prize in Physics** for their discovery of the electroweak theory. This theory was important not only in its own right, but because it pointed the way toward **further unification**, with the goal of finding **one set of equations** to describe all four forces.

« SEE ALSO
pp.302–03 Theories of Relativity
pp.314–17 Quantum Revolution
pp.404–405 Grand Unified Theory

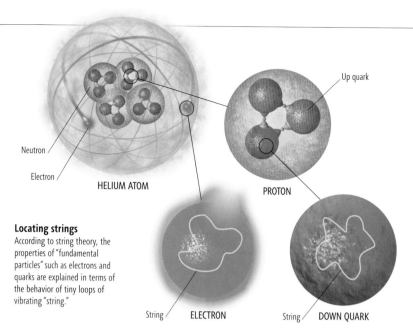

Neutron

Electron

HELIUM ATOM

Up quark

PROTON

Locating strings
According to string theory, the properties of "fundamental particles" such as electrons and quarks are explained in terms of the behavior of tiny loops of vibrating "string."

String **ELECTRON**

String **DOWN QUARK**

The search for a unified theory of physics is the quest to unite the two great theories of 20th-century physics: the general theory of relativity (see p.302) and quantum mechanics (see p.314). The general theory of relativity relates gravity to the structure of space and time, treating them as a single four-dimensional entity, space-time. Quantum mechanics describes the other three forces of nature: electromagnetism and the strong and weak nuclear forces (see pp.404–405).

A unified theory must describe gravity as well as the other three forces in quantum terms. This idea implies that space-time itself must be quantized (packaged) into discrete lumps. From the

String Theory

In particle physics, fundamental entities such as electrons are regarded as mathematical points. In string theory this idea is replaced by the concept of one-dimensional entities ("strings") or multidimensional entities ("membranes"), whose vibrations produce the properties formerly associated with "particles."

weakness of gravity compared with the other three forces, physicists know that the scale on which this happens is tiny—only around 10^{-35}m. The motivation for string theory came from attempts to mathematically model the behavior of space-time on this miniscule scale.

Loops and threads
Although physicists had previously considered the possibility of treating fundamental entities as strings, string theory only took off in the mid 1980s, when John Schwarz of the US and Michael Green of the UK set out to include gravity in the package. They developed the idea that fundamental entities such as electrons and quarks are outward manifestations of "strings" vibrating on the quantum gravity scale. Properties such as mass are explained as different vibrations of the same kind of string. An electron is a string vibrating in a certain way, while a quark is an identical string vibrating in a different way.

At first, the physicists thought of the strings as open-ended, like a wiggly line; but they soon developed an improved version of the theory, in which the strings are tiny loops, like stretched elastic bands. It would take about 10^{20} such strings, side by side, to span the diameter of a proton.

Good vibrations
One of the motivations for studying string theory is that it is free from infinities, which plague theories of fundamental particles. The problem is essentially that forces such as the electromagnetic force

are inversely proportional to distance; this means that the "self interaction" of a particle with zero size would be 1 divided by zero, which is infinity. Since strings do not exist at a point, they don't have this problem.

String theory caused great excitement in the mid 1980s because it turned out that the mathematical description of these tiny loops of vibrating string automatically includes a description of gravity. The equations were set up to

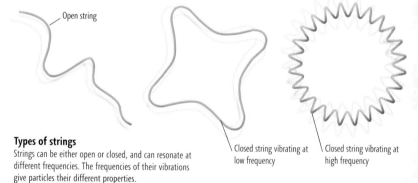
Open string

Types of strings
Strings can be either open or closed, and can resonate at different frequencies. The frequencies of their vibrations give particles their different properties.

Closed string vibrating at low frequency

Closed string vibrating at high frequency

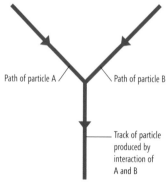

Particle physics interaction
If two particles with zero size meet and interact at a point, the forces operating between them become infinite, because these forces are divided by zero. This is a sign that there is something wrong with this description of particle interactions.

Path of particle A

Path of particle B

Track of particle produced by interaction of A and B

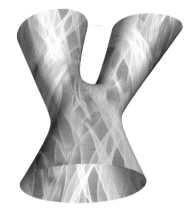

String interaction
Unlike two particles with zero size, two loops of vibrating string can meet and merge with one another to make a loop vibrating in a different way (a different particle), without any problems arising with infinities.

THEORETICAL PHYSICIST (1963–)

BRIAN GREENE

American physicist Brian Greene is one of the leading string theorists today. Greene was born in New York City in 1963, and has worked at Harvard, Oxford, Cornell, and Columbia universities. His special interest is applying ideas from string theory to cosmological problems, including the origin of the Universe and the nature of black holes. Greene is the most eloquent spokesman for the idea of string theory.

His book *The Elegant Universe* was nominated for the Pulitzer Prize in nonfiction in 2000, and was later made into a television special.

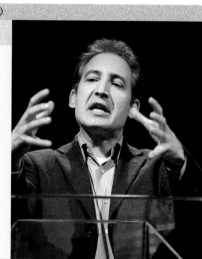

COMPACTIFICATION

"Extra" dimensions are essential in defining the properties of strings, but they are thought of as hidden by "compactification." The usual example of compactification is to think of a hose pipe. Viewed from a distance, it looks like a one-dimensional line; but look closer and you see that it is made of a two-dimensional sheet (which might ripple in interesting ways) wrapped around the third dimension. The same trick can be used mathematically to wrap up any number of dimensions, shrinking them and leaving only the familiar four dimensions visible.

Hidden dimensions
A flat sheet is rolled, made into a ring and shrunk to the size of a full stop, concealing its original form.

describe the other forces and particles of nature, and physicists were surprised to find that a description of another kind of particle, technically known as a spin-2 boson, fell out of the equations. At first they were baffled; then they realized that this was the particle associated with the gravitational field, in the same way that the photon is associated with the electromagnetic field (see pp.382–83).

Beyond the fourth dimension

The snag about string theory is that it requires many more dimensions than the familiar four of space-time. In order to produce all the variety of forces and particles from the vibrations of a single kind of string, the vibrations have to take place in many different dimensions. In

The latest development of string theory, known as M-theory, unites different versions of string theory in one package.

WHEN SIX BECOME ONE

In the mid 1990s string theory seemed to be in trouble. There were **five different versions** of string theory, plus a **sixth theory called supergravity**, and they seemed unrelated. But in 1995 **Edward Witten**, of the Institute for Advanced Study in New Jersey, caused a stir when he proposed that all six of these theories could be described in **one mathematical package**, which he called **M-theory**. (The "M" has variously been said to stand for Matrix, Mystery, Master, Mother, Magic, or even Murky.) It is as if the six string theories were the six sides of a die. Which one you see depends on how you roll the die. Could this be the long-sought **theory of everything**?

NOT EVEN WRONG?

Some scientists are skeptical about string theory. One criticism is that strings would be undetectable by any human-made particle accelerator, and the vast number of possible universes described by the theory means it cannot make predictions that can be tested. This would make it useless for predicting anything, and arguably unscientific. This has led some critics of string theory to describe it as "not even wrong."

A multidimensional world
A 10-dimensional Calabi-Yau manifold—shown here in cross section—is the kind of higher-dimensional space that can be "compactified" (see above) mathematically to leave three dimensions of space plus one of time.

one version of string theory, a closed loop of string has 10-dimensional waves— which describe fermions (see pp.382– 83)—running around it one way, and 26-dimensional waves—which describe bosons—running around it the other way.

So why is it that we only perceive three dimensions of space and one of time? The answer may be that the extra dimensions are rolled up very small, or "compactified" (see above). Strings may be thought of as loops embedded in higher-dimensional membranes, with harmonics on the vibrating string translating into properties such as mass and charge.

An alternative view
Some physicists worry that too much effort goes into string theory, compared with other areas of research. American physicist Lee Smolin speaks forcefully for this minority in *The Trouble with Physics*, published in 2006.

The first internal image of the body dates from 1895, when Wilhelm Röntgen made an X-ray photo of his wife's hand (see p.294). Within weeks, X-ray images made their debut in medical diagnostics, a role that they have maintained ever since.

X-ray technology is constantly improving and it is ideally suited for imaging solid structures within the body, such as teeth and bones. They can also be used to identify small regions of unusually dense tissue, which may be a sign of cancer. However, standard X-rays have one major limitation: because they reduce three-dimensional objects to a flat image, dense parts of the body, such as the skull and rib cage

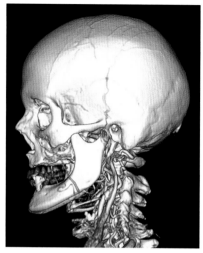

◁ **CT skull**
The astounding clarity of CT scans is demonstrated by this 3-D image of a man's skull. In this mode, the image has been processed by edge-detection software, giving the bones a lifelike appearance.

Body Imaging

By gathering data in different ways, modern imaging techniques have revolutionized our ability to look inside the human body. In addition to revealing internal organs, they can often show body processes at work—an important aid in diagnosing disorders and disease.

« BEFORE

Before medical imaging became available, doctors relied mainly on observation of symptoms and signs to make diagnoses.

SYMPTOMS AND SIGNS

A **symptom** is a **subjective** condition noticed by the patient. A **sign** is something **noticed by a doctor**, possibly visual such as change in skin color, but not by the patient.

DOCTORS RELIED ON OBSERVATIONS

ENDOSCOPY

Light-conducting tubes, or **endoscopes**, were the earliest means of looking inside the body. The first recorded use was in 1822, by **William Beaumont**, an American army surgeon. In the early 20th century **laparoscopes** with their own light source were used in **abdominal surgery**.

X-RAYS

The earliest form of **noninvasive body** imaging, **X-rays** were discovered by accident in 1895 by **Wilhelm Röntgen**. His breakthrough led to a new medical discipline, **radiology**.

« SEE ALSO
pp.294–95 THE DISCOVERY OF X-RAYS

mask less dense tissue inside them. The solution to this problem was first sketched out in the 1920s, although it did not become a practical possibility until a series of advancements in electronic design in the late 1960s. Known as computerized tomography, or CT scanning, it uses a rotating X-ray source that circles the body, passing X-rays through it to a detector on the opposite side. Instead of producing direct X-ray images, the scanner builds up a bank of data that can be processed to show specific regions in 3-D form. Using digital techniques, superficial structures can be subtracted, revealing the tissues underneath.

Imaging blood vessels (angiography)
This colored MRI angiogram shows a blocked artery in the neck of a stroke patient. Blood vessels can also be viewed using CT scanning, if they are first injected with a contrasting medium to highlight them on the CT scan.

Magnetic imagery

CT scanning is one of several 3-D imaging techniques that are now widely used in medicine. The most widespread alternative, magnetic resonance imaging, or MRI, also coevolved with the computer. Unlike CT scanning, MRI exposes the body to an intense magnetic field, created by a giant superconducting magnetic coil. Smaller electromagnets apply a much weaker variable field, and the scanner also emits pulses of radio waves. These radio pulses act on the body's hydrogen atoms in water, making them precess, or resonate. When each radio pulse ends, these atoms give off signals that the scanner detects. The pattern of these signals varies from one tissue type to another, because each contains different amounts of water.

Like CT scans, MRI scans provide a snapshot of the body at a single moment in time. A more recent development, called functional MRI, records real-time changes

Timeline of human evolution

Multiple genera have been identified in the hominin lineage, each often including several species (a selection of which is shown here). Although the overall evolutionary lineage leading to *Homo sapiens* is clear, direct ancestral relationships are difficult to establish, particularly in the earlier phases.

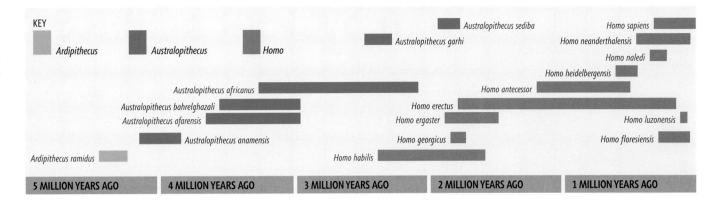

KEY

Ardipithecus Australopithecus Homo

Australopithecus sediba Homo sapiens
Australopithecus garhi Homo neanderthalensis
Homo naledi
Homo heidelbergensis
Australopithecus africanus Homo antecessor
Australopithecus bahrelghazali Homo erectus
Australopithecus afarensis Homo ergaster Homo luzonensis
Australopithecus anamensis Homo georgicus Homo floresiensis
Ardipithecus ramidus Homo habilis

| 5 MILLION YEARS AGO | 4 MILLION YEARS AGO | 3 MILLION YEARS AGO | 2 MILLION YEARS AGO | 1 MILLION YEARS AGO |

were recovered from the Philippine island of Luzon and date to 67,000 years ago; in 2019 these were distinguished as *Homo luzonensis*. Both species show a mixture of modern and archaic human features, and their ancestry remains unclear—they may have descended from local groups of *H. erectus*, or perhaps from even earlier hominin species. Modern humans arriving in Southeast Asia may have overlapped with these species, but as yet there is no evidence they interbred.

Resetting modern humans

Our understanding of the evolution of *Homo sapiens* has been revolutionized by recent archaeological and biomolecular discoveries. Previously it was thought our species evolved in East Africa some 200,000 years ago and spread across Eurasia 90,000–45,000 years ago, encountering only one other relative, the Neanderthals. However, new finds have identified modern humans in northwest Africa at least 315,000 years ago and in the eastern Mediterranean by 200,000 years ago. While they did not extend further into

Europe until after 45,000 years ago, they may have reached China as early as 120,000 years ago and Australia by 65,000–45,000 years ago. We now know, too, that several other closely related species were still present across Africa and Eurasia at this time, including *Homo naledi*, which shows an intriguing mix of modern and archaic

traits and dates to around 335,000–236,000 years ago. In Europe and western Asia, *Homo sapiens* encountered and interbred with Denisovans and Neanderthals, preserving traces of these hominins' genes in our genome today, but is it not fully understood whether or how they interacted with other species alive at the time.

PALEOANTHROPOLOGIST (1913–1996)

MARY LEAKEY

British paleoanthropologist Mary Leakey's work at the Olduvai site in Tanzania unearthed several key fossils that helped unravel the human origin story. Often working with her husband, Louis Leakey, she discovered and excavated many new species, including the potential human ancestor *Homo habilis*. In 1976, she discovered the Laetoli footprints—humanlike footsteps preserved in volcanic ash from 3.7 million years ago. Their discovery pushed back the suspected date of the origin of bipedalism to long before the development of tools, settling what had been a heated debate.

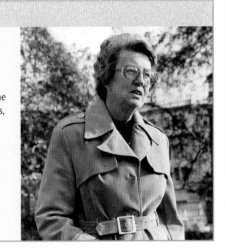

AFTER

Since the 1950s, advances in molecular biology and the study of the role of DNA in the inheritance of genetic traits have revolutionized our understanding of human evolution.

OUR AFRICAN ANCESTRY

Following the discovery of the helical structure of the DNA molecule by Crick and Watson in 1953 « 346–47, Rebecca Cann and colleagues collected and analyzed the mitochondrial DNA of a global sample of living humans in 1987 and found that they were descended from a common population that lived in Africa around 170,000 years ago. Just a couple of decades later, the first fragments of ancient DNA recovered from 50,000-year-old Neanderthal bones by Svaante Paabo and his colleagues revealed this species's distinctiveness.

GENE FLOW

In 2016, analysis of the Neanderthal genome revealed that Neanderthals contributed genetically to modern humans beyond Africa through interbreeding 65,000–47,000 years ago.

Generally flatter face than earlier species

Homo erectus
Although different from modern humans, this species is thought to be more "modern" looking than its ancestors. *Homo erectus* lived about 1.9–0.1 million years ago and was the first human relative to spread beyond Africa and across Asia.

Homo neanderthalensis
The large-brained human *Homo neanderthalensis* lived in Europe and western Asia between 400,000 and 40,000 years ago and shared a common ancestor with *Homo sapiens*.

Heavily worn teeth suggest their use in preparing skins as well as eating

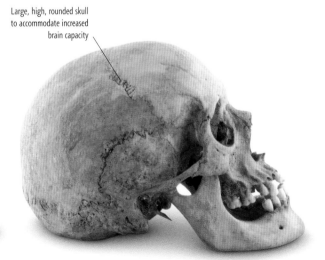

Large, high, rounded skull to accommodate increased brain capacity

Homo sapiens
Of all hominin species, only *Homo sapiens* survives today. The skull of a modern human is distinguished by a high forehead, large brain, vertical face, small teeth, and prominent chin.

BEFORE

Earth system science has developed out of a number of allied topics, some Earth-related, others largely theoretical.

HUMBOLDT EXPLORES THE ORINOCO RIVER

UNIFYING DISCIPLINES
The German naturalist **Alexander von Humboldt** (1769–1859) strove to unify the diverse branches of science. During his travels in South America he gathered huge amounts of data from different phenomena and tried to induce unifying principles and connections between them.

GAIA HYPOTHESIS
In a series of publications in the 1970s, **James Lovelock** and **Lynn Margulis** proposed the Gaia theory, which suggests that Earth's organisms and inorganic surroundings evolve together as a single, self-regulating planetary system.

SYSTEMS SCIENCE
From the 1940s systems science grew in its application to many fields: robotics and intelligent machines, electronic circuitry, communications, industrial production, ecosystems of nature, and many others.

《 SEE ALSO
pp.206–07 **ATMOSPHERIC MOVEMENT**
pp.278–79 **CYCLES IN THE BIOSPHERE**

Earth System Science

One of the grandest of scientific endeavors is the relatively new discipline of Earth system science. It applies system science, with its study of component parts and how they relate and interact as a whole, to the entire Earth with its constituent parts, cycles, and spheres, involving land, water, air, life, and human activity.

Earth system science is ambitious in scope and vast in the scale of its complexities. Its status in relation to our planet has been compared to the position of cosmology with respect to the Universe. The discipline grew out of systems science, which developed in the 1950s as a way to analyse how parts interrelate to make a functioning whole, how these workings can be disturbed by various factors, and how the whole relates to its wider environment. For example, the relationships between some parts of a system may be robust enough to cope with disruption, whereas the interaction between two particular components may be so critical that even if it is perturbed slightly, the whole system may descend into chaos.

As an interdisciplinary field, systems science has been incorporated into many areas, and the 1970s saw its application to Earth sciences—the whole Earth together with its wider environment. British chemist and physician James Lovelock and American biologist Lynn Margulis developed the Gaia hypothesis, which attempted to explain how life could have thrived continually on Earth

for more than three billion years in the face of changes in Earth's atmosphere and energy balance. Margulis believed that cooperative interaction is important in driving evolution. She suggested that living organisms and the inanimate components of Earth—oceans, landmasses, and atmosphere—interact to make the planet habitable. Lovelock and Margulis proposed that the biosphere itself, together with the atmosphere and other components, acted as a self-regulating system, controlling the environment to favor the continued existence of life.

This inspired scientists to view how the world works in a more holistic way—a view supported by the images of our planet returned by a fleet of Earth-observing satellites. In the 1980s, a NASA-based committee chaired by Francis Bretherton set out to "obtain a scientific understanding of the entire Earth System on a global scale by describing how its component parts and their interactions have evolved, how they function, and how they may be expected to continue to evolve at all time scales." Their report was published as *Earth System Science: A Closer View*.

Spheres of influence
Our planet can be regarded as having constituent parts or "spheres"—water, air, land, and life. The hydrosphere involves all things watery; the atmosphere is the gaseous sphere surrounding Earth; the geosphere comprises the pedosphere (soil cover) and lithosphere (Earth's solid outer layers); and the biosphere is plants, animals, and other living matter. Each of these can be viewed as a planetary subsystem that works with, and depends on, the others. This interdependency is central to Earth system science, which conceptualizes Earth's processes, mechanisms and cycles as a "system of systems." Physics, chemistry, biology, mathematics, and other sciences combine to portray Earth as one integrated holistic system.

For example, traditional Earth sciences describe how water, carbon, nitrogen, and other substances are cycled by natural processes through the atmosphere, rocks, soil and living things (see pp.278–79). In the past, such studies often focused on the cycle itself—that is, they primarily looked inward, with much less emphasis on how the cycle

Spaceship Earth
Launched in 2015, NASA's Deep Space Climate Observatory satellite is in orbit around 1 million miles (1.6 million km) from our planet. It has a continuous view of Earth's sunlit side and its cameras take full Earth images (as shown here) every two hours, allowing—for the first time—study of daily variations over the entire globe.

connected to other systems. Earth system science focuses on interconnectedness—how the various cycles and spheres interact and respond to one another. Since the 1950s, a revolution has been brought about in the understanding of the linkage between internal and external processes. This has been due to the development of international science, projects such as the International Geophysical Year (1957–58), and the application of sciences such as geophysics to Earth processes, from the internal driving force of core heat to the ocean-atmosphere system, and climatology.

Linked systems
The four "spheres" of water, air, land, and life are well established, but others too can be envisaged. The cryosphere is wherever water is frozen into ice. It includes ice caps, ice sheets, glaciers, permafrost, and even snowflakes and hailstones. Frozen and semifrozen water relate to the surroundings in a very different way from liquid water, for

The exosphere
One of Earth's interconnected spheres, the exosphere includes the Sun and all of its influences on our planet, such as solar flares (pictured), which can cause geomagnetic storms on Earth. It also includes all the comets, asteroids, and similar space bodies that have impacted Earth through the ages, bringing minerals, some water, and possibly even the building-blocks of life itself.

example, harbouring the powerful greenhouse gas methane as methane clathrates (see p.423).

Other interlinked spheres include the exosphere—beyond our planet's atmosphere—and the anthroposphere. The latter encompasses parts of the planet that have been modified or made by humans, including farmland, mines, quarries, reservoirs, cities, transportation systems, retail and leisure facilities, and urban ecosystems of plants and animals.

The Anthropocene

A related concept is the Anthropocene, a label suggested in the 1980s by US ecologist Eugene Stoerme, and popularized by Dutch atmospheric chemist Paul Crutzen in 2000, for the pervasive human impact upon Earth. The Anthropocene would follow the Holocene, which is the current officially accepted epoch of geologic time, stretching from the waning of the last major ice age 11,700 years ago to today. The scientific community is currently debating the validity of the concept for understanding geologic history but, more broadly, the Anthropocene is a way of defining human activity as an integral part of Earth's systems instead of separating Earth systems into "natural" and "human" categories.

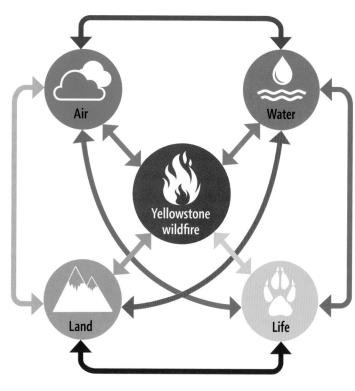

System analysis of a wildfire

Earth system science provides a new, holistic way of analyzing events in our environment. The relationships between an event and Earth's "spheres"—Air, Water, Land, and Life—can be envisaged graphically, as shown above. In this example, the event is the record-breaking wildfire that swept through Yellowstone National Park in 1988. The analysis takes into account how the event affects each sphere; how the spheres affect the event; and how the spheres affect one another.

Ash carried to distant watercourses by wind; airborne particles can act as nuclei for raindrops.

Soil erosion increases because foliage cover has been destroyed; blackened soil absorbs more heat from sunlight.

Burnt plant matter fertilizes soil; loss of existing vegetation cover favours pioneer plant species.

Temperature of ponds and other waterways increases as shade cover is destroyed.

Ash is deposited in waterways, choking aquatic life; rivers act as natural firebreaks.

Soil is dried out, so rain is needed for germination; wind speeds are higher, so wind-dispersed plants colonize.

Smoke particles affect air quality and visibility; land ahead of the fire is dried out by air heated by the fire.

Topsoil is damaged by scorching; humus burns off, making plant re-establishment more difficult.

Plants and slower moving animals killed or displaced; food sources destroyed; some seeds activated.

Turbidity, pH, and temperature of waterways rises, disturbing aquatic ecosystems.

AFTER

As a relatively new discipline, Earth system science is establishing itself in various ways.

EARTH MONITORING

Satellites and high-technology sensors on land and at sea provide scientists with powerful tools to view Earth as an integrated system. There are currently around **200 Earth observing satellites** in orbit, operated by 45 nations.

AWARENESS AND EDUCATION

In 1998 the **American Geosciences Institute** began Earth Science Week, usually held each October, to raise awareness of the interconnected phenomena that affect our planet. The **Earth System Science Education Alliance** formed in 2000 to foster its teaching, development and understanding.

Yellowstone wildfire

In September 1988, wildfires burned more than 30 percent of the area of Yellowstone National Park. Such intense firestorms occur when large fires burn so intensely that they create their own wind systems that draw in more air to sustain the inferno.

BEFORE

Suspicions of a link between changes in atmospheric chemistry and temperature have existed for over 150 years.

GREENHOUSE GAS INVESTIGATION
In 1859 Irish physicist **John Tyndall** set out to find which gases play a part in **trapping heat** in the atmosphere. He found the most important were water vapor and carbon dioxide (CO_2).

ICE AGES
In 1837 Swiss geologist **Louis Agassiz** was the first to propose that a great **Ice Age** had once affected Earth's high latitudes, and that the entire planet's climate was subject to rapid change.

« SEE ALSO
pp.206–07 ATMOSPHERIC MOVEMENT
pp.208–09 PREDICTING THE WEATHER
pp.210–11 STRUCTURE OF THE ATMOSPHERE

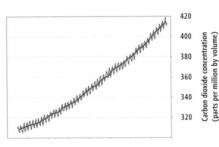

The Keeling curve: rising carbon dioxide levels
Keeling's graph shows that the level of carbon dioxide in the atmosphere has risen steadily since at least 1958. A small annual fluctuation (shown in red) occurs due to seasonal changes in carbon dioxide uptake by plants.

GEOCHEMIST (1931–2019)

WALLACE BROECKER

American geochemist Wallace Broecker is best known for his discovery of the role played by the oceans in triggering the sudden climate change in the past. He developed chemical tracer techniques for studying the behavior of carbon dioxide (CO_2) in the oceans and its interactions with the atmosphere. In 1975 he coined the term "global warming." A pioneer of geo-engineering, he believes in capturing and storing CO_2 as a possible solution.

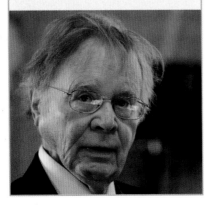

Global Warming

The long-term heating of our planet's surface—global warming—is creating massive and potentially catastrophic problems for life and the environment. Human activities, especially the burning of fossil fuels, have been releasing heat-trapping greenhouse gases, such as carbon dioxide, into Earth's atmosphere and climate system.

Life on Earth depends on a physical phenomenon called "the greenhouse effect." This describes the way that gases in the our atmosphere—notably water vapor, carbon dioxide (CO_2), methane, and ozone—absorb some of the infrared radiation (heat) emitted from Earth's surface that would otherwise be lost to space. Like the glass in a greenhouse, these "greenhouse gases" help to warm our planet's surface; without them, average temperatures on Earth's surface would be around 0°F (– 18°C)—so low that all water on Earth would freeze.

As early as 1896, Swedish scientist Svante Arrhenius realized that carbon dioxide emitted by burning fossil fuel (then mainly coal) could have profound impacts on climate, calculating that a doubling of CO_2 levels could increase surface temperatures by as much as 9–11°F (5–6°C). The relatively low human outputs of CO_2 at the time—added to a belief that any impacts of CO_2 emission would either be swamped by natural change or be buffered by the oceans—meant that the significance of Arrhenius's work would not be realized for another 50 years.

In the 1950s, scientists began to realize that relatively small changes were capable of destabilizing Earth's climate, and rising CO_2 levels could trigger disastrous warming. In 1958 American scientist Charles Keeling began accurate year-by-year measurements of the gas in the atmosphere, and found that CO_2 levels were rising steadily at the rate of a few percent per decade. Taking advantage of the processing power offered by the new digital computers of the 1970s, scientists constructed models to predict what effect the rising greenhouse gas concentrations would have on Earth's temperature. The results predicted a steady warming.

Ice record
Evidence for a link between fluctuating CO_2 levels and climate change has come from the analysis of ice formed hundreds of thousands of years ago in the Antarctic. Measurements from air trapped in the ice can reveal historic levels of CO_2 in the atmosphere, and other clues give an indication of air temperature around the time the ice originally formed.

Gathering evidence
Scrutiny of historical temperature records going back decades—and studies of the growth rings of trees, which gave information about temperature trends over centuries—seemed to indicate that this warming had already started, and that a distinct acceleration was occurring. By the late 1970s most scientists became convinced that rising CO_2 levels and global warming were, at least partly, the result of human

activities, particularly the burning of fossil fuels, which had increased rapidly in the 19th century.

Current warming
Human activities on Earth have increased the levels of greenhouse gases in our atmosphere by about 40 percent since preindustrial times; and more than half of this increase has occurred since 1970. This has been paralleled by an increase of about 1.4°F (0.8°C) in global

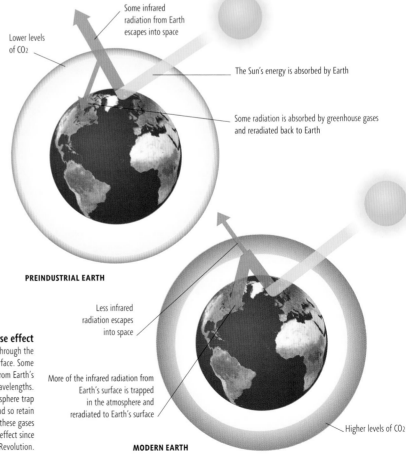

Some infrared radiation from Earth escapes into space

Lower levels of CO_2

The Sun's energy is absorbed by Earth

Some radiation is absorbed by greenhouse gases and reradiated back to Earth

PREINDUSTRIAL EARTH

The greenhouse effect
Light from the Sun passes through the atmosphere, warming Earth's surface. Some of this heat is radiated back from Earth's surface at longer (infrared) wavelengths. Greenhouse gases in the atmosphere trap some of this infrared radiation and so retain the heat. Rises in the levels of these gases have enhanced the greenhouse effect since the Industrial Revolution.

Less infrared radiation escapes into space

More of the infrared radiation from Earth's surface is trapped in the atmosphere and reradiated to Earth's surface

Higher levels of CO_2

MODERN EARTH

woodland would be needed to compensate for human activity. There are also ways of artificially capturing CO₂ from the chimney stacks of power stations using "scrubbers," but the technology currently requires huge amounts of energy to work: up to one third of a power station's output is needed to remove the CO₂ it emits.

Once trapped, CO₂ needs to be safely stored. Proposed methods include storing the CO₂ as a gas in geological formations or as solid carbonate at the bottom of the oceans. It could also be converted naturally into minerals by peridotite rock. In theory, the peridotite deposits in Oman (in the Arabian peninsula) alone could absorb several billion tons of CO₂ every year; however, such storage sites would soon reach capacity.

Alternatively, captured CO₂ could be turned into fuel, using photosynthesizing devices known as "artificial leaves."

Other more radical ways to sequester CO₂ include "fertilizing" the oceans with powdered iron (see box, left), or farming certain marine algae in huge ponds. These algae use CO₂ dissolved in the water to make calcium carbonate (limestone), which they then incorporate into their exoskeletons. After locking up the CO₂, the algae could be harvested, compacted into bricks, and used as a building material.

> ## 1 MILLION fewer tons of carbon dioxide would be emitted if just 10 percent of the US's and Europe's workforce telecommuted once a week for a year.

Low-energy house
Many improvements can be made to homes to reduce energy usage. These may be financially more expensive initially but over the long term would pay off and help the environment.

Carbon stores
The carbon sequestered by 1.2 acres (half a hectare) of conifer woodland in one growing cycle can compensate for the carbon dioxide emitted by a car over one average driver's lifetime.

Apart from solutions based on carbon sequestration (capturing and storage), various other geo-engineering options have been suggested and may be employed in future.

AEROSOLS
Tiny sulfate particles could be injected into the stratosphere at relatively low cost to **reflect the Sun's rays.** However, there would be a risk of ozone depletion and unknown consequences for the weather.

REFLECTIVE CROPS
Some scientists believe that planting arable land with crop varieties whose **leaves reflect more light back** into the atmosphere could significantly cut regional summer temperatures.

CLOUD SYNTHESIS
There are suggestions that **seawater could be atomized to create extra clouds,** which would reflect the Sun's rays. Again, there would be a risk of unknown weather effects.

Wind power
Wind turbines can produce electricity to supplement the supply from the local grid.

Loft insulation
A highly effective way of saving energy, loft insulation is particularly eco-friendly if it uses natural materials.

Active solar power
Solar energy collectors can be used to generate hot water and electricity.

Rafter insulation
In a lived-in roof space, ensuring that the ceiling is sufficiently insulated can be a key energy-saving measure.

Green living roof
This type of structure provides good insulation and a habitat for wildlife.

Ground-source heating
Extracting a small amount of heat from the ground can provide hot water and heating for the house.

Biomass boilers
A carbon-neutral option that offers a viable alternative to conventional boilers.

Rainwater storage
Rainwater can be stored in tanks and used for purposes such as garden maintenance.

Cavity wall insulation
Insulating these walls can help reduce energy requirements further.

REFERENCE

The following section provides quick access to facts and figures on the major fields of science covered in this book. Tables, laws, formulas, illustrations, and maps are included.

Measurement

Historically, different people used independently formulated, and often imprecise, units of measurement—such as a grain of wheat. Standard measuring systems developed when commerce spread beyond local boundaries. International trade led to the introduction of units defining basic quantities, such as weight and length, that meant the same everywhere. Worldwide, metric measurement is now the most commonly used system.

Base SI units

The SI (Système International d'Unités) metric system of measurement is founded on seven base units of independent physical quantities. All other physical units are obtained from these units.

Unit	Symbol	Definition
meter	m	The meter is the length of a path traveled by light in a vacuum during a time interval of $1/299\,792\,458$ of a second.
kilogram	kg	The kilogram is the unit of mass equal to the mass of the international prototype kilogram.
second	s	The second is the duration of 9,192,631,770 periods of the radiation corresponding to the transition between the two hyperfine levels of the ground state of the cesium-133 atom.

Unit	Symbol	Definition
ampere	A	The ampere is that constant electric current which, if maintained in two straight parallel conductors of infinite length, of negligible cross section, and placed 1 meter apart in vacuum, would produce between these conductors a force equal to 2×10^{-7} newton per meter.
kelvin	K	The kelvin, unit of thermodynamic temperature, is the fraction $1/273.16$ of the thermodynamic temperature of the triple point of water.
candela	cd	The candela is the luminous intensity, in a given direction, of a source that emits monochromatic radiation of frequency 540×10^{-12} hertz and that has a radiant intensity in that direction of $1/683$ watt per steradian.
mole	mol	The mole is the amount of substance that contains as many elementary units as there are carbon atoms in 0.012 kilogram of carbon-12.

Supplementary and derived SI units

Two supplementary units are used in addition to the seven SI base units. Many other units, some examples of which are listed here, are derived from the SI base units using a system of quantity equations.

Supplementary unit	Symbol	Definition
radian	rad	The unit of measurement of angle; it is the angle subtended at the center of a circle by an arc equal in length to the circle radius.
steradian	sr	The unit of measurement of solid angle; it is the solid angle subtended at the center of a circle by a spherical cap equal in area to the square of the circle radius.

Derived unit	Symbol	Definition
coulomb	C	A unit of electrical charge equal to the amount of charge transferred by a current of 1 ampere in 1 second.
farad	F	The capacitance of a capacitor that has an equal and opposite charge of 1 coulomb on each plate and a voltage difference of 1 volt between the plates.
henry	H	A unit of inductance in which an induced electromotive force of 1 volt is produced when the current is varied at the rate of 1 ampere per second.
hertz	Hz	The unit of frequency; 1 hertz has a periodic interval of 1 second.
joule	J	The energy exerted by a force of 1 newton acting to move an object through a distance of 1 meter.
lumen	lm	A unit of luminous flux equal to the amount of light given out through a solid angle of 1 steradian by a point source of 1 candela intensity radiating uniformly in all directions.
newton	N	A unit of force equal to the force that imparts an acceleration of 1m/sec/sec to a mass of 1 kilogram.
ohm	Ω	A unit of electrical resistance equal to the resistance between two points on a conductor when a potential difference of 1 volt between them produces a current of 1 ampere.
pascal	Pa	A unit of pressure equal to 1 newton per square meter.
volt	V	A unit of potential equal to the potential difference between two points on a conductor carrying a current of 1 ampere when the power dissipated between the two points is 1 watt; equivalent to the potential difference across a resistance of 1 ohm when 1 ampere of current flows through it.
watt	W	A unit of power equal to 1 joule per second; the power dissipated by a current of 1 ampere flowing across a resistance of 1 ohm.

SI conversion factors

The table below gives the SI equivalents and conversion factors for units of measurement from other systems that are still in use. "SI equivalent" refers to one unit of the type named in the first column; the column headed "Reciprocal" gives the reverse conversion factors. For example: 1yd = 0.914m; 1m = 1.093yd.

Unit	Symbol	Quantity	SI equivalent	SI unit	Reciprocal
acre		area	0.405	hm^2	2.471
ångström	Å	length	0.1	nm	10
astronomical unit	AU	length	0.150	Tm	6.684
atomic mass unit	amu	mass	1.661×10^{-27}	kg	6.022×10^{26}
bar	bar	pressure	0.1	MPa	10
barrel (US) = 42 US gal	bbl	volume	0.159	m^3	6.290
British thermal unit	btu	energy	1.055	kJ	0.948
calorie	cal	energy	4.187	J	0.239
cubic foot	cu ft	volume	0.028	m^3	35.315
cubic inch	cu in	volume	16.387	cm^3	0.061
cubic yard	cu yd	volume	0.765	m^3	1.308
curie	Ci	activity of radionuclide	37	GBq	0.027
degree = $1/90$ right angle	°	plane angle	0.0175	rad	57.296
degree Celsius	°C	temperature	1	K	1
degree Fahrenheit	°F	temperature	0.556	K	1.8
degree Rankine	°R	temperature	0.556	K	1.8
dyne	dyn	force	10	µN	0.1
electronvolt	eV	energy	0.160	aJ	6.241
erg	erg	energy	0.1	µJ	10
fathom (6 ft)		length	1.829	m	0.547
fermi	fm	length	1	fm	1
foot	ft	length	30.48	cm	0.033
foot per second	ft s^{-1}	velocity	0.305	m s^{-1}	3.281
			1.097	km h^{-1}	0.911
gallon (US) = 231cu in	gal	volume	3.785	dm^3	0.264
gallon (UK)	gal	volume	4.546	dm^3	0.220

Continued »

>> Continued

Unit	Symbol	Quantity	SI equivalent	SI unit	Reciprocal
gauss	Gs, G	magnetic flux density	100	μT	0.01
grade = 0.01 rt angle	rt angle	plane angle	0.0157	rad	63.662
grain	gr	mass	0.065	g	15.432
hectare	ha	area	1	hm²	1
horsepower	hp	power	0.746	kW	1.341
inch	in	length	2.54	cm	0.394
kilogram-force	kgf	force	9.807	N	0.102
knot		velocity	1.852	km h⁻¹	0.540
light year	ly	length	9.461×10^{15}	m	1.057×10^{-16}
liter	l	volume	1	dm³	1
Mach number	Ma	velocity	1193.3	km h⁻¹	8.380×10^{-4}
maxwell	Mx	magnetic flux	10	nWb	0.1
micron	μ	length	1	μm	1
mile (nautical)		length	1.852	km	0.540
mile (statute)		length	1.609	km	0.621
miles per hour (mph)	mile h⁻¹	velocity	1.609	km h⁻¹	0.621
minute =(1/60)°	'	plane angle	2.91×10^{-4}	rad	3437.75
oersted	Oe	magnetic field strength	0.079	kA m⁻¹	12.6
ounce (avoirdupois)	oz	mass	28.349	g	0.035
ounce (troy) = 480 gr		mass	31.103	g	0.032
parsec	pc	length	30857	Tm	0.0000324
phot	ph	illuminance	10	klx	0.1
pint (UK)	pt	volume	0.568	dm³	1.760
poise	P	viscosity	0.1	Pa s	10
pound	lb	mass	0.454	kg	2.205
pound force	lbf	force	4.448	N	0.225
pound force/in		pressure	6.895	kPa	0.145
poundal	pdl	force	0.138	N	7.233
pounds per square inch	psi	pressure	6.895×10^3	kPa	0.145
rad	rad	absorbed dose	0.01	Gy	100
rem	rem	dose equivalent	0.01	Sv	100
right angle = □/2 rad		plane angle	1.571	rad	0.637
röntgen	R	exposure	0.258	mC kg⁻¹	3.876
second = (1/60)'	"	plane angle	4.85×10^{-6}	mrad	2.063×10^{-5}
slug		mass	14.594	kg	0.068
solar mass	M	mass	1.989×10^{30}	kg	5.028×10^{-31}
square foot	sq ft	area	9.290	dm²	0.108
square inch	sq in	area	6.452	cm²	0.155
square mile (statute)	sq mi	area	2.590	km²	0.386
square yard	sq yd	area	0.836	m²	1.196
standard atmosphere	atm	pressure	0.101	MPa	9.869
stere	st	volume	1	m³	1
stilb	sb	luminance	10	kcd m⁻²	0.1
stokes	St	viscosity	1	cm² s⁻¹	1
therm = 10⁵ btu		energy	0.105	GJ	9.478
ton = 2240 lb		mass	1.016	Mg	0.984
ton-force	tonf	force	9.964	kN	0.100
ton-force/sq in		pressure	15.444	MPa	0.065
tonne	t	mass	1	Mg	1
torr, or mmHg	torr	pressure	0.133	kPa	7.501
X unit		length	0.100	pm	10
yard	yd	length	0.914	m	1.093

SI prefixes

Factor	Prefix	Symbol	Factor	Prefix	Symbol	Factor	Prefix	Symbol
10^{18}	exa-	E	10^2	hecto-	h	10^{-9}	nano-	n
10^{15}	peta-	P	10^1	deca-	da	10^{-12}	pico-	p
10^{12}	tera-	T	10^{-1}	deci-	d	10^{-15}	femto-	f
10^9	giga-	G	10^{-2}	centi-	c	10^{-18}	atto-	a
10^6	mega-	M	10^{-3}	milli-	m			
10^3	kilo-	k	10^{-6}	micro-	μ			

Converting common measures

IMPERIAL TO METRIC

Length	Multiply by
inches→millimeters	25.4
inches→centimeters	2.54
feet→meters	0.3048
yards→meters	0.9144
statute miles→kilometers	1.6093
nautical miles→kilometers	1.852

Area	Multiply by
square inches→square centimeters	6.4516
square feet→square meters	0.0929
square yards→square meters	0.8361
acres→hectares	0.4047
square miles→square kilometers	2.5899

Volume	Multiply by
cubic inches→cubic centimeters	16.3871
cubic feet→cubic meters	0.0283
cubic yards→cubic meters	0.7646

Capacity	Multiply by
US fluid ounces→liters	0.0296
US pints→liters	0.4732
US gallons→liters	3.7854
UK fluid ounces→liters	0.0284
UK pints→liters	0.5682
UK gallons→liters	4.546

Weight	Multiply by
ounces (avoirdupois)→grams	28.3495
ounces (troy)→grams	31.1035
pounds→kilograms	0.4536
US tons→tonnes	1.1023

METRIC TO IMPERIAL

Length	Multiply by
millimeters→inches	0.0394
centimeters→inches	0.3937
meters→feet	3.2806
meters→yards	1.0936
kilometers→statute miles	0.6214
kilometers→nautical miles	0.54

Area	Multiply by
square centimeters→square inches	0.155
square meters→square feet	10.764
square meters→square yards	1.196
hectares→acres	2.471
square kilometers→square miles	0.386

Volume	Multiply by
cubic centimeters→cubic inches	0.061
cubic meters→cubic feet	35.315
cubic meters→cubic yards	1.308

Capacity	Multiply by
liters→US fluid ounces	33.8150
liters→US pints	2.1134
liters→US gallons	0.2642
liters→UK fluid ounces	35.1961
liters→UK pints	1.7598
liters→UK gallons	0.2199

Weight	Multiply by
grams→ounces (avoirdupois)	0.0353
grams→ounces (troy)	0.0322
kilograms→pounds	2.2046
tonnes→US tons	0.9072

TEMPERATURE

To convert	Equation
°Fahrenheit→°Celsius	−32, × 5, ÷ 9
°Celsius→°Fahrenheit	× 9, ÷ 5, + 32
Kelvin→°Celsius	− 273.16
°Celsius→Kelvin	+ 273.16

Astronomy and Space
The Solar System

The expanse of space known as the solar system contains the Sun together with the planets and their attendant moons, and countless small objects—such as asteroids and comets—that fall within its gravitational influence (see pp.182–83). This immense area measures 9,000 billion miles (15,000 billion km) in diameter and is divided roughly into three regions: the inner and outer planet groups and, in the farthest reaches, a vast comet cloud.

The planets

Eight planets orbit the Sun. In the inner region of the solar system there are four rocky planets—Mercury, Venus, Earth, and Mars—each comprising a metallic core and a rocky mantle and crust. The four planets of the outer region are the giants Jupiter, Saturn, Uranus, and Neptune, which are predominantly liquid surrounded by gas. A ninth planet, Pluto, on the outer reaches of the solar system, was reclassified as a dwarf planet in 2006.

Planet	MERCURY	VENUS	EARTH	MARS	JUPITER	SATURN	URANUS	NEPTUNE
Distance from Sun millions of miles (km)	36.0 (57.9)	67.2 (108.2)	93 (149.6)	141.5 (227.9)	483.3 (778.3)	886 (1,427)	1,782 (2,870)	2,774 (4,497)
Diameter at equator miles (km)	3,033 (4,879)	7,523 (12,104)	7,928 (12,756)	4,222 (6,786)	88,784 (142,984)	74,914 (120,536)	31,770 (51,118)	30,757 (49,528)
Mass (Earth = 1)	0.06	0.82	1	0.107	318	95	14.5	17
Volume (Earth = 1)	0.056	0.86	1	0.15	1,319	744	67	57
Surface temperature °F (°C)	−356 to +800 (−180 to +430)	+896 (+480)	−158 to +133 (−70 to +55)	−248 to +77 (−120 to +25)	−238 (−150)	−292 (−180)	−353 (−214)	−364 (−220)
Surface gravity (Earth = 1)	0.38	0.9	1	0.38	2.64	0.925	0.79	1.12
Time to orbit Sun ("year")	87.97 days	224.7 days	365.26 days	686.98 days	11.86 years	29.46 yrs	84.01 yrs	164.8 yrs
Time to turn 360° ("day")	58.65 days	243.01 days	23 h 56 m 4 s	24 h 37 m 23 s	9 h 55 m 30 s	10 h 39 m	17 h 14 m	16 h 7 m
Orbital speed miles/s (km/s)	29.7 (47.9)	21.8 (35)	18.5 (29.8)	15 (24.1)	8.1 (13.1)	6 (9.6)	4.2 (6.8)	3.4 (5.4)
Number of observed moons	0	0	1	2	79	82	27	14

The sizes of the Sun and planets

The Sun is by far the largest object in the Solar System, containing about 99.8 percent of the mass in the entire system. The largest planets are the four giants, led by Jupiter, which has 2.5 times the combined mass of the other seven planets. At the lower end of the size scale are the four rocky planets, of which Earth is the largest. The smallest planet, Mercury, is smaller than the largest moons of Jupiter and Saturn.

SUN

Main Belt asteroids

Asteroids are small, mostly irregularly shaped bodies that orbit the Sun. The majority, around a billion, congregate in the Main Belt, an area between Mars and Jupiter. Most are only a few kilometers across. Ceres, the first asteroid to be discovered, in 1801, is now classified as a dwarf planet. Listed below are some asteroids that have been imaged in detail.

Name	Size miles (km)	Orbital period (years)
Vesta	Diameter: 348 (560)	3.63
Mathilde	Length: 41 (66)	4.31
Ida	Length: 37 (60)	4.84
Eros	Length: 19.25 (31)	1.76
Gaspra	Length: 11.2 (18)	3.29
Annefrank	Length: 3.7 (6)	3.29

Kepler's laws of planetary motion

Formulated by German astronomer Johannes Kepler, these laws state that the orbits of the planets are elliptical, not circular, and that orbital speed is slower the further away a planet is from the Sun (see pp.76–78).

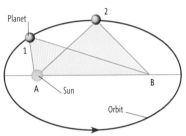

First law
This states that a planet orbits the Sun in a path known as an ellipse, with the Sun at one focus. There are two focuses within an ellipse (A and B). On any particular ellipse, the total distance from one focus (A) to any point on the ellipse (1 or 2) and back to the other focus (B) is always the same.

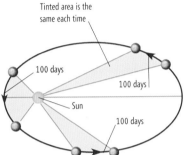

Second law
This describes how the speed of a planet changes as it orbits the Sun. The law states that a line from the Sun to a planet sweeps out equal areas in equal periods of time. This means that the planet moves faster when closer to the Sun and slower when farther away.

Third law
Kepler's third law describes the mathematical relationship between the distances of planets from the Sun and their orbital periods. It states that the square of each planet's orbital period is directly proportional to the cube of its average distance from the Sun. This allows orbital period and distance to be calculated accurately for all planets.

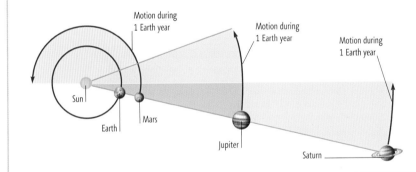

Meteor showers

Meteoroids are pieces of debris from comets or asteroids. As the Earth orbits the Sun, it passes through streams of meteoroids. These burn up in the upper atmosphere (see pp.118–19), becoming visible as trails of light, or meteors. Showers of meteors are seen at fixed times of year and are named after the constellations from which they radiate.

Name	Date	Constellation
Quadrantids	Jan 1–6	Boötes
April Lyrids	April 19–24	Lyra
Eta Aquarids	May 1–8	Aquarius
Delta Aquarids	July 15–Aug 15	Aquarius
Perseids	July 25–Aug 18	Perseus
Orionids	Oct 16–27	Orion
Taurids	Oct 20–Nov 30	Taurus
Leonids	Nov 15–20	Leo
Geminids	Dec 7–15	Gemini

Main planetary moons

Besides Earth, most of the other solar system planets have their own orbiting moons. The exceptions are Mercury and Venus. Listed here are the major planetary moons.

PLANET/ moons	Orbital period (days)	Distance from planet in miles (km)		Diameter in miles (km)	
EARTH					
Moon	27.32	238,000	(384,000)	2,155	(3,476)
MARS					
Phobos	0.31	5,827	(9,377)	22	(14)
Deimos	1.26	14,580	(23,460)	12	(7)
JUPITER					
Metis	0.29	79,000	(128,000)	27	(43)
Adrastea	0.29	80,000	(129,000)	10	(16)
Amalthea	0.49	112,000	(181,000)	104	(167)
Thebe	0.67	138,000	(222,000)	60	(100)
Io	1.76	262,000	(422,000)	2,263	(3,642)
Europa	3.55	417,000	(671,000)	1,940	(3,122)
Ganymede	7.15	665,000	(1,070,000)	3,270	(5,260)
Callisto	16.68	1,170,000	(1,883,000)	2,995	(4,820)
Leda	238.72	6,936,988	(11,164,000)	12	(20)
Himalia	250.57	7,134,000	(11,480,000)	87	(140)
Lysithea	259.22	7,283,000	(11,720,000)	26	(42)
Elara	259.65	7,295,000	(11,740,000)	50	(80)
Ananke	631	13,222,779	(21,280,000)	19	(30)
Carme	734	14,540,086	(23,400,000)	28	(46)
Pasiphae	744	14,664,360	(23,600,000)	37	(60)
Sinope	758	14,850,771	(23,900,000)	22	(36)
SATURN					
Pan	0.57	83,000	(134,000)	12	(28)
Atlas	0.6	86,000	(138,000)	19	(30)
Prometheus	0.61	86,000	(139,000)	53	(86)
Pandora	0.62	88,000	(142,000)	50	(81)
Epimetheus	0.69	94,000	(151,000)	72	(116)
Janus	0.69	94,000	(151,000)	111	(179)
Mimas	0.94	116,000	(186,000)	246	(396)
Enceladus	1.37	148,000	(238,000)	310	(500)
Calypso	1.88	183,000	(295,000)	14	(22)
Telesto	1.88	183,000	(295,000)	15	(25)
Tethys	1.88	183,000	(295,000)	660	(1,060)
Dione	2.73	234,000	(377,000)	700	(1,120)
Helene	2.73	234,000	(377,000)	22	(35)
Rhea	4.51	527,000	(327,000)	950	(1,530)
Titan	15.94	759,000	(1,222,000)	3,200	(5,150)
Hyperion	21.27	932,678	(1,501,000)	168	(270)
Iapetus	14.72	2,212,000	(3,560,000)	913	(1,470)
Phoebe	550.48	8,047,000	(12,950,000)	132	(213)
URANUS					
Miranda	1.41	81,000	(130,000)	293	(472)
Ariel	2.52	119,000	(191,000)	720	(1,160)
Umbriel	4.14	165,000	(266,000)	730	(1,170)
Titania	8.7	271,000	(436,000)	980	(1,580)
Oberon	13.46	362,000	(583,000)	947	(1,524)
NEPTUNE					
Triton	5.87	221,000	(355,000)	1,681	(2,705)
Nereid	360.16	3,424,000	(5,510,000)	210	(340)
Proteus	1.12	73,322	(118,000)	261	(420)

Earth, Moon, and Sun

Earth is the third closest planet to the Sun after Mercury and Venus, and is the only planet known to support living organisms. It is orbited by one natural satellite, the Moon, which is one of the largest satellites in the solar system in comparison to its parent planet. Interactions between the Earth, Moon, and Sun are responsible for variations in Earth's atmosphere, night and day cycles, climate, seasons, and tidal patterns.

Earth at a glance

Largest of the inner planets, Earth is unique in the solar system for having an oxygen-rich atmosphere, liquid as well as frozen surface water, and a crust of moving plates. Earth orbits the Sun at an average speed of 18.5 miles/s (29.8 km/s)—about 67,000 mph (108,000 kph).

STRUCTURE

- Crust
- Mantle of silicate rock
- Outer core (liquid iron)
- Inner core (solid iron–nickel)

SCALE

Earth is about four times the size of its moon

TILT, SPIN, AND ORBIT

- Line at right angles to plane of orbit
- Orbits Sun in 365.25 days
- Sun
- Axis tilted at 23.5°
- Spins on its axis once every 23.93 h

ATMOSPHERE

- Water vapor (1%) and trace gases
- Oxygen (20.9%)
- Nitrogen (78.1%)

Earth: vital statistics

Diameter: miles (km)	7,928 (12,756)
Average distance from Sun: miles (km)	93 million (149.6 million)
Orbital speed around Sun: miles/s (km/s)	18.5 (29.8)
Mean solar day	24 hours
Mass	5.98×10^{24} kg
Volume	1.08321×10^{12} km³
Average density (water=1)	5.52
Surface gravity	9.8m/s²
Average surface temperature	59° F (15° C)
Number of moons	1

Earth's chemical composition

The chemical compounds that make up Earth occur in different proportions in each of its layers (see pp.352–53). Gravitational separation of material means that the deeper the layer is beneath the surface, the richer the concentrations of denser materials such as iron.

- Silicon dioxide
- Aluminum oxide
- Iron and iron oxides
- Calcium oxide
- Magnesium oxide
- Nickel oxide
- Others

CONTINENTAL CRUST | OCEANIC CRUST | MANTLE | CORE

Moon at a glance

The Moon is a barren, dusty sphere with a similar structure to that of the rocky planets. It has no atmosphere or liquid water. In the 27.32 days the Moon takes to orbit Earth, it makes a complete rotation on its axis.

STRUCTURE

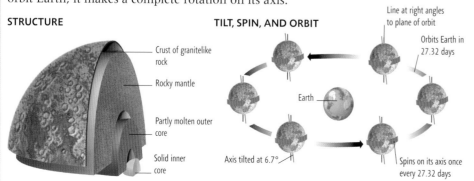

- Crust of granitelike rock
- Rocky mantle
- Partly molten outer core
- Solid inner core

TILT, SPIN, AND ORBIT

- Line at right angles to plane of orbit
- Orbits Earth in 27.32 days
- Earth
- Axis tilted at 6.7°
- Spins on its axis once every 27.32 days

Moon: vital statistics

Diameter: miles (km)	2,160 (3,476)
Average distance from Earth: miles (km)	238,900 (384,400)
Orbital speed around Earth: miles/s (km/s)	0.63 (1.02)
New Moon to new Moon	29.53 days
Mass (Earth=1)	0.01
Volume (Earth=1)	0.02
Average density (water=1)	3.34
Surface gravity (Earth=1)	0.17
Average surface temperature	-4° F(-20° C)

Phases of the Moon

As the Moon orbits, the proportion of its Earth-facing hemisphere that is lit by the Sun changes in a cycle of phases. During the 29.5 days of the lunar month, the Moon's appearance ranges from a thin crescent to quarter, gibbous (more than half illuminated), and full face. When the Moon passes between the Sun and Earth, sunlight falls on its far side; the side facing Earth is in darkness and cannot be seen at all.

CYCLE OF THE MOON'S PHASES

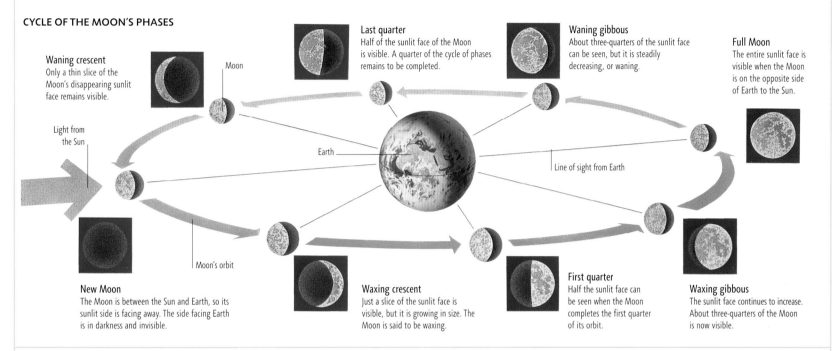

Waning crescent
Only a thin slice of the Moon's disappearing sunlit face remains visible.

Moon

Last quarter
Half of the sunlit face of the Moon is visible. A quarter of the cycle of phases remains to be completed.

Waning gibbous
About three-quarters of the sunlit face can be seen, but it is steadily decreasing, or waning.

Full Moon
The entire sunlit face is visible when the Moon is on the opposite side of Earth to the Sun.

Light from the Sun

Earth

Line of sight from Earth

Moon's orbit

New Moon
The Moon is between the Sun and Earth, so its sunlit side is facing away. The side facing Earth is in darkness and invisible.

Waxing crescent
Just a slice of the sunlit face is visible, but it is growing in size. The Moon is said to be waxing.

First quarter
Half the sunlit face can be seen when the Moon completes the first quarter of its orbit.

Waxing gibbous
The sunlit face continues to increase. About three-quarters of the Moon is now visible.

Lunar eclipses 2021–2027

A lunar eclipse occurs when the orbiting Moon passes through Earth's shadow and its brightness is dimmed to a varying degree (see p.79). In a penumbral eclipse, the Moon passes through the lighter, outer shadow; when part of the Moon passes through full shadow, that is a partial eclipse; in a total eclipse, the entire Moon passes through full shadow.

Date	Type	Region of visibility	Date	Type	Region of visibility
November 19, 2021	Partial	Americas, Europe, Asia, Australia, Pacific	March 14, 2025	Total	Pacific, Americas, Europe, Africa
May 16, 2022	Total	America, Europe, Africa	September 7, 2025	Total	Europe, Africa, Asia, Australia
November 8, 2022	Total	Asia, Australia, Pacific, Americas	March 3, 2026	Total	East Asia, Australia, Pacific, Americas
May 5, 2023	Penumbral	Africa, Asia, Australia	August 28, 2026	Partial	Eastern Pacific, Americas, Europe, Africa
October 28, 2023	Partial	Americas, Europe, Africa, Australia	February 20, 2027	Penumbral	Americas, Europe, Africa, Asia
March 25, 2024	Penumbral	Americas	July 18, 2027	Penumbral	East Africa, Asia, Australia, Pacific
September 18, 2024	Partial	Americas, Europe, Africa	August 17, 2027	Penumbral	Pacific, Americas

The Sun: vital statistics

Mass (Earth = 1)	333,000
Surface temperature	9,900° F (5,500° C)
Core temperature	27 million ° F (15 million ° C)
Diameter: miles (km)	864,900 (1,392,000)
Distance from Earth: miles (km)	93 million (149.6 million)
Luminosity	3.9×10^{26} watts
Rotation period	25.4 days (at equator)
Age	4.6 billion years

Solar composition

The Sun's outer layers are 73 percent hydrogen, 25 percent helium, and two percent other elements. In the superheated core, where more than 660 million tons of hydrogen are converted into helium every second, the amount of hydrogen is only about 34 percent, while the amount of helium is about 64 percent.

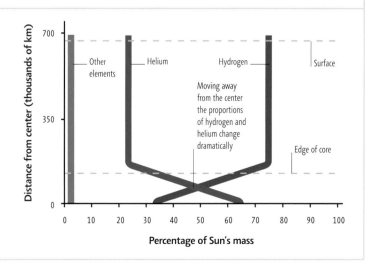

Distance from center (thousands of km)

Other elements

Helium

Hydrogen

Surface

Moving away from the center the proportions of hydrogen and helium change dramatically

Edge of core

Percentage of Sun's mass

Total solar eclipses 2021–2034

In a total eclipse, the Moon completely obscures the Sun from viewers within a limited region (see pp.78–79). This region, known as the path of totality, can be predicted precisely.

Date	Duration	Region of totality
December 4, 2021	1 min 54 sec	Antartica
April 20, 2023	1 min 16 sec	Indonesia, Australia, Papua New Guinea
April 8, 2024	4 min 28sec	Mexico, central US, eastern Canada
August 12, 2026	2 min 18 sec	Arctic, Greenland, Iceland, Spain
August 2, 2027	6 min 23 sec	Morocco, Spain, north Africa, Saudi Arabia, Yemen, Somalia
July 22, 2028	5 min 10 sec	Australia, New Zealand
November 25, 2030	3 min 44 sec	Botswana, South Africa, Australia
March 30, 2033	2 min 37 sec	Eastern Russia, Alaska
March 20, 2034	4 min 9 sec	Nigeria, Cameroon, Chad, Sudan, Egypt, Saudi Arabia, Iran, Afghanistan, Pakistan, India, China

Solar heating

Heat from the Sun in the form of infrared, UV, and visible radiation warms the Earth's atmosphere, oceans, and land. Solar heating is more intense in the tropical regions than at the poles.

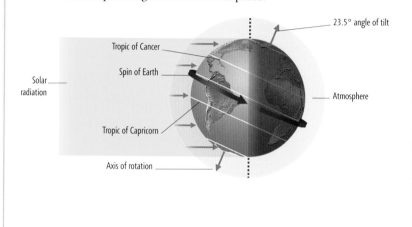

The stars

A star is a body formed when a cloud of gas and dust shrinks and collapses under its own gravity (see pp.330–31). The collapsed cloud heats up, setting off nuclear reactions that create energy and light. Stars are held together by gravity in groups called galaxies (see pp.340–41). Between galaxies lie the immensities of empty space. Our home galaxy, the Milky Way, is made up of 200–500 billion stars of varying types and brightness, of which the Sun is one.

The magnitude scale

Astronomers measure the luminosity, or brightness, of stars as well as other celestial objects in magnitudes. The smaller the magnitude, the greater the brightness. The very brightest objects have negative magnitudes. Each step on the scale shown below represents an increase or decrease in brightness of 2.5 times. The scale includes both stars and Solar System objects.

MAGNITUDE SCALE

Spectral types

When a star's light is split into different wavelengths, it produces a spectrum containing dark lines (absorption lines), the pattern of which reveals the star's composition, temperature, and color. Astronomers classify stars into seven spectral types: O, B, A, F, G, K, and M. O, blue-white, stars are the hottest; M, orange-red, the coolest. Each type has 10 subdivisions, numbered 0 to 9 (hotter to cooler). The Sun is type G2.

| **Type O** (40,000–29,000° C) | **Type B** (28,000–9,700° C) | **Type A** (9,600–7,200° C) | **Type F** (7,100–5,800° C) | **Type G** (5,700–4,700° C) | **Type K** (4,600–3,300° C) | **Type M** (3,200–2,100° C) |

Average sizes of stars

There are enormous differences in the sizes of stars, which range from supergiants (300 times larger than the Sun) to giant and dwarf stars. A star may end its life as a very small, dense neutron star, or as a black hole very much smaller than Earth.

Supergiant: 10 times size of red giant

Red giant: 30 times size of Sun

Hydrogen-burning type B star: 7 times size of Sun

Sun

Sun: 100 times size of white dwarf

White dwarf: 500 times size of neutron star

Neutron star: 5 times size of black hole

Red giant

Sun

Hydrogen-burning type M star: 1/10 size of Sun

White dwarf

Neutron star

Black hole

Hertzsprung–Russell (H–R) diagram

Named after the astronomers who devised it, the H–R diagram is a graph that plots the luminosity of stars against their temperature (see pp.328–29). Most stars lie in a band called the main sequence, and are at roughly midlife. Older, cooler stars—red giants and supergiants—occupy the upper right side of the graph. Dying stars shrink to become tiny, hot white dwarfs and move to the bottom left of the graph.

Main sequence stars
The main sequence runs diagonally from the brightest stars at top left to the faintest at bottom right. Main sequence stars burn hydrogen in nuclear reaction and convert it to helium. Stars spend about 90 percent of their lives on the main sequence, changing very little in luminosity or temperature while they are there. The Sun is a typical main sequence star.

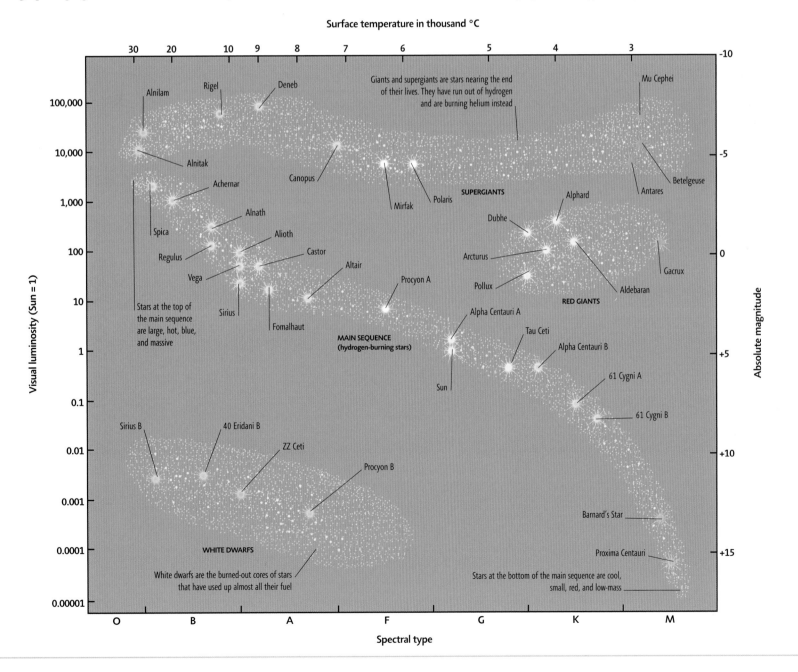

Brightest stars

A star's brightness, or magnitude, is described in two ways. The apparent magnitude of a star is its brightness as seen from Earth. Absolute magnitude is the apparent magnitude it would have if it was 32.6 light-years away. The magnitudes of variable stars (followed by "v") are average values.

| Name | Magnitude | | Distance from |
	Apparent	Absolute	Sun (light years)
Sirius	-1.46	+1.4	860
Canopus	-0.73	5.7	310
Alpha Centauri A & B	-0.27	+4.1	4.38
Arcturus	-0.06	-0.3	36
Vega	+0.0v	+0.5v	26
Capella	+0.08	-0.5	42
Rigel	+0.10v	-7.8v	860
Procyon	+0.35	+2.6	11.4
Achernar	+0.46v	-1.5	139
Betelgeuse	+0.49v	-5.9v	550
Hadar	+0.61	-4.9	390
Altair	+0.77	+2.3	16
Aldebaran	+0.85v	-0.7v	69
Acrux A & B	+0.80	-3.7	325
Antares	+0.96v	5.2v	550
Spica	+0.96v	-3.6v	250
Pollux	+1.15	+1.0	35
Fomalhaut	+1.16	+1.7	25
Deneb	+1.25v	-9.5--10.8	1,500–2,600
Mimosa	+1.25v	-3.4	280
Regulus	+1.39	-0.6	79
Adhara	+1.50	-6.5	430

Nearest stars and star systems

Stellar distances are measured in light-years (the distance light travels in a year: 9,461 billion km). The following stars and multiple star systems are among the Sun's nearest neighbors.

Name of star or star system	Apparent magnitude	Spectral type	Distance (light-years)
Alpha Centauri A, B, C	0.1, 1.4, 11.0	G2, K1, M5	4.4
Barnard's Star	9.5	M4	6.0
Lalande 21185	7.5	M2	8.3
Sirius A, B	-1.4, 8.5	A0, white dwarf	8.6
Ross 154	10.4	M4	9.7
Epsilon Eridani	3.7	K2	10.5
HD 217987	7.4	M0	10.7
Ross 128	11.1	M4	11
61 Cygni A, B	5.2, 6.1	K5, K7	11.4
Procyon A, B	0.4, 10.7	F5, white dwarf	11.4

The constellations

Earlier people associated distinctive star patterns with stories, naming them accordingly. Usually, the stars in a constellation are chance groupings and are not related. Astronomers officially recognize 88 constellations, now as set regions of sky rather than celestial images.

Latin name	Common name	Latin name	Common name
Andromeda	Andromeda	Lacerta	Lizard
Antlia	Air Pump	Leo	Lion
Apus	Bird of Paradise	Leo Minor	Little Lion
Aquarius	Water Carrier	Lepus	Hare
Aquila	Eagle	Libra	Scales
Ara	Altar	Lupus	Wolf
Aries	Ram	Lynx	Lynx
Auriga	Charioteer	Lyra	Lyre
Boötes	Herdsman	Mensa	Table Mountain
Caelum	Chisel	Microscopium	Microscope
Camelopardalis	Giraffe	Monoceros	Unicorn
Cancer	Crab	Musca	Fly
Canes Venatici	Hunting Dogs	Norma	Level
Canis Major	Great Dog	Octans	Octant
Canis Minor	Little Dog	Ophiuchus	Serpent Bearer
Capricornus	Sea Goat	Orion	Orion, Hunter
Carina	Keel	Pavo	Peacock
Cassiopeia	Cassiopeia	Pegasus	Winged Horse
Centaurus	Centaur	Perseus	Perseus
Cepheus	Cepheus	Phoenix	Phoenix
Cetus	Whale	Pictor	Painter's Easel
Chamaeleon	Chameleon	Pisces	Fishes
Circinus	Compasses	Piscis Austrinus	Southern Fish
Columba	Dove	Puppis	Poop Deck
Coma Berenices	Berenice's Hair	Pyxis	Mariner's Compass
Corona Australis	Southern Crown	Reticulum	Net
Corona Borealis	Northern Crown	Sagitta	Arrow
Corvus	Crow	Sagittarius	Archer
Crater	Cup	Scorpius	Scorpion
Crux	Southern Cross	Sculptor	Sculptor
Cygnus	Swan	Scutum	Shield
Delphinus	Dolphin	Serpens	Serpent
Dorado	Swordfish	Sextans	Sextant
Draco	Dragon	Taurus	Bull
Equuleus	Foal	Telescopium	Telescope
Eridanus	River	Triangulum	Triangle
Fornax	Furnace	Triangulum Australe	Southern Triangle
Gemini	Twins	Tucana	Toucan
Grus	Crane	Ursa Major	Great Bear
Hercules	Hercules	Ursa Minor	Little Bear
Horologium	Clock	Vela	Sails
Hydra	Water Snake	Virgo	Virgin
Hydrus	Little Water Snake	Volans	Flying Fish
Indus	Indian	Vulpecula	Fox

Star maps

Around 6,000 stars are visible to the naked eye. These star maps show the main constellations in the Northern and Southern Hemispheres of the sky. As the earth rotates, some stars appear to rise and set in the sky.

The constellations visible differ according to a viewer's latitude and vary with date and time. In mid-latitudes, the stars closest to the pole never set and are visible all year round.

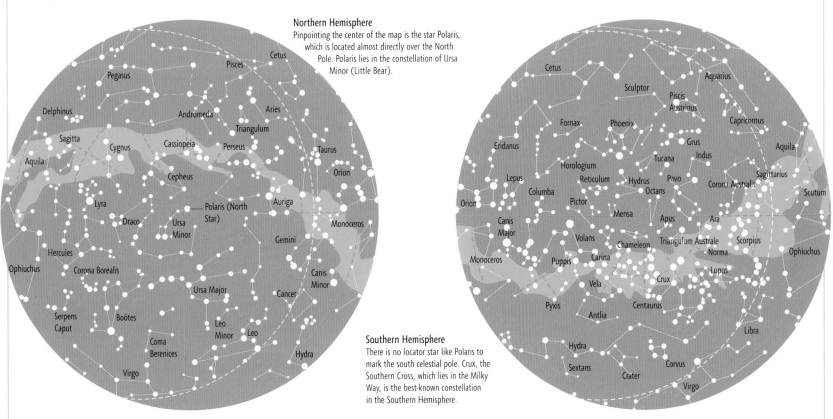

Northern Hemisphere
Pinpointing the center of the map is the star Polaris, which is located almost directly over the North Pole. Polaris lies in the constellation of Ursa Minor (Little Bear).

Southern Hemisphere
There is no locator star like Polaris to mark the south celestial pole. Crux, the Southern Cross, which lies in the Milky Way, is the best-known constellation in the Southern Hemisphere.

Local Group galaxies

The Local Group is a cluster of more than 50 galaxies that includes our home galaxy, the Milky Way (see p.341). Although the members of the group are scattered over a region of space measuring approximately 3 million light years, they are bound together by powerful gravitational forces. By far the largest galaxies in the Local Group are the Milky Way and the Andromeda Galaxy (which is just discernible to the naked eye).

Name	Distance in light-years	Diameter in light-years	Absolute magnitude	Type	Name	Distance in light-years	Diameter in light-years	Absolute magnitude	Type
Milky Way	0	150,000	–20.8	Barred spiral	Cetus Dwarf	2,460,000	5,000	–10.2	Dwarf spheroidal
SagDEG	81,000	10,000	–12.7	Dwarf spheroidal	M32	2,490,000	6,500	–16.0	Dwarf elliptical
LMC	163,000	14,000	–17.9	Magellanic spiral	And XXXI	2,520,000	4,750	–13.0	Dwarf spheroidal
SMC	206,000	7,000	–16.4	Dwarf Irregular	And XXXII	2,520,000	9,500	–11.4	Dwarf
Sculptor Dwarf	290,000	3,500	–10.1	Dwarf spheroidal	NGC 147	2,530,000	10,000	–14.9	Dwarf spheroidal
Fornax Dwarf	460,000	2,300	–11.5	Dwarf elliptical	Andromeda (M31)	2,538,000	220,000	–21.6	Spiral
Leo I	820,000	2,500	–11.0	Dwarf spheroidal	And VI	2,550,000	3,000	–10.8	Dwarf spherical
Phoenix Dwarf	1,440,000	2,200	–10.2	Dwarf irregular	And XXXIII	2,560,000	2,600	–10.3	Dwarf
NGC 6822	1,630,000	7,000	–15.2	Irregular	And VII	2,580,000	2,000	–11.7	Dwarf spheroidal
NGC 185	2,050,000	7,000	–14.8	Dwarf spheroidal	M110 (NGC 205)	2,670,000	17,500	–16.2	Dwarf elliptical
IC 10	2,200,000	4,600	–15.6	Dwarf irregular	Triangulum (M33)	2,730,000	60,000	–18.9	Spiral
And II	2,220,000	2,400	–12.6	Dwarf spheroidal	PegDIG	3,000,000	4,500	–11.5	Dwarf irregular
IC 1613	2,380,000	11,400	–14.5	Irregular dwarf	WLM	3,040,000	10,000	–14.1	Irregular
And I	2,400,000	1,800	–12.0	Dwarf spheroidal	Aquarius Dwarf	3,200,000	2,000	–11.1	Dwarf irregular
And III	2,440,000	2,750	–10.1	Dwarf spheroidal	SagDIG	3,390,000	3,000	–11.5	Irregular
Leo III	2,600,000	3,900	–11.7	Irregular	UGC 4879	4,180,00	3,000	–11.5	Irregular

Messier objects

The French astronomer and comet-hunter Charles Messier compiled a list of 110 nebulous objects to distinguish them from comets, with which they could be readily confused. Most Messier objects belong to one of three major categories: galaxies, star clusters, and nebulae. All of them are identified by the prefix M and a number; a few Messier objects have been given common names.

Number/Name	Constellation	Type
M1 Crab Nebula	Taurus	Supernova remnant
M2	Aquarius	Star cluster
M3	Canes Venatici	Star cluster
M4	Scorpius	Star cluster
M5	Serpens	Star cluster
M6 Butterfly Cluster	Scorpius	Star cluster
M7 Ptolemy Cluster	Scorpius	Star cluster
M8 Lagoon Nebula	Sagittarius	Nebula
M9	Ophiuchus	Star cluster
M10	Ophiuchus	Star cluster
M11 Wild Duck Cluster	Scutum	Star cluster
M12	Ophiuchus	Star cluster
M13 Great Globular Cluster	Hercules	Star cluster
M14	Ophiuchus	Star cluster
M15 Pegasus Cluster	Pegasus	Star cluster
M16 Eagle Nebula	Serpens	Nebula
M17 Omega Nebula	Sagittarius	Nebula
M18	Sagittarius	Star cluster
M19	Ophiuchus	Star cluster
M20 Trifid Nebula	Sagittarius	Nebula
M21	Sagittarius	Star cluster
M22	Sagittarius	Star cluster
M23	Sagittarius	Star cluster
M24	Sagittarius	Star cluster
M25	Sagittarius	Star cluster
M26	Scutum	Star cluster
M27 Dumbbell Nebula	Vulpecula	Nebula
M28	Sagittarius	Star cluster
M29	Cygnus	Star cluster
M30	Capricornus	Star cluster
M31 Andromeda Galaxy	Andromeda	Galaxy
M32	Andromeda	Galaxy
M33 Triangulum Galaxy	Triangulum	Galaxy
M34	Perseus	Star cluster
M35	Gemini	Star cluster
M36	Auriga	Star cluster

Number/Name	Constellation	Type
M37	Auriga	Star cluster
M38	Auriga	Star cluster
M39	Cygnus	Star cluster
M40	Ursa Major	Double star
M41	Canis Major	Star cluster
M42 Orion Nebula	Orion	Nebula
M43 De Mairan's Nebula	Orion	Nebula
M44 Beehive Cluster	Cancer	Star cluster
M45 Pleiades	Taurus	Star cluster
M46	Puppis	Star cluster
M47	Puppis	Star cluster
M48	Hydra	Star cluster
M49	Virgo	Galaxy
M50	Monoceros	Star cluster
M51 Whirlpool Galaxy	Canes Venatici	Galaxy
M52	Cassiopeia	Star cluster
M53	Coma Berenices	Star cluster
M54	Sagittarius	Star cluster
M55	Sagittarius	Star cluster
M56	Lyra	Star cluster
M57 Ring Nebula	Lyra	Nebula
M58	Virgo	Galaxy
M59	Virgo	Galaxy
M60	Virgo	Galaxy
M61	Virgo	Galaxy
M62	Ophiuchus	Star cluster
M63 Sunflower Galaxy	Canes Venatici	Galaxy
M64 Black Eye Galaxy	Coma Berenices	Galaxy
M65	Leo	Galaxy
M66	Leo	Galaxy
M67	Cancer	Star cluster
M68	Hydra	Star cluster
M69	Sagittarius	Star cluster
M70	Sagittarius	Star cluster
M71	Sagitta	Star cluster
M72	Aquarius	Star cluster
M73	Aquarius	Star cluster

Number/Name	Constellation	Type
M74	Pisces	Galaxy
M75	Sagittarius	Star cluster
M76 Little Dumbbell Nebula	Perseus	Nebula
M77	Cetus	Galaxy
M78	Orion	Nebula
M79	Lepus	Star cluster
M80	Scorpius	Star cluster
M81 Bode's Galaxy	Ursa Major	Galaxy
M82 Cigar Galaxy	Ursa Major	Galaxy
M83 Southern Pinwheel Galaxy	Hydra	Galaxy
M84	Virgo	Galaxy
M85	Coma Berenices	Galaxy
M86	Virgo	Galaxy
M87 Virgo A Galaxy	Virgo	Galaxy
M88	Coma Berenices	Galaxy
M89	Virgo	Galaxy
M90	Virgo	Galaxy
M91	Coma Berenices	Galaxy
M92	Hercules	Star cluster
M93	Puppis	Star cluster
M94	Canes Venatici	Galaxy
M95	Leo	Galaxy
M96	Leo	Galaxy
M97 Owl Nebula	Ursa Major	Nebula
M98	Coma Berenices	Galaxy
M99	Coma Berenices	Galaxy
M100	Coma Berenices	Galaxy
M101 Pinwheel Galaxy	Ursa Major	Galaxy
M102	not fully identified	–
M103	Cassiopeia	Star cluster
M104 Sombrero Galaxy	Virgo	Galaxy
M105	Leo	Galaxy
M106	Canes Venatici	Galaxy
M107	Ophiuchus	Star cluster
M108	Ursa Major	Galaxy
M109	Ursa Major	Galaxy
M110	Andromeda	Galaxy

Galaxy classification

Galaxies are classified into three main groups: elliptical (E), spiral (S), and barred spiral (SB). Variations are denoted by a, b, c, and so on. Some galaxies are irregular (Irr) in shape. Examples are listed here.

Name	Constellation	Type	Distance in millions of light-years
M105	Leo	E1	37
M59	Virgo	E5	50
Sombrero	Virgo	Sa	31
NGC 2841	Ursa Major	Sa	46
Andromeda	Andromeda	Sb	2.5
Pinwheel	Ursa Major	Sc	20
Triangulum	Triangulum	Sc	3.2
Whirlpool	Canes Venatici	Sc	31
NGC 2859	Leo Minor	SBa	72
NGC 5850	Virgo	SBb	130
NGC 7479	Pegasus	SBc	110
M82	Ursa Major	Irr	12
NGC 2337	Lynx	Irr	25

Clusters of galaxies

Galaxies tend to form clusters, pulled together by huge gravitational forces (see p.341). Such clusters differ in density and pattern, some having a regular structure while others are more random groupings.

Name	Distance in millions of light-years	Brightest galaxy	Estimated number of members
Virgo	50	M49	2,000
Fornax	65	NGC 1316	60
Eridanus	75	NGC 1232	200
Antlia	130	NGC 3271	250
Centaurus	170	NGC 4696	100+
Hydra	190	NGC 3311	1,000+
Perseus	240	NGC 1275	1,000+
Coma	300	NGC 4874	3,000+
Hercules	500	NGC 6041	200
Corona Borealis	1,000	PCG 54881	400+

Space exploration

Since 1957, scientists have explored the Universe from beyond Earth's atmosphere. Spacecraft have returned data from every planet in the solar system; people live on space stations for months at a time; and observatories both on Earth and in orbit provide information about near and deep space. Although space exploration has been dominated by Russia and the United States, many other countries now actively participate.

Artificial satellites

Thousands of artificial satellites orbit Earth (see pp.368–69) and most are in one of the four main orbit types. A low-Earth orbit is just above Earth's atmosphere. Satellites in polar orbit are often up to 500 miles (800 km) above Earth. In a highly elliptical orbit, a satellite is at much higher altitude at one end of its orbit than the other.

In geostationary orbit, 23,500 miles (36,000 km) high, a satellite synchronizes with Earth's rotation and appears fixed in the sky.

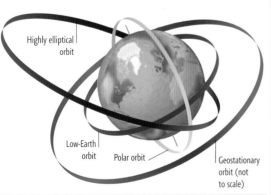

Highly elliptical orbit

Low-Earth orbit

Polar orbit

Geostationary orbit (not to scale)

Type	Use
Low-Earth orbit	Mobile communications, reconnaissance, astronomy
Polar orbit	Weather, navigation
Highly elliptical orbit	Communications at northern latitudes
Geostationary orbit	Communications, weather

Major events in space exploration

Mission	Date	Country	First
Sputnik 1	October 1957	USSR	Artificial satellite in space
Vostok 1	April 1961	USSR	Human in space (Yuri Gagarin)
Mercury	February 1962	US	US manned space flight (John Glenn)
Voskhod 2	March 1965	USSR	Space walk (Alexsey Leonov)
Luna 9	February 1966	USSR	Soft landing on the Moon
OAO-2	December 1968	US	Space telescope launched into orbit
Apollo 11	July 1969	US	Human on the Moon (Neil Armstrong)
Venera 7	December 1970	USSR	Soft landing on Venus
Salyut 1	April 1971	USSR	Space station launched
Mars 3	December 1971	USSR	Soft landing on Mars
Pioneer 10	December 1973	US	Flyby of Jupiter
Voyager 2	August 1977	US	Spacecraft launched to fly by Jupiter, Saturn, Uranus, Neptune
Rosetta	August 2014	ESA	Probe goes into orbit around a comet
New Horizons	July 2015	USA	Fly-by of Pluto
Hayabusa 2	September 2018	Japan	Operational rover on an asteroid (Ryugu)
Chang'e 4	January 2019	China	Soft landing on far side of Moon

(ESA = European Space Agency)

Earth Sciences
Structure of Earth

Earth has taken about 4.6 billion years to develop. It is a remarkably varied world, especially on the thin outer crust that covers its rock mantle and metallic core (see pp.352–53). The processes that shaped Earth are ongoing. The influence of forces such as shifting tectonic plates is usually imperceptible, except over geological time, but occasionally tectonic movement causes sudden, violent events, including earthquakes.

Geological timescale

Earth's timescale covers thousands of millions of years. Study of the rocks and fossils of the planet's crust has made it possible to divide this time into segments (see p.191). The divisions are constantly being refined.

Era	Period	Epoch	Start (million years ago)
Cenozoic	Neogene	Holocene	0.01
		Pleistocene	2.58
		Pliocene	5.33
		Miocene	23
	Paleogene	Oligocene	33.9
		Eocene	56
		Paleocene	66
Mesozoic	Cretaceous		145
	Jurassic		201
	Triassic		252
Paleozoic	Permian		299
	Carboniferous (divided into Lower Mississippian and Upper Pennsylvanian in United States)		359
	Devonian		419
	Silurian		444
	Ordovician		485
	Cambrian		541
Precambrian	Formation of Earth		4,554

Earth's minerals

Minerals are the basic materials of Earth's rocks. All of them have an internal crystal structure composed of atoms arranged in a regular, repeating, three-dimensional pattern. A free-growing crystal forms a geometric shape with flat planes, or faces, that are often arranged symmetrically. Patterns of crystal symmetry are classified into six main groups, or crystal systems, which are defined by the specific angles of their faces. These systems are: cubic; hexagonal/trigonal; monoclinic; orthorhombic; tetragonal; and triclinic.

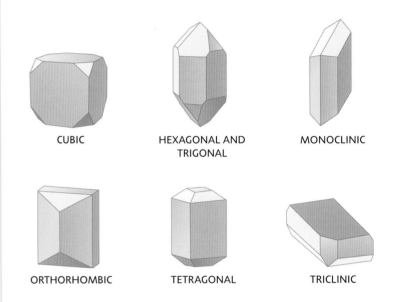

CUBIC

HEXAGONAL AND TRIGONAL

MONOCLINIC

ORTHORHOMBIC

TETRAGONAL

TRICLINIC

Mineral classification

There are more than 4,000 known natural minerals, although only about 100 of these are abundant. Minerals are classified according to their chemical composition, and are commonly divided into the major groups listed below. Each grouping contains many minerals.

Group	Approx. no. of minerals	Examples
Sulfides	600	Pyrite, galena
Silicates	500	Olivine, quartz, feldspar, garnet
Oxides and hydroxides	400	Chromite, hematite
Phosphates and vanadates	400	Apatite, carnotite
Sulfates	300	Anhydrite, barite, gypsum
Carbonates	200	Calcite, aragonite, dolomite
Halides	140	Fluorite, halite, sylvite
Borates and nitrates	125	Borax, colemanite, kernite, nitratine
Molybdates and tungstates	42	Wulfenite, wolframite
Native elements	20	Gold, platinum, copper, sulfur, carbon

Earth's rocks

Rocks are grouped according to the way they have formed. There are three main types—igneous, sedimentary, and metamorphic—common examples of which are listed here. Igneous rocks form from molten magma; sedimentary rocks accumulate from layers of sediment; metamorphism occurs when heat or pressure converts one rock type into another.

Igneous	Sedimentary	Metamorphic
Syenite	Limestone	Slate
Trachyte	Dolomite	Phyllite
Granite	Evaporite	Schist
Rhyolite	Breccia	Gneiss
Diorite	Conglomerate	Hornfels
Andesite	Sandstone	Marble
Gabbro	Siltstone	Quartzite
Basalt	Mudstone	Migmatite
Peridotite	Shale	Amphibolite
Picrite	Coal	Mylonite

The "rock cycle"

All types of rock—igneous, metamorphic, and sedimentary—form continuously both within Earth and on its surface. Many of these rock materials are also "recycled" over geological time, with igneous and metamorphic rocks from Earth's interior being broken down by surface processes, deposited as sediments, and buried. The "rock cycle" (see below and pp.184–185) is a simplified illustration of some of these interlinked processes and transformations.

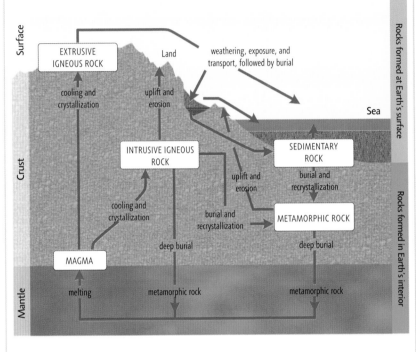

Tectonic plates

Earth's outermost rocks are broken into seven major tectonic plates and about twelve smaller ones (see pp.356–57). Over time the plates have moved apart (diverged) and joined together (converged) to shift continents, open and close oceans, and form mountains. At transform boundaries, plates slide past one another along fault planes.

PLATE NAMES:

1 North American plate
2 Pacific plate
3 Nazca plate
4 South American plate
5 African plate
6 Arabian plate
7 Eurasian plate
8 Antarctic plate
9 Indo-Australian plate

KEY TO PLATE BOUNDARIES:

Converging
Diverging
Transform
Uncertain
Boundary between African and Arabian plates

Earthquake zones

Movement of Earth's crust and tectonic plates produces stress and fracture of its rocks along fault planes. The buildup of stress in these brittle crustal rocks results in sudden failure and the release of massive amounts of energy as earthquakes (see pp.359–61). The shock waves are transmitted through the rocks, often over great distances; those of major quakes reach the other side of the world. Earth's major earthquake zones follow the boundaries where tectonic places move relative to one another.

KEY TO PLATE BOUNDARIES:

Converging
Diverging
Transform
Uncertain
Earthquake zone

Earthquake severity scales

The size of an earthquake can be measured by magnitude and by intensity (see pp.360–61). Two traditionally used scales are the Mercalli intensity scale (which measures the effects of an earthquake on the surface, at 12 levels) and the Richter scale (which measures the amount of energy released). However, in scientific use these have largely been superseded by the moment magnitude scale. Hundreds of earthquakes are generated weekly at plate boundaries around the world, mostly on the ocean floor. Few are large enough to cause widespread damage.

THE MERCALLI INTENSITY SCALE		
I	Instrumental	No Earth movement felt; detectable only by seismograph
II	Feeble	Movement felt only by people at rest or on upper floors of buildings
III	Slight	Noticeable indoors, especially on upper floors; hanging objects swing
IV	Moderate	Felt indoors by most people and outdoors by a few; hanging objects swing; dishes, doors, and windows rattle; sensation of a heavy truck hitting the walls; parked cars rock
V	Rather strong	Felt by nearly all; sleeping people waken; doors swing; small objects move; dishes and glassware break; liquids may spill
VI	Strong	Felt by all; difficult to walk; furniture moves; objects fall from shelves and pictures from walls; plaster may crack; little structural damage
VII	Very strong	Difficult to stand; drivers feel cars shaking; furniture may break; bricks or chimney pots may fall; some structural damage depending on construction of building
VIII	Destructive	Car drivers cannot steer; considerable damage in poorly constructed buildings or tall structures such as factory chimneys; branches broken from trees; wet or sloping ground may crack; water levels in wells may be affected
IX	Ruinous	Considerable structural damage to all building types; houses may shift off their foundations; the ground cracks and reservoirs are damaged seriously
X	Disastrous	Most buildings destroyed; bridges and dams seriously damaged; large cracks in ground; major landslides, rail tracks bend
XI	Very disastrous	Few structures remain standing; pipelines and rail tracks destroyed
XII	Catastrophic	Total destruction; upheavals of ground and rock

RICHTER SCALE: TYPICAL RANGES	
‹3.5	Recordable, but not generally felt
3.5–5.4	Felt by many, but causes little or no damage
5.5–6.0	May be slight damage to well-constructed buildings; more severe damage to poorly constructed buildings
6.1–6.9	Can cause severe and widespread damage up to a range of 62.5 miles (100 km)
7.0–7.9	Major effects, causing severe damage over larger areas
›8.0	Can cause devastation over several hundred miles (kilometers)

The land and oceans

The topography of Earth's thin outer layer, or crust, ranges over 12 miles (20 km) in elevation, from high mountains across plains into deep valleys on both land and sea. Although significant for life and the environment, the variation is small on a global scale. The upstanding landmasses and their surrounding shelf seas are formed of continental crust, while the deep oceans are floored by relatively dense oceanic crust. Gravity allows water to fill the ocean basins, but water is constantly recycled from the sea through the atmosphere onto land and back to the sea.

Earth's surface

About 70 percent of Earth's total surface is covered in water. More than half (53.5 percent) of this water covered area lies at depths ranging between approximately 1¾–3¾ miles (3 and 6 km). About 20.9 percent of the land surface lies between sea level and 0.6 miles (1 km) in altitude. The highest point on Earth's surface is Mount Everest at 29,032 ft (8,849 m) above sea level; the lowest point is the bottom of the Mariana deep-sea trench, located in the floor of the Pacific Ocean, at a depth of 36,201 ft (11,034 m) below sea level.

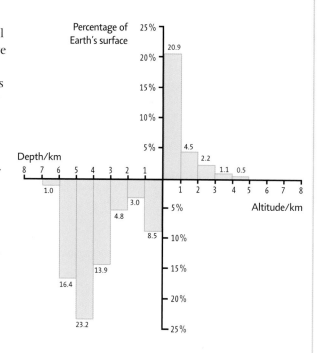

Continents of the world

Most geographers recognize the seven continents named here, although some combine Europe and Asia as Eurasia, and regard North and South America as one American continent.

Continent	Size sq miles (km²)	Estimated population
Asia	17,179,000 (44,493,000)	4,667,000,000
Africa	11,690,000 (30,970,000)	1,363,000,000
North America	9,442,000 (24,454,000)	370,000,000
Central/South America	6,887,000 (17,838,000)	614,000,000
Antarctica	6,000,000 (15,500,000)	4,000
Europe	3,956,000 (10,245,000)	748,000,000
Australia/Oceania	3,454,000 (8,945,000)	43,000,000

The world's longest rivers

The lengths given here are approximations, as exact measurement of a river is difficult. Multiple tributaries often make a river's source hard to pinpoint and the outflow may follow numerous channels.

River	Length miles (km)	Location
Nile	4,160 (6,690)	North and East Africa
Amazon	4,080 (6,570)	South America
Mississippi–Missouri	3,740 (6,020)	United States of America
Chang Jiang (Yangtze)	3,720 (5,980)	China
Yenisey	3,650 (5,870)	Russia
Amur	3,590 (5,780)	Northeast Asia
Ob–Irtysh	3,360 (5,410)	Russia
Parana	3,030 (4,880)	South America
Huang He (Yellow)	3,010 (4,840)	China
Congo	2,880 (4,630)	Central Africa
Lena	2,730 (4,400)	Russia
Mackenzie	2,630 (4,240)	Northwest Canada
Mekong	2,600 (4,180)	Southeast Asia
Niger	2,550 (4,100)	West Africa

The world's highest mountain peaks

Fourteen peaks rise over 26,000 ft (8,000 m) above sea level, all of them located in Asia. The majority are in the Himalayas, the highest and one of the most recently formed mountain ranges on Earth.

Mountain	Range	Location	Height feet (meters)
Everest	Himalayas	Nepal/Tibet	29,032 (8,849)
K2	Karakoram	Pakistan/China	28,250 (8,611)
Kangchenjunga	Himalayas	Nepal/India	28,169 (8,586)
Lhotse	Himalayas	Nepal/Tibet	27,940 (8,516)
Makalu	Himalayas	Nepal/Tibet	27,766 (8,463)
Cho Oyu	Himalayas	Nepal/Tibet	26,906 (8,201)
Dhaulagiri	Himalayas	Nepal	26,795 (8,167)
Manaslu	Himalayas	Nepal	26,781 (8,163)
Nanga Parbat	Himalayas	Pakistan	26,660 (8,125)
Annapurna	Himalayas	Nepal	26,545 (8,091)
Gasherbrum I	Karakoram	Pakistan/China	26,470 (8,068)
Broad Peak	Karakoram	Pakistan/China	26,400 (8,047)
Gasherbrum II	Karakoram	Pakistan/China	26,360 (8,035)
Shishna Pangma	Himalayas	Tibet	26,289 (8,013)

Ocean currents

Ocean water is in constant circulation and transfers heat from the surface to the depths. Through interaction with the atmosphere, it greatly influences global climates (see pp.212–13). Surface currents are driven by winds but their patterns are modified by the effects of Earth's rotation on its axis, which causes circular water movements called gyres. These gyres rotate clockwise in the Northern Hemisphere and counterclockwise in the Southern. Specific components of these gyres are called boundary currents. The boundary currents on the eastern side of oceans are predominantly cold and move toward the equator; those on the western sides tend to be warm and move away from the equator. The present global circulation pattern is also partly controlled by the configuration of the ocean basins and surrounding coastlines. In the past, the distribution of oceans and continents, and therefore current circulation, was very different.

Warm ocean currents

Cold ocean currents

CURRENTS:

1 North Pacific gyre
2 South Pacific gyre
3 Humboldt current
4 Gulf Stream

5 North Atlantic gyre
6 South Atlantic gyre
7 Antarctic circumpolar current
8 Aghulas current

9 South Indian gyre
10 North Pacific gyre

Major mountain ranges

The world's major mountain ranges, such as the Andes, Himalayas, Alps, and Rockies, are comparatively young, having formed within the last few hundred million years. They are situated along the boundaries of tectonic plates that collided in the geological past (see pp.358–59). The ranges were formed as immense pressures caused by the convergence of the plates deformed and uplifted the rocks high above the surrounding land. This process of uplift is still continuing.

RANGES:

1 Alaska Range	9 Ethiopian Highlands
2 Rocky Mountains	10 Caucasus
3 Appalachians	11 Ural Mountains
4 Andes	12 Tien Shan
5 Pyrenees	13 Himalayas
6 Atlas Mountains	14 Great Dividing Range
7 Alps	15 Transantarctic Mountains
8 Drakensberg	

Climate and weather

Earth's atmosphere, surface waters, and landmasses combine to produce the characteristic climates and weather patterns of regions around the world. In particular, oceans, which cover about 70 percent of Earth's surface, have a major role in the regulation of climates, transporting heat from the equatorial regions to higher, colder latitudes and releasing water vapor into the atmosphere.

Climate regions

Earth's landmasses can be divided into climate regions, based on average temperatures and rainfall, and type of vegetation. Tropical areas are hot all year round and often very wet, while polar regions and the tops of high mountains are characterized by extreme cold. In the temperate regions between the poles and the tropics, climates are more moderate. Areas classified as desert usually have an annual rainfall of less than 10in (25cm). Climate is influenced not only by latitude but also by factors such as height above sea level and distance from the ocean.

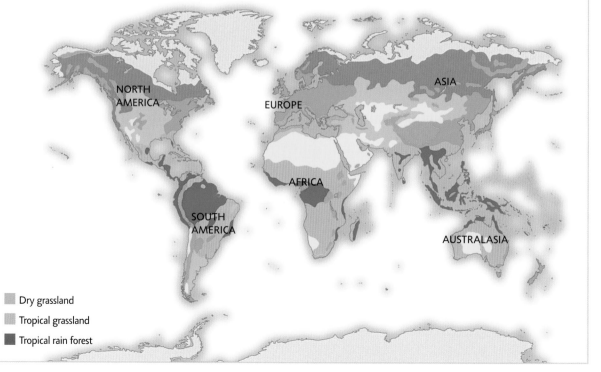

CLIMATE REGIONS:

Polar and tundra	Temperate forest	Dry grassland
Boreal forest	Mediterranean	Tropical grassland
Mountain	Desert	Tropical rain forest

Weather map symbols

The symbols shown below are an internationally recognized system used for classifying types of weather. They were devised by the World Meteorological Organization and represent varying intensities of precipitation and other conditions.

Symbol	Description	Symbol	Description
●	Intermittent light rain	∿	Freezing rain
●●	Continuous light rain	∿	Freezing drizzle
● ●	Intermittent moderate rain	℞	Thunderstorm
●● ●	Continuous moderate rain	⚇	Tornado
●●● ●	Continuous heavy rain	≡	Fog
▽	Light rain shower	∞	Haze
▽	Moderate rain shower	▲▲	Stationary front
▽	Heavy rain shower	▼▼▼	Cold front
✳	Snow	▲▲▲	Occluded front
❟	Drizzle	▲▲▲	Warm front

The Beaufort wind force scale

Beaufort's scale was originally devised for sailors as a means of judging wind strength by observations of sea disturbance and the way sails, and therefore steerage, were affected. The scale was later modified for use on land, substituting features such as trees and cars.

Beaufort no.	Wind speed mph (kph)	Description
0	0–1 (0–2)	Calm—smoke rises vertically, air feels still
1	1–3 (2–6)	Light air—smoke drifts
2	4–7 (7–11)	Slight breeze—wind detectable on face, some leaf movement
3	8–12 (12–19)	Gentle breeze—leaves and twigs move gently
4	13–18 (20–29)	Moderate breeze—loose paper blows around
5	19–24 (30–39)	Fresh breeze—small trees sway
6	25–31 (40–50)	Strong breeze—difficult to use an umbrella
7	32–38 (51–61)	High wind—whole trees bend
8	39–46 (62–74)	Gale—twigs break off trees, walking into wind is difficult
9	47–54 (75–87)	Severe gale—roof tiles blow away
10	55–63 (88–101)	Whole gale—trees break and are uprooted
11	64–74 (102–119)	Storm—damage is extensive, cars overturn
12	75+ (120+)	Hurricane—widespread devastation

Where the wind blows

At Earth's poles, the air is at high pressure and low temperature; at the equator, air is at low pressure but higher temperature. This, together with the rotation of Earth on its axis, creates a predictable pattern of warm and cool winds around the globe (see pp.206–207). Continental landmasses and high mountains also influence wind patterns, such as the monsoon winds over southern Asia. On the equator there is an area known as the doldrums, once a trap for sailing ships, where there is very little wind. In certain latitudes, a combination of disturbances in wind speed and direction and high sea surface temperatures can create tropical cyclones, also known as hurricanes or typhoons. In a cyclone, the winds move in a spiral pattern and may reach speeds of more than 125mph (200kph).

Symbol	Description	No.		No.	
🌀	Tropical cyclones	1	Northeast trades	4	Westerlies
→	Warm winds	2	Southeast trades	5	Southwest monsoon
→	Cool winds	3	Polar easterlies	6	Doldrums

Biology
Variety of Life

Living things exist in many forms. The simplest are microscopic single cells, but others are multicelled and can grow to giant size—such as trees or whales. All organisms need energy to survive, and the basic categories of life mark out fundamental differences in the way different organisms achieve this—for example, plants make food by photosynthesis, and animals consume food acquired from their surroundings.

Scientific classification

Scientists classify organisms into groups according to similarities that reflect evolutionary relationships. All life on Earth is classified into major categories called kingdoms—such as plants and animals. Each kingdom contains subdivisions, or phyla, which in turn are split further.

The chart below shows the hierarchy from the rank of kingdom to the rank of species—the basic unit of classification. Not all life-forms are categorized easily and scientists may prefer to classify organisms in different ways according to the evidence available.

Classification levels

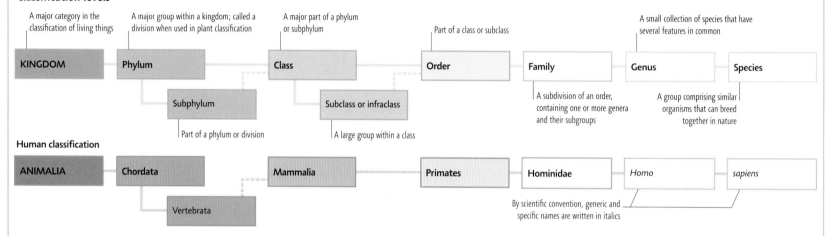

- **KINGDOM** — A major category in the classification of living things
- **Phylum** — A major group within a kingdom; called a division when used in plant classification
- **Subphylum** — Part of a phylum or division
- **Class** — A major part of a phylum or subphylum
- **Subclass or infraclass** — A large group within a class
- **Order** — Part of a class or subclass
- **Family** — A subdivision of an order, containing one or more genera and their subgroups
- **Genus** — A group comprising similar organisms that can breed together in nature
- **Species** — A small collection of species that have several features in common

Human classification

ANIMALIA — Chordata — Vertebrata — Mammalia — Primates — Hominidae — *Homo* — *sapiens*

By scientific convention, generic and specific names are written in italics

Kingdoms of Life

Modern biology has revealed that microbial organisms are far more diverse than previously supposed; microbes that look similar under a microscope can, in fact, be far more distantly related than, for instance, plants and animals. Biologists now think that there are more than a dozen kingdoms of living organisms—only two of which are plants and animals. The remainder include many kinds of microbes, such as algae and amoebas. Two of these groups—the Archaea and Bacteria—are so fundamentally different from other life forms that biologists have erected a new classification rank above the kingdom level—called the domain—to properly illustrate their relationships.

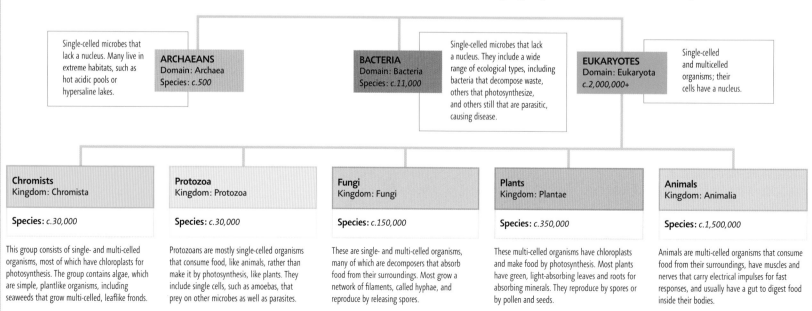

ARCHAEANS
Domain: Archaea
Species: *c.500*

Single-celled microbes that lack a nucleus. Many live in extreme habitats, such as hot acidic pools or hypersaline lakes.

BACTERIA
Domain: Bacteria
Species: *c.11,000*

Single-celled microbes that lack a nucleus. They include a wide range of ecological types, including bacteria that decompose waste, others that photosynthesize, and others still that are parasitic, causing disease.

EUKARYOTES
Domain: Eukaryota
c.2,000,000+

Single-celled and multicelled organisms; their cells have a nucleus.

Chromists
Kingdom: Chromista
Species: *c.30,000*

This group consists of single- and multi-celled organisms, most of which have chloroplasts for photosynthesis. The group contains algae, which are simple, plantlike organisms, including seaweeds that grow multi-celled, leaflike fronds.

Protozoa
Kingdom: Protozoa
Species: *c.30,000*

Protozoans are mostly single-celled organisms that consume food, like animals, rather than make it by photosynthesis, like plants. They include single cells, such as amoebas, that prey on other microbes as well as parasites.

Fungi
Kingdom: Fungi
Species: *c.150,000*

These are single- and multi-celled organisms, many of which are decomposers that absorb food from their surroundings. Most grow a network of filaments, called hyphae, and reproduce by releasing spores.

Plants
Kingdom: Plantae
Species: *c.350,000*

These multi-celled organisms have chloroplasts and make food by photosynthesis. Most plants have green, light-absorbing leaves and roots for absorbing minerals. They reproduce by spores or by pollen and seeds.

Animals
Kingdom: Animalia
Species: *c.1,500,000*

Animals are multi-celled organisms that consume food from their surroundings, have muscles and nerves that carry electrical impulses for fast responses, and usually have a gut to digest food inside their bodies.

The plant kingdom

Plants are organisms that produce their food by photosynthesis, using the pigment chlorophyll in their leaves to absorb energy from sunlight (see pp.154–55). Chemical reactions use this energy to combine water and carbon dioxide to make glucose, which fuels growth. The plant kingdom contains over 400,000 species. It can be divided into two major groups: vascular plants, which contain tissues specialized to carry water and food, and nonvascular plants, which do not. Most commercially valuable species are member of one phylum—flowering plants.

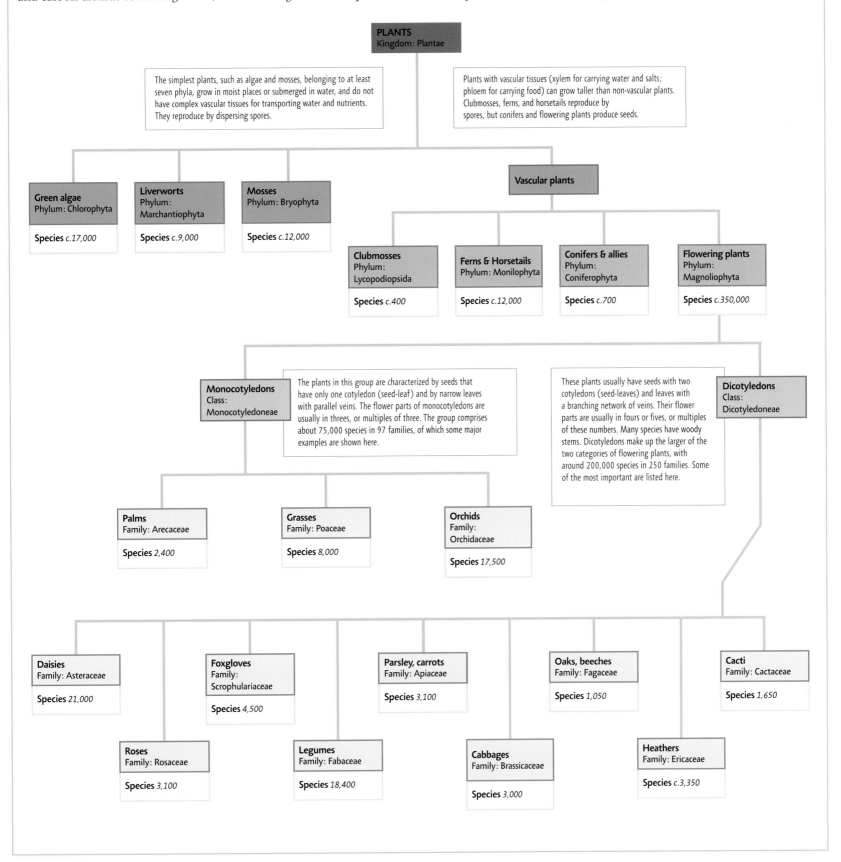

The animal kingdom

Animals are multicellular life-forms that obtain energy by eating food, including other living organisms or their remains. One of the defining features of an animal is mobility. Most are able to move from one place to another, or to move part of their body to feed or protect themselves. More than one million animal species have been catalogued, but a 2011 study estimated the true total to be around 7.7 million. There are more than 30 recognized major animal groupings, or phyla, whose members share a basic body plan, but the vast majority of animal species belong to those phyla described below. Any of the groupings is liable to change in response to shifts in current scientific opinion.

ANIMALS
Kingdom: Animalia

All animals other than Chordates belong to an informal category called Invertebrates, which includes over nine-tenths of all animal species. Invertebrates lack a backbone; many are soft-bodied and live in water, or in damp habitats. One phylum, the arthropods, has been outstandingly successful both in water and on land.

Cnidarians
Phylum: Cnidaria

Species 10,000
Mainly marine

Corals; jellyfish; sea anemones; hydras

Annelid worms
Phylum: Annelida

Species 19,000
Aquatic and terrestrial

Earthworms; leeches; lugworms and other marine worms

Sponges
Phylum: Porifera

Species 8,000
Mainly marine

Nematode worms
Phylum: Nematoda

Species 25,000

Small phyla
Ctenopora, Rotifera, Nemertea

Species 3,000+

Comb jellies; rotifers; nemertean worms

Echinoderms
Phylum: Echinodermata

Species 21,000
Marine

Brittle stars; sea urchins; sea cucumbers; starfish; sea lilies and feather stars

Bryozoans
Phylum: Bryozoa

Species 6,000
Mainly marine

Colony-forming animals

Flatworms
Phylum: Platyhelminthes

Species 30,000

Free-living flatworms; flukes; tapeworms

Mollusks
Phylum: Mollusca

Species 117,000
Aquatic and terrestrial

Chitons; slugs and snails; clams, scallops; tusk or tooth shells; octopuses, squids, and cuttlefish

Arthropods
Phylum: Arthropoda

This large phylum contains animals that have a jointed body covered by an external skeleton and divided into segments. The skeleton supports and protects the body and, on land, it helps to prevent it from drying out.

Millipedes
Class: Diplopoda

Species 12,000

Centipedes
Class: Chilopoda

Species 3,000

Crustaceans
Class: Crustacea

Species 65,000
Mainly marine

Water fleas; wood lice; crabs and lobsters

Arachnids
Class: Arachnida

Species 70,000
Mainly terrestrial

Scorpions; spiders; mites; harvestmen

Insects
Class: Insecta

Species 1,000,000+
Mainly terrestrial

Springtails; silverfish and bristletails; dragonflies and damselflies; butterflies and moths; grasshoppers and crickets; earwigs; stick insects; cockroaches; mantids; termites; lice; true bugs; thrips; lacewings; beetles; flies; fleas; bees, wasps, and ants

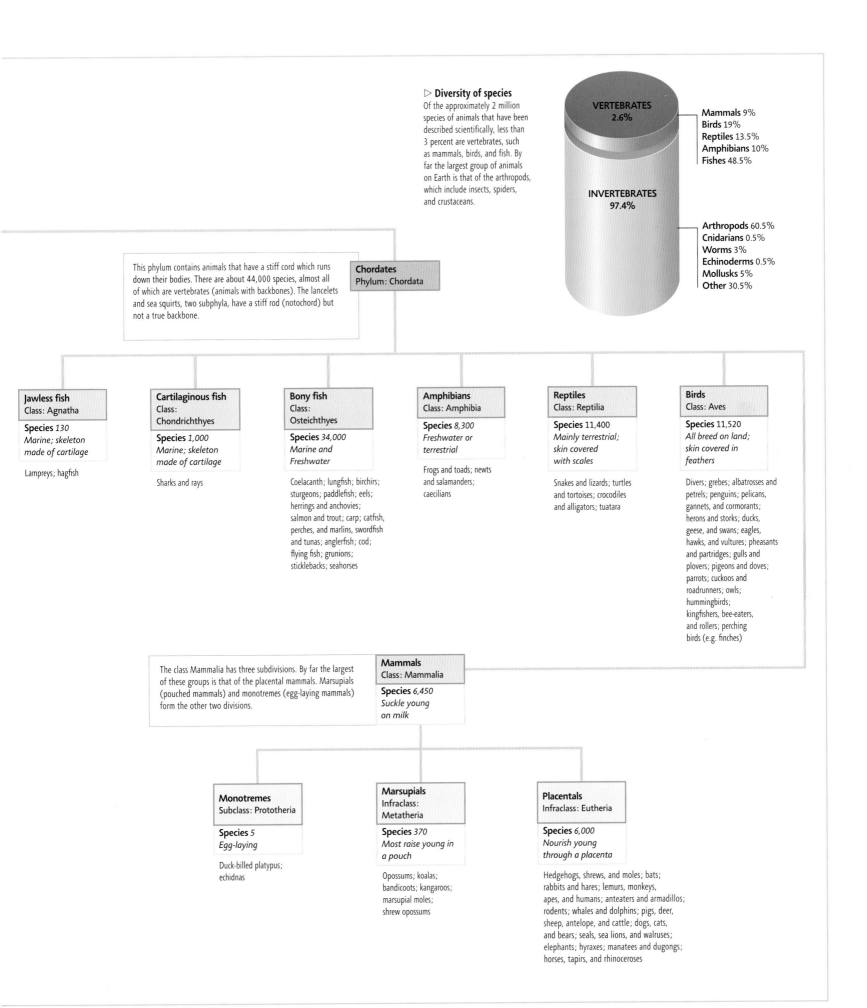

▷ **Diversity of species**
Of the approximately 2 million species of animals that have been described scientifically, less than 3 percent are vertebrates, such as mammals, birds, and fish. By far the largest group of animals on Earth is that of the arthropods, which include insects, spiders, and crustaceans.

VERTEBRATES 2.6%

Mammals 9%
Birds 19%
Reptiles 13.5%
Amphibians 10%
Fishes 48.5%

INVERTEBRATES 97.4%

Arthropods 60.5%
Cnidarians 0.5%
Worms 3%
Echinoderms 0.5%
Mollusks 5%
Other 30.5%

This phylum contains animals that have a stiff cord which runs down their bodies. There are about 44,000 species, almost all of which are vertebrates (animals with backbones). The lancelets and sea squirts, two subphyla, have a stiff rod (notochord) but not a true backbone.

Chordates
Phylum: Chordata

Jawless fish
Class: Agnatha

Species 130
Marine; skeleton made of cartilage

Lampreys; hagfish

Cartilaginous fish
Class: Chondrichthyes

Species 1,000
Marine; skeleton made of cartilage

Sharks and rays

Bony fish
Class: Osteichthyes

Species 34,000
Marine and Freshwater

Coelacanth; lungfish; birchirs; sturgeons; paddlefish; eels; herrings and anchovies; salmon and trout; carp; catfish, perches, and marlins, swordfish and tunas; anglerfish; cod; flying fish; grunions; sticklebacks; seahorses

Amphibians
Class: Amphibia

Species 8,300
Freshwater or terrestrial

Frogs and toads; newts and salamanders; caecilians

Reptiles
Class: Reptilia

Species 11,400
Mainly terrestrial; skin covered with scales

Snakes and lizards; turtles and tortoises; crocodiles and alligators; tuatara

Birds
Class: Aves

Species 11,520
All breed on land; skin covered in feathers

Divers; grebes; albatrosses and petrels; penguins; pelicans, gannets, and cormorants; herons and storks; ducks, geese, and swans; eagles, hawks, and vultures; pheasants and partridges; gulls and plovers; pigeons and doves; parrots; cuckoos and roadrunners; owls; hummingbirds; kingfishers, bee-eaters, and rollers; perching birds (e.g. finches)

The class Mammalia has three subdivisions. By far the largest of these groups is that of the placental mammals. Marsupials (pouched mammals) and monotremes (egg-laying mammals) form the other two divisions.

Mammals
Class: Mammalia

Species 6,450
Suckle young on milk

Monotremes
Subclass: Prototheria

Species 5
Egg-laying

Duck-billed platypus; echidnas

Marsupials
Infraclass: Metatheria

Species 370
Most raise young in a pouch

Opossums; koalas; bandicoots; kangaroos; marsupial moles; shrew opossums

Placentals
Infraclass: Eutheria

Species 6,000
Nourish young through a placenta

Hedgehogs, shrews, and moles; bats; rabbits and hares; lemurs, monkeys, apes, and humans; anteaters and armadillos; rodents; whales and dolphins; pigs, deer, sheep, antelope, and cattle; dogs, cats, and bears; seals, sea lions, and walruses; elephants; hyraxes; manatees and dugongs; horses, tapirs, and rhinoceroses

The human species

The human body took a few million years to evolve once we branched away from our closest relatives, but essentially it works in the same way as that of any other mammal. The body's systems are made up of tissues and organs designed for specific functions. A system cannot work alone: each one depends on the others. For example, the skeletal system needs muscles for mobility and blood to provide oxygen and nutrients.

Human ancestors

The evolution of humans follows a complicated, many-branched family tree. We did not simply advance in an unbroken line, progressing from the first hominid to stand upright and walk on two legs to modern man. At one time in the distant past, several closely related species of hominids co-existed. All of them except our direct ancestors died out, leaving *Homo sapiens* as the sole representative of the human species.

Years ago	Key event	Evidence	Discovery site
7 million	Existence of a possible hominid	The "Chad" skull (*Sahelanthropus*)	Chad, Central Africa
6 million	Possible earliest date of upright, bipedal walking hominid	Leg bones (*Orrorin*)	Tugen Hills, Kenya
5.8 million	Earliest evidence of first known bipedal hominid	Toe bone and other fragments (*Ardipithecus kadabba*)	Middle Awash, Ethiopia
4.4 million	First evidence for an omnivorous diet in hominids	Numerous skeletal remains	Middle Awash, Ethiopia
4.2 million	Hominid (*Australopithecus anamensis*) capable of climbing and bipedal walking	Teeth and limb fragments	Lake Turkana, Kenya
3.85 million	Abundant fossils and footprints confirm bipedalism and climbing abilities	Footprints of *Australopithecus afarensis*	Laetoli, northern Tanzania
3.2 million	"Lucy" (*Australopithecus afarensis*) found	Partial skeleton of bipedal hominid	Hadar, Ethiopia
3.2 million	First australopithecine to be found, possible ancestor for genus *Homo*	Skull and other fragments	4 sites Southern Africa
2.6 million	Earliest robust plant-eating australopithecine (*Australopithecus aethiopicus*)	Skulls and other fragments	Ethiopia; northern Tanzania
2.3 million	One of the earliest humans (*Homo habilis*) appears and may have made stone tools	Skull and other fragments	Olduvai Gorge, Tanzania
2 million	Bipedal *Australopithecus sediba* combines human and australopithecine characters	Near complete skeleton	Malapa, South Africa
1.9 million	Oldest humans (*Homo erectus*) outside Africa		Dmanisi, Georgia
1.9 million	*Homo erectus*, first hominid to leave Africa, emerges	Nearly complete skeleton	Kenya
1.8 million	Appearance of plant-eating robust australopithecine (*Australopithecus robustus*)	Jaw and skull fragments	South Africa
1.6 million	Earliest evidence of *Homo* (*H. erectus*) in Asia	Skull and limb bone	Java, Indonesia
1.5 million	"Turkana Boy," most complete specimen of *Homo erectus*, found	Nearly complete skeleton	Southern Africa
1.2 million	*Australopithecus robustus* known to be extinct in Africa		
900,000	Oldest European human remains (*Homo antecessor* or *Homo heidelbergensis*)	Bones of 6 individuals	Gran Dolina, northern Spain
700,000	Emergence of *Homo heidelbergensis*	Jawbone	near Heidelberg, Germany
335,000	Appearance of a new human species (*Homo naledi*)	Bones of some 15 individuals	Rising Star Cave, South Africa
300,000	Earliest *Homo sapiens* in Africa	Partial remains of 5 individuals	Jebel Irhoud, Morocco
400,000	Emergence of *Homo neanderthalensis* in Europe	Bones of 29 individuals	Sima de los Huesos, northern Spain
195,000	Oldest *Homo sapiens* in East Africa	Two skull caps	Omo Kibish, Ethiopia
260,000	Possible *Homo sapiens* in southern Africa	Skull	Florisbad, South Africa
210,000	Possible arrival of *Homo sapiens* in southern Europe	Skull fragment (Apidima 1)	Apidima Cave, southern Greece
100,000	*Homo sapiens* reaches Near East	Skeletons at grave site	Jebel Qafzeh, Israel
100,000	Appearance of small human species (*Homo floresiensis*—the "Hobbit")	Near complete skeleton, bones, and stone tools	Flores, Indonesia
90,000	Appearance of a new human species, the Denisovans	DNA from finger bone	Denisova Cave, Siberia
60,000	*Homo sapiens* reaches China	Skull and teeth	Shaanxi and Guizhou Provinces, China
55,000	*Homo sapiens* reaches Australia	Skull	Lake Mungo, South Australia
46,000	Oldest DNA evidence for modern humans in Europe	Artifacts and DNA from a tooth	Bacho Kiro Cave, Bulgaria
40,000	*Homo neanderthalensis* extinct as a discrete species	Reanalysis of radiocarbon dates	40 sites from Spain to Russia
24,000	*Homo sapiens* crosses into American continent	Artifacts	Bluefish Caves, Yukon, Canada
14,500	*Homo sapiens* reaches South America	Artifacts	Monte Verde, Chile
770	*Homo sapiens* reaches New Zealand	Artifacts	Papatowai, South Island, New Zealand

Skeletal system

The skeleton is a strong but flexible framework that provides support for all other body parts, and protects delicate organs such as the brain and heart (see pp.224–25). Bone is made of living tissue; it is a reservoir of important minerals, especially calcium and phosphorus, and also makes new cells for the blood. The bones are arranged symmetrically on either side of the body into two main divisions: the axial and appendicular skeletons. The axial skeleton consists of 80 bones, including the skull, spine, ribs, and breastbone. Attached to these bones is the appendicular skeleton, which has 126 bones and provides the body with mobility. Bones in this division, which include those of the arms and legs, have many joints. Also part of the appendicular skeleton are the shoulder blades and pelvis, the linking structures between the limbs and the core of the body.

COMPONENTS Skull, spine, ribs, and breastbone (axial skeleton); limb bones, shoulders, and hips (appendicular skeleton); ligaments

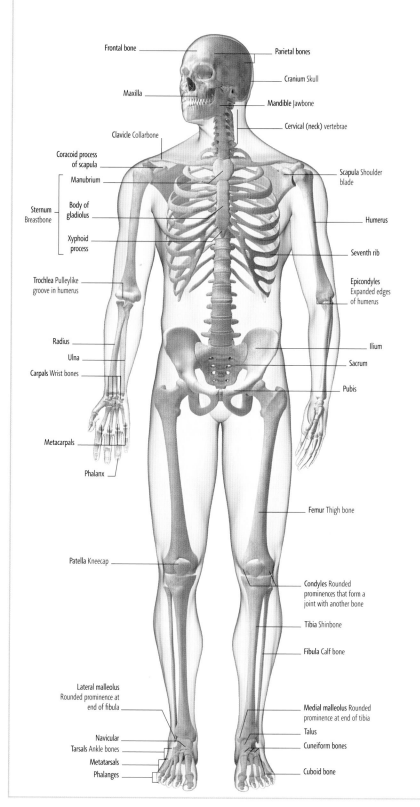

Frontal bone
Parietal bones
Cranium Skull
Maxilla
Mandible Jawbone
Cervical (neck) vertebrae
Clavicle Collarbone
Coracoid process of scapula
Manubrium
Scapula Shoulder blade
Sternum Breastbone
Body of gladiolus
Xyphoid process
Humerus
Seventh rib
Trochlea Pulleylike groove in humerus
Epicondyles Expanded edges of humerus
Radius
Ulna
Carpals Wrist bones
Ilium
Sacrum
Pubis
Metacarpals
Phalanx
Femur Thigh bone
Patella Kneecap
Condyles Rounded prominences that form a joint with another bone
Tibia Shinbone
Fibula Calf bone
Lateral malleolus Rounded prominence at end of fibula
Medial malleolus Rounded prominence at end of tibia
Talus
Navicular
Tarsals Ankle bones
Cuneiform bones
Metatarsals
Phalanges
Cuboid bone

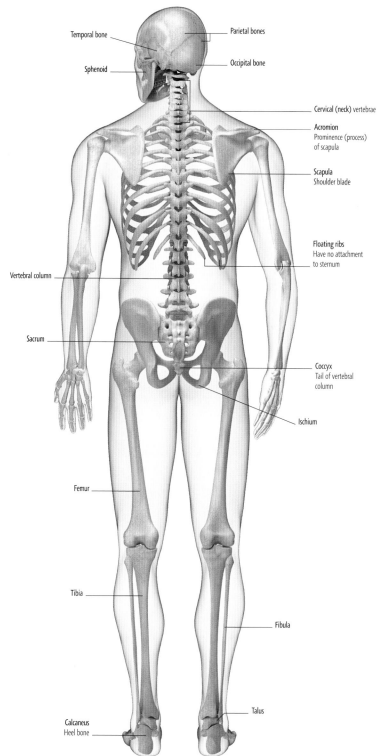

Temporal bone
Parietal bones
Sphenoid
Occipital bone
Cervical (neck) vertebrae
Acromion Prominence (process) of scapula
Scapula Shoulder blade
Floating ribs Have no attachment to sternum
Vertebral column
Sacrum
Coccyx Tail of vertebral column
Ischium
Femur
Tibia
Fibula
Talus
Calcaneus Heel bone

Muscular system

Muscles (see pp.224–25) account for about half of the body's total weight. There are three types: skeletal, cardiac, and smooth muscle. Over 600 skeletal muscles make up the flesh of the body. They come in a variety of sizes—from large triangular slabs such as the deltoid in the shoulder, to thin slivers, such as the sartorius that runs from the hip to the inside knee. The muscles cover the skeleton in overlapping layers. Those just below the skin and its underlying fat are described as superficial and those beneath them are the deep muscles. They enable the body to move by contracting quickly and powerfully to pull on the bones, to which they are linked by cordlike tendons. In addition, they maintain a steady tension that sustains posture. The skeletal muscles work in groups, usually under voluntary control. Both the other types of muscles are involuntary and work automatically: cardiac muscle controls heart rhythm; and smooth muscle controls such actions as movement of food through the digestive tract.

COMPONENTS Skeletal muscles (attached to bones); smooth muscles within organs; tendons; cardiac muscle of heart

Occipitofrontalis Raises eyebrows

Zygomaticus major Raises corners of mouth

Orbicularis oculi Closes eye

Mentalis Raises lower lip and wrinkles chin

Orbicularis oris Narrows mouth and purses lips

Scalenus Aids breathing and neck flexion

Sternocleidomastoid Tilts and twists neck

Pectoralis major Draws arm in toward body; rotates upper arm inward

Deltoid Raises arm away from body

Long head of triceps Straightens arm

Pectoralis minor Moves shoulder blade

External intercostal Elevates ribs

Biceps brachii Flexes forearm at elbow

Internal intercostal Pulls adjacent ribs together

Rectus abdominis Flexes spine and draws pelvis forward

Internal oblique abdominal Flexes and rotates trunk

External oblique abdominal Flexes and rotates trunk

Flexor carpi radialis Flexes hand at wrist

Brachioradialis Flexes arm at elbow

Iliopsoas Flexes thigh at hip

Flexor digitorum superficialis Flexes joints of hand and wrist

Pectineus Flexes and draws thigh in toward body

Abductor pollicis brevis Pulls thumb in to center palm

Adductor brevis Rotates thigh to center body

Sartorius Flexes thigh at hip and leg at knee joint, and turns thigh outward

Adductor longus Rotates thigh

Rectus femoris Flexes thigh at hip

Gracilis Flexes and rotates leg

Vastus lateralis Plays a part in extending knee

Peroneus brevis Flexes foot downward

Peroneus longus Flexes foot downward and turns it outward

Tibialis anterior Flexes foot upward and inward

Gastrocnemius Flexes foot downward

Soleus Flexes foot downward

Extensor digitorum longus Extends outer toes and helps flex foot upward

Extensor hallucis brevis Helps extend big toe

Retinaculum (ligamentous band) Stabilizes ankle joint

Extensor digitorum brevis Helps extend middle three toes

Extensor hallucis longus Extends big toe

Abductor hallucis Flexes big toe

Tendon of extensor hallucis longus

Tendons of extensor digitorum longus

Semispinalis capitis Extends head and neck and flexes them from side to side

Rhomboideus minor Helps retract shoulder blade

Splenius capitis Moves head, twists neck

Rhomboideus major Helps to retract shoulder blade

Levator scapulae Lifts and twists shoulder

Trapezius Rotates, elevates, and retracts shoulder blade

Infraspinatus Rotates arm and stabilizes shoulder

Deltoid Raises arm away from body

Long head of triceps Moves arm down to center body

Teres minor Lifts and twists arm, stabilizes shoulder

Lateral head of triceps Extends elbow

Teres major Lifts and twists arm, stabilizes shoulder

Latissimus dorsi Largest surface area of any muscle in body; moves arm, pulls shoulder back

Erector spinae Elevates and straightens spine — Spinalis / Longissimus / Iliocostalis

External oblique abdominal Supports abdominal wall and helps flex and rotate trunk

Internal abdominal oblique Supports abdominal wall and helps flex and rotate trunk

Gluteus minimus Raises thigh away from body at hip, rotates thigh; when walking, tilts pelvis

Extensor carpi ulnaris Bends and pulls wrist in to center body

Piriformis

Gemellus superior

All rotate and stabilize hip

Internal obturator

Flexor carpi ulnaris Flexes and pulls wrist in to center body

Extensor digitorum Extends finger joints

Gemellus inferior

Extensor retinaculum Ligamentous band supporting wrist

Quadratus femoris Rotates and stabilizes hip

Gluteus maximus Bulkiest muscle in body; pulls thigh back

Vastus lateralis Extends and stabilizes knee

Biceps femoris

Semitendinosus

Hamstrings Extend thigh at hip, flex and rotate leg

Plantaris Assists in knee flexion

Semimembranosus

Popliteus Flexes and turns leg

Gastrocnemius Main calf muscle; contracts to flex ankle and pull up heel

Tibialis posterior Main muscle in turning foot inward

Flexor digitorum longus Flexes and turns in foot, helps toes to grip

Soleus Flexes foot

Flexor hallucis longus The "push-off" muscle in walking

Achilles (calcaneal) tendon

Fibularis longus Flexes and turns foot outward

Abductor digiti minimi Moves little toe outward

Fibularis brevis Flexes and turns foot outward

Nervous system

The body's prime communication and coordination network, the nervous system (see pp.220–21), has two main parts. The central nervous system (CNS) comprises the brain and the spinal cord. Nerves branch from the CNS to form the peripheral nervous system (PNS), which extends to every organ and tissue of the body. A component of this, the autonomic nervous system (ANS), controls involuntary body functions such as heart rate.

COMPONENTS Brain; spinal cord; peripheral nerves; sense organs

Brain

Auriculotemporal nerve

Facial nerve

Axillary nerve

Phrenic nerve
Extends to diaphragm

Spinal ganglion One of many
nodules that send sensory
information to brain via
spinal cord

Spinal cord Part of central
nervous system, protected
by vertebral column

Sympathetic ganglia chain
Part of sympathetic
nervous system; conveys
stress signals to body

Gluteal nerve

Pudendal nerve

Common palmar
digital nerve

Deep branch of
ulnar nerve

Sciatic nerve

Common peroneal nerve

Tibial nerve

Cutaneous branch
of saphenous nerve

Deep peroneal nerve
Saphenous nerve
Interosseous nerve

Superficial peroneal nerve

Intermediate dorsal
cutaneous nerve

Lateral plantar nerve

Supraclavicular nerve

Brachial plexus

Vagus nerve

Lateral pectoral nerve

Musculo-cutaneous nerve

Intercostal nerves

Subcostal nerve

Median nerve
Radial nerve
Ulnar nerve
Obturator nerve
Iliohypogastric nerve

Ilioinguinal nerve

Filum terminale Fibrous
tissue connecting
spinal cord to coccyx

Femoral nerve

Muscular branch
of femoral nerve

Sciatic nerve

Common peroneal nerve

Tibial nerve

Deep peroneal nerve

Saphenous nerve

Interosseous nerve

Superficial peroneal nerve

Medial dorsal cutaneous nerve

Medial plantar nerve

Endocrine system

The body's endocrine system is composed of bodies of glandular tissue, such as the thyroid, and glands within certain organs, such as the testis, ovary, and heart. These glands are ductless, secreting hormones into the bloodstream; the hormones are chemical messengers that regulate many body processes—including fluid balance and urine production, growth and development, and sexual reproduction (see pp.274–75).

COMPONENTS Pituitary gland; pineal gland; hypothalamus; thyroid gland; thymus gland; heart; stomach; pancreas; intestines; adrenal glands; kidney; ovaries (female); testes (male)

Pineal gland Pea-sized gland in middle of brain; makes melatonin, a hormone important in body rhythms such as sleep–wake cycle; influences sexual development

Hypothalamus Cluster of nerve cells serving as main link between nerves and hormones; produces regulatory hormones that travel to pituitary gland

Pituitary gland Called the "master gland"; controls many other endocrine glands

Thyroid gland Controls rates of metabolism and energy use, and heart rate

Thymus gland Produces hormones involved in development of white blood cells that function in the immune system

Heart Produces hormone called atriopeptin, which reduces blood volume and pressure and helps to regulate fluid balance

Adrenal gland Cortex (outer layer) manufactures steroid hormones that help regulate metabolism and maintain fluid balance; medulla (inner layer) produces adrenaline

Kidney Secretes erythropoietin, which stimulates production of red blood cells in bone marrow

Stomach Makes hormones that stimulate production or release of digestive enzymes

Pancreas Bears clusters of cells, known as islets of Langerhans, which produce hormones (insulin and glucagon) that control blood glucose levels

Intestines Makes hormones that stimulate production or release of digestive enzymes

Ovary Manufactures female sex hormones estrogen and progesterone, which stimulate egg ripening and thickening of the uterine wall respectively

Testes Produce male sex hormones, including testosterone, that stimulate development of male sexual organs and production of sperm

Cardiovascular system

Also known as the circulatory system, the cardiovascular system comprises the heart, blood vessels—arteries, veins, capillaries, and many smaller vessels—and blood (see pp.90–91). The heart pumps oxygen-rich blood (red, below) around the body to all organs and tissues through a network of arteries. Deoxygenated blood (blue, below) returning to the heart through the veins carries away waste produced by cell processes.

COMPONENTS Heart; blood; major blood vessels (arteries and veins); minor blood vessels (arterioles, venules, and capillaries)

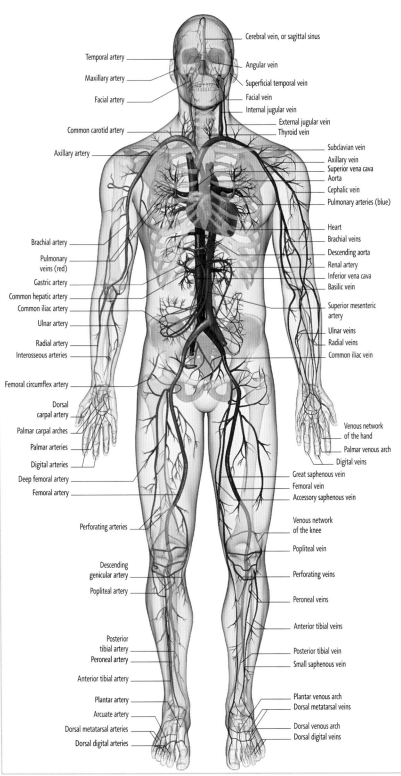

Cerebral vein, or sagittal sinus
Temporal artery
Angular vein
Maxillary artery
Superficial temporal vein
Facial artery
Facial vein
Internal jugular vein
External jugular vein
Thyroid vein
Common carotid artery
Subclavian vein
Axillary artery
Axillary vein
Superior vena cava
Aorta
Cephalic vein
Pulmonary arteries (blue)
Heart
Brachial artery
Brachial veins
Pulmonary veins (red)
Descending aorta
Renal artery
Gastric artery
Inferior vena cava
Common hepatic artery
Basilic vein
Common iliac artery
Ulnar artery
Superior mesenteric artery
Radial artery
Ulnar veins
Interosseous arteries
Radial veins
Common iliac vein
Femoral circumflex artery
Dorsal carpal artery
Palmar carpal arches
Venous network of the hand
Palmar arteries
Palmar venous arch
Digital arteries
Digital veins
Deep femoral artery
Great saphenous vein
Femoral artery
Femoral vein
Accessory saphenous vein
Perforating arteries
Venous network of the knee
Popliteal vein
Descending genicular artery
Perforating veins
Popliteal artery
Peroneal veins
Anterior tibial veins
Posterior tibial artery
Posterior tibial vein
Peroneal artery
Small saphenous vein
Anterior tibial artery
Plantar artery
Plantar venous arch
Arcuate artery
Dorsal metatarsal veins
Dorsal metatarsal arteries
Dorsal venous arch
Dorsal digital arteries
Dorsal digital veins

Respiratory system

The respiratory system allows gases to be exchanged between air and blood. With each inhalation, air passes through passages of varying sizes until it reaches tiny air sacs deep inside the lungs. Here, vital oxygen diffuses across the thin walls of the sacs and into the bloodstream. In exchange, waste carbon dioxide passes into the lungs, from where it is exhaled. The respiratory system is powered by breathing muscles—the diaphragm and the intercostal muscles between the ribs. On inhalation, these contract to increase the space inside the chest cavity, which allows the lungs to expand and take in air. On exhalation, the breathing muscles return to their resting positions.

COMPONENTS Nasal passage; throat (pharynx); windpipe (trachea); lungs; major and minor lung airways (bronchi and bronchioles); diaphragm and other respiratory muscles

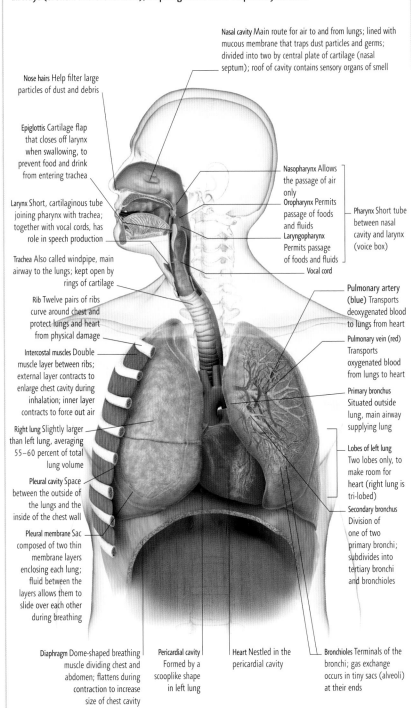

Nasal cavity Main route for air to and from lungs; lined with mucous membrane that traps dust particles and germs; divided into two by central plate of cartilage (nasal septum); roof of cavity contains sensory organs of smell
Nose hairs Help filter large particles of dust and debris
Epiglottis Cartilage flap that closes off larynx when swallowing, to prevent food and drink from entering trachea
Nasopharynx Allows the passage of air only
Oropharynx Permits passage of foods and fluids
Pharynx Short tube between nasal cavity and larynx (voice box)
Larynx Short, cartilaginous tube joining pharynx with trachea; together with vocal cords, has role in speech production
Laryngopharynx Permits passage of foods and fluids
Trachea Also called windpipe, main airway to the lungs; kept open by rings of cartilage
Vocal cord
Rib Twelve pairs of ribs curve around chest and protect lungs and heart from physical damage
Pulmonary artery (blue) Transports deoxygenated blood to lungs from heart
Intercostal muscles Double muscle layer between ribs; external layer contracts to enlarge chest cavity during inhalation; inner layer contracts to force out air
Pulmonary vein (red) Transports oxygenated blood from lungs to heart
Primary bronchus Situated outside lung, main airway supplying lung
Right lung Slightly larger than left lung, averaging 55–60 percent of total lung volume
Lobes of left lung Two lobes only, to make room for heart (right lung is tri-lobed)
Pleural cavity Space between the outside of the lungs and the inside of the chest wall
Secondary bronchus Division of one of two primary bronchi; subdivides into tertiary bronchi and bronchioles
Pleural membrane Sac composed of two thin membrane layers enclosing each lung; fluid between the layers allows them to slide over each other during breathing
Diaphragm Dome-shaped breathing muscle dividing chest and abdomen; flattens during contraction to increase size of chest cavity
Pericardial cavity Formed by a scooplike shape in left lung
Heart Nestled in the pericardial cavity
Bronchioles Terminals of the bronchi; gas exchange occurs in tiny sacs (alveoli) at their ends

Lymphatic and immune system

The body has an internal defense mechanism, the immune system, which is based on specialized white blood cells and helps provide protection against disease. The lymphatic system is an integral part of the immune system. Its active element is lymph, a watery fluid that circulates in the body and drains through lymph nodes that filter out harmful organisms.

COMPONENTS White blood cells; antibodies; spleen; tonsils and adenoids; thymus gland; lymph fluid; lymph vessels, nodes, and ducts

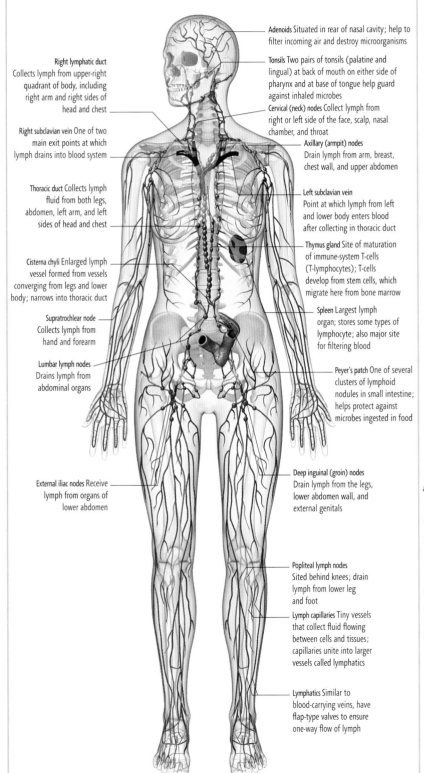

Right lymphatic duct Collects lymph from upper-right quadrant of body, including right arm and right sides of head and chest

Right subclavian vein One of two main exit points at which lymph drains into blood system

Thoracic duct Collects lymph fluid from both legs, abdomen, left arm, and left sides of head and chest

Cisterna chyli Enlarged lymph vessel formed from vessels converging from legs and lower body; narrows into thoracic duct

Supratrochlear node Collects lymph from hand and forearm

Lumbar lymph nodes Drains lymph from abdominal organs

External iliac nodes Receive lymph from organs of lower abdomen

Adenoids Situated in rear of nasal cavity; help to filter incoming air and destroy microorganisms

Tonsils Two pairs of tonsils (palatine and lingual) at back of mouth on either side of pharynx and at base of tongue help guard against inhaled microbes

Cervical (neck) nodes Collect lymph from right or left side of the face, scalp, nasal chamber, and throat

Axillary (armpit) nodes Drain lymph from arm, breast, chest wall, and upper abdomen

Left subclavian vein Point at which lymph from left and lower body enters blood after collecting in thoracic duct

Thymus gland Site of maturation of immune-system T-cells (T-lymphocytes); T-cells develop from stem cells, which migrate here from bone marrow

Spleen Largest lymph organ; stores some types of lymphocyte; also major site for filtering blood

Peyer's patch One of several clusters of lymphoid nodules in small intestine; helps protect against microbes ingested in food

Deep inguinal (groin) nodes Drain lymph from the legs, lower abdomen wall, and external genitals

Popliteal lymph nodes Sited behind knees; drain lymph from lower leg and foot

Lymph capillaries Tiny vessels that collect fluid flowing between cells and tissues; capillaries unite into larger vessels called lymphatics

Lymphatics Similar to blood-carrying veins, have flap-type valves to ensure one-way flow of lymph

Digestive system

The long passageway known as the digestive tract starts at the mouth and continues through the esophagus, stomach, and intestines to the anus. Along its length, food is broken down and nutrients absorbed, while waste is formed into feces to be eliminated from the body. The liver, gallbladder, and pancreas are also associated with digestion (see pp.216–17).

COMPONENTS Mouth; throat (pharynx); esophagus; stomach; pancreas; liver; gallbladder; small intestine (duodenum, jejunum, and ileum); large intestine (colon and rectum); anus

Parotid duct Tube that conveys parotid secretions to mouth

Mouth Entry point of food to digestive system

Tooth

Tongue

Sublingual (left) and submandibular salivary glands Secrete saliva, which lubricates food and contains enzymes that start digestion

Trachea

Esophagus Also known as the gullet; thick-walled muscular tube that connects pharynx with stomach

Parotid salivary gland Largest pair of salivary glands

Pharynx Passage from the mouth to the esophagus

Epiglottis Flap of cartilage covering larynx opening during swallowing

Stomach J-shaped muscular bag that churns, digests, and stores food

Liver Large organ that processes absorbed nutrients, detoxifies harmful substances, and produces bile

Gallbladder Stores bile produced by liver

Pancreas Secretes digestive enzymes, which pour along a duct into first part of small intestine

Small intestine The major site of digestion and absorption of nutrients

Large intestine Absorbs water from food residue, and forms and stores feces

Appendix Dead-end tube with no apparent function in humans

Rectum Stores waste matter until it can be eliminated conveniently

Anus Short, tubelike muscular valve that relaxes to let out digestive waste

Urinary system

The urinary tract is the body's filtering system. Circulating blood passes through the kidneys, which remove waste products and excess water and convert them into urine. Long tubes, called ureters, convey urine from the kidneys to the bladder for storage and excretion. As the urinary system filters the blood, it regulates the body's levels of fluids and chemicals.

COMPONENTS Kidneys; ureters; bladder; urethra

Aorta

Inferior vena cava

Kidney Each is about 4–5in (10–12.5cm) long, and contains about 1 million filtering units

Renal pelvis Funnel-shaped chamber in which urine collects before passing down the ureter

Renal artery

Renal vein

Ureters Convey urine from kidneys to bladder; their walls have three layers: the outer layer, composed of connective and adipose (fat) tissue; the middle layer of muscular fibers, which contract to propel urine to the bladder; the inner layer secretes mucus to prevent its cells from coming into contact with urine

Opening of ureter

Bladder lining Secretes mucus to protect body tissues from urine; folds in lining stretch out as bladder fills

Bladder wall Contains three layers of muscle fiber, together called the detrusor muscle

Femoral artery

Urethra

Bladder outlet

Prostate gland Involved in semen production as part of the reproductive system; encircles the urethra

Membranous part of urethra

Spongy part of urethra

Penis

Male urethra
In males, the urethra passes along the length of the penis and conveys semen as well as urine.

Reproductive system

Of the body's major systems, the reproductive system (see pp.226–27) is the one that differs most between males and females. In both sexes the system starts to function at puberty, enabling the production of offspring by bringing together male and female sex cells. Hormones manufactured by the ovaries in women and the testes in men influence the development of sexual characteristics such as the shape of the body.

COMPONENTS Female: ovaries; fallopian tubes; uterus; vagina and external genitalia; breasts.
Male: testes; spermatic ducts, seminal vesicles; urethra; penis; prostate and bulbourethral glands

Female reproductive organs
The internal female reproductive organs are well protected within the girdle of the pelvic bones. Eggs released from the ovaries pass along the fallopian tubes into the muscular, thick-walled uterus.

Fallopian tube Also called oviduct, or egg tube; carries ripe eggs from ovary to uterus

Fimbriae Fingerlike projections that oscillate to guide an ovulated egg into a fallopian tube

Ovary Produces and stores eggs and makes female sex hormones

Uterus Also called the womb; protects and nourishes the developing baby before birth

Bladder

Pubic symphysis Cartilage forming the junction of the two pubic bones

Clitoris Sensitive tissue that enlarges during sexual arousal

Urethra Carries urine from bladder to outside; opens into front part of vulva

Pelvic floor muscles Hold and support pelvic organs

Rectum | Vagina Elastic wall stretches during intercourse and childbirth

Cervix Narrow, protruding, collarlike neck of the uterus

Male reproductive organs
The male genitals (the penis and scrotum) hang outside the body. In the complex internal system of ducts, tubes, and glands, sperm develop and are stored before being ejaculated in semen.

Ureters Carry urine from kidneys to bladder

Bladder

Vas deferens Thick-walled tube that carries sperm to ejaculatory duct

Cartilage of pubic symphysis Joint at midpoint of front of pelvis

Corpus cavernosum Spongy erectile tissue within penis

Corpus spongiosum Spongy erectile tissue within penis

Urethra Conveys sperm from testes, or urine from bladder

Glans penis Sensitive end of the penis

Prepuce (foreskin) Loosely fitting skin sheath that protects glans penis

Testis Produces sperm

Scrotum Skin pouch that suspends testes away from body and keeps them cool

Anus | Rectum

Seminal vesicle Produces bulk of seminal fluid, including nutrients for sperm

Ejaculatory duct Conveys sperm and seminal fluid to urethra

Prostate gland Surrounds ejaculatory ducts and first portion of urethra; produces fluid for semen

Epididymis Coiled tube in which sperm mature for about 1–3 weeks

Skin, hair, and nails

The body's outer coverings of skin, hair, and nails (together known as the integumentary system) provide first-line defense against physical injury and harmful microorganisms. Skin has two main layers. The top layer, or epidermis, is a tough waterproof tissue with a surface of dead cells that are constantly sloughed off. Beneath the epidermis is the dermis. This layer contains sweat glands, sebaceous glands that produce oily sebum to lubricate the skin, and sensory organs that keep the body informed about pressure, heat, and cold. The dermis also contains blood vessels that regulate body temperature. Below the epidermis and dermis is the subcutaneous fat layer, which acts as an insulator and shock absorber.

COMPONENTS Skin; hair; nails; subcutaneous fat layer

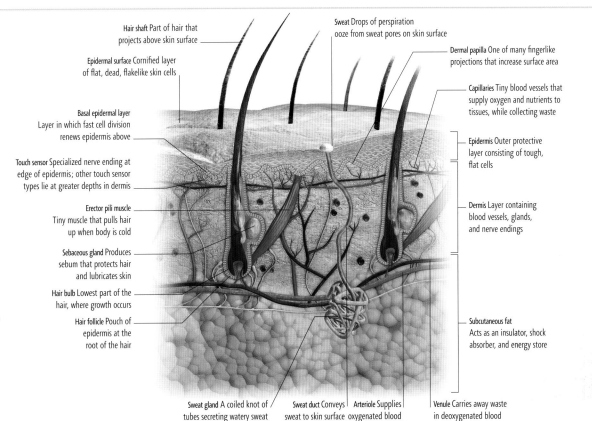

Hair shaft Part of hair that projects above skin surface

Epidermal surface Cornified layer of flat, dead, flakelike skin cells

Basal epidermal layer Layer in which fast cell division renews epidermis above

Touch sensor Specialized nerve ending at edge of epidermis; other touch sensor types lie at greater depths in dermis

Erector pili muscle Tiny muscle that pulls hair up when body is cold

Sebaceous gland Produces sebum that protects hair and lubricates skin

Hair bulb Lowest part of the hair, where growth occurs

Hair follicle Pouch of epidermis at the root of the hair

Sweat Drops of perspiration ooze from sweat pores on skin surface

Dermal papilla One of many fingerlike projections that increase surface area

Capillaries Tiny blood vessels that supply oxygen and nutrients to tissues, while collecting waste

Epidermis Outer protective layer consisting of tough, flat cells

Dermis Layer containing blood vessels, glands, and nerve endings

Subcutaneous fat Acts as an insulator, shock absorber, and energy store

Sweat gland A coiled knot of tubes secreting watery sweat | Sweat duct Conveys sweat to skin surface | Arteriole Supplies oxygenated blood | Venule Carries away waste in deoxygenated blood

Chromosome set

Each body cell contains genetic material (DNA) arranged on structures called chromosomes (see pp.308–09). In humans, there are 46 chromosomes in a full set. Of these, 22 are matching pairs, one chromosome of which is derived from the mother and one from the father. The 23rd pair are the sex chromosomes: XX signifying female and XY (shown here) male. When stained by chemicals, each chromosome shows a banding pattern that makes it possible for the locations of genes to be mapped (see pp.414–15).

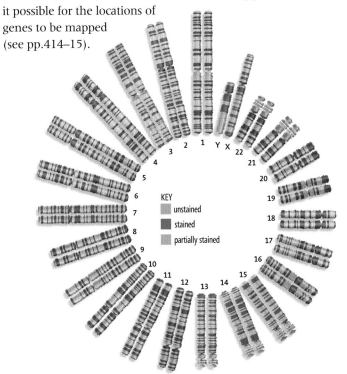

KEY
■ unstained
■ stained
■ partially stained

Chromosome No.	Genes	Examples of gene functions
1	3,100	Pancreatic secretion; tumor suppression; type of collagen; coagulation factor V (F5)
2	1,900	Red hair color; growth of bone and cartilage (BMPR2)
3	2,000	Rhodopsin retinal pigment (RHO); smell (olfactory receptors); DNA repair (MLH1)
4	1,200	Red hair color (HCL2); coagulation factor XI; dentin (DSPP)
5	1,300	Taste receptor; growth hormone receptor; embryonic development (NIPBL)
6	1,500	Control of light-sensing pigments (RDS); iron balance (HFE); immune system (HLA-B)
7	1,500	Type of collagen; color blindness (blue/yellow); growth-rate controlling factor
8	1,000	Fibroblast growth factor receptor 1; fat transportation and breakdown (LPL)
9	1,100	Blood group; removal of excess cholesterol (ABCA1)
10	1,100	Fibroblast growth factor receptor 2; tumor suppression (PTEN)
11	1,800	Albinism (OCA1); beta hemoglobin (HBB); parathyroid hormone; calcitonin
12	1,400	Interferon; type of collagen; regulate muscle proteins (PPP1R12A)
13	550	Cholesterol-lowering factor; coagulation factors VII and X
14	1,300	Thyroid-stimulating hormone factor; coagulation factor C
15	950	Brown and blue eye color (Bey 1/Bey 2); brown hair color (HC13); albinism (OCA2)
16	1,100	Alpha hemoglobin; skin color—melanin (MC1R); breakdown of fatty acids (MLYCD)
17	1,500	Growth hormone; tumor suppression (FLCN); urea production in liver (NAGS)
18	400	Controlling inflammation (MEFV); tumor suppression (SMAD4)
19	1,700	Green eye color (Gey 1); brown hair color (HCL1); protection of nerve cells (PRX)
20	700	Embryonic development (JAG1); nervous system function (PRNP)
21	350	Calcium channels in cardiac muscle and ears (KCNE1)
22	700	Tumor suppression (CHEK2); normal nerve function (NEFH)
X	1,300	Color blindness (red/green); maternal–fetal interface role; brain development (CDKL5)
Y	300	Sex determination; skeletal development (SHOX), also present on X chromosome

Sources and effects of hormones

The various glands of the hormonal system constitute a control and communications network that is complementary to the nervous system. However, instead of using nerve impulses, the glands secrete chemical messages (in the form of hormones) to affect other glands and tissues in various parts of the body. Hormones are carried in the bloodstream to their targets, where they exert their specific effects. The table below lists the hormones that are secreted by different parts of the body and describes their wide-ranging actions.

Gland or hormone-secreting tissue	Hormone secreted	Effects
Hypothalamus	Releasing or inhibiting hormones	Stimulate or suppress hormone secretion by pituitary gland
Pituitary gland	Growth hormone	Stimulates growth and metabolism
	Prolactin	Stimulates milk production after childbirth
	ACTH (adrenocorticotrophic hormone)	Stimulates hormone production by adrenal glands
	TSH (thyroid-stimulating hormone)	Stimulates hormone production by thyroid gland
	FSH (follicle-stimulating hormone); LH (luteinizing hormone)	Stimulate gonads (ovaries or testes)
	ADH (antidiuretic hormone)	Acts on kidneys to conserve water
	Oxytocin	Stimulates contractions of uterus during labor and milk letdown reflex in breast-feeding
	MSH (melanocyte-stimulating hormone)	Acts on skin to promote production of skin pigment (melanin)
Brain	Endorphins; enkephalins	Alleviate pain; boost mood
Thyroid gland	Thyroid hormones	Increase metabolic rate; affect growth
	Calcitonin	Lowers level of calcium in blood
Parathyroid glands	Parathyroid hormone	Increases level of calcium in blood
Thymus	Thymic hormone	Stimulates lymphocyte development
Heart	Atrial natriuretic factor	Lowers blood pressure
Adrenal glands	Adrenaline (epinephrine); noradrenaline (norepinephrine)	Prepare body for physical and mental stress
	Hydrocortisone	Affects metabolism
	Aldosterone	Regulates sodium and potassium excretion by kidneys
	Androgens	Affect growth and sex drive (in both males and females)
Kidneys	Renin	Regulates blood pressure
	Erythropoietin	Stimulates production of red blood cells
Pancreas	Insulin	Lowers blood sugar level
	Glucagon	Raises blood sugar level
Placenta	Chorionic gonadotrophin (HCG); estrogens; progesterone	Maintain pregnancy
Gastrointestinal tract	Gastrin; secretin; cholecystokinin	Regulate secretion of some digestive enzymes
Testes	Testosterone	Affects development of male secondary sexual characteristics and genital organs
Ovaries	Estrogens; progesterone	Affect development of female secondary sexual characteristics and genital organs; control menstrual cycle; maintain pregnancy

Functions of vitamins and minerals

Vitamins are organic substances that are mostly incorporated into coenzymes—molecules that assist and support enzymes in the control of metabolic processes. Regular vitamin intake is required because only a few vitamins can be manufactured in the body. Minerals are simple inorganic substances such as calcium, iron, chloride, and iodide. They are needed both for general metabolism and specialized uses, such as iron for hemoglobin in red blood cells.

Function	Vitamins and minerals needed
Blood clotting	Vitamin K Calcium Iron
Blood cell formation and functioning	Vitamins B_6 and B_{12} Vitamin E Folic acid Copper Iron Cobalt
Healthy skin and hair	Vitamin A Vitamin B_2 (riboflavin) Vitamin B_3 (niacin) Vitamin B_6 Vitamin B_{12} Biotin Sulfur Zinc
Heart functioning	Vitamin B_1 (thiamine) Vitamin D Inositol Calcium Potassium Magnesium Selenium Sodium Copper
Healthy teeth	Vitamins C and D Calcium Phosphorus Fluorine Magnesium Boron
Healthy eyes	Vitamin A Zinc
Bone formation	Vitamins A, C, and D Fluorine Calcium Copper Phosphorus Magnesium Boron
Muscle functioning	Vitamin B_1 (thiamine) Vitamin B_6 Vitamin B_{12} Vitamin E Biotin Calcium Potassium Sodium Magnesium

Medical landmarks

The history of medicine is punctuated by breakthroughs that have radically changed our understanding of the human body and the processes of disease. In the 21st century the development of new treatments and technologies is accelerating rapidly, as scientists and the chemical industry make advances on an almost daily basis. The following are some of the medical events that have made a significant contribution to health.

Date	Developments in diagnosis and prevention of disease
c. 400 BCE	Disease concept introduced by the Greek physician Hippocrates.
c. 1666	Single-lens light microscope developed by the Dutch naturalist Antonie van Leeuwenhoek, who discovered microorganisms using it.
1725	Medical thermometer with a scale devised by Daniel Gabriel Fahrenheit.
1796	First smallpox vaccination given by the English physician Edward Jenner. The first true vaccine (comprising weakened microorganisms), against chicken cholera, was developed in 1880 by the French scientist Louis Pasteur.
1816	Stethoscope invented by the French physician René Laënnec.
1850–1900	The germ theory of disease proposed by the French scientist Louis Pasteur and developed by the German bacteriologist Robert Koch.
1851	Ophthalmoscope invented by the German scientist Herman von Helmholtz.
1881	The Czech–Austrian physician Samuel von Basch invented the sphygmomanometer to measure blood pressure.
1887	First electrocardiograph (ECG) described by the English physiologist Augustus Waller at St. Mary's Hospital, London. Dutch physiologist Willem Einthoven went on to develop the five phases of the ECG and won the Nobel Prize in 1924.
1891	Baby incubator introduced by the French doctor Alexandre Lion.
1895	X-rays discovered by the German physicist Wilhelm Röntgen.
1898	British doctor Ronald Ross proved that malaria is transmitted by mosquitoes.
1901	Hearing aid (electric) developed by the American inventor Miller Reese Hutchinson.
c. 1932	Transmission electron microscope (TEM) constructed by the German scientists Max Knoll and Ernst Ruska.
1938	First cardiac catheterization performed by George Peter Robb and Israel Steinberg in New York.
1946	Franz Greiter developed sunscreen after severe sunburn while climbing Piz Buin, a mountain on the Swiss–Austrian border.
1950	Amniocentesis to detect fetal disease developed by English doctor Douglas Bevis, but its significance went unrecognized for 10 years.
1951	Link discovered between smoking and lung cancer demonstrated by British scientists Richard Doll and Austin Bradford Hill.
1957	Fiberoptic endoscopy pioneered by the South African-born doctor Basil Hirschowitz at the University of Michigan.
1958	Ultrasound first used on pregnant women to examine the fetus by obstetrician and gynecologist Ian Donald, working in Scotland.
1972	CT scanner invented by the British engineer Godfrey Hounsfield and the South African-born physicist Alan Cormack of Tufts University, Massachusetts.
1976	Chorionic villus sampling developed by Chinese gynecologists to aid the early diagnosis of genetic disorders.
1977	First MRI scanner completed by Raymond Damadian and team in the US.
1979	The global eradication of smallpox was formally certified, following an intensive campaign launched by the World Health Organization in 1967.
2003	Completion of the Human Genome Project by the International Human Genome International Consortium, following first draft produced in 2001.
2006	Cervical cancer vaccine developed by Scottish-born doctor Ian Frazer, and Chinese molecular virologist Jian Zhou, at the University of Queensland, Australia, is the first vaccine to target a cause of cancer.

Date	Developments in surgery and drug treatment
1545	Basic surgical principles established by the French surgeon Ambroise Paré.
1666	Quinine (in form of Jesuits' bark) popularized by the British physician Thomas Sydenham for treating malaria.
1785	Use of digitalis to treat heart failure first described by the British physician William Withering.
1805	Morphine, extracted from opium, first used as pain relief by the German pharmacist Friedrich Sertürner.
1842	First surgery using general anesthesia performed by the American surgeon Crawford Long, who used ether. In 1845 the American dentist Horace Wells used nitrous oxide (laughing gas) as an anesthetic. In 1847 the British obstetrician James Simpson introduced chloroform anesthesia.
1870	Antiseptic surgery pioneered by the British surgeon Joseph Lister, who used a carbolic acid (phenol) spray to help prevent infection.
1878	Acetaminophen discovered by American chemistry professor Harmon Morse; its use was not realized until 70 years later.
1899	Aspirin first manufactured by German company Bayer.
1901	ABO blood groups discovered by the Austrian pathologist Karl Landsteiner, thereby establishing the basis for safe transfusions.
1911	Salvarsan introduced by the German bacteriologist Paul Ehrlich to treat syphilis.
1920	Adhesive bandages invented by Earle Dickson while working for Johnson & Johnson.
1928	British bacteriologist Alexander Fleming first recognized penicillin's antibacterial action.
1935	Sulfonamides' antibacterial action discovered by the German pharmacologist Gerhard Domagk.
1943	Kidney dialysis machine developed by the Dutch surgeon Willem Kolff.
1951	Oral contraceptive pill developed by the American doctors Gregory Pincus and John Rock, and the Austrian-born American chemist Carl Djerassi.
1955	First successful kidney transplant (between identical twins) performed by a team of American surgeons, led by Joseph Murray, of Harvard Medical School.
1962	First beta-blocker heart drug, nethalide (pronethalol), developed by scientists at Imperial Chemical Industries, England.
1967	First coronary artery bypass graft performed by Argentinian doctor René Géronimo Favaloro working in Cleveland, Ohio.
1967	First human heart transplant performed by the South African surgeon Christiaan Barnard at the Groote Schur Hospital, Cape Town.
1969	Scottish scientist David Jack discovered salbutamol to treat asthma.
1976	Coronary angioplasty introduced by Swiss surgeon Andreas Grüntzig.
1978	First "test-tube baby" Louise Brown, resulting from in vitro fertilization, born in the UK. British gynecologist Patrick Steptoe and the embryologist Robert Edwards developed the IVF techniques.
1980	German gynecologist Kurt Semm performed the first keyhole surgery.
1982	First genetically engineered medicine, human insulin, developed at Genentech, California.
1986	Zidovudine (originally called AZT) developed to treat HIV/AIDS by scientists at Burroughs Wellcome Research Laboratories, North Carolina.
1998	American James Thomson isolated the first human embryonic stem cell line.
2001	French doctor Jacques Marescaux performed the first telesurgery.
2005	First face transplant performed by French team led by Bernarad Devauchelle and Jean-Michel Dubernard on 38-year-old Isabelle Dinoire.
2013	Japanese researchers grew human liver tissue from stem cells.
2012	American biochemist Jennifer Doudna and French microbiologist Emmanuelle Charpentier develop the CRISPR-Cas9 technique for editing genes, paving the way for cures for genetic disorders.
2020	Vaccines are approved for controlling the COVID-19 virus less than a year after the World Health Organization declared the global outbreak a pandemic.

Chemistry

Chemistry is the study of the composition, structure, and properties of substances and the changes they undergo during chemical reactions. Organic chemistry studies the structures and reactions of carbon (organic) compounds; inorganic chemistry deals with the properties of all other elements; physical chemistry investigates the physical properties of chemical reactions; and biochemistry studies the compounds and reactions integral to life processes.

Periodic table

The periodic table currently contains 118 elements, of which around 90 occur natually on Earth. The elements are structured in order of increasing atomic number, and are arranged in vertical "groups" and horizontal "periods" (see pp.232–33). Groupings of elements demonstrate a close family resemblance, with reactive metallic elements on the left, through less reactive metals to reactive nonmetals, to barely reactive gases on the far right. Two rows of "inner transition elements" are positioned at the bottom of the table to avoid making the periods too long.

Atomic number
The number of protons in the nuclei of the element's atoms.

Chemical symbol
Used as a form of shorthand in chemical equations (in this case for aluminum).

13
Al
Aluminum
27

Relative atomic mass
See below. Numbers are rounded to the nearest integer for simplicity.

Key
The elements fall into several color-coded categories or blocks containing similar elements. Nine categories are shown here, some of them corresponding closely or exactly to groups (columns) of the table.

- Alkali metals
- Alkaline-earth metals
- Transition metals
- Inner transition elements
- Other metals
- Metalloids
- Other nonmetals
- Noble gases
- Halogens

Relative atomic mass

Relative atomic mass (RAM) is a measure of the mass of one atom of an element compared to a "standard mass" given as one-twelfth of the mass of an atom of carbon-12. However, elements exist in more than one form, or isotope; their relative atomic mass is a weighted average of the mass of their naturally occurring isotopes.

Calculating RAM
All atoms of silicon have 14 protons. There are three stable isotopes of the element, silicon-28, silicon-29, and silicon-30, which all have different numbers of neutrons. The relative atomic mass of silicon is 28.09; this is the average mass of the isotopes, weighted according to their natural abundance.

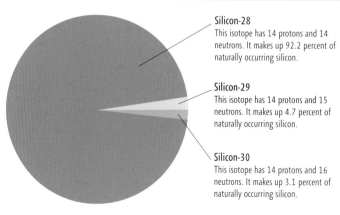

Silicon-28
This isotope has 14 protons and 14 neutrons. It makes up 92.2 percent of naturally occurring silicon.

Silicon-29
This isotope has 14 protons and 15 neutrons. It makes up 4.7 percent of naturally occurring silicon.

Silicon-30
This isotope has 14 protons and 16 neutrons. It makes up 3.1 percent of naturally occurring silicon.

Table of Elements

This chart lays out essential information for each of the elements, apart from the five most massive and recently discovered. It is sorted by atomic number—the number of protons in the nucleus of an element's atoms—and includes the commonly known symbol, atomic mass (number of protons and neutrons in the nucleus) and valency (the number of chemical bonds formed by the atoms) for each element.

Atomic number	Element	Symbol	Atomic mass	Melting point °F	Melting point °C	Boiling point °F	Boiling point °C	Valency
1	Hydrogen	H	1	-434	-259	-253	-423	1
2	Helium	He	4	-458	-272	-269	-452	0
3	Lithium	Li	7	354	179	2,440	2,990	1
4	Beryllium	Be	9	2,341	1,283	5400	3,660	2
5	Boron	B	11	4,170	2,300	6620	4,827	3
6	Carbon	C	12	6,332	3,500	8721	4,827	2,4
7	Nitrogen	N	14	-346	-210	-321	-183	3,5
8	Oxygen	O	16	-362	-219	-297	-188	2
9	Fluorine	F	19	-364	-220	-306	-246	1
10	Neon	Ne	20	-416	-249	-410	890	0
11	Sodium	Na	23	208	98	1,634	1,105	1
12	Magnesium	Mg	24	1,202	660	2,021	2,467	2
13	Aluminum	Al	27	1,220	1,420	4,473	2,355	3
14	Silicon	Si	28	2,588	44	4,271	280	4
15	Phosphorus	P	31	111	113	536	445	3,5
16	Sulfur	S	32	235	-101	832	-34	2,4,6
17	Chlorine	Cl	35	-150	-189	-29	-186	1,3,5,7
18	Argon	Ar	40	-308	64	-303	754	0
19	Potassium	K	39	147	848	1389	1487	1
20	Calcium	Ca	40	1558	1541	2709	2831	2
21	Scandium	Sc	45	2806	1677	5128	3277	3
22	Titanium	Ti	48	3051	1917	5931	3377	3,4
23	Vanadium	V	51	3483	1903	6111	2642	2,3,4,5
24	Chromium	Cr	52	3457	1244	4788	2041	2,3,6
25	Manganese	Mn	55	2271	1539	3706	2750	2,3,4,6,7
26	Iron	Fe	56	2802	1495	4980	2877	2,3
27	Cobalt	Co	59	2723	1455	5211	2730	2,3
28	Nickel	Ni	59	2651	1083	4950	2582	2,3
29	Copper	Cu	64	1981	420	4680	907	1,2
30	Zinc	Zn	65	788	30	1665	2403	2
31	Gallium	Ga	70	86	937	4357	2355	2,3
32	Germanium	Ge	73	1719	817	4271	613	4
33	Arsenic	As	75	1503	217	1135	685	3,5
34	Selenium	Se	79	423	-7	1265	59	2,4,6
35	Bromine	Br	80	19	-157	138	-152	1,3,5,7
36	Krypton	Kr	84	-251	39	-242	688	0
37	Rubidium	Rb	85	102	769	1270	1384	1
38	Strontium	Sr	88	1416	1522	2523	3338	2
39	Yttrium	Y	89	2772	1852	6040	4377	3
40	Zirconium	Zr	91	3366	2467	7911	4742	4
41	Niobium	Nb	93	4473	2610	8568	5560	3,5
42	Molybdenum	Mo	96	4730	2172	8811	4877	2,3,4,5,6
43	Technetium	Tc	98	3942	2310	8811	3900	2,3,4,6,7
44	Ruthenium	Ru	101	4190	1966	7052	3727	3,4,6,8
45	Rhodium	Rh	103	3571	1554	6741	2970	3,4
46	Palladium	Pd	106	2829	962	5378	2212	2,4
47	Silver	Ag	108	1764	321	4014	767	1
48	Cadmium	Cd	112	610	156	1413	2028	2
49	Indium	In	115	313	232	3680	2270	1,3
50	Tin	Sn	119	450	631	4118	1635	2,4
51	Antimony	Sb	122	1168	450	2975	990	3,5
52	Tellurium	Te	128	842	114	1814	184	2,4,6
53	Iodine	I	127	237	-112	363	184	1,3,5,7
54	Xenon	Xe	131	-170	29	-161	671	0
55	Cesium	Cs	133	84	725	1240	1640	1
56	Barium	Ba	137	725	1337	1640	2984	2

Atomic number	Element	Symbol	Atomic mass	Melting point °F	Melting point °C	Boiling point °F	Boiling point °C	Valency
57	Lanthanum	La	139	1,690	921	6,255	3,457	3
58	Cerium	Ce	140	1,470	799	6,199	3,426	3,4
59	Praseodymium	Pr	141	1,708	931	6,354	3,512	3
60	Neodymium	Nd	144	1,870	1,021	5,554	3,068	3
61	Promethium	Pm	145	2,134	1,168	4,892	2,700	3
62	Samarium	Sm	150	1,971	1,077	4,892	1,791	2,3
63	Europium	Eu	152	1,512	822	3,256	1,597	2,3
64	Gadolinium	Gd	157	2,395	1,313	2,907	3,266	3
65	Terbium	Tb	159	2,473	1,356	5,911	3,123	3
66	Dysprosium	Dy	163	2,574	1,412	5,653	2,562	3
67	Holmium	Ho	165	2,685	1,474	4,644	2,695	3
68	Erbium	Er	167	2,784	1,529	4,883	2,863	3
69	Thulium	Tm	169	2,813	1,545	5,185	1,947	2,3
70	Ytterbium	Yb	173	1,506	819	3,537	1,194	2,3
71	Lutetium	Lu	175	3,025	1,663	2,181	3,395	3
72	Hafnium	Hf	178	4,041	2,227	6,143	4,602	4
73	Tantalum	Ta	181	5,425	2,996	8,316	5,427	3,5
74	Tungsten	W	184	6,170	3,410	9,801	5,660	2,4,5,6
75	Rhenium	Re	186	5,756	3,180	10,220	5,627	1,4,7
76	Osmium	Os	190	5,510	3,045	10,161	5,090	2,3,4,6,8
77	Iridium	Ir	192	4,370	2,410	9,190	4,130	3,4
78	Platinum	Pt	195	3,222	1,772	7,466	3,827	2,4
79	Gold	Au	197	1,947	1,064	6,921	2,807	1,3
80	Mercury	Hg	201	-38	-39	5,080	357	1,2
81	Thallium	Tl	204	577	303	675	1,457	1,3
82	Lead	Pb	207	622	328	2,655	1,744	2,4
83	Bismuth	Bi	209	520	271	3,171	1,560	3,5
84	Polonium	Po	209	489	254	2,840	962	2,3,4
85	Astatine	At	210	572	300	1,764	370	1,3,5,7
86	Radon	Rn	222	-96	-71	698	-62	0
87	Francium	Fr	223	81	27	-80	677	1
88	Radium	Ra	226	1292	700	1251	1200	2
89	Actinium	Ac	227	1922	1050	2190	3200	3
90	Thorium	Th	232	3182	1750	5792	4787	4
91	Protactinium	Pa	231	2907	1597	8649	4027	4,5
92	Uranium	U	238	2070	1132	7281	3818	3,4,5,6
93	Neptunium	Np	237	1179	637	6904	4090	2,3,4,5,6
94	Plutonium	Pu	244	1184	640	7394	3230	2,3,4,5,6
95	Americium	Am	243	1821	994	5850	2607	2,3,4,5,6
96	Curium	Cm	247	2444	1340	4724	3190	2,3,4
97	Berkelium	Bk	247	1922	1050	5774	710	2,3,4
98	Californium	Cf	251	1652	900	1310	1470	2,3,4
99	Einsteinium	Es	252	1580	860	2678	966	2,3
100	Fermium	Fm	257	unknown	unknown	unknown	unknown	2,3
101	Mendelevium	Md	258	unknown	unknown	unknown	unknown	2,3
102	Nobelium	No	259	unknown	unknown	unknown	unknown	2,3
103	Lawrencium	Lr	262	unknown	unknown	unknown	unknown	3
104	Rutherfordium	Rf	267	unknown	unknown	unknown	unknown	unknown
105	Dubnium	Db	268	unknown	unknown	unknown	unknown	unknown
106	Seaborgium	Sg	271	unknown	unknown	unknown	unknown	unknown
107	Bohrium	Bh	272	unknown	unknown	unknown	unknown	unknown
108	Hassium	Hs	277	unknown	unknown	unknown	unknown	unknown
109	Meitnerium	Mt	276	unknown	unknown	unknown	unknown	unknown
110	Darmstadtium	Ds	281	unknown	unknown	unknown	unknown	unknown
111	Roentgenium	Rg	280	unknown	unknown	unknown	unknown	unknown
112	Copernicium	Cn	285	unknown	unknown	unknown	unknown	unknown

Atoms and molecules

An atom is the smallest part of an element that can exist as a stable entity. Atoms are the basic units of matter, and many bond together to form larger particles called molecules. Atoms within a molecule may be of the same chemical element, as in oxygen, or of different elements, as in water. Atoms are held together by strong chemical bonds (see pp.138–39) formed as electrons are shared or exchanged between the atoms.

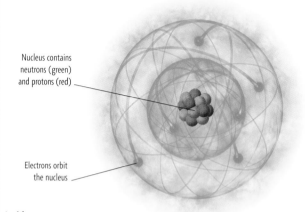

Nucleus contains neutrons (green) and protons (red)

Electrons orbit the nucleus

Inside an atom
Inside each atom are a number of subatomic particles: the nucleus, a tightly bound cluster of protons and neutrons, and electrons that orbit the nucleus. Protons have a positive charge and neutrons have no charge. The positively charged protons would normally repel each other, but the nucleus is held together by a powerful force called the strong nuclear force.

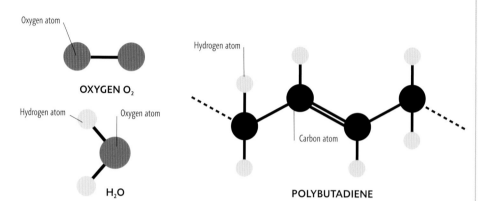

Oxygen atom

OXYGEN O$_2$

Hydrogen atom   Oxygen atom

H$_2$O

Hydrogen atom

Carbon atom

POLYBUTADIENE

Simple molecules
Elements may form simple (small) molecules by bonding with themselves or other elements. Above, the oxygen molecule is made up of two oxygen atoms bonded together, and the water molecule (H_2O) consists of one oxygen atom bonded to two hydrogen atoms.

Complex molecules
Some molecules consist of thousands of atoms joined together in long chains called polymers (see pp.468–69). They are possible because carbon forms stable bonds with other carbon atoms. Polybutadiene is a synthetic polymer used in the manufacture of tires.

Allotropes

The atoms of some elements can bond in a number of different ways to create forms called allotropes. For example, the element carbon has two common allotropes: diamond, where the carbon atoms are bonded together in a tetrahedral lattice arrangement; and graphite, where the carbon atoms are bonded together in hexagonal sheets. Allotropy refers only to different forms of the same element; allotropes do not exist in compounds.

Shared electrons form covalent bonds between atoms

Carbon atom bonded to four other carbon atoms

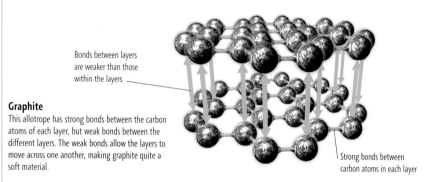

Bonds between layers are weaker than those within the layers

Graphite
This allotrope has strong bonds between the carbon atoms of each layer, but weak bonds between the different layers. The weak bonds allow the layers to move across one another, making graphite quite a soft material.

Strong bonds between carbon atoms in each layer

Diamond
The hardest known mineral, diamond has carbon atoms tightly bound to each other in an extremely rigid structure called a diamond lattice. Even though it is immensely tough, diamond will burn in air to make carbon dioxide.

The pH scale for acids and bases

The pH scale measures the strength of acids and bases (also known as alkalis). "pH" stands for potential of hydrogen, and it measures the concentration of hydrogen ions in a liquid. The scale ranges from pH0 (most acid) to pH14 (most basic), with pure water defining the neutral point in the scale at pH7. Each value produces a characteristic color in response to a universal indicator solution (see pp.238–39).

pH	pH0	pH1	pH2	pH3	pH4	pH5	pH6	pH7	pH8	pH9	pH10	pH11	pH12	pH13	pH14
Acid/alkali	stronger acid			weak acid				neutral		weak alkali				stronger alkali	
Substance	Battery acid = pH0		Orange juice = pH3		Tomato juice = pH4			Water = pH7	Seawater = pH8	Toothpaste = pH9		Ammonia = pH11			Drain cleaner = pH14

Chemical reactions

In any chemical reaction, the starting materials undergo bond breaking and/or bond making. The products of the reaction are different chemicals with different properties. There are several different types of chemical reaction (see pp.234–35).

Hydrogen molecule made up of two atoms

Ammonia made up of three hydrogen atoms and one nitrogen atom

Nitrogen molecule with a triple bond

HYDROGEN (H) + NITROGEN (N) = AMMONIA (NH₃)

△ **Synthesis**
A synthesis reaction is when two or more simple compounds combine to form a more complicated one. In the above example, molecules of hydrogen and nitrogen bond together to form ammonia.

Carbonic acid molecule

Carbon dioxide molecule

Water molecule

CARBONIC ACID (H₂CO₃) = CARBON DIOXIDE (CO₂) + WATER (H₂O)

△ **Decomposition**
A decomposition reaction is the opposite of a synthesis reaction (see top), whereby a complex molecule breaks down into simpler ones. In the above reaction, carbonic acid (a weak acid) breaks down into carbon dioxide (CO₂) and water (H₂O) when warmed in water.

Factors affecting reactions	
Concentration of reactants	A higher concentration of particles increases the chance of them colliding with one another, hence causing a reaction.
Pressure of reactants	In gases, particles at high pressure are brought closer together than particles at low pressure, so they are more likely to collide.
Temperature of reactants	As temperature increases, particles move faster and so collide more frequently than they do at low temperatures.
Light levels	Some particles can absorb light energy, making them more reactive at high light levels than they are at low light levels.
Size of reactants	Smaller particles of solids react faster than bigger particles because they have a larger combined surface area over which reactions can occur.

Fission and fusion

Fission occurs when the nuclei of radioactive atoms are split into fragments, emitting alpha (α) and beta (β) particles, gamma (γ) radiation, and masses of energy (see pp.298–99). Fusion occurs when, at very high temperatures (millions of degrees centigrade), some of the hydrogen atoms lose their electron shells and the unprotected nuclei collide with each other producing helium (see pp.324–25). When the collision occurs, huge amounts of energy are released. The most striking natural example of nuclear fusion is the Sun.

▽ **How fission works**
Fission is usually done with uranium-235 because it is the only fissile isotope found in any quantity. The large number of neutrons over protons in the nucleus of a U-235 atom makes it highly unstable. If bombarded with neutrons, it cannot absorb them but splits to start a chain reaction.

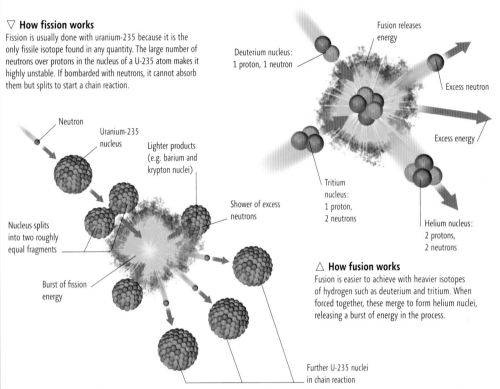

Neutron

Uranium-235 nucleus

Lighter products (e.g. barium and krypton nuclei)

Nucleus splits into two roughly equal fragments

Burst of fission energy

Shower of excess neutrons

Further U-235 nuclei in chain reaction

Deuterium nucleus: 1 proton, 1 neutron

Fusion releases energy

Excess neutron

Excess energy

Tritium nucleus: 1 proton, 2 neutrons

Helium nucleus: 2 protons, 2 neutrons

△ **How fusion works**
Fusion is easier to achieve with heavier isotopes of hydrogen such as deuterium and tritium. When forced together, these merge to form helium nuclei, releasing a burst of energy in the process.

Radioactive half-life

Radioactive elements decay at different rates. A half-life is the time it takes for half of the original amount of a radioactive element to decay. Different elements emit different types of radiation when they decay—alpha (α) particles, beta (β) particles, and gamma (γ) rays. After two half-lives, a quarter the original sample will remain; after three half-lives one-eighth the original sample will remain, and so on. Carbon dating (see pp.190–91) uses carbon-14 to determine the age of organic material. Half lives of other materials are given in the table below.

Element	Decay rate	Particle
Uranium-238	4,500 million years	α
Plutonium-239	24,400 years	α
Carbon-14	5,700 years	β
Radium-226	1,600 years	α
Strontium-90	28 years	β
Hydrogen-3	12.3 years	β
Cobalt-60	5.3 years	γ
Phosphorus-32	14.3 days	β
Iodine-131	8.1 days	β
Radon-222	3.8 days	α
Lead-214	26.8 minutes	β
Astatine-215	0.0001 seconds	α

Chemical bonds

The atoms within chemical compounds are held together by bonds, which form when atoms come into contact with each other and their electron shells interact (exchange electrons). The strongest bonds are ionic, covalent, or metallic (see pp.288–89), and they vary in strength according to the atoms involved; for example hydrogen forms strong bonds with many elements, but helium forms no bonds at all.

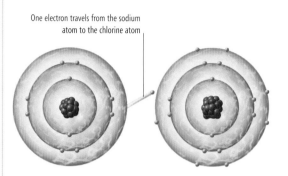

One electron travels from the sodium atom to the chlorine atom

Two oxygen atoms share four electrons

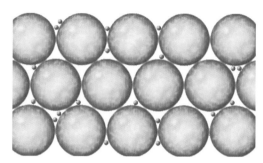

Ionic bonds
In ionic bonds, electrons are transferred from one atom to another. When sodium and chlorine combine to form sodium chloride (salt), sodium loses an electron and becomes positively charged; chlorine takes that electron and becomes negatively charged. Ionic bonds are difficult to break. Ionic compounds are usually solids with high melting points.

Covalent bonds
In a covalent bond, electrons are shared between two atoms. When two oxygen atoms bond together to form an oxygen molecule, they share four electrons—two from each oxygen atom. Other examples of covalent bonding are water (H_2O), and carbon dioxide (CO_2). Covalent compounds are usually liquids or gases with low melting points.

Metallic bonds
Metal atoms are bonded to each other through metallic bonding. In this type of bonding, all the atoms lose electrons, which float around in a common pool. The electrons in this pool can move around freely, which is why metals can transfer heat or electricity so well. If one part of the metal is heated, the electrons carry the heat quickly to other parts.

Chemical nomenclature

The system of naming of chemical compounds is referred to as chemical nomenclature. Specific rules and conventions are used to avoid ambiguity between compounds, and to convey information about the structure of the compound. Prefixes are used to denote the number of atoms present, mono- indicating one atom, di- two atoms, and so on, while suffixes -ite and -ate indicate the presence of oxygen.

Compound ending in	Description	Example
–ide	Contains just the two elements in the name	Iron sulfide FeS
–ite	Contains oxygen as well as the other elements in the name	Iron sulfite $FeSO_3$
–ate	Contains oxygen as well as the other elements in the name. Contains more oxygen than is found in –ites	Iron sulfate $FeSO_4$

Prefix	Number of atoms in prefix	Example
Mono-	1	Nitrogen monoxide NO
Di-	2	Nitrogen dioxide NO_2
Tri-	3	Dinitrogen trioxide N_2O_3
Tetra–	4	Dinitrogen tetroxide N_2O_4
Penta-	5	Dinitrogen pentoxide N_2O_5

Alkanes

Alkanes contain only hydrogen and carbon atoms, which are linked together by single bonds. Each carbon atom has 4 bonds and each hydrogen atom is linked to a carbon atom. Alkanes always have the suffix –ane. A selection of alkanes and their properties is listed below.

Number of carbon atoms	Name	Physical state at 77° F (22° C)	Molecular formula
1	Methane	gas	CH_4
2	Ethane	gas	C_2H_6
3	Propane	gas	C_3H_8
4	Butane	gas	C_4H_{10}
5	Pentane	liquid	C_5H_{12}
6	Hexane	liquid	C_6H_{14}
7	Heptane	liquid	C_7H_{16}
8	Octane	liquid	C_8H_{18}
9	Nonane	liquid	C_9H_{20}
10	Decane	liquid	$C_{10}H_{22}$

Alkenes

Like alkanes, alkenes also contain only hydrogen and carbon atoms, but all contain at least one carbon–carbon double bond. This means that they are more reactive than alkanes. Alkenes always have the suffix –ene. A selection of alkenes and their properties is listed below.

Carbon atoms in chain	Name	Physical state at 77° F (22° C)	Molecular formula
2	Ethene	gas	C_2H_4
3	Propene	gas	C_3H_6
4	Butene	gas	C_4H_8
5	Pentene	liquid	C_5H_{10}
6	Hexene	liquid	C_6H_{12}
7	Heptene	liquid	C_7H_{14}
8	Octene	liquid	C_8H_{16}
9	Nonene	liquid	C_9H_{18}
10	Decene	liquid	$C_{10}H_{20}$
11	Undecene	liquid	$C_{11}H_{22}$

Properties of common metals

About three-quarters of the known elements are metals. Most have high melting points and are good conductors of heat and electricity. Most are strong but malleable. Properties of common metals are listed below, with electronegativity referring to the tendency of the atoms in a stable molecule to attract electrons within covalent bonds.

Metal	Electronegativity (0.7 to 4.0)	Atomic number	Melting point		Boiling point	
			(°F)	(°C)	(°F)	(°C)
Aluminum (Al)	1.6	13	1220	660	4473	2467
Barium (Ba)	0.9	56	1337	725	2984	1640
Beryllium (Be)	1.6	4	2341	1283	5400	2990
Cesium (Cs)	0.8	55	84	29	1240	671
Calcium (Ca)	1.0	20	1558	848	2709	1487
Chromium (Cr)	1.7	24	3457	1903	4788	2642
Cobalt (Co)	1.9	27	2723	1495	5211	2877
Copper (Cu)	1.9	29	1981	1083	4680	2582
Iron (Fe)	1.8	26	2802	1539	4980	2750
Lead (Pb)	1.8	82	622	328	3171	1744
Lithium (Li)	1.0	3	354	179	2440	1340
Magnesium (Mg)	1.3	12	1202	650	2021	1105
Manganese (Mn)	1.6	25	2271	1244	3706	2401
Nickel (Ni)	1.9	28	2651	1455	4950	2730
Tin (Sn)	2.0	50	232	119	4118	2270
Zinc (Zn)	1.7	30	788	420	1665	907

Salt formation

When metals come into contact with acids, salts are formed along with hydrogen. This table describes what happens in a number of metal-acid reactions.

Metal	Acid	Salt
Sodium	Hydrochloric acid	Sodium chloride
Potassium	Sulfuric acid	Potassium sulfate
Copper	Hydrochloric acid	Copper chloride
Zinc	Sulfuric acid	Zinc sulfate
Calcium	Hydrochloric acid	Calcium chloride
Sodium	Sulfuric acid	Sodium sulfate

Mohs scale of hardness

The mohs scale is a measure of the hardness (resistance to scratching) of minerals. It lists 10 minerals in ascending order: the higher the number, the harder the mineral. Each mineral in the scale scratches all those above it. All other materials can be compared to these minerals.

Mineral	Composition	Simple Hardness Test	Hardness
Talc	$Mg_3Si_4O_{10}(OH)_2$	Crushed by finger nail	1
Gypsum	$CaSO_4.2H_2O$	Scratched by finger nail	2
Calcite	$CaCO_3$	Scratched by copper coin	3
Fluorite	CaF_2	Scratched by glass	4
Apatite	$Ca_5(PO_4)_3F$	Scratched by penknife	5
Feldspar	$KA1Si_3O_8$	Scratched by quartz	6
Quartz	SiO_2	Scratched by steel file	7
Topaz	$A1_2SiO_4F_2$	Scratched by corundum	8
Corundum	$A1_2O_3$	Scratched by diamond	9
Diamond	C		10

Organic compounds

Organic chemistry is concerned with compounds whose molecules contain carbon, the foundation of all living things. Carbon can combine with itself and other elements to form a huge variety of molecules (see pp.150–51).

SIMPLE HYDROCARBON (METHANE)

CARBON CHAINS (BUTANE)

CARBON RING (BENZENE)

Carbon compounds

Carbon atoms usually have a tetrahedral (pyramidal) arrangement of bonds, which means they can form complex branching chains and rings. This structural complexity allows carbon to form the structural and functional building blocks of all living cells. The typical shapes of a simple hydrocarbon (methane), a carbon chain (butane), and a carbon ring (benzene) are shown above.

ETHANE

ETHENE

ETHANOL

ETHANOIC ACID

ETHANAL

Organic molecules

The most common elements to bond with carbon are hydrogen, oxygen, and nitrogen. These elements form arrangements of atomic clusters called functional groups, which are attached to the carbon atoms and determine the chemical properties of each molecule. A selection of two-carbon molecules is shown above.

Synthetic polymers

Polymers are large molecules made from thousands of identical smaller molecules, called "repeating units" or "monomers." They can be synthesized, in a process called polymerization. Synthetic polymers can be designed to have different features, such as flexibility, hardness, or stickiness, and used in substances such as plastics (see pp.336–37). The chart below shows the monomers, properties, and applications of some common synthetic polymers, although it does not faithfully represent their structure—usually long, twisting three-dimensional chains.

Key
The molecules comprising each of these polymers, together with the types of bonds, are shown here.

- Hydrogen
- Chlorine
- Carbon
- Oxygen
- Nitrogen
- Fluorine
- Benzene ring
- Single bond
- Double bond

Polymer	Properties and uses	Polymer	Properties and uses
Polyamides Nylon-6, 6	Textile fiber, threads, ropes; molded gears and electrical insulation	**Polyurethanes** Polyurethane	Foam rubber, synthetic leather
Nylon-6, 10	Sports equipment; bristles for brushes	Lycra	Expanded foam rubber, carpet underlays, clothing
Polyesters Terylene, Dacron	Textile fiber; basis for magnetic tape and photographic film	**Polyalkenes** Neoprene	Adhesive golf-ball covers, liquid seals
Polycarbonates Lexan	Tough and transparent. Used in bullet-proof windows, safety glass, food containers, and car components	Polythene (polyethene)	Tough plastic used in fibers, thin films, extrusion-molded objects, toys, and bottles
Polyethers Polyglycol 166	Making urethanes and speciality elastomers	Butyl rubber (polyisobutylene)	Tire inner tubes, raincoats, seals
Polyoxymethylene Delrin, Celcon	Tough plastic for gears, pipes, and pens	PTFE polytetrafluoroethene, Teflon	Highly water-repellent. Used in nonstick cooking ware; also has industrial uses where very low friction needed
Phenol–based Bakelite	Hard thermosetting polymer used in items such as telephones, buttons, and electrical insulators	Polystyrene	Transparent and glasslike. Used in a wide variety of molded and expanded objects, packing, and insulation
Poly (melamine formaldehyde), Melamine, Formica	Laminated surfaces such as table tops and cupboards	Perspex poly (methylmethacrylate)	Transparent and glasslike. Used in windows, fiber optics, and illuminated signs
		PVC (polyvinyl chloride)	Inflexible polymer. With plasticizer, used in thin films, tubing, car seat covers, and floor tiles

Natural polymers

Polymers are abundant in the natural world. The most common is cellulose, which is formed from long chains of glucose repeating units. Other natural polymers, or biopolymers, include amber and starch. Polymers are even found in the human body, in the form of proteins, made from amino acids (see p.348), and DNA, made from nucleic acids. Some natural polymers can be adapted for different uses—for example, rubber can be modified, by a process called vulcanization, to make it tougher and more elastic.

This oxygen–hydrogen (OH) group points horizontally in beta-glucose. The OH group is also found in alpha-glucose (better known as blood sugar), in which it points downward.

Adjacent glucose building blocks are joined by a type of bond called a glycosidic bond.

The beta-glucose unit is a form of glucose unique to cellulose. The orientation of its atoms allows parallel chains to link together by hydrogen bonds.

Structure of cellulose
Cellulose is the most abundant organic molecule on Earth and makes up the cell walls of all plants. Its molecules consist of thousands of glucose (sugar) building blocks (see right), bonded to form long chains.

Hexagonal ring arranged in a "chair" shape

Carbon atom

Oxygen atom

Hydrogen atom

Basic glucose unit
The fundamental building block of cellulose is the glucose unit, which consists of a six-sided (hexagonal) ring containing five carbon atoms and one oxygen atom, usually arranged in a "chair" shape.

States of matter

Matter normally exists as a solid, liquid, or gas. A fourth state, plasma, occurs on Earth in phenomena such as ball lightning and aurorae, and also in fluorescent lights, and naked flames. Plasma is a hot gas made up of positively charged ions and electrons. A highly exotic fifth state, called a Bose-Einstein condensate, was discovered in 1995. It occurs at a few billionths of a degree above zero K, when atoms lose nearly all their energy; with no energy to transfer, they clump together into one "super atom."

THE FOUR COMMONLY OBSERVED STATES OF MATTER

State	Features
Solid	Molecules are fixed in position by chemical bonds. Energy from heat makes them vibrate, but they stay in place. Their shape can be changed only by physical force.
Liquid	Molecules are more widely spaced and can move around (flow), but they are still held together by inter-particle attraction; as a result, liquids have a fixed volume.
Gas	Molecules move independently of one another; they can flow and scatter. Volume is not fixed.
Plasma	Particles comprise flowing negative electrons and positive ions. Plasma has no fixed shape or volume. It can conduct electricity, and responds strongly to magnetic fields.

Changes in state
The changes of matter from one state to another are called phase transitions. They are usually brought about by the transfer of heat (see p.471), although the change from gas to plasma involves the breakdown of gas atoms.

PLASMA

Deionization

Ionization

Vaporization

Condensation

GAS

Deposition

Sublimation

LIQUID

SOLID

Freezing

Melting

FREEZING AND BOILING POINTS

Substance	Freezing point		Boiling point	
	(°F)	(°C)	(°F)	(°C)
Water	32	0	212	100
Saltwater (brine	-5.8	-21	226.4	108
Alcohol (ethanol)	-173.2	-114	172.9	78.3
Gasoline	140	<-60	100–400	40–200
Methane	296.5	-182.5	258.9	-161.6
Carbon dioxide	-68.1	-55.6	109.3	-78.5
Acetic acid	62.6	17	224.6	118.1
Benzene	41.9	5.5	176.4	80.2
Acetone	-138.1	-94.5	133.7	56.5
Camphor	355.6	179.8	399.2	204
Chloroform	-82.3	-63.5	143.6	62
Nitrogen dioxide	11.84	-11.2	69.9	21.1
Ammonia	-107.9	-77.7	-28.3	-33.5
Magnesium oxide	5,165.6	2,852	6,512	3,600
Sodium chloride	1,473.8	801	2,669	1,465
Calcium chloride	1,421.6	772	3,515	1,935

Physics

Physics is the study of the physical forces and qualities that govern the Universe, from the subatomic to the interstellar level. It provides definitions of matter, energy, force, and motion, and includes laws and principles that demonstrate the ways in which these phenomena relate to each other. There are various branches, including classical mechanics, electromagnetism, optics, and more recent disciplines such as quantum mechanics.

Energy, work, and power

According to the First law of Thermodynamics (see opposite), energy cannot be created or destroyed. It can, however, be converted into other forms—for example, to do work. There are two main types of energy: potential energy, which is stored in an object, and kinetic energy, which an object possesses due to its motion.

TYPES OF ENERGY

Potential energy	Motion energy
Magnetic	Kinetic
Gravitational	Heat
Chemical	Light
Electrical	Sound
Elastic	

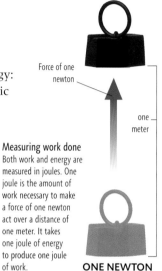

Force of one newton

one meter

Measuring work done
Both work and energy are measured in joules. One joule is the amount of work necessary to make a force of one newton act over a distance of one meter. It takes one joule of energy to produce one joule of work.

ONE NEWTON

ENERGY FORMULAS

Quantity	Description	Formula
Change in gravitational potential energy	mass × gravitational field strength × height difference	$\Delta E_p = mg\Delta h$
Kinetic energy	$1/2$ mass × square of velocity	$E_k = 1/2\,mv^2$
Weight	mass × gravitational field strength	$W = mg$
Work done	force × distance moved in direction of force	$W = Fs$
Power	$\dfrac{\text{work done (or energy transferred)}}{\text{time taken}}$	$P = \dfrac{W}{t}$
Efficiency	$\dfrac{\text{work output}}{\text{work input}} \times 100\%$	$\dfrac{W_o}{W_i} \times 100\%$

Mechanical forces

A force is a push or pull that makes an object move in a straight line or turn (see pp.112–13). Forces have a magnitude (size) and a direction, and are called vector quantities (unlike speed, which has only magnitude and is a scalar). They may act alone or together. We can harness forces by using machines to make work easier.

MOTION FORMULAS

Quantity	Description	Formula
Speed	$\dfrac{\text{distance}}{\text{time}}$	$s = \dfrac{d}{t}$
Time	$\dfrac{\text{distance}}{\text{speed}}$	$t = \dfrac{d}{s}$
Distance	speed × time	$d = st$
Velocity	$\dfrac{\text{displacement (distance in a given direction)}}{\text{time}}$	$v = \dfrac{s}{t}$
Acceleration	$\dfrac{\text{change in velocity}}{\text{time taken for change}}$	$a = \dfrac{(v-u)}{t}$
Resultant force	mass × acceleration	$F = ma$
Momentum	mass × velocity	$p = mv$

TURNING FORCES

Force	Description	Formula	Key
Moment of inertia	The equivalent of mass for an object rotating around an axis	$I = mr^2$	I = moment of inertia m = mass r^2 = square of distance from axis
Angular velocity	The velocity of an object rotating around an axis	$\omega = \dfrac{\Delta\theta}{\Delta t}$	ω = angular velocity $\Delta\theta$ = angular displacement Δt = change in time
Angular momentum	The momentum of an object rotating around an axis	$L = I\omega$	L = angular momentum I = moment of inertia ω = angular velocity

EQUATIONS OF MOTION UNDER CONSTANT ACCELERATION

These four equations are used to express constant acceleration in different ways:

$$v = u + at$$

$$s = \frac{(u+v)t}{2}$$

$$v^2 = u^2 + 2as$$

$$s = ut + \tfrac{1}{2}at^2$$

s = displacement
u = original velocity
v = final velocity
a = acceleration
t = time taken

HOOKE'S LAW

$$F_s = -kx$$

F_s = force of spring
k = spring constant (indication of spring's stiffness)
x = extension

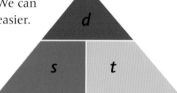

Mnemonic triangle
The relationship between distance, speed, and time can be shown in various ways, as seen in the equations on the left. This triangle can help you remember the relationships. The top element, d, can be divided by s or t, and s and t can be multiplied together.

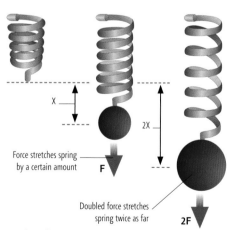

X

2X

Force stretches spring by a certain amount **F**

Doubled force stretches spring twice as far **2F**

Stretching force
Hooke's law states that the extension of a spring (or other stretchy object) is directly proportional to the force acting on it. This law is only true if the elastic limit of the object has not been reached. If the elastic limit has been reached, the object will not return to its original shape and may eventually break.

Newton's laws

Newton's laws of motion demonstrate how forces make objects move, remain at rest, or interact with other objects and forces (see pp.104–05). Newton also formulated the law of universal gravitation to define the force of gravity, which holds together everything in the Universe.

FIRST LAW	Unless disturbed by a force, an object will either stay still or travel at a constant speed in a straight line.
SECOND LAW	The force acting on a body is equal to its rate of change of momentum, which is the product of its mass and its acceleration.
THIRD LAW	For every action there is an equal and opposite reaction.

LAW OF UNIVERSAL GRAVITATION

$$F = \frac{Gm_1m_2}{r^2}$$

F = force
G = universal gravitational constant
m_1, m_2 = masses
r^2 = square of distances between masses

Laws of thermodynamics

These laws are fundamental physical statements that describe what happens when a thermodynamic system goes through an energy change (see pp.180–81).

ZEROTH LAW	If two bodies (distinct systems) are each in thermal equilibrium with a third body, these two bodies will also be in thermal equilibrium with each other.
FIRST LAW	Energy can be neither created nor destroyed.
SECOND LAW	The total entropy of an isolated system increases over time.
THIRD LAW	There is a theoretical minimum temperature at which the motion of the particles of matter would cease.

Heat transfer

Heat is a form of kinetic energy (see opposite) that causes atoms in a substance to vibrate. The more heat energy a substance contains, the more vigorously the atoms will vibrate. The vibrating atoms collide with each other, transferring heat from one to the next.

If two objects of different temperatures come into contact, heat will be transferred from the warmer to the cooler one until the two hold the same amount of heat. There are three main forms of heat transfer— conduction, convection, and radiation (see p.179).

HEAT FORMULAS

Quantity	Description	Formula	Key
Specific heat capacity	The amount of energy (measured in joules or calories) needed per unit mass (gram or kilogram) to raise the temperature of a given mass of a material by 1°Celsius.	$Q = cm\Delta T$	Q = energy c = specific heat capacity m = mass ΔT = temperature change
Specific latent heat	The amount of energy absorbed or lost as a material changes from one state to another (e.g. from liquid to solid).	$E = mL$	E = energy m = mass L = specific latent heat

TYPES OF SPECIFIC LATENT HEAT

Type	Description
Specific latent heat of melting	The energy (in joules) needed to convert 1kg of a material from solid to liquid without changing its temperature.
Specific latent heat of fusion	The energy needed to convert 1kg of a material from liquid to solid without changing its temperature.
Specific latent heat of boiling (or vaporization)	The energy needed to convert 1kg of a material from liquid to gas without changing its temperature.

Hottest atoms vibrate most vigorously

Heat energy is transferred to adjacent atoms, making them vibrate more actively

Conduction
Conduction occurs mainly in solid materials. Metals, such as iron bars, are very good conductors of heat energy, unlike nonmetals and gases. Atoms vibrate more when heated; if one end of an iron bar is heated, the atoms pass this heat energy on toward the cold end of the bar.

Convection
Convection is the transfer of heat through fluids (moving liquids or gases). A sea breeze is a convection current. Air heated by land becomes less dense and rises; to replace it, cooler air is drawn in from over the sea. At night, the direction of the current is reversed.

Radiation
Radiation is the transfer of heat through electromagnetic waves, such as from the Sun, or from the heating element in a toaster. Radiation is the only form of heat transfer that allows heat to pass through empty space.

373K	100°C	212°F
300K	27°C	81°F
273K	0°C	32°F
255K	−18°C	0°F
200K	−73°C	−99°F
100K	−173°C	−279°F
Absolute zero 0K	−273°C	−460°F
Kelvin	Celsius	Fahrenheit

Temperature
Temperature is a measure of how hot or cold an object or substance is. It actually shows how much kinetic (heat) energy something contains. Different scales are used to measure temperature, including the kelvin scale, on which 0K is absolute zero—the point at which atoms in a substance have no heat energy, and do not vibrate at all.

Pressure

Pressure is defined as the force applied to an object divided by the area over which the force is applied (see pp.100–01). It is a scalar quantity (one that has only magnitude or size), usually expressed in pascals (Pa), newtons (N) per m² (see p.477) or atmospheres (1 atmosphere being equivalent to 101.33kPa). Pressure varies according to the weight of the body exerting the force. For example, atmospheric pressure on a climber decreases as he or she nears the summit of a mountain because there is less atmosphere the higher from sea level the climber ascends. In contrast, pressure increases with increasing depth in the ocean. For every 33ft (10m) of depth, the pressure increases by one atmosphere. Pressure in fluids also acts equally in all directions and can be transmitted to act in another place (see p.180). If a fluid is moving (for example, water flowing in a river), pressure is reduced if the rate of flow increases. This principle, named after Bernoulli, applies to all fluids.

Relative density is the ratio of the density of a substance to the density of a reference substance. For liquids or solids it is the ratio of the density to the density of freshwater at 4°C, which is given as 1g per cm³; substances less dense than this will float in water.

PRESSURE AND DENSITY

Law	Description	Formula
Pressure	$\dfrac{\text{force}}{\text{area}}$	$P = \dfrac{F}{A}$
Density	$\dfrac{\text{mass}}{\text{Volume}}$	$\rho = \dfrac{m}{V}$
Volume	$\dfrac{\text{mass}}{\text{density}}$	$V = \dfrac{m}{\rho}$
Mass	volume × density	$m = V\rho$

Mnemonic triangle
The relationship between mass, volume, and pressure can be shown in various ways. This triangle can help you to remember the relationships. The top element, m, can be divided by v or ρ, and v and ρ can be multiplied together.

Pascal's principle
This states that a change in the pressure of an enclosed incompressible fluid is conveyed equally and undiminished to every part of the fluid and to the surfaces of its container. For a fluid in an enclosed container, therefore, the pressure increases toward the base of the container, because of the increased mass of the fluid above. For information on hydraulics, see diagram on p.145.

water under low pressure

water under high pressure

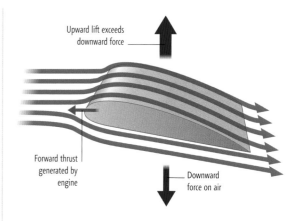

Ship displaces large amount of water, so can support large load

Archimedes' Principle
Any object wholly or partly immersed in a fluid is buoyed up by a force (upthrust) equal to the weight of the fluid displaced by the object.

Upward lift exceeds downward force

Forward thrust generated by engine

Downward force on air

Bernoulli's Principle
This principle describes the relationship between velocity and pressure in a moving fluid—a slow-flowing fluid exerting more pressure than a fast-flowing fluid. This principle also applies to airflow. One of the most dramatic everyday examples of Bernoulli's principle can be found in the airplane.

ATMOSPHERIC AND WATER PRESSURE

The pressure of the air in our atmosphere, and of the water in lakes and oceans, varies with height and depth.

Altitude/depth	Pressure (kPa)*
65,000ft (20,000m)	5.46
36,000ft (11,000m)	22.69
24,600ft (7,500m)	38.30
Sea level: 0ft (0m)	101.29
-400ft (-120m)	1,114.21
-21,300ft (-6,500m)	65,839.80
-35,800ft (-10,912m)	107,000.00

*kPa = kilopascals

GAS LAWS

Law	Description	Formula	Key
Avogadro's Law	At a constant temperature and pressure, volume is proportional to number of molecules. Equal volumes of different gases, in identical conditions of temperature and pressure, will contain equal numbers of molecules.	$V \propto n$	V = volume n = number of molecules $\propto$ = proportional to
Boyle's Law	For a given mass of gas at a constant temperature, volume is inversely proportional to pressure. So, for example, if volume doubles, pressure halves.	PV = constant	P = pressure V = volume
Charles's Law	For a given mass of gas at a constant pressure, volume is directly proportional to absolute temperature (measured in Kelvin).	V/T = constant	V = volume T = (absolute) temperature
Gay-Lussac's Law	For a given mass of gas at a constant volume, pressure is directly proportional to absolute temperature (measured in Kelvin).	P/T = constant	P = pressure T = (absolute) temperature
Ideal gas law	An "ideal" gas is a hypothetical gas in which particles may collide but do not have attractive forces between them. The ideal gas law is a good approximation of the behavior of many gases in various conditions.	$pV = nRT$	n = number of moles R = universal gas constant T = absolute temperature
Dalton's Law of Partial Pressures	Applies to mixtures of two or more gases. States that total pressure of gaseous mixture equals sum of partial pressures of individual component gases. This law is used to figure out gas mixtures breathed by scuba divers.	$P = \Sigma p$ or $P_{total} = p1 + p2 + p3 \dots$ etc	P = total pressure p = individual partial pressures

Sound

Mechanical vibrations cause sound waves (see pp.256–57). These can travel through solids, liquids or gases, but cannot move through a vacuum. Sound has three elements: pitch, which increases with the frequency of the sound waves; loudness, which increases with the amplitude; and quality, which depends on the wave form. The speed of the waves depends on the medium through which they travel. The speed of sound varies with air density and temperature; in dry air at sea level, its speed is 1,125ft (343m) per second.

SOUND-INTENSITY LEVEL

Source	Sound-intensity level (dB)*
Jet aircraft	120
Heavy machinery	90
Busy street	70
Conversation	50
Whisper	20

*dB = decibels

PROPERTIES OF SOUND WAVES

Longitudinal

Cannot travel through empty space (vacuum)

Can travel through solids, liquids, and gases

Comprise back-and-forth oscillations of particles in a medium, e.g. air

Travel at about 1,125 ft/s (343 m/s) through air at sea level

WAVE SPEED

$$v = f\lambda$$

v = speed f = frequency
λ = wavelength

Frequency and amplitude

Wavelength is related to frequency: the lower the wave's frequency, the longer the wavelength. The energy of a sound wave is indicated by its amplitude: the higher the intensity, the greater the amplitude.

1/100 second   1/100 second

short wavelength   long wavelength

High amplitude

Low amplitude

HIGH–FREQUENCY WAVE   **LOW–FREQUENCY WAVE**

Light

Light is a form of electromagnetic radiation that is visible to the human eye (see pp.114–17). It can only pass through certain solids and liquids, such as glass and water; however, unlike sound, it can travel through a vacuum. Light can be reflected from polished or shiny surfaces, such as water or a mirror, or it can be refracted through a medium such as glass—a quality used in the correction of near- and farsightedness. White light can also be split to reveal its constituent parts—red, orange, yellow, green, blue, indigo, and violet light—by passing it through a prism.

REFRACTION OF LIGHT

Feature	Description	Formula	Key
Refractive index	The ability of a material to refract (bend) light.	$n = \dfrac{c}{v}$	c = speed of light in vacuum v = speed of light in medium
Snell's Law	Expresses relationship between angles of incidence and refraction in different media; shows what happens when light passes from one medium to another.	$n = \dfrac{\sin i}{\sin r}$	n = refractive index i = angle of incidence r = angle of refraction

Reflecting light

There are two laws that govern reflection. The first is that the incident ray (striking the boundary), the reflected ray (returning from the boundary), and the "normal" (the line drawn at right angles to the boundary) are in the same plane. The second is that the angle of incidence equals the angle of reflection. Images seen by reflection appear to be the same distance behind the reflecting surface as the object is in front of it.

PROPERTIES OF ELECTROMAGNETIC WAVES

Transverse

Can travel through empty space (vacuum)

Can travel through some solids, liquids, and gases

Comprise side-to-side oscillations of electric and magnetic fields

Travel at about 186,000 miles/s (300,000km/s) in a vacuum, slightly slower in air

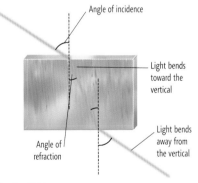

Angle of incidence

Light bends toward the vertical

Light bends away from the vertical

Angle of refraction

Refracting light

Light bends toward the vertical when slowed on entering a denser medium, and bends away from it on exiting. The amount of bending is described quantitatively by Snell's Law.

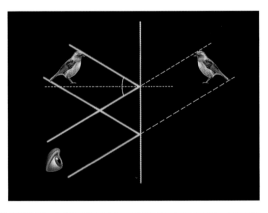

Ray of light   Point of focus

Point of focus

Refracted light

Refracted light

Splitting light

White light passing through a prism is refracted to form a spectrum. This is because each of the colors is refracted at a slightly different angle. The blue light is "bent" most and the red light least. If the spectrum is passed through a second prism, white light can be restored (see p.116–17).

Convex and concave lenses

Parallel light rays converge on passing through a convex lens and diverge on passing through a concave lens. The distance between the point of focus and the lens is the focal length. A highly curved lens has a smaller focal length and is more powerful than a less curved lens.

EXAMPLES OF REFRACTIVE INDICES

Substance	Refractive index	Speed of light (m/s)
Air	1.0	300,000,000
Water	1.33	225,000,000
Perspex	1.5	200,000,000
Glass	1.5	200,000,000
Diamond	2.4	120,000,000

Magnetism and electricity

A magnet is any object or device that produces a magnetic field (see pp.80–81)—an area comprising lines of force that emanate from one end of a magnet, called the "north-seeking" (N) pole, and re-enter the magnet at the other end, the "south-seeking" (S) pole. This field can attract other objects, such as iron filings, with the potential to be magnetized. Certain substances, such as magnetite (rock containing iron oxide), are strongly magnetic, and Earth itself has a magnetic field. Magnetism has a close interrelationship with electricity—magnetic fields can be used to generate electric currents, and vice versa (see pp.166–67).

Similar poles repel each other

Opposing poles attract each other

REPULSION   **ATTRACTION**

Magnetic repulsion and attraction
If two magnets are placed with their N poles close together, the fields will repel each other. If the magnets are placed with opposing poles close together, the field from the N pole of one will be attracted to the S pole of the other.

ELECTRICITY FORMULAS

Quantity	Description	Formula
Electric charge	current × time	$q = I\,t$
Electrical energy	potential difference (or voltage) × current × time	$E = V I\,t$
Electrical power	potential difference (or voltage) × current	$P = V I$

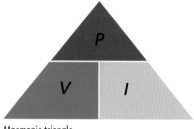

Mnemonic triangle
The relationship between electrical power, current, and potential difference (voltage) can be shown in various ways, as seen here. The top element, P, can be divided by V or I, and V and I can be multiplied together.

Maxwell's equations

These four equations, developed from the work of James Clerk Maxwell, comprise a full description of classical (nonquantum) electromagnetism, showing how electric and magnetic fields are produced, and how the rates of change of these fields are related to their sources (see p.167). For this reason, they are some of the most important equations used in science. In the process of figuring out these laws, Maxwell derived the electromagnetic wave equation, which demonstrated that light was a form of wave with interlinked electric and magnetic elements. This finding led to the discovery of radio waves and the understanding of other parts of the electromagnetic spectrum (see below).

THE FOUR EQUATIONS

Law	Statement	Application
Gauss's law for electricity	The electric flux through any closed surface is proportional to the total charge contained within that surface.	Used to calculate electric fields around charged objects.
Gauss's law for magnetism	For a magnetic dipole with any closed surface, the magnetic flux drawn inward toward the south pole will equal the flux directed outward from the north pole; the net flux will always be zero.	Describes sources of magnetic fields and shows that they will always be closed loops.
Faraday's law of induction	The induced electromotive force (emf) around any closed loop equals the negative of the rate of change of the magnetic flux through the area enclosed by the loop.	Describes how a changing magnetic field can generate an electric field; this is the operating principle for electric generators, inductors, and transformers.
Ampère's law with Maxwell's correction	In static electric field, the line integral of the magnetic field around any closed loop is proportional to the electric current that is flowing through the loop.	Used to calculate magnetic fields. Shows that magnetic current can be generated in two ways: by electric current and by changing electric fields.

Electromagnetic spectrum

The electromagnetic spectrum is a continuum comprising various forms of radiation wave. The waves travel at the same speed (the speed of light), but have different wavelengths. Their frequency, and the energy that they carry, increases as the wavelength decreases. Visible light is the part of the spectrum that we can perceive naturally; other forms can be detected by instruments. Some forms are blocked by Earth's atmosphere.

Radio waves
The longest radio waves are absorbed in Earth's upper atmosphere by the electrically charged ionosphere. Intermediate waves, with wavelengths between a few centimeters and about 10 meters, pass through the atmosphere and reach Earth's surface, while those with shorter wavelengths are absorbed by atmospheric carbon dioxide and water vapor.

Microwaves
The shortest form of radio waves, with wavelengths between 1cm and 100μm, are known as microwaves. These waves are widely used in communications as well as in microwave ovens.

RADIO WAVES   MICROWAVES

1Km   100m   10m   1m   10cm   1cm   1mm

Electric currents and circuits

Electric charge can be made to travel in a flow, or current (see pp.164–65). This current can be generated by an electromotive force (emf) such as a battery cell, and directed around wire loops, or circuits, to provide power for electrical devices. There are two main ways to drive the flow around a circuit: by creating an "alternating" current (AC), which flows to and fro; or by producing a "direct" current (DC), which flows in one direction only (see p.252). The way current flows around a circuit is governed by particular laws (see below).

Mnemonic triangle
Ohm's Law can be illustrated by a triangle such as this one, which shows the relationship between electrical current, resistance, and potential difference (voltage). The top element, V, can be divided by I or R, and I and R can be multiplied together.

Series circuit
If a circuit has components connected in a single loop, the components are said to be "in series." In a series circuit, the same current flows through all of the components.

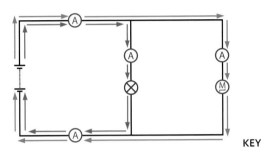

Parallel circuit
If a circuit includes two or more branches, the components are said to be "in parallel." In a parallel circuit, the components are connected so that the same voltage is supplied to each component.

KEY

Path of current around single circuit

Path of current around parallel branch

ELECTRICITY AND CIRCUIT LAWS

Law	Description	Formula	Key
Coulomb's Law	The force of attraction or repulsion between two charged particles is directly proportional to the product of the charges and inversely proportional to the distance between them.	$F = k\dfrac{q_1 q_2}{r^2}$	k = Coulomb's constant q_1 and q_2 = electric (point) charges r^2 = square of distance
Ohm's Law	This law expresses the relationships between voltage, resistance and current, which can be expressed in several ways.	$I = \dfrac{V}{R}$ $\quad R = \dfrac{V}{I}$ $V = IR$	R = resistance I = current V = potential difference (voltage)
Kirchhoff's Current Law	The sum of the electric currents entering any junction in a circuit is equal to the sum of those leaving the junction.	$\Sigma I = 0$	Σ = summation symbol I = current
Kirchhoff's Voltage Law	The sum of the voltage changes around any closed loop path is zero.	$\Sigma V = 0$	Σ = summation symbol V = potential difference (voltage)

KEY TO CIRCUIT SYMBOLS

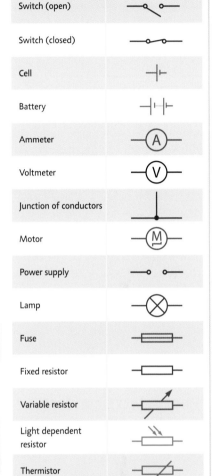

Switch (open)	
Switch (closed)	
Cell	
Battery	
Ammeter	
Voltmeter	
Junction of conductors	
Motor	
Power supply	
Lamp	
Fuse	
Fixed resistor	
Variable resistor	
Light dependent resistor	
Thermistor	
Diode	
Generator	
AC power supply	

Infrared
Emitted by warm objects, infrared radiation is mostly absorbed in Earth's atmosphere by gases such as carbon dioxide and water vapor.

Optical window
Wavelengths measured in billionths of a foot pass through Earth's atmosphere. Animal eyesight has evolved to be sensitive to radiations in this "window."

Utraviolet
The wavelengths of ultraviolet rays are shorter than visible light. Most ultraviolet rays from the Sun are blocked by the atmospheric ozone layer.

X-rays
With wavelengths from 10 nanometers (nm) down to 0.01nm, X-rays carry large amounts of harmful energy. Fortunately Earth's atmosphere acts as a good barrier to them.

Gamma rays
Even shorter than X-rays, gamma rays have an extreme potential for penetrating materials and damaging tissues. As with X-rays, our atmosphere prevents them from reaching Earth's surface.

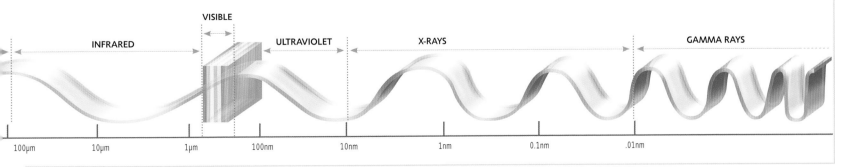

VISIBLE

INFRARED — ULTRAVIOLET — X-RAYS — GAMMA RAYS

100μm · 10μm · 1μm · 100nm · 10nm · 1nm · 0.1nm · .01nm

Quantum mechanics

The branch of science that deals with minute, indivisible units of energy is known as quantum mechanics. Since its early theoretical development at the beginning of the 20th century, it has provided a means to answer questions that were beyond the realm of classical mechanics and contributed to many new ideas, including string theory and grand unified theory. Quantum mechanics has also been used to help explain fundamental forces.

The four fundamental forces

Ordinary matter is subject to four basic forces—gravity, electromagnetism, and strong and weak nuclear forces (see pp.404–05). Each is associated with a particular particle.

FORCE-CARRIER PARTICLES

Matter particles affected by a particular force produce and absorb specific force carriers. For example, strong forces are linked with gluons.

Particle	Force	Relative strength	Range in (m)
*Graviton	Gravity	10^{-41}	Infinite
Photon	Electromagnetic	1	Infinite force
Gluon	Strong force	25	10^{-15}
W, Z bosons	Weak interaction	0.8	10^{-18}

* hypothetical force particle

Subatomic particles

Elementary particles are the constituents of everything in the Universe, except dark matter (see pp.382–83). Quarks and leptons are the basic components of matter, while gauge bosons are the particles that transmit the fundamental forces. These particles may combine to form larger particles. Scientists think that every type of particle has an opposing "antiparticle." If paired, they can annihilate each other, producing particles of light (photons).

SUBATOMIC PARTICLES: TYPES

	Elementary particles	Composite particles (hadrons)
Fermions (matter particles)	**Quarks** Particles that make up protons and neutrons. There are six "flavors": Up, Down, Charm, Strange, Top, Bottom. **Leptons** A group of six particles, comprising electron, muon, and tau particles and their associated neutrinos.	**Baryons** Particles made up of three quarks. The best known are the proton (two Up and one Down particle) and the neutron (one Up and two Down particles)
Bosons (force-carrier particles)	**Gauge bosons** Particles associated with the four fundamental forces (see left). No gauge boson has yet been found for gravity, although the existence of a "graviton" has been hypothesized.	**Mesons** Particles made from a quark with an antiquark. There are many types, including the positive pion (Up quark with Down antiquark) and the negative kaon (Strange quark with Up antiquark).

Photoelectric effect

In 1905 Albert Einstein proposed the idea that light exists as tiny packets, or particles, called photons. The photoelectric effect demonstrates its particle-like properties: when light is shone on a metal (or photoemissive) target, photons can "push" electrons off the surface of the metal, causing a current to flow. Increasing the intensity, or shortening the wavelength, of the light increases the energy of the ejected electrons. Changing the material of the target also affects the current produced.

Red, blue, and ultraviolet light
When red light (left) is shone on to a metal surface, no electrons are ejected. Shorter-wavelength blue light (center) is more energetic than red, so it does cause electrons to be ejected. Ultraviolet light (right) causes many electrons to be ejected at very high energy.

Planck's constant

In 1900 Max Planck explained the peaks in "black body" curves (see p.314) by relating the temperature of a black body to the intensity of energy radiated by it at a given wavelength. He proposed that the peaks were areas where there were lots of atoms producing particles of light that, when added together, were greater in number than at either end. He divided the light into "packets," or quanta, and assigned each an energy (E) related to its frequency (f). This worked provided that $E = hf$, h being a new constant in nature now called Planck's constant.

Power radiated at each wavelength

Wavelength (μm)

Black body curves
Plotted on a graph, black body curves at different temperatures have strong peaks with little radiation to either side. This phenomenon could not be explained by classical physics and so led to the quantum revolution.

Einstein's theories of relativity

Albert Einstein challenged Newton's accepted theory of gravitation with the publication of his General Theory of Relativity in 1915 (see pp.302–03). This showed that Newton's three Laws of Motion were only approximately correct, breaking down particularly as the speed of light was approached. Similarly, Einstein showed Newtonian gravity to be an approximation that works well for weak gravitational fields but less well for strong fields.

THEORY	PROPOSITION
Special Theory of Relativity	1) All physical laws are the same in all frames of reference in uniform motion with respect to one another.
	2) The speed of light is a constant, regardless of the motions of the light source and the observer.
General Theory of Relativity	Spacetime is curved: gravity causes distortions of time and mass, and massive objects (such as stars) warp spacetime around them.

Common equations

The following equations are commonly used in physics. Some of the units used to calculate these equations can be found in the metric and imperial measurement table on p.431.

EQUATIONS		
Quantity	**Statement**	**Formula**
Gravitational force	universal gravitational constant × mass$_1$ × mass$_2$ / distance^2	$\dfrac{Gm_1m_2}{r^2}$
Change in gravitational potential energy	mass × gravitational field strength × height difference	$\Delta E_p = mg\Delta h$
Kinetic energy	½mass × velocity^2	$E_k = \frac{1}{2}mv^2$
weight	mass × gravitational field strength	$W = mg$
Work done	force × distance moved in the direction of the force	$W = Fs$
Power	$\dfrac{\text{work done}}{\text{time taken}}$ or $\dfrac{\text{energy transferred}}{\text{time taken}}$	$P = \dfrac{W}{t}$
Efficiency	$\dfrac{\text{work output}}{\text{work input}} \times 100\%$	$\dfrac{W_o}{W_i} \times 100\%$
Speed	$\dfrac{\text{distance moved}}{\text{time taken}}$	$s = \dfrac{d}{t}$
Velocity	$\dfrac{\text{displacement}}{\text{time}}$	$v = \dfrac{s}{t}$
Acceleration	$\dfrac{\text{change in velocity}}{\text{time taken for this change}}$	$a = \dfrac{(v-u)}{t}$
Resultant force	mass × acceleration	$F = ma$
Momentum	mass × velocity	$m\,v$
Moment of inertia	mass × radius from the axis^2	$I = mr^2$
Density	$\dfrac{\text{mass}}{\text{volume}}$	$\rho = \dfrac{m}{V}$
Volume	$\dfrac{\text{mass}}{\text{density}}$	$V = \dfrac{m}{\rho}$
Mass	volume × density	$m = V\rho$
Pressure	$\dfrac{\text{force}}{\text{area}}$	$P = \dfrac{F}{A}$
Refractive index	$\dfrac{\text{speed of light in a vacuum}}{\text{speed of light in a medium}}$	$n = \dfrac{c}{v}$
Wave speed	frequency × wavelength	$v = f\lambda$
Electric charge	current × time	$q = It$
Potential difference (voltage)	current × resistance or $\dfrac{\text{energy transferred}}{\text{charge}}$	$V = IR$ $V = \dfrac{W}{q}$
Resistance	$\dfrac{\text{voltage}}{\text{current}}$	$R = \dfrac{V}{I}$
Electrical energy	potential difference (or voltage) × current × time	$E = VIt$
Electrical power (DC)	potential difference (or voltage) × current	$P = VI$ $V = \dfrac{P}{I}$ $I = \dfrac{P}{V}$

Common physical qualities

The following table shows some common physical properties and their symbols. The units by which they are measured are indicated by the SI units and their symbols.

Physical quality	Symbol	SI unit	SI unit symbol
Acceleration, deceleration	a	meter/second^2 kilometer/hour/second	m s^{-2} km h^{-1} s^{-1}
Angular velocity	ω	radian/second	rad s^{-1}; s^{-1}
Density	ρ	kilogram/meter^3 kilogram/milliliter	kg m^{-3} kg ml^{-1}
Electric charge	Q, q	coulomb	C
Electric current	I, i	ampere (coulomb/second)	A (Cs^{-1})
Electrical energy	–	megajoule kilowatt-hour	MJ kWh
Electrical power	P	watt (joule/second)	W (J s^{-1})
Electromotive force (emf)	E	volt (watt/ampere)	V (W A^{-1})
Electrical conductance	S	siemen (ohm^{-1})	A V^{-1}
Electrical resistance	R	ohm (volt/ampere)	Ω (V A^{-1})
Frequency	f	hertz (cycles/second)	Hz (s^{-1})
Force	F	newton (kilogram meter/second^2)	N (kg m s^{-2})
Gravitational intensity, field strength	–	newton/kilogram	N kg^{-1}
Magnetic field strength	H	ampere/meter	A m^{-1}
Magnetic flux	Φ	weber	Wb
Magnetic flux density	B	tesla (weber/meter^2)	T (Wb m^{-2})
Mass	m	kilogram	kg
Mechanical power	P	watt (joule/second)	W (J s^{-1})
Moment of inertia	I	kilogram meter^2	Kg m^2
Momentum	p	kilogram meter/second	Kg m s^{-1}
Pressure	P	pascal (newton/meter^2)	Pa (N m^{-2})
Quantity of substance	n	mole	mol
Specific heat capacity	C or c	joule/kilogram/kelvin	J kg^{-1} K^{-1}
Specific latent heats of fusion, vaporization	L	joule/kilogram	J kg^{-1}
Torque	τ	newton meter	N m
Velocity, speed	u, v	meter/second kilometer/hour	m s^{-1} km h^{-1}
Volume	V	meter^3 milliliter	m^3 ml
Wavelength	λ	meter	m
Weight	W	newton	N
Work, energy	W	joule (newton meter)	J (N m)

Mathematics

Mathematics is a huge subject, encompassing the study of numbers, quantities, shapes, space, and patterns, as well as relationships between entities and operations that can be performed on them. Mathematics plays a fundamental role in all the sciences, and over the millennia it has developed its own set of symbols, theorems, and other key concepts, a selection of which are described in the following pages.

Counting systems

Tallies have been used since prehistoric times to represent amounts and numbers, and from these, various counting systems have been developed. The Roman numeral system is based on seven letters of the alphabet whose combined values give a total. Our current system, based on Hindu–Arabic numerals, uses ten digits that can be arranged to represent larger numbers, while the binary system, which is used mainly in computing, uses just two digits: 1 and 0. Numerical equivalents for the different counting systems are shown here.

Hindu–Arabic	Roman	Binary	Hindu–Arabic	Roman	Binary	Hindu–Arabic	Roman	Binary
1	I	1	13	XIII	1101	70	LXX	1000110
2	II	10	14	XIV	1110	80	LXXX	1010000
3	III	11	15	XV	1111	90	XC	1011010
4	IV	100	16	XVI	10000	100	C	1100100
5	V	101	17	XVII	10001	200	CC	11001000
6	VI	110	18	XVIII	10010	300	CCC	100101100
7	VII	111	19	XIX	10011	400	CD	110010000
8	VIII	1000	20	XX	10100	500	D	111110100
9	IX	1001	30	XXX	11110	1,000	M	1111101000
10	X	1010	40	XL	101000	5,000	$\bar{V}$	1001110001000
11	XI	1011	50	L	110010	10,000	$\bar{X}$	10011100010000
12	XII	1100	60	LX	111100	100,000	$\bar{C}$	11000011010100000

Numerical equivalents

Percentages, decimals, and fractions are different ways of presenting a numerical value as a proportion of a given amount. For example, 10 percent (10%) has the equivalent value of the decimal 0.1 and the fraction $\frac{1}{10}$. However, some values that are expressed precisely by fractions have unresolved decimal equivalents that contain recurring decimal numbers. For example, the fraction $\frac{1}{3}$ has the decimal equivalent of 0.333 recurring, where the "3" repeats infinitely. In these instances decimal values are given to the first three decimal places below.

%	Decimal	Fraction	%	Decimal	Fraction	%	Decimal	Fraction	%	Decimal	Fraction	%	Decimal	Fraction
1	0.01	$^1/_{100}$	12½	0.125	$^1/_8$	24	0.24	$^6/_{25}$	36	0.36	$^9/_{25}$	49	0.49	$^{49}/_{100}$
2	0.02	$^1/_{50}$	13	0.13	$^{13}/_{100}$	25	0.25	$^1/_4$	37	0.37	$^{37}/_{100}$	50	0.50	$^1/_2$
3	0.03	$^3/_{100}$	14	0.14	$^7/_{50}$	26	0.26	$^{13}/_{50}$	38	0.38	$^{19}/_{50}$	55	0.55	$^{11}/_{20}$
4	0.04	$^1/_{25}$	15	0.15	$^3/_{20}$	27	0.27	$^{27}/_{100}$	39	0.39	$^{39}/_{100}$	60	0.60	$^3/_5$
5	0.05	$^1/_{20}$	16	0.16	$^4/_{25}$	28	0.28	$^7/_{25}$	40	0.40	$^2/_5$	65	0.65	$^{13}/_{20}$
6	0.06	$^3/_{50}$	16⅔	0.166	$^1/_6$	29	0.29	$^{29}/_{100}$	41	0.41	$^{41}/_{100}$	66⅔	0.666	$^2/_3$
7	0.07	$^7/_{100}$	17	0.17	$^{17}/_{100}$	30	0.30	$^3/_{10}$	42	0.42	$^{21}/_{50}$	70	0.70	$^7/_{10}$
8	0.08	$^2/_{25}$	18	0.18	$^9/_{50}$	31	0.31	$^{31}/_{100}$	43	0.43	$^{43}/_{100}$	75	0.75	$^3/_4$
8⅓	0.083	$^1/_{12}$	19	0.19	$^{19}/_{100}$	32	0.32	$^8/_{25}$	44	0.44	$^{11}/_{25}$	80	0.80	$^4/_5$
9	0.09	$^9/_{100}$	20	0.20	$^1/_5$	33	0.33	$^{33}/_{100}$	45	0.45	$^9/_{20}$	85	0.85	$^{17}/_{20}$
10	0.10	$^1/_{10}$	21	0.21	$^{21}/_{100}$	33⅓	0.333	$^1/_3$	46	0.46	$^{23}/_{50}$	90	0.90	$^9/_{10}$
11	0.11	$^{11}/_{100}$	22	0.22	$^{11}/_{50}$	34	0.34	$^{17}/_{50}$	47	0.47	$^{47}/_{100}$	95	0.95	$^{19}/_{20}$
12	0.12	$^3/_{25}$	23	0.23	$^{23}/_{100}$	35	0.35	$^7/_{20}$	48	0.48	$^{12}/_{25}$	100	1.00	1

Mathematical signs and symbols

This table presents a selection of signs and symbols commonly used in mathematics and other branches of science. These generally represent abbreviations for entities, relationships between entities and/or values, and operations. Using signs and symbols, mathematicians and scientists can express complex equations and formulas in a standardized way that is universally understood.

Symbol	Definition
$+$	plus; positive
$-$	minus; negative
$\pm$	plus or minus; positive or negative; degree of accuracy
$\mp$	minus or plus; negative or positive
$\times$	multiplied by (6 x 4)
$\cdot$	multiplied by (6·4); scalar product of two vectors (A·B)
$\div$	divided by (6 ÷ 4)
$/$	divided by; ratio of (6/4)
$—$	divided by; ratio of (⁶/₄)
$!$	factorial (4! = 4 x 3 x 2 x 1)
$=$	equals
$\neq$	not equal to
$\equiv$	identical with; congruent to
$\not\equiv$	not identical with
$\triangleq$	corresponds to
$:$	ratio of (6:4)
$::$	proportionately equals (1:2 :: 2:4)
$\approx$	approximately equal to; equivalent to; similar to
$>$	greater than
$\gg$	much greater than
$\ngtr$	not greater than
$<$	less than

Symbol	Definition
$\ll$	much less than
$\nless$	not less than
$\geqslant, \geqq, \geqq$	equal to or greater than
$\leqslant, \leqq, \leqq$	equal to or less than
$\propto$	directly proportional to
$( )$	parentheses
$[ ]$	brackets
$\{ \}$	braces
$-$	vinculum: division (a-b); chord of circle or length of line (AB); arithmetic mean (X)
$\exists$	there exists
$\forall$	for all
∞	infinity
$\rightarrow$	approaches the limit
$\sqrt{\ }$	square root
$^3\sqrt{\ } \ ^4\sqrt{\ }$	cube root, fourth root, etc.
$\%$	percent
$'$	prime; minute(s) of arc; foot/feet
$''$	double prime; second(s) of arc; inch/inches
$\cap$	arc of circle
$°$	degree of arc
$\angle, \angle^s$	angle(s)
	equiangular
$\perp$	perpendicular

Symbol	Definition
$\parallel$	parallel
$\bigcirc$	circle
$\triangle$	triangle
$\square$	square
$\square$	rectangle
$\square$	parallelogram
$\cong$	congruent to; identical with
$\therefore$	therefore
$\because$	because
$\underset{\_}{m}$	measured by
$\mathbf{\Delta}$	increment
$\sum$	summation
$\prod$	product
$\int$	integral
∇	del: differential operator
$\cup$	union
$\cap$	intersection
$\in$	is an element of
$\subset$	strict inclusion
$\supset$	contains
$\Rightarrow$	implies
$\Leftarrow$	implied by
$\Leftrightarrow$	implies and is implied by

Platonic solids

Also known as convex regular polyhedrons, the five Platonic solids were known by the ancient Greeks and are named after the Greek philosopher Plato (see p.34). All of their faces are identical regular polygons. The same number of faces meet at every vertex (point where edges meet), and all angles at each vertex are equal. Their names come from their respective number of faces: 4, 6, 8, 12, 20.

Regular tetrahedron
Composed of equilateral triangles, the regular tetrahedron has 4 faces, 6 edges, and 4 vertices.

Regular hexahedron (cube)
The most easily recognizable of the Platonic solids, the cube consists of 6 square faces, 12 edges, and 8 vertices.

Regular octahedron
Composed of equilateral triangles, the regular octahedron has 8 faces, 12 edges, and 6 vertices.

Regular dodecahedron
Composed of regular pentagons, the regular dodecahedron has 12 faces, 30 edges, and 20 vertices.

Regular icosahedron
Composed of equilateral triangles, the regular icosahedron has 20 faces, 30 edges, and 12 vertices.

Areas and volumes

The illustrations below show various two- and three-dimensional shapes, and the formulas for calculating their areas and volumes. In the formulas, two letters together means that they are multiplied together (for example, "ah" means "a" multiplied by "h"); the superscript ^{2} means squared (for example, "a^2" is "a" multiplied by "a"); the superscript ^{3} means cubed (for example, "a^3" is "a" multiplied by "a" multiplied by "a"); the symbol $\sqrt{}$ means square root; and the symbol π represents pi, which is 3.142 to three decimal places.

AREA OF TWO-DIMENSIONAL SHAPES

TRIANGLE
Area = $\dfrac{ah}{2}$

SQUARE
Area = a^2

RECTANGLE
Area = ab

RHOMBUS
Area = $\dfrac{ab}{2}$

TRAPEZIUM
Area = $\dfrac{(a+b)}{2}$ h

PARALLELOGRAM
Area = ah

REGULAR POLYGON WITH N SIDES
Area = $\dfrac{(abN)}{2}$

CIRCLE
Area = πr^2

SURFACE AREA AND VOLUME OF THREE-DIMENSIONAL SHAPES

CONE
Surface area
(including base) = $\pi rs + \pi r^2$
Volume = $\dfrac{\pi r^2 h}{3}$

SPHERE
Surface area = $4\pi r^2$
Volume = $\dfrac{4\pi r^3}{3}$

TORUS
Surface area = $4\pi^2 Rr$
Volume = $2\pi^2 Rr^2$

CYLINDER
Surface area = $2\pi r^2 + 2\pi rh$
Volume = $\pi r^2 h$

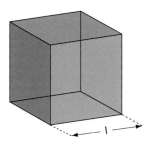

CUBE
Surface area = $6l^2$
Volume = l^3

RECTANGULAR PRISM
Surface area = 2(lh + lw + wh)
Volume = lwh

REGULAR SQUARE PYRAMID
Surface area = $2bs + b^2$
Volume = $\dfrac{b^2 h}{3}$

REGULAR TETRAHEDRON
Surface area = $a^2\sqrt{3}$
Volume = $\dfrac{\sqrt{2}}{12} a^3$

Factorials

The factorial of a number (denoted by the symbol !) is the product of all positive integers up to and including that number. For example, 4 factorial (written as 4!) = 4 x 3 x 2 x 1 = 24.

N	N!	N	N!
0	1	11	39916800
1	1	12	479001600
2	2	13	6227020800
3	6	14	87178291200
4	24	15	1307674368000
5	120	16	20922789888000
6	720	17	355687428096000
7	5040	18	6402373705728000
8	40320	19	121645100408832000
9	362880	20	2432902008176640000
10	3628800	21	51090942200000000000

Compound interest

Compound interest is interest that is paid on both the principal amount of money invested and on any accumulated interest. The formula for calculating the total value of an invested sum after a period of compound interest if the interest is paid once a year is:

$$A = P(1+i)^n$$

A = final amount of money accumulated, including interest
P = principal amount of money invested
i = rate of interest (expressed as a decimal) per period
(per year for interest paid annually)
n = number of periods for which the money is invested

For example, if $1,000 was invested for three years at an interest rate of 5% per year:

   P = $1,000
   i = 5% per year; expressed as a decimal, 5% = 5/100 = 0.05
   n = 3
Therefore the amount after three years, A = 1,000 x $(1 + 0.05)^3$
   = 1,000 x 1.05^3
   = 1,000 x 1.157625
   = $1,157.62 (rounded down)
If the same amount ($1,000) was invested for 5 years at 5% per year, the amount after that period would be:
   1,000 x $(1 + 0.05)^5$ = $1,276.28

Coefficients

In algebra a coefficient is a number that is the multiple of a variable in an equation. For example, in the equation $3x^2 + 7y - 4 = 0$, the coefficient of the variable x^2 is 3, and the coefficient of the variable y is 7; in the equation $3(x + y) = 0$, the coefficient of the variable $(x + y)$ is 3. If there is no number in front of a variable, the coefficient is 1; for example, in the equation x + 2 = 5, the coefficient of x is 1.

Pascal's triangle

A triangular arrangement of numbers, Pascal's triangle is named after the 17th-century French mathematician Blaise Pascal (see p.196). In Pascal's triangle, each row has 1 as the first and last number. The other numbers are created by adding the two numbers that are to the left and right in the row above. Pascal's triangle is used mainly in algebra and probability.

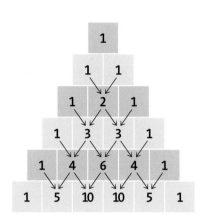

Types of algebraic equations

Equations can be classified according to the highest power to which one or more of their variables is raised (see p.49). For example, in a quadratic equation at least one variable is squared (raised to the power of two).

Equation	Common mathematical form	Description/uses
Linear	$ax + by + c = 0$	No variable is raised to a power greater than one. Used in simple problems of addition, multiplication, and division.
Quadratic	$ax^2 + bx + c = 0$	At least one of the variables is squared. Used in calculations involving area.
Cubic	$ax^3 + bx^2 + cx + d = 0$	At least one of the variables is cubed. Used in calculations involving volume.
Quartic	$ax^4 + bx^3 + cx^2 + dx + e = 0$	At least one of the variables is to the power of four. Used in computer graphics.
Quintic	$ax^5 + bx^4 + cx^3 + dx^2 + ex + f = 0$	At least one of the variables is to the power of five. Solution of quintic equations was important to the development of abstract algebra.

Averages

An average is a single number that represents a group of numbers. The term "average" is usually used to refer to the arithmetic mean (often called simply the mean) but it may also refer to the median or mode.

Measure	Definition	Example
Arithmetic mean (mean)	The sum of all the values in a group divided by the number of values.	To find the mean of 2, 4, 5, 5, 6, add together all of the values: 2 + 4 + 5 + 5 + 6 = 22. There are 5 values so the mean is 22 ÷ 5 = 4.4.
Median	The middle value of a group when the values are put in order. If there are two values in the middle, the median is the mean of those two values.	To find the median of 6, 5, 2, 4, 5, put the values in order: 2, 4, 5, 5, 6. The middle value—the median—is the third value along, which is 5.
Mode	The value that occurs most often in a group. It is possible to have more than one mode if more than one value appears "the most."	In the group of values 2, 4, 5, 5, 6, the value that appears most—the mode—is 5. In the group 2, 4, 4, 5, 5, 6, the values 4 and 5 both appear "the most" and so both are modes.

Basic trigonometry

The key formulas used in trigonometry are the sine (sin), cosine (cos), and tangent (tan) ratios. They can be used to find the angles in a right-angled triangle when at least two sides are known, or conversely, if an angle and the length of a side are known, these can be used to figure out lengths of the other sides. A useful way of recalling the formulas is by remembering the word "SOH-CAH-TOA."

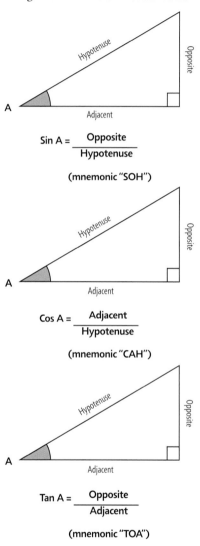

$$\text{Sin A} = \frac{\text{Opposite}}{\text{Hypotenuse}}$$

(mnemonic "SOH")

$$\text{Cos A} = \frac{\text{Adjacent}}{\text{Hypotenuse}}$$

(mnemonic "CAH")

$$\text{Tan A} = \frac{\text{Opposite}}{\text{Adjacent}}$$

(mnemonic "TOA")

Conic sections

Also known as conics, conic sections are curves or shapes produced when a cone or double cone is intersected by a plane. The diagrams below show the shapes produced (in red) from various intersections.

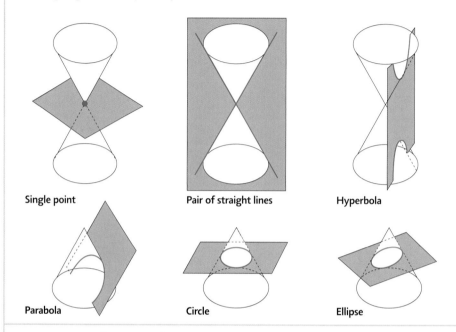

Single point
Pair of straight lines
Hyperbola
Parabola
Circle
Ellipse

Squares, cubes, and roots

Number	Square	Cube	Square root	Cube root	Number	Square	Cube	Square root	Cube root
1	1	1	1.000	1.000	13	169	2,197	3.606	2.351
2	4	8	1.414	1.260	14	196	2,744	3.742	2.410
3	9	27	1.732	1.442	15	225	3,375	3.873	2.466
4	16	64	2.000	1.587	16	256	4,096	4.000	2.520
5	25	125	2.236	1.710	17	289	4,913	4.123	2.571
6	36	216	2.449	1.817	18	324	5,832	4.243	2.621
7	49	343	2.646	1.913	19	361	6,859	4.359	2.668
8	64	512	2.828	2.000	20	400	8,000	4.472	2.714
9	81	729	3.000	2.080	25	625	15,625	5.000	2.924
10	100	1,000	3.162	2.154	30	900	27,000	5.477	3.107
11	121	1,331	3.317	2.224	40	1,600	64,000	6.325	3.420
12	144	1,728	3.464	2.289	50	2,500	125,000	7.071	3.684

The square roots and cube roots are rounded to three decimal places.

Pythagoras's theorem

Pythagoras's theorem states that in any right-angled triangle, the square of the length of the hypotenuse is equal to the sum of the squares of the other two sides (see p.35). It can be used to calculate the length of any side of such a triangle if the length of the other two sides are known. The formula is written $c^2 = a^2 + b^2$, where c is the length of the hypotenuse, and a and b are the lengths of the other two sides.

Although known to the Babylonians, tradition ascribes to Pythagoras himself the first proof. There are hundreds of proofs of Pythagoras's theorem. The one shown here uses algebra.

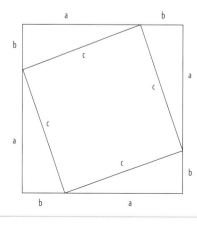

Algebraic proof
The formula for Pythagoras's theorem, $c^2 = a^2 + b^2$, can be proved using this diagram, which features four right-angled triangles created when a square is positioned at an angle within a larger square. The following algebraic equations provide the proof:

The area of each of the the four triangles = $\frac{1}{2}ab$

Therefore, the total area of the four triangles = $4(\frac{1}{2}ab) = 2ab$

The area of the inner square = c^2

Therefore, the total area of the large outer square = $2ab + c^2$

But the total area of the outer square also = $(a + b)^2 = a^2 + 2ab + b^2$

Therefore, $2ab + c^2 = a^2 + 2ab + b^2$

Subtracting 2 ab from both sides gives $c^2 = a^2 + b^2$

Prime numbers

A prime number is any number that can only be exactly divided by 1 and itself without leaving a remainder. By definition, 1 is not a prime. The Greek mathematician Euclid proved that there is an infinite number of primes. There is no one formula for yielding every prime, and the largest prime numbers discovered by supercomputers run to many million of digits. Shown here are the first 918 prime numbers.

2	3	5	7	11	13	17	19	23	29	31	37	41	43	47	53	59	61	67	71	73	79	83	89	97	101	103
107	109	113	127	131	137	139	149	151	157	163	167	173	179	181	191	193	197	199	211	223	227	229	233	239	241	251
257	263	269	271	277	281	283	293	307	311	313	317	331	337	347	349	353	359	367	373	379	383	389	397	401	409	419
421	431	433	439	443	449	457	461	463	467	479	487	491	499	503	509	521	523	541	547	557	563	569	571	577	587	593
599	601	607	613	617	619	631	641	643	647	653	659	661	673	677	683	691	701	709	719	727	733	739	743	751	757	761
769	773	787	797	809	811	821	823	827	829	839	853	857	859	863	877	881	883	887	907	911	919	929	937	941	947	953
967	971	977	983	991	997	1009	1013	1019	1021	1031	1033	1039	1049	1051	1061	1063	1069	1087	1091	1093	1097	1103	1109	1117	1123	1129
1151	1153	1163	1171	1181	1187	1193	1201	1213	1217	1223	1229	1231	1237	1249	1259	1277	1279	1283	1289	1291	1297	1301	1303	1307	1319	1321
1327	1361	1367	1373	1381	1399	1409	1423	1427	1429	1433	1439	1447	1451	1453	1459	1471	1481	1483	1487	1489	1493	1499	1511	1523	1531	1543
1549	1553	1559	1567	1571	1579	1583	597	1601	1607	1609	1613	1619	1621	1627	1637	1657	1663	1667	1669	1693	1697	1699	1709	1721	1723	1733
1741	1747	1753	1759	1777	1783	1787	1789	1801	1811	1823	1831	1847	1861	1867	1871	1873	1877	1879	1889	1901	1907	1913	1931	1933	1949	1951
1973	1979	1987	1993	1997	1999	2003	2011	2017	2027	2029	2039	2053	2063	2069	2081	2083	2087	2089	2099	2111	2113	2129	2131	2137	2141	2143
2153	2161	2179	2203	2207	2213	2221	2237	2239	2243	2251	2267	2269	2273	2281	2287	2293	2297	2309	2311	2333	2339	2341	2347	2351	2357	2371
2377	2381	2383	2389	2393	2399	2411	2417	2423	2437	2441	2447	2459	2467	2473	2477	2503	2521	2531	2539	2543	2549	2551	2557	2579	2591	2593
2609	2617	2621	2633	2647	2657	2659	2663	2671	2677	2683	2687	2689	2693	2699	2707	2711	2713	2719	2729	2731	2741	2749	2753	2767	2777	2789
2791	2797	2801	2803	2819	2833	2837	2843	2851	2857	2861	2879	2887	2897	2903	2909	2917	2927	2939	2953	2957	2963	2969	2971	2999	3001	3011
3019	3023	3037	3041	3049	3061	3067	3079	3083	3089	3109	3119	3121	3137	3163	3167	3169	3181	3187	3191	3203	3209	3217	3221	3229	3251	3253
3257	3259	3271	3299	3301	3307	3313	3319	3323	3329	3331	3343	3347	3359	3361	3371	3373	3389	3391	3407	3413	3433	3449	3457	3461	3463	3467
3469	3491	3499	3511	3517	3527	3529	3533	3539	3541	3547	3557	3559	3571	3581	3583	3593	3607	3613	3617	3623	3631	3637	3643	3659	3671	3673
3677	3691	3697	3701	3709	3719	3727	3733	3739	3761	3767	3769	3779	3793	3797	3803	3821	3823	3833	3847	3851	3853	3863	3877	3881	3889	3907
3911	3917	3919	3923	3929	3931	3943	3947	3967	3989	4001	4003	4007	4013	4019	4021	4027	4049	4051	4057	4073	4079	4091	4093	4099	4111	4127
4129	4133	4139	4153	4157	4159	4177	4201	4211	4217	4219	4229	4231	4241	4243	4253	4259	4261	4271	4273	4283	4289	4297	4327	4337	4339	4349
4357	4363	4373	4391	4397	4409	4421	4423	4441	4447	4451	4457	4463	4481	4483	4493	4507	4513	4517	4519	4523	4547	4549	4561	4567	4583	4591
4597	4603	4621	4637	4639	4643	4649	4651	4657	4663	4673	4679	4691	4703	4721	4723	4729	4733	4751	4759	4783	4787	4789	4793	4799	4801	4813
4817	4831	4861	4871	4877	4889	4903	4909	4919	4931	4933	4937	4943	4951	4957	4967	4969	4973	4987	4993	4999	5003	5009	5011	5021	5023	5039
5051	5059	5077	5081	5087	5099	5101	5107	5113	5119	5147	5153	5167	5171	5179	5189	5197	5209	5227	5231	5233	5237	5261	5273	5279	5281	5297
5303	5309	5323	5333	5347	5351	5381	5387	5393	5399	5407	5413	5417	5419	5431	5437	5441	5443	5449	5471	5477	5479	5483	5501	5503	5507	5519
5521	5527	5531	5557	5563	5569	5573	5581	5591	5623	5639	5641	5647	5651	5653	5657	5659	5669	5683	5689	5693	5701	5711	5717	5737	5741	5743
5749	5779	5783	5791	5801	5807	5813	5821	5827	5839	5843	5849	5851	5857	5861	5867	5869	5879	5881	5897	5903	5923	5927	5939	5953	5981	5987
6007	6011	6029	6037	6043	6047	6053	6067	6073	6079	6089	6091	6101	6113	6121	6131	6133	6143	6151	6163	6173	6197	6199	6203	6211	6217	6221
6229	6247	6257	6263	6269	6271	6277	6287	6299	6301	6311	6317	6323	6329	6337	6343	6353	6359	6361	6367	6373	6379	6389	6397	6421	6427	6449
6451	6469	6473	6481	6491	6521	6529	6547	6551	6553	6563	6569	6571	6577	6581	6599	6607	6619	6637	6653	6659	6661	6673	6679	6689	6691	6701
6703	6709	6719	6733	6737	6761	6763	6779	6781	6791	6793	6803	6823	6827	6829	6833	6841	6857	6863	6869	6871	6883	6899	6907	6911	6917	6947
6949	6959	6961	6967	6971	6977	6983	6991	6997	7001	7013	7019	7027	7039	7043	7057	7069	7079	7103	7109	7121	7127	7129	7151	7159	7177	7187

Multiplication table

A multiplication table is a quick-reference tool that can be used to multiply two numbers. A range of numbers is aligned down the left hand side of the chart, starting a series of rows; another along the top of the chart, heads a series of columns. At the intersections of the rows and columns lie the products of the numbers at the far left and top respectively.

	2	3	4	5	6	7	8	9	10	11	12	13	14	15	16	17	18	19	20	21	22	23	24	25
2	4	6	8	10	12	14	16	18	20	22	24	26	28	30	32	34	36	38	40	42	44	46	48	50
3	6	9	12	15	18	21	24	27	30	33	36	39	42	45	48	51	54	57	60	63	66	69	72	75
4	8	12	16	20	24	28	32	36	40	44	48	52	56	60	64	68	72	76	80	84	88	92	96	100
5	10	15	20	25	30	35	40	45	50	55	60	65	70	75	80	85	90	95	100	105	110	115	120	125
6	12	18	24	30	36	42	48	54	60	66	72	78	84	90	96	102	108	114	120	126	132	138	144	150
7	14	21	28	35	42	49	56	63	70	77	84	91	98	105	112	119	126	133	140	147	154	161	168	175
8	16	24	32	40	48	56	64	72	80	88	96	104	112	120	128	136	144	152	160	168	176	184	192	200
9	18	27	36	45	54	63	72	81	90	99	108	117	126	135	144	153	162	171	180	189	198	207	216	225
10	20	30	40	50	60	70	80	90	100	110	120	130	140	150	160	170	180	190	200	210	220	230	240	250
11	22	33	44	55	66	77	88	99	110	121	132	143	154	165	176	187	198	209	220	231	242	253	264	275
12	24	36	48	60	72	84	96	108	120	132	144	156	168	180	192	204	216	228	240	252	264	276	288	300
13	26	39	52	65	78	91	104	117	130	143	156	169	182	195	208	221	234	247	260	273	286	299	312	325
14	28	42	56	70	84	98	112	126	140	154	168	182	196	210	224	238	252	266	280	294	308	322	336	350
15	30	45	60	75	90	105	120	135	150	165	180	195	210	225	240	255	270	285	300	315	330	345	360	375
16	32	48	64	80	96	112	128	144	160	176	192	208	224	240	256	272	288	304	320	336	352	368	384	400
17	34	51	68	85	102	119	136	153	170	187	204	221	238	255	272	289	306	323	340	357	374	391	408	425
18	36	54	72	90	108	126	144	162	180	198	216	234	252	270	288	306	324	342	360	378	396	414	432	450
19	38	57	76	95	114	133	152	171	190	209	228	247	266	285	304	323	342	361	380	399	418	437	456	475
20	40	60	80	100	120	140	160	180	200	220	240	260	280	300	320	340	360	380	400	420	440	460	480	500
21	42	63	84	105	126	147	168	189	210	231	252	273	294	315	336	357	378	399	420	441	462	483	504	525
22	44	66	88	110	132	154	176	198	220	242	264	286	308	330	352	374	396	418	440	462	484	506	528	550

Who's Who

Agassiz, Louis 1835–1910
See p.195.

Al-Biruni 973–1048 CE
Persian scholar and scientist who excelled in astronomy, mathematics, physics, medicine, and history, writing about them prolifically. Al-Biruni traveled to India, making a comprehensive study of its culture in *A History of India*. In his studies on astronomy, he supported the theory of the Earth's rotation on its axis and made accurate calculations of latitude and longitude. Al-Biruni's work in physics included an explanation of how natural springs were governed by the laws of hydrostatics, while in geography he radically suggested that the Indus valley had once been a sea basin.

Alhazen 965–1039
See pp.60–61.

Al-Idrisi c.1099–1165
Arab geographer, cartographer, and traveler, Al-Idrisi is best known for his *Tabula Rogeriana*, a remarkably accurate map of the world as it was then known. He prepared the map and its commentary in 1154 for King Roger II of Sicily after collating information from merchants and explorers. The map, which has south at the top, extends across the whole of Europe and Asia, although only northern Africa is shown. Al-Idrisi also created a huge, two-sided planisphere: one side showed the positions of the constellations, while the other showed where the countries of the world were.

Al-Kwharizmi c.820
See p.48.

Alvarez, Luis Walter 1911–1988
US experimental physicist who won the 1968 Nobel Prize for Physics for developing the liquid-hydrogen bubble chamber, which he used to identify many resonances (very short-lived subatomic particles that occur only in high-energy nuclear collisions). During World War II, Alvarez invented the radar guidance system for aircraft landings and designed the detonators for the spherical implosives used on the Trinity and Nagasaki atomic bombs. In 1947, he helped put together the first proton linear accelerator. In 1980, with his son, Walter, he proposed the asteroid-impact theory to explain the extinction of the dinosaurs.

Amontons, Guillaume 1663–1705
French inventor and physicist who invented and perfected instruments for measuring temperature and pressure. In 1687, he invented an improved hygrometer, a device used to measure humidity. He improved the barometer and thermometer (1695), particularly for use at sea. He was the first researcher to discuss the theory of an absolute zero of temperature and proposed theories on the laws of friction.

Ampère, André-Marie 1775–1836
French mathematician and physicist who developed the mathematical theory of electromagnetism. Ampère's theory served as the basis for other fundamental ideas of the 19th century regarding electricity and magnetism and inspired discoveries by scientists such as Faraday, Weber, Thomson, and Maxwell. The ampere, the unit of electric current, was named after him.

Anaximander 611–547 BCE
Greek philosopher. He drew a map of the entire known world and is credited as being the first scientific geographer. He invented the gnomon, used to determine the solstices, which became the basis for the development of sundials, and suggested that the heavens are a perfect sphere, with the Earth floating freely at its center. He was also the first to try to determine the magnitude of the celestial bodies and how they are positioned in space.

Anderson, Carl David 1905–1991
American physicist who opened up the field of particle physics and the study of the atom. He discovered the positron and muon, for which he was jointly awarded the Nobel Prize in Physics in 1936.

Angstrom, Anders Jonas 1814–1874
Swedish physicist and spectroscopist who deduced in 1855 that a hot gas emits light at the same wavelengths at which it absorbs light when cooler. In 1861, Angstrom began to study the solar spectrum, proving the existence of hydrogen in the sun and recording his measurements in units, subsequently named in his honor. He was also the first to examine the spectra of the aurora borealis.

Anning, Mary 1799–1847
See p.187.

Appleton, Edward Victor 1892–1965
English physicist who proved the existence of the ionosphere. From 1924 to 1936, Appleton was professor of physics at King's College, London, where he worked on experiments using radio waves. He determined the height at which the waves were reflected back to earth—the ionosphere—and discovered another reflecting layer at approximately 150 miles (250km), now called the Appleton layer. This work led directly to the development of radar technologies.

Archimedes c.287–212 BCE
Greek philosopher and inventor, and arguably the greatest mathematician of the ancient world. He wrote influential works on arithmetic, geometry, and mechanics, and famously, while taking a bath, solved a tricky problem concerning the king's crown—realizing that it would displace its own volume of water. This led him to investigate hydrostatics and discover Archimedes's Principle, which states that any object immersed in water will experience an upthrust equal to the weight of water displaced. He gave mathematical descriptions of levers and pulleys, devised machines to repel invading Roman ships, and may have invented the pump known as the Archimedes screw.

Aristarchus of Samos c.310–230 BCE
Greek astronomer noted as being the first to suggest that the Earth revolves around the sun. His only surviving treatise, *On the Magnitudes and Distances of the Sun and Moon*, details his geometric argument, based on observation, whereby he calculated that the sun was about 20 times as far from the Earth as the moon, and 20 times the moon's size. Although these estimates proved inaccurate, the fault lay in the inadequacies of the instruments he was using rather than in his reasoning, and his pioneering methods provided a platform for future astronomers and mathematicians.

Aristotle 384–322 BCE
See pp.36–37.

Arkwright, Richard 1733–1792
English inventor and industrialist, who in 1769 patented an automatic cotton-spinning machine that allowed an unskilled teenager to spin 96 threads simultaneously. Powered by a water wheel, it came to be called the water frame. He installed many of these machines in a purpose-built factory in Derbyshire in 1771, thus introducing the factory system and mass production, and driving forward the Industrial Revolution. He had to borrow money to build that first factory, but when he died in 1792, he left several factories, a town house in London, a castle in Derbyshire, and half a million pounds.

Arrhenius, Svante August 1859–1927
See p.234.

Aston, Francis William 1877–1945
English physicist and chemist who won the 1922 Nobel Prize for chemistry for his work on isotopes. He developed the mass spectrograph, which separates atoms or molecular fragments of different mass and measures those masses. This led to his discovery of a large number of naturally occurring isotopes, and the formulation of his Whole Number Rule, which became important in the development of nuclear energy.

Avicenna c.979–1037
See p.25.

Avogadro, Amedeo 1776–1856
Italian physicist and mathematician most noted for his

work on molecular theory and Avogadro's law. In 1811, he published work that made the distinction between the molecule and the atom. He suggested that equal volumes of all gases at the same temperature and pressure contain the same number of molecules. In honor of Avogadro's contributions to molecular theory, the number of molecules in one mole in the molecular weight in grams was named Avogadro's number.

Babbage, Charles 1791–1871
See p.174.

Bacon, Francis 1561–1626
English statesman who became a leader in natural philosophy and is regarded as the father of modern scientific methodology. Rejecting ancient beliefs that scientific truth could be reached only by authoritative discussion, Bacon tried to put natural science on a sound empirical foundation in *Novum Organum Scientiarum* (1620) and *The Advancement of Learning* (1623). His classification of the sciences later inspired the 18th-century French encyclopedists, while his empiricism influenced 19th-century British philosophers of science. He died following an experiment on how freezing affected meat preservation: having braved a blizzard to stuff a chicken with snow, Bacon developed pneumonia and died.

Bacon, Roger 1214–1294
See p.57.

Baekeland, Leo Hendrik 1863–1944
Belgian-American chemist and inventor of Bakelite, whose first invention was Velox, an improved photographic paper that freed photographers from having to use sunlight for developing images. In 1899, George Eastman bought full rights to Velox for $1 million, and with this windfall, Baekeland was able to invest in research and set up a company in 1909 to produce Bakelite, the first completely synthetic plastic, which could flow and be poured into molds before hardening.

Bahcall, Neta 1942–
See p.400.

Baird, John Logie 1888–1946
Scottish television pioneer. Baird began his television experiments in 1922 and obtained the world's first moving television image in 1925. He demonstrated the world's first color transmission in 1928 and made the first television program for the BBC. His mechanical scanning system, however, was less versatile than Marconi-EMI's electronic system, which was adopted by the BBC in 1936.

Banks, Joseph 1743–1820
English explorer and botanist who introduced many new plants to the West, including the eucalyptus, and helped establish the Royal Botanic Gardens at Kew, London. In 1768, Banks joined a Royal Society expedition to the South Pacific, led by Captain James Cook aboard HMS *Endeavour*. He collected numerous specimens en route, later writing a scientific account of the voyage and its discoveries. Banks was interested in the practical and commercial uses of plants and in introducing them into other countries. Around 75 species are named after him, as is a group of Pacific Islands. There was even a suggestion to name Australia "Banksia." He was a long-time president (1778–1820) of the Royal Society.

Bardeen, John 1908–91
US physicist known for his work on semiconductors; he co-invented the transistor (1947), paving the way for modern electronics, and later helped develop a theory of superconductivity (1957). The transistor enabled the development of thousands of different electronic devices, while his developments in superconductivity are used in medical advances such as MRI. Bardeen won the Nobel Prize for each of these achievements in 1956 and 1972, respectively. He worked at Bell Laboratories from 1945–1951 and was a professor of electrical engineering and physics at the University of Illinois until 1975.

Barnard, Christian Neething 1922–2001
See p.406.

Barré-Sinoussi, Françoise 1947–
See p.412.

Bath, Patricia 1942–2019
See p.365.

Baudot, Jean Maurice Emile 1845–1903
French telegraph engineer and inventor of the Baudot code. The telegraph service trained him in the Morse telegraph and the Hughes printing telegraph method, which inspired his own system. Baudot used synchronous distributors running at a constant speed, allowing several messages to be sent simultaneously. In June 1874, he patented his first printing telegraph. His apparatus was shown at the Paris Exposition Universelle (1878) and won him the gold medal.

Baumé, Antoine 1728–1804
French chemist best known as the inventor of the Baumé scale hydrometer. He became professor of chemistry at the École de Pharmacie in 1752. He retired in 1780 but was ruined by the French Revolution and forced to return to commercial life. He improved many scientific instruments, among them the hydrometer, used to measure the specific gravity of liquids. Baumé devised an improved scale for the hydrometer that now bears his name.

Becquerel, Antoine Henri 1852–1908
See p.299.

Beg, Ulugh 1394–1449
See p.29.

Bell, Alexander Graham 1847–1922
See p.260.

Benz, Karl 1844–1929
German inventor who, with Gottlieb Daimler, made the first practical gas-engined motor vehicle. In June 1886, Benz road-tested a three-wheeled car powered by a four-stroke single-cylinder engine with electric ignition. Two years later, he was the first to offer gas-driven cars to sell to the public, so founding the motor industry.

Berg, Paul 1926–
American biochemist and molecular biologist who shared the 1980 Nobel Prize for Chemistry for studies on recombinant DNA. Berg developed recombinant DNA techniques that enable DNA to be spliced from different organisms and recombined. An important leap forward for scientific research, this has valuable commercial applications in fields such as agriculture, medicine, and industry. Berg has long been a leading figure in genetic engineering; he also played a key role in addressing its potential risks and devising guidelines for self-regulation.

Bergius, Friedrich 1884–1949
German chemist who made lubricants and synthetic fuels, such as gasoline, from coal. In 1910, Bergius established a private laboratory where he researched the hydrogenating effect of hydrogen on coal and heavy oils under high pressure. After WWI, he managed to develop the hydrogenation of coal on a large scale in what is know as the Bergius process. In 1931, he shared the Nobel Prize in Chemistry with Carl Bosch for their invention of chemical high-pressure methods.

Bernard, Claude 1813–1878
See p.274.

Berners-Lee, Tim 1955–
See p.378.

Bernoulli, Daniel 1700–1782
Swiss mathematician and physicist who discovered that the pressure in a fluid decreases as the velocity of fluid flow increases. This became known as Bernoulli's principle and is fundamental to many engineering applications. Bernoulli also formulated a law enabling the velocity of a liquid to be measured and contributed to probability theory. His publication *Hydrodynamica* (1738), in which he developed the theory

of water mills, windmills, water pumps, and water propellers, was a milestone in the kinetic theory of fluids and gases. Bernoulli was the first to differentiate between hydrostatic and hydrodynamic pressure.

Berthelot, Marcellin Pierre Eugène 1827–1907
French chemist who, working on the synthesis of acetylene and other hydrocarbons, demonstrated that living processes are not essential to the production of organic chemicals. His research and publications made a significant contribution to the development of chemistry in the late 19th century.

Berzelius, Jöns Jakob 1779–1848
Swedish chemist considered one of the founders of modern chemistry, Berzelius is best known for his system of electrolysis. He produced the first accurate list of atomic weights and developed modern chemical symbols. He discovered selenium, cerium, and thorium, isolated the elements of zirconium, silicon, and titanium, developed classical analytical techniques, and investigated isomerism and catalysis. From 1807 to 1832, Berzelius was a professor of medicine and pharmacy at the Karolinska Institute, and he became a member of the Royal Swedish Academy of Sciences in 1808.

Best, Charles Herbert 1899–1978
Canadian medical scientist who helped discover insulin in 1921, after which it became a standard treatment for diabetes. Best worked closely with Frederick Banting and missed out on winning the Nobel Prize in Physiology or Medicine in 1923 to Banting and physiology professor John Macleod. Best went on to develop heparin, recognizing that it could be an important anticoagulant drug for preventing blood clotting.

Bethe, Hans Albrecht 1906–2005
German-born American physicist who, in 1938, made the first detailed calculations concerning the nuclear reactions that power the stars. He theorized that hydrogen is turned into helium by a series of nuclear reactions known as the carbon cycle. For his work, Bethe won the 1967 Nobel Prize in Physics. In the 1950s, he joined a project to develop the hydrogen bomb. He later campaigned, alongside Albert Einstein, against nuclear testing.

Biot, Jean Baptiste 1774–1862
French physicist, mathematician, and astronomer who confirmed the reality of meteorites. He also worked on the relationship between electrical currents and magnetism. He discovered that the intensity of the magnetic field set up by a current flowing through a wire varies inversely with the distance from the wire. Known as Biot-Savart's law, it became fundamental to modern electromagnetic theory. A crater on the moon is named after him.

Bjerknes, Vilhelm 1862–1951
See p.208.

Black, Joseph 1728–1799
See pp.148–49.

Blackett, Patrick Maynard Stuart 1897–1974
English physicist best known for his work on cloud chambers, rock magnetism, and cosmic rays. In 1932, working with Giuseppe Occhialini, he devised a system of Geiger counters to investigate cosmic rays. In 1937, he went to the University of Manchester where he created a major international research laboratory. He played an important role in World War II, advising on military strategy and in 1948 was awarded the Nobel Prize in Physics.

Boas, Franz 1858–1942
German-born US anthropologist known for applying scientific methods to the study of human cultures and societies, Boas is credited with playing a pivotal role in developing modern cultural anthropology. His concept was that all human races are equally capable of developing cultural forms; where differences exist it is because of culture rather than race or genetics. At Columbia University (1899–1942), Boas developed one of the leading departments of anthropology in the US. He specialized in Native American culture and languages and founded the *International Journal of American Linguistics*.

Bode, Johann Elert 1747–1826
German astronomer best known for Bode's law, or the Titus-Bode law, a rough rule that predicts the spacing of the planets in the solar system. As director of the Berlin Observatory, he discovered several new star clusters and a comet. In 1801, he published the *Uranographia*, a celestial atlas that showed the positions of the stars and gave the planet Uranus its name.

Bohr, Niels 1885–1962
Danish physicist who carried out research into the atom and was the first to apply quantum theory, as established by Max Planck. In 1913, on the basis of Rutherford's theories, he published his model of atomic structure that still serves as the basis of the physical and chemical properties of the elements. He was awarded the Nobel Prize for Chemistry in 1922. During World War II, he joined the Manhattan Project but later concentrated on the peaceful application of atomic physics and the political conflicts caused by the development of atomic weapons.

Boole, George 1815–1864
English teacher, mathematician, and philosopher who invented symbolic knowledge and formulated the rules that govern it, known as Boolean algebra. His most important work, *The Laws of Thought*, was published in 1854. Boole's ideas became the basis for the development of much modern abstract mathematics and computer science.

Borlaug, Norman 1914–2009
American agronomist who is widely considered to be the founding father of the Green Revolution, which has seen the development of disease-resistant varieties of wheat. Following a period of research in plant pathology, Borlaug combined the results of selective breeding with improved agricultural techniques to increase food productivity in Mexico, India, and Pakistan, averting famine in these countries. In 1970, he received the Nobel Peace prize for his humanitarian efforts, and in 1986 he established the World Food Prize for individuals working to the same ends.

Bosch, Carl 1874–1940
German industrial chemist who developed the Haber-Bosch process for high-pressure synthesis of ammonia. This process became the main industrial procedure for nitrogen fixation, and in 1931 he was jointly awarded the Nobel Prize for Chemistry with Friedrich Bergius. Bosch then studied how to make ammonia available for use in agriculture and industry, developing a means for the production of nitrogen fertilizers. This had a global impact on agriculture.

Bose, Satyendra Nath 1894–1974
See p.305.

Boyle, Robert 1627–1691
See pp.98–99.

Tycho 1546–1601
Danish nobleman and astronomer known for his work in developing astronomical instruments and for his accurate and comprehensive planetary and astronomical observations. These included a thorough study of the solar system, proof that the orbit of the 1577 comet lay beyond the moon, and accurate positions of more than 777 fixed stars. Incredibly, all were achieved before the invention of the telescope, and his work provided the gateway to future discoveries. Tycho is also known for his contributions to medicine. He famously lost the tip of his nose during a duel and wore a metal prosthetic.

Brahmagupta c.596–668 CE
See p.31.

Bramah, Joseph 1748–1814
See p.145.

Broca, Paul 1824–1880
See p.223.

Broecker, Wallace 1931–2019
See p.422.

Brunel, Isambard Kingdom 1806–1859
English engineer whose first project was helping his father to dig the first tunnel under a major river—the Thames in London. He designed the Clifton Suspension Bridge across the Avon Gorge in Bristol and constructed the Great Western Railway from London all the way to Cornwall in the southwest, including all its tunnels, bridges, and viaducts. He also built three great ships—the *Great Western*, the first steamship built to carry passengers to America; the *Great Britain*, made of iron and fitted with the newly invented screw propeller; and the *Great Eastern*, which in 1866 laid the first successful cable across the Atlantic.

Budd, William 1811–1880
English doctor and epidemiologist whose studies on infectious diseases, such as typhoid, advanced contagion theory. Budd also advocated the use of disinfectant and the prevention of sewage entering the water supplies. Such preventative methods enabled Budd to significantly reduce the spread of cholera in Bristol.

Burbank, Luther 1849–1926
American botanist, horticulturalist, and pioneer in agricultural science noted for his plant-breeding experiments. Burbank developed more than 800 strains and varieties of plants. He helped transform plant breeding into a modern science, aiding the later study of genetics. His objective was to improve plant quality, so increasing the world's food supply. His Burbank potato was introduced to Ireland to counteract the bight epidemic. Burbank settled in Santa Rosa, California, where he established a nursery garden, greenhouse, and experimental farms that became internationally renowned.

Cabrera, Blas c.1942–
Spanish physicist at Stanford University, best known for his experiment in search of magnetic monopoles. In 1982 his detector recorded an event which bore the perfect signature hypothesized for a magnetic monopole. No similar event has since been recorded. The Cabrera Group at Stanford is currently involved in several research projects including the Cryogenic Dark Matter Search (CDMS) experiment. This seeks to identify the dominant form of matter in and around our galaxy, under the hypothesis that it comprises weakly interfacing massive particles (WIMPS). If WIMPS and therefore supersymmetry were discovered, the most important problems in elementary particle physics and cosmology would be simultaneously solved.

Calmette, Albert (Leon Charles) 1863–1933
French bacteriologist and immunologist best known for his work in public health with Camille Guérin. They developed the BCG (*bacillus* Calmette-Guérin) vaccine against tuberculosis, first trialed in 1922. He also produced the first antivenin for snake venom, the Calmette's serum.

Cajal, Santiago Ramón y 1852–1934
Spanish histologist, physician, and pathologist, Cajal is noted for his work on the structure of the central nervous system. Using a histological staining technique based on silver chromate developed by Camillo Golgi, Cajal identified the structure of individual neurons, which led him to conclude that nervous tissue is made of independent cells rather than a complex connecting network. He was joint-winner of the 1905 Nobel Prize for Physiology or Medicine.

Cannizzaro, Stanislao 1826–1910
An Italian chemist, Cannizzaro is remembered for his discovery of cyanamide, for obtaining alcohols from aldehydes (a process known as Cannizarro's reaction), and for distinguishing between molecular and atomic weights. He supported Avogadro's law—the idea that equal volumes of gas at the same pressure and temperature held equal numbers of molecules or atoms, and that equal volumes of gas could be used to calculate atomic weights—which led to a new understanding of chemistry. He held posts at various Italian universities, including Alessandria, Palermo, and Rome, and was also politically active.

Cannon, Walter Bradford 1871–1945
American neurologist and physiologist who studied the regulation of hunger and thirst in animals and developed the concept of homeostasis (the ability of an organism to maintain its internal environment), popularizing it in *The Wisdom of the*

Body (1932). With Philip Bard, he developed the Cannon-Bard theory to explain why people feel emotions first and then act upon them. Cannon was also the first to use X-rays in physiological studies; the barium meal is a modern derivative of this research. He investigated hemorrhagic and traumatic shock, worked on methods of storing blood, and in 1931 discovered sympathin, an adrenaline-like substance. He became a leading authority in the emerging field of psychosomatic medicine. Cannon was professor and chairman of the Department of Physiology at Harvard Medical School.

Carnot, Nicolas Leonard Sadi 1796–1832
French physicist and military engineer who gave the first successful theoretical account of heat engines, now known as the Carnot Cycle. His work laid the foundation for thermodynamics. Carnot's most important work, *Réflexions sur la Puissance Motrice du Feu* (Reflections on the Driving Power of Fire) published in 1824, provided the theoretical basis for the development of the steam engine and introduced the idea of the second law of thermodynamics. The Carnot Cycle demonstrates the law of entropy: an engine cannot convert all of its heat energy into mechanical energy; some always stays unused.

Carson, Rachel 1907–1964
See p.333.

Carver, George Washington 1864–1943
See p.363.

Cassini, Giovanni Domenico 1625–1712
Italian-born engineer and astronomer. He discovered a dark gap, 1,696 miles (2,730 km) wide, in Saturn's rings, caused by the gravitational influence of Saturn's inner satellites. The gap was named Cassini's division. He also discovered four satellites of Saturn and the Great Red Spot on Jupiter. The NASA *Cassini* spacecraft, which orbited Saturn, was named after him.

Cauchy, Augustin Louis, Baron 1789–1857
French mathematician who laid the groundwork for rigor in analysis and researched differential equations, probability, and mathematical physics. A prolific science writer and one of the greatest modern mathematicians, he was elected to the Academy of Sciences in Paris aged 27.

Cavendish, Henry 1731–1810
See p.210.

Caventou, Joseph Bienaimé 1795–1877
French pharmacist and one of the earliest investigators of alkaloids. Working alongside Pierre-Joseph Pelletier, he isolated a number of active ingredients from plants, including chlorophyll, strychnine, caffeine, and quinine sulfate, which became an important treatment for malaria. The moon crater Caventou is named after him.

Chadwick, James 1891–1974
English physicist who won the Nobel Prize for Physics in 1935 for discovering the neutron. Chadwick worked on radioactivity with Ernest Rutherford and in 1932 proved Rutherford's prediction that a particle without electric charge, which Chadwick named a neutron, existed in the nucleus of an atom; he also calculated its mass. Chadwick's discovery advanced scientific experiment internationally, and scientists began to bombard all types of materials with neutrons; one of these was uranium, which resulted in nuclear fission. During World War II, Chadwick worked on the Manhattan Project to develop the atomic bomb. He was knighted in 1945.

Chamberlain, Owen 1920–2006
American physicist who, with Emilio Segrè, won the Nobel Prize in Physics in 1959 for their discovery of the antiproton, the antimatter equivalent and negatively charged mirror image of the proton. This discovery opened up a new field of physics and expanded the understanding of particle physics. He worked on the Manhattan Project, helping to develop the first atomic bomb.

Chandrasekhar, Dr. Subrahmanyan 1910–1995
See p.331.

Chappe, Claude 1763–1805
French inventor who, with his brother Ignace, developed a mechanical telegraph system. Their practical semaphore design was demonstrated in 1792, with a message passed between Paris and Lille, and eventually spanned France and parts of Europe.

Chaptal, Jean-Antoine 1756–1832
French technical chemist, industrialist, and statesman who discovered that adding sugar to unfermented wine increased the final alcohol level; this became known as chaptalization. Some of his achievements included improvements in the manufacture of sulfuric acid, saltpeter for gunpowder, wine, and dyeing. Focusing on technical chemistry, Chaptal supplemented Lavoisier's theoretical work, and his writings pioneered the application of chemical principles to industry and agriculture.

Charles, Jaques Alexandre Cesar 1746–1823
French chemist and physicist who invented the hydrogen balloon. On August 27, 1783, his first balloon ascended to a height of nearly 3,000 ft (914m). In 1783, Charles and Ainé Roberts covered over 27 miles (43km) in La Charlière. He is also known for his formulation of Charles's law (1787), which states that, at constant pressure, the volume occupied by a fixed mass of gas is directly proportional to the absolute temperature.

Charpentier, Emmanuelle 1968–
See p.417.

Cherenkov, Pavel Alekseyevich 1904–1990
Soviet physicist who shared the 1959 Nobel Prize for Physics with Igor Tamm and Illya Frank for the discovery and theoretical interpretation of Cherenkov radiation. As a research student, Cherenkov had observed that electrons produce a blue glow when passing through a transparent liquid at high velocity. This Cherenkov radiation, correctly interpreted by Tamm and Frank in 1937, led to the development of the Cherenkov Detector that became so useful in experimental nuclear and particle physics.

Clarke, Arthur 1917–2008
See p.369.

Claude, Albert 1899–1983
Belgian-American cell biologist and co-winner of the 1974 Nobel Prize for Physiology or Medicine for discoveries concerning the structural and functional organization of the living cell. During the 1930s and '40s Claude helped pioneer the use of an electron microscope for cellular study and was involved in developing differential centrifugation to separate various cell components. He discovered that the cell interior comprised a highly organized membranous network (the endoplasmic reticulum), dispelling the belief that it was merely a chaotic mass. Claude joined the Rockefeller Institute in 1929 and remained there for the rest of his career.

Cockcroft, John Douglas 1897–1967
English physicist and co-winner with Ernest Walton of the 1951 Nobel Prize for Physics for pioneering the use of particle accelerators to study atomic nuclei. While a professor at Cambridge, he and Walton designed the Cockcroft-Walton generator and used it to split lithium atoms by bombarding them with protons. This was one of the earliest experiments to artificially change the atomic nucleus of one element to a different nucleus. They then researched the splitting of other atoms, establishing the importance of accelerators like their generator as a tool for nuclear research. In 1946, he set up the British Atomic Energy Research Establishment with the aim of developing Britain's atomic power program. He was master of Churchill College, Cambridge, UK, from 1959 to 1967.

Cohen, Stanley Norman 1935–
American microbiologist known for pioneering new methods of combining and transplanting genes, which led to the field of genetic engineering. At Stanford University, from 1968, Cohen began researching how the genes of plasmids could make bacteria resistant to antibiotics. In 1972, he collaborated with Herbert Boyer and Paul Berg to devise ways of combining and transplanting genes, received the first patent for gene splicing, and in 1988 was awarded the National Medal of Science. Cohen remains at Stanford as professor of genetics and medicine; his primary field of research is molecular genetics.

Compton, Arthur Holly 1892–1962
American physicist and joint winner with CTR Wilson of the 1927 Nobel Prize in Physics for his discovery of the Compton Effect in which wavelengths of X-rays increase when the rays collide with electrons in metals. This discovery provided the first evidence that X-rays can act like particles, confirming the theory that electromagnetic radiation can act as both a wave and a particle and helping legitimize quantum theory. Compton worked on the Manhattan Project.

Coolidge, William David 1873–1975
US engineer, physical chemist, and inventor famous for his work on incandescent electric lighting and X-ray development. Coolidge conducted experiments that led to the use of tungsten filaments in light bulbs and invented the X-ray vacuum tube or Coolidge tube (1916). A major breakthrough in the field of radiology, it became the prototype of the modern X-ray tube. Coolidge also helped develop the first successful submarine-detection system. He was the director of the General Electric Research Laboratory and a vice president of the corporation, joining as a researcher in 1905.

Cope, Edward Drinker 1840–1897
American paleontologist and evolutionary theoretician who helped define American paleontology. Cope's contribution to knowledge of vertebrates of the Tertiary Period was invaluable. He discovered some 1,000 species of extinct vertebrates in the US, including the dinosaurs Camarasaurus and Coelophysis. Much of Cope's time was spent on expeditions across the American West. A competition with fellow paleontologist, Othniel Marsh, to discover the most American fossil dinosaurs, famously degenerated in to an ongoing feud dubbed the Bone Wars. Cope was one of the founders of the Neo-Lamarckian school of evolutionary thought, believing in the inheritance of acquired characteristics, rather than in natural selection.

Copernicus, Nicolaus 1473–1543
Polish polymath of the Renaissance, Copernicus combined studies as a mathematician, astronomer, and physician with a career as a scholar, cleric, civic leader, magistrate, diplomat, military commander, translator, and economist. In 1543, the year he died, he published his masterwork, *De Revolutionibus Orbium Coelestium* (On the Revolutions of the Celestial Spheres), in which he set out a fully predictive mathematical model of a heliocentric solar system, with Earth and all the planets revolving around the sun, and all the bodies in the universe moving in circular motions. The book had a crucial influence on the European Scientific Revolution.

Cori, Gerty 1896–1957
See p.217.

Coriolis, Gaspard-Gustave de 1792–1843
French mathematician, mechanical engineer, and scientist, Coriolis dedicated his career to the study of friction, hydraulics, machine performance, and ergonomics. Teaching at the École Polytechnique, Paris, in 1829, he published his first paper, *Du Calcul de l'Effet des Machines* (On the Calculation of Mechanical Action), in which he introduced the important terms "work" and "kinetic energy." He is now best known for the Coriolis force, an apparent deflection of moving objects when viewed from a rotating reference frame, such as the surface of Earth.

Cousteau, Jacques-Yves 1910–1997
French explorer, ecologist, filmmaker, scientist, photographer, author, and nature presenter, Cousteau became known to millions through the television series he made from 1966 to 1994, such as *The Undersea World of Jacques Cousteau* (1968–1976). In the 1940s, he improved the aqualung underwater breathing apparatus, enabling his prolonged explorations underwater. His beautifully photographed and soundly researched television documentaries and over 50 books laid an increasing emphasis on the need to conserve marine life and habitats, and the televised explorations aboard his boat *Calypso* were pioneering in raising mass awareness of ecological issues.

Crick, Francis 1916–2004
English molecular biologist, physicist, and neuroscientist who, along with colleague James Watson, worked out the structure of deoxyribonucleic acid, or DNA when they were both carrying out research at Cambridge University. They discovered its double helix form, one that could duplicate itself, confirming theories that it carried life's heredity information. He then went on to find the link between DNA and the genetic code. His research spawned an entire industry of biotechnology and earned Crick, Watson, and physicist Maurice Wilkins the Nobel Prize for Physiology or Medicine in 1962. In 1977, he became professor at the Salk Institute in San Diego and branched into neuroscience.

Crookes, William 1832–1919
See p.292.

Curie, Marie 1867–1934
See pp.296–97.

Cuvier, Georges 1769–1832
See p.186.

Da Vinci, Leonardo 1452–1519
Italian artist, musician, botanist, mathematician, and architect, Leonardo was the archetypal Renaissance man and his brilliant inventiveness was expressed in hundreds of detailed technical drawings. Leonardo set out to write the first systematic explanations of how machines work, but many of his inventions progressed no further than plans on paper. His submarine, clockwork road vehicle, hydraulic pumps, armored tank, helicopter and flapping-wing aircraft were never made. However, he greatly advanced the ideas of civil engineering and hydrodynamics for future generations

Daguerre, Louis 1787–1851
See p. 262.

Dalton, John 1766–1844
See p.138.

Daoyuan, Li 469 or 476–527 CE
A geographer during the Northern Wei Dynasty, Daoyuan traversed China's mountains and streams analyzing the topography. He discovered that the landscape had changed over time and set out to record a detailed geographical log, which he called *Shui Jing Zhu* (Commentary on the Waterways Classic), describing hills, lakes, cities, as well as hydrology and local climate.

Darwin, Charles 1809–1882
See pp.202–03.

Davy, Humphry 1778–1829
See p.170.

Dawkins, Richard 1941–
English evolutionary biologist, Dawkins taught zoology at the Universities of Oxford and California. As a student, he was mentored by Niko Tinbergen, famous for his pioneering research on animal behavior. He was also influenced by the work of Crick and Watson on DNA and genetics and pushed his theories of evolutionary biology, the "selfish gene" and the term "meme" into mainstream scientific culture. He is a Fellow of the Royal Society and remains an influential and controversial figure.

de Betancourt, Agustín 1758–1824
Spanish engineer whose work ranged from steam engines to balloons, urban planning, and structural engineering. De Betancourt was the first director of the Madrid School of Civil Engineering. On a visit to England in 1788, he met James Watt and designed a steam-powered pump and mechanical loom. He built an optical telegraph connection between Madrid and Aranjuez, and in Russia he constructed Saint Petersburg's first bridge across Malaya Nevka.

de Fermat, Pierre 1601–1665
See p.196.

de Gimbernat, Anthoni 1734–1816
Spanish anatomist and surgeon. In 1765, he became professor of anatomy at the Royal School of Surgery in Barcelona. He was a pioneer in ophthalmology, urology, and vascular surgery and developed techniques for mending diaphragmatic hernias. He entered the king's medical cabinet in 1798, serving as personal physician to King Carlos III. In 1801, he was appointed First Royal Surgeon and president of all surgical schools in Spain.

de la Cierva, Juna 1895–1936
Spanish civil engineer and pilot, de la Cierva made a major contribution toward the practical takeoff and landing of aircraft. His autogyro of the 1920s had a freely turning rotor and could take off and land in a short distance. He influenced the design of helicopters and is credited with establishing an understanding of rotor control and stalling.

de la Place, Pierre 1749–1827
See p.183.

Delbrück, Max 1906–81
German-American biologist who pioneered the study of molecular biology. Originally trained in astrophysics Delbrück became interested in biology after meeting Wolfgang Pauli and Neils Bohr. In the US, Delbrück teamed up with Salvador Luria, and, with Alfred Day Hershey, they won the Nobel Prize for Physiology or Medicine in 1969 for their work on the genetic structure of viruses.

Descartes, René 1596–1650
See p.103.

de Graaf, Regnier 1641–1673
Dutch physician and anatomist who discovered the development of the ovarian follicles, later called Graafian follicles. He also gave exact descriptions of the testicles and isolated and collected secretions of the gallbladder and pancreatic juices.

de Vries, Hugo 1848–1935
Dutch botanist and geneticist who is best known for his studies on plant mutation. De Vries rediscovered George Mendel's laws of heredity, hypothesized on the existence of pangenes, and experimented with osmosis, coining the term "isotonic." He became professor at the University of Amsterdam in 1878 and was awarded the Darwin Medal in 1906.

Diesel, Rudolph 1858–1913
See p.255.

Diophantus of Alexandria c.200–c.284 CE
Greek mathematician best known for his *Arithmetica*. He studied at the University of Alexandria in Egypt and wrote a series of books that set out to explore algebraic equations. He is known as the father of algebra, though some claim this title should be shared with the Persian mathematician Al-Khwarizimi.

Dirac, Paul 1902–1984
Influential English theoretical physicist who won the Nobel Prize for physics in 1933 for his work on quantum theory. He applied Einstein's relativity theory to quantum mechanics and deduced that electrons must be spinning around their own axes and behaving like small magnets. Dirac predicted the existence of positrons and formulated the Dirac equation, a relativistic quantum mechanical wave equation that is important in modern theoretical physics. The equation demands the existence of antiparticles and actually predated their experimental discovery. Dirac held the Lucasian Chair of Mathematics at the University of Cambridge and spent the last years of his life at the University of Florida.

Dolland, John 1706–1761
English optician, the son of a Huguenot silk-weaver, who found how to construct achromatic lenses by a combination of crown and flint glasses. He also developed the heliometer, a telescope used to find the angular distance between two stars. Awarded the Copley Medal from the Royal Society in 1758, he was elected as a Fellow in 1761 and was appointed optician to King George III.

Doppler, Christian Johann 1803–1853
Austrian mathematician, famous for conceiving the Doppler Effect, which he presented at the Royal Bohemian Society in Prague, 1842. He theorized on the apparent change in the frequency of a light or sound wave when the observer and the source are moving relative to each other. In 1850, he was appointed as the first director of the new Institute of Physics at the Imperial University in Vienna.

Doudna, Jennifer 1964–
See p.417.

Duchenne, Guillaume 1806–1875
See p.225.

Eastman, George 1854–1932
American philanthropist, inventor, and industrialist. Founder of

the Eastman Kodak, he is remembered as the pioneer of snapshot photography. The most significant innovations under Eastman's direction were the Kodak snapshot box camera (1888), the celluloid-based roll film (1889), and 16mm ciné film (1923).

Eddington, Arthur 1882–1944
English astrophysicist best known for his confirmation of Einstein's general theory of relativity. In 1919, he took measurements on the Atlantic island of Principe during a total solar eclipse, showing that massive objects such as the sun distort space-time around them. He developed the first detailed models of the internal structure of stars and supported models of an expanding universe but virulently attacked suggestions of "extreme" stars such as white dwarfs.

Edison, Thomas 1847–1931
See p.265.

Ehrlich, Paul 1854–1915
German immunologist who discovered the blood-brain barrier—a sheet of cells that selectively shields the brain and spinal cord. Ehrlich used synthetic stains to study blood cells and realized that these stains might lead the way to antibacterial drugs. In 1909, Ehrlich and his Japanese assistant, Sahachiro Hata, found that their compound number 606 (Salvarsan) was deadly to syphilis bacteria. The discovery led to the development of the first effective treatment for syphilis.

Einstein, Albert 1879–1955
See pp.304–05.

Empedocles c.495–c.435 BCE
See p.22.

Entralgo, Pedro Laín 1908–2001
Spanish medical researcher and humanist, who wrote on the anthropology of hope. He obtained the chair of the history of medicine at the University of Madrid in 1942, founded the Instituto Arnau de Vilanova of the History of Medicine in 1943, and was elected a member of the Royal National Academy of Medicine in 1946.

Erastothenes 276–194 BCE
See p.38.

Esaki, Leo 1925–
Japanese physicist who invented the Esaki tunnel diode, the first quantum electron device. In 1973, he was awarded the Nobel Prize in Physics for his work on electron tunneling in solids. Esaki was elected a Fellow of the American Academy of Arts and Sciences in May 1974, a Foreign Associate of the National Academy of Engineering (US) in 1977, and a foreign member of the American Philosophical Society in 1991.

Euclid c.330–c.260 BCE
See p.34.

Euler, Leonhard 1703–1783
Swiss mathematician and physicist who created new areas of mathematical study, such as topology. He was a leading pioneer in astronomy, mechanics, and optics. Although totally blind toward the end of his life, his vast output was undiminished and he carried on performing astronomical calculations mentally.

Eustachio, Bartolomeo c.1500–1574
Italian founder of the science of human anatomy. He was the first to describe the anatomy of the teeth, identified the adrenal glands, and extended the knowledge of the inner ear. The eustachian tube linking the ear to the throat, to balance pressure on the eardrum, is named after him.

Fabricius, Hieronymus 1537–1619
Italian surgeon and anatomist. He was educated at the University of Padua where he studied under and succeeded Gabriel Fallopius as professor of anatomy in 1565. Fabricius's pioneering work on veins led to later knowledge of blood circulation. He identified the larynx as a vocal organ and elevated embryology to an independent science.

Fahrenheit, Gabriel Daniel 1686–1736
German physicist, engineer, and instrument maker who made the first reliable thermometers. Fahrenheit used mercury successfully and introduced cylindrical bulbs instead of spherical ones. When

Europe switched to Celsius, the Fahrenheit temperature scale was still widely used in the rest of the world.

Fallopius, Gabriello 1523–1562
Italian anatomist considered to be one of the most important anatomists of the 16th century. He spent much of his time working on the anatomy of the head but is best known for identifying the reproductive organs of both men and women. The fallopian tube, which connects an ovary to the uterus, is named after him. In 1551, he took the chair of anatomy and surgery at the University of Padua and held a professorship of botany.

Faraday, Michael 1791–1867
See pp.170–71.

Fermi, Enrico 1901–1954
Italian physicist most noted for his development of the first nuclear reactor and for his work on quantum theory and statistical mechanics. Fermi was elected professor of theoretical physics at the University of Rome in 1927, became the leading expert on neutrons, and was awarded the Nobel Prize for physics in 1938. He became professor at the Institute for Nuclear Studies of the University of Chicago in 1946, a position he held until his death.

Ferrel, William 1817–1891
See p.207.

Feynman, Richard 1918–1988
See pp.326–327.

Fibonacci, Leonardo c.1170–c.1250
See p.63.

Fitzroy, Robert 1805–1865
English captain of HMS Beagle and meteorologist. He was awarded a Gold Medal by the Royal Geographic Society in 1837, and his achievement in nautical surveying and scientific navigation resulted in his election to the Fellowship of the Royal Society in 1851. He was Governor of New Zealand from 1843 to 1845 and became head of the British Meteorological Department in 1854. Promoted to Vice Admiral in 1863, he suffered from depression and committed suicide, leaving substantial debts.

Fleming, Alexander 1881–1955
Scottish bacteriologist who, in 1928, discovered the antibiotic penicillin. Working at St. Mary's Hospital in London, he noticed that penicillin mold growing on cultures of bacteria prevented the bacteria from multiplying. Penicillin has since saved millions of lives. He was knighted in 1944 and in 1945 shared the Nobel Prize for Physiology or Medicine with Ernest Chain and Sir Howard Florey, who developed the commercial production of penicillin.

Fleming, Walter 1843–1905
See p.306.

Florey, Howard Walter 1898–1968
Australian biochemist who, along with Ernst Chain, took up Alexander Fleming's neglected work on penicillium. They successfully isolated and purified penicillin. With mass production of the drug achieved in 1943, it was used in treating casualties of war, saving countless lives. He shared the Nobel Prize for Physiology or Medicine with Chain and Fleming in 1945 and raised to the British peerage in 1965.

Ford, Henry 1863–1947
American inventor and industrialist who introduced the moving assembly line into motor manufacture, enabling full-scale mass production. He founded the Ford Motor Company in 1903. His first vehicle was the Model T Ford, introduced in 1908. The Model T remained in production almost unaltered for 15 years, during which time 15 million were built.

Foucault, Jean Bernard Léon 1819–1868
French physicist who was the first to show how a pendulum can demonstrate Earth's rotation. He is credited with naming the gyroscope and measured the speed of light, showing how it travels more slowly through water than air. In 1855, he received the Copley Medal of the Royal Society and was made a Fellow in 1864.

Fowler, William 1911–1955
American nuclear astrophysicist who formulated a theory of element generation. Fowler won the Henry Norris Russell Lectureship of the American Astronomical Society in 1963 and the Eddington Medal in 1978. He shared the Nobel Prize for Physics in 1983 for his theoretical and experimental studies of nuclear reactions in the formation of the chemical elements in the universe. He received the national Medal of Science in 1974 and was awarded the Legion of Honour in 1989.

Franklin, Benjamin 1706–1790
See pp.160–61.

Franklin, Rosalind 1920–1958
English scientist, who, along with Maurice Wilkins, took X-ray diffraction photographs of DNA and found that the DNA molecule had a corkscrew-like shape known as a helix. These results were put together by Francis Crick and James Watson to become the double helix model. Though less well known than her colleagues, Franklin's crucial role in the discovery of DNA structure is now widely acknowledged.

Fresnel, Augustin-Jean 1788–1827
French physicist who built on the work of Thomas Young to extend the wave theory of light to a large range of optical uses. He studied the behavior of light and is best known for the Fresnel lens, which used compound lenses instead of mirrors in lighthouses. He received scant recognition for his work during his lifetime but became a member of the Royal Society in 1825 and was awarded the Rumford Medal by the Society in 1827.

Gabor, Dennis 1900–1979
See p.365.

Gagarin, Yuri 1934–1968
See p.372.

Galen, Claudius c.130 CE–c.200 CE
Greek physician, born in Pergamum in Asia Minor, who settled in Rome in 164. He was physician to five Roman emperors and wrote extensively on physiology and anatomy. He based his writing on his knowledge of early medicine and his own experiments. Of the 500 treatises he wrote, 180 survive. Galen's arguments dominated medicine throughout the Middle Ages.

Galilei, Galileo 1564–1642
See pp.82–83.

Galvani, Luigi 1737–1798
See p.221.

Gamow, George 1904–1968
American physicist, cosmologist, and champion of the Big Bang theory of the universe. He also predicted the existence of a background warmth in the universe, left over from the Big Bang's fireball, a prediction confirmed in 1965. Renowned as a popular science writer, he joined the University of Colorado, in 1956 as professor and worked there until his death.

Gassendi, Pierre 1592–1655
French priest who achieved distinction as a scientist, astronomer, mathematician, and philosopher. He tried to revive the philosophy of Epicurus and reconcile it with Christianity. He is best known for his opposition to the ideas of Descartes. The lunar crater Gassendi is named after him.

Gauss, Carl Friedrich 1777–1855
German mathematician whose major work concerned number theory or higher arithmetic. One of history's most influential mathematicians, Gauss worked on complex numbers, non-Euclidean geometry, electromagnetism, and the theory of errors, which led to the concept of the normal distribution curve. He was awarded the Royal Society's Copley Medal in 1838 and has a moon crater, ship, asteroid, and extinct volcano named after him.

Gay-Lussac, Joseph-Louis 1778–1850
See p.101.

Geber, Jabir c.721–815 CE
Islamic alchemist, astronomer, physicist, and philosopher. Also known by the full name of Abu Musa Jabir ibn Hayyan, he is described as the father of Arab chemistry. He turned alchemy

into a science, invented many pieces of laboratory equipment, and identified chemical substances such as hydrochloric acid and nitric acid.

Geiger, Hans 1882–1945
German physicist best known for developing the Geiger counter, an instrument for measuring and detecting nuclear radiation. In 1909, while working under the direction of Ernest Rutherford, Geiger and Ernest Marsden set up their famous gold foil experiment, the Geiger-Marsden experiment and discovered that atoms have a nucleus. In 1928, Geiger and Walther Müller created an improved Geiger counter with increased speed and sensitivity.

Gilbert, William 1544–1603
See p.81.

Goddard, Robert H 1882–1945
See p.339.

Goeppert-Meyer, Maria 1906–1972
See p.287.

Golgi, Camillo 1843–1926
Italian scientist, pathologist, and physician who spent much of his career studying the central nervous system. In 1873, he published *On the Structure of the Brain Grey Matter*, in which he described the discovery of the "black reaction." This revolutionary tissue staining is still in use today and is named after him (Golgi staining or Golgi impregnation). Golgi established the Institute of General Pathology in 1881 and was jointly awarded the Nobel Prize for Physiology or Medicine in 1906 with Santiago Ramón y Cajal.

Goodall, Jane 1934–
See p.277.

Gould, Stephen Jay 1941–2002
American paleontologist and evolutionary biologist. He is best known for his work with Niles Eldridge and their theory of punctuated equilibrium. He spent most of his career at Harvard University, where he campaigned against creationism and argued that science and religion should be kept as two distinct fields. He was posthumously awarded the prestigious Linnean Society of London's Darwin-Wallace Medal in 2008.

Greene, Brian 1963–
See p.402.

Grew, Nehemiah 1641–1712
English botanist, vegetable anatomist, and physiologist. His most popular and influential work, *The Anatomy of Plants* (1682) shows the microscopic detail of plant tissue. His work was contemporaneous with that of Marcello Malpighi, and the two shared much of their research. He was elected Fellow of the Royal Society in 1671 and in 1684 was the first to describe sweat pores on the hands and feet.

Guangqi, Xu 1562–1633
Chinese astronomer, mathematician, and agricultural scientist who experimented with western irrigation methods and new agricultural practices. Working with Italian Jesuit Matteo Ricci, he translated several important western texts, including the first parts of Euclid's *Elements*, and began to reform the Chinese calendar. His collaboration with western scientists was hailed as the start of Chinese enlightenment.

Gutenberg, Johannes 1400–1468
See p.58.

Haber, Fritz 1868–1934
German chemist who in 1908–1909 devised a method of making synthetic ammonia, a raw material essential in making explosives and fertilizers. His discovery led to a great expansion of the chemical industry, and for this he won the Nobel Prize in Chemistry in 1918. A patriotic German, he was given the rank of captain by the Kaiser and developed the use of poisonous gases in trench warfare.

Hadamard, Jacques 1865–1963
French mathematician best known for proving the prime number theorem and the Hadamard three-circle theorem. He published *Psychology of Invention in the Mathematical Field* in 1945,

demonstrating the use of introspection to describe mathematical thought processes. Hadamard is also associated with the Dreyfus Affair, a French political scandal in which Alfred Dreyfus, a Jewish army captain, was falsely convicted of treason. Dreyfus was Hadamard's brother-in-law, and Hadamard worked tirelessly to clear his name.

Hadley, George 1685–1768
English physicist and meteorologist who advanced the first accurate theory describing the trade winds and explained why they blew from the northeast in the northern hemisphere and the southeast in the southern hemisphere. Although a partial explanation had previously been offered by Edmond Halley (1686), he was unable to explain the associated north-south circulation pattern. In 1735, Hadley suggested that heated air at the equator rises and flows toward the poles, where it cools, descends, and flows back toward the equator; this pattern became known as the Hadley cell. However, Hadley's theory remained unacknowledged until 1793, when it was rediscovered by English scientist John Dalton.

Haeckel, Ernst 1834–1919
German biologist and artist best known for his now-discounted "biogenetic law" or theory of recapitulation. He was a supporter of Charles Darwin's theories and produced studies of marine organisms. Haeckel made the basic division of the animal kingdom into single-celled and multicellular creatures. He described and named thousands of new species and mapped a genealogical tree relating all forms of life. He was awarded the Linnean Society of London's Darwin-Wallace Medal in 1908.

Hahn, Otto 1879–1968
German chemist credited as the first person to demonstrate the deliberate fission of uranium atoms in 1938. Once nuclear fission had been accepted, Hahn was able to demonstrate its use in warfare and energy production. He was awarded the Nobel Prize in Chemistry in 1945 but could not attend the ceremony as he was interned at Farm Hall near Cambridge, suspected of working on the German nuclear energy project. Hahn had never worked on the program, and after the Americans dropped atomic bombs on Hiroshima and Nagasaki, Hahn became an outspoken advocate against nuclear weapons. In 1961, Pope John XXIII awarded him the Gold medal of the Papal Academy.

Hales, Stephen 1677–1761
See p.155.

Halley, Edmond 1656–1742
See p.119.

Harrison, John 1693–1776
English carpenter and clockmaker who invented the marine chronometer crucial in establishing longitude at sea. Built for a prize offered by the British Government in 1714, his design solved one of the greatest navigation problems. Harrison built a succession of three chronometers, but it was with his fourth timepiece (H4) that he intended to claim the prize. Despite its accuracy, the Government was unconvinced, and Harrison had to prove the design could be replicated. Harrison eventually won some of the prize money but remained bitter at his treatment.

Harvey, William 1578–1657
See p.90.

Hawking, Stephen 1942–2018
English physicist and expert on black holes, who developed a single theory on the beginning and end of the universe. Hawking is regarded as one of the leading contributors to the science of cosmology and his work is widely recognized. In 1979, he took the post of Lucasian Professor of Mathematics at Cambridge. Author of the bestselling *A Brief History of Time*, he suffered from ALS (amyotrophic lateral sclerosis) from a young age. Confined to a wheelchair, he used an electronic voice synthesizer to communicate. His many accolades include 12 honorary degrees, Fellowship of the Royal Society, and Membership of the US National Academy of Sciences.

Heisenberg, Werner 1901–1976
See p.316.

Henry, Joseph 1797–1878
See p.168.

Herschel, Caroline 1750–1848
See p.119.

Herschel, William 1738–1822
German-born English astronomer. He discovered Uranus in 1781 and developed a theory of nebulae and the evolution of stars. He also showed that the solar system moved through space, discovered infrared radiation, and constructed over 400 telescopes. Herschel coined the term "asteroid," and controversially believed that every planet was inhabited. He received the Copley Medal in 1781, was made a Fellow of the Royal Society, and knighted in 1816. Herschel helped found the Astronomical Society of London in 1820, which later became the Royal Astronomical Society.

Hertz, Heinrich 1857–1894
German physicist whose experiments with electromagnetic waves led to the development of the wireless telegraph and the radio. All frequencies are measured in Hertz, meaning cycles per second. He progressed the electromagnetic theory of light formulated by English physicist James Clerk Maxwell in 1884 and proved that electricity can be transmitted in electromagnetic waves that travel at the speed of light and possess various other properties of light. Hertz was the first person to broadcast and receive radio waves and to show that light was a form of electromagnetic radiation (1885–1889).

Hertzsprung, Henry Russell 1873–1967
See p.329.

Hilbert, David 1862–1943
German mathematician whose many outstanding contributions span number theory, mathematical logic, differential equations, and the three-body problem. He invented a space-filling curve known as the Hilbert Curve. In his *Foundations of Geometry*, he stated the first rigorous set of geometrical axioms. In a world meeting in Paris in 1900, he posed a list of 23 problems, now called Hilbert problems; this was considered the most influential speech ever given to mathematicians.

Hipparchus 190–120 BCE
See p.28.

Hippocrates c.460–377 BCE
Greek physician and teacher regarded as the father of clinical medicine. Hippocrates is credited with being the first physician to reject the idea that supernatural or divine forces caused illness. He based his medical practice on observations, study of the human body, and documentation of symptoms and treatments. Hippocrates was the first physician to accurately describe many diseases and conditions, categorize illnesses according to severity (acute, chronic), and to use terms such as "relapse" and "convalescence." He founded the Hippocratic school of medicine on the Greek island of Cos. With his followers, Hippocrates taught medicine and set out rigorous codes of behavior, ethics, and good practice through the Hippocratic oath and some 60 medical treatises (the Hippocratic corpus).

Hodgkin, Dorothy 1910–1994
See pp.384–385.

Hooke, Robert 1635–1703
See pp.92–93.

Hooker, Joseph Dalton 1817–1911
See p.153.

Hoover, Erna Schneider 1926–
See p.378.

Hopper, Grace 1906–1922
American computer scientist who devised some of the first computer programs. Notably, Hopper was one of the first programmers of the Harvard Mark I computer. Her work on compilers led ultimately to the development of the business language COBOL. She was a United States Naval Officer and rose to the rank of Rear Admiral; the US Battleship USS Hopper was named after her. In 1973, she became the first woman to be made a Distinguished Fellow of the British Computer Society.

Hoyle, Fred 1915–2001
See p.321.

Hubble, Edwin 1889–1953
See p.340.

Hutton, James 1726–1797
See p.184.

Huygens, Christiaan 1629–1695
See p.114.

Ingenhousz, Jan 1730–1799
Dutch-born English physician and scientist who discovered photosynthesis, whereby sunlight enables green plants to absorb carbon dioxide and release oxygen. Ingenhousz also invented an improved device for generating large quantities of static electricity (1766), made the first quantitative measurements of heat conduction in metal rods (1789), and was a discoverer of Brownian motion, describing the erratic movement of coal dust on the surface of alcohol (1785). He was a physician to the Austrian Empress, Maria Theresa, and worked in various European countries ranging from Britain to Switzerland.

Jackson, Mary 1921–2005
See p.371.

Jenner, Edward 1749–1823
English scientist and medical practitioner, Edward Jenner demonstrated how inoculation with cowpox virus could safely provide immunity from the deadly viral disease of smallpox. Before Jenner's time, Europe had become aware that in the East people protected themselves from smallpox by treating themselves with material from smallpox survivors, but Jenner's "vaccination" with cowpox virus produced dependable immunization that carried no risk of smallpox infection from the vaccine itself. Thanks in large part to universal adoption of cowpox inoculation, smallpox was declared an eradicated disease by the World Health Organization in 1980.

Johnson, Katherine 1918–2020
See p.371.

Joliot-Curie, Irène 1897–1956
See p.297.

Joule, James Prescott 1818–1889
See p.177.

Julien, Percy 1899–1975
See p.241.

Jun, Heo 1546–1615
Korean court physician during the reign of King Seonjo of the Joseon Dynasty, Heo Jun based his herbal medicine on Chinese principles but rewrote the prescriptions, simplifying them and replacing expensive Chinese ingredients with native Korean equivalents. The greatest of his texts is *Dongui bogam*, in which he made remedies understandable to commoners by using simple hangui characters rather than Chinese hanja characters. Heo Jun was a holistic practitioner who would promote training of the mind and body and mind in preference to other forms of treatment.

Kamil, Abu c.850–c.930
Egyptian mathematician Abu Kamil, also known as al-Hasib al-Misri, the "calculator from Egypt," specialized in algebra. He was familiar with the works of Al-Khwarizmi, whom he acknowledged as inventor of all the principles of algebra. In his own *Book on Algebra*, Abu Kamil combined Greek geometry with Al-Khwarizmi's algebra and Babylonian methodology to solve geometrical problems. In his later *Book of Rare Things in the Art of Calculation*, Abu Kamil addressed solutions to indeterminate equations. Abu Kamil's books gained importance in the 13th century as the basis of Fibonacci's introduction of algebra to Europe.

Kant, Immanuel 1724–1804
German philosopher Immanuel Kant had a profound effect not only on mathematical and scientific thinking but also on other areas of philosophy, such as ethics and aesthetics. By addressing the question, "What can we know?" he worked to reconcile empiricism (we know only what our senses reveal) and rationalism (we know only what our minds construct). Kant concluded that objective experience is actively constructed by the functioning of the human mind. As a result, he placed the human mind at the center of the acquisition of knowledge, drawing a definite boundary between physics (what we can know) and metaphysics (about which we can only speculate).

Kapany, Narinder Singh 1927–2020
Indian physicist whose 1955 doctoral research opened the gateway to fiber optics, with its myriad uses from data transmission to imaging. In addition to fiber-optic communications, his research and inventions involved biomedical instrumentation, pollution monitoring, lasers, and solar energy; he had more than 120 patents. Long based in the US, his eclectic career has also encompassed business, publishing, lecturing, farming, and academia.

Keeling, Charles 1928–2005
American geochemist who developed a method for measuring levels of carbon dioxide (CO_2) in the air, while working at the California Institute of Technology in the 1950s. His research alerted the world to the relentless rise in atmospheric CO_2 and he continued to measure atmospheric CO_2 for over 40 years. In 2002, he received the National Medal of Science, the highest award in the US for lifetime achievement in scientific research.

Kekulé, Friedrich August 1829–1896
See p.288.

Kelvin, William Thompson, Lord 1824–1907
See p.181.

Kepler, Johannes 1571–1630
See p.77.

Khayyám, Omar 1048–1123 BCE
Persian scholar, poet, and mathematician. In mathematics, he developed a geometric method for solving cubic equations by intersecting a parabola with a circle. Khayyam also contributed to calendar reform, compiled astronomical tables, and constructed important critiques of Euclid's theories of proportion and parallels. He is also known for his poem, *The Rubáiyát of Omar Khayyám*.

Khorana, Har Gobind 1922–2011
Indian-born US biochemist who shared the 1968 Nobel Prize in Physiology or Medicine with Marshall Nirenberg and Robert Holley for work on the interpretation of the genetic code and its role in protein synthesis. He later developed the first artificial gene. Such genes are used in laboratories for sequencing, cloning, and other forms of genetic engineering. In 1970, he became a professor of chemistry and biology at the Massachusetts Institute of Technology, remaining there until retirement in 2007.

Koch, Robert 1843–1910
German physician, whose discoveries facilitated the development of the first chemicals designed to attack specific bacteria. Koch isolated the bacilli for anthrax in 1877, tuberculosis in 1882, and cholera in 1883. He also developed Koch's postulates of germ theory that sets out the criteria to be met to establish whether specific bacteria cause specific diseases. Koch won the 1905 Nobel Prize in Physiology or Medicine for his tuberculosis discoveries. He was appointed professor of hygiene at the University of Berlin in 1885 and was director of the Institute for Infectious Diseases, Berlin from 1891 to 1904.

Kuhne, Wilhelm 1857–1945
See p.216.

LaPlace, Pierre-Simon 1749–1827
See p.183.

Lamarck, Jean-Baptiste 1749–1823
French biologist and an early evolutionist, who believed that acquired characteristics are inheritable. Lamarck recognized that the environment helps shape the species that inhabit it and noted that the Earth's surface undergoes constant transformation, creating significant changes over time. His Lamarckian theory of evolution suggested that living things evolved continuously upward, from dead matter to simple and then increasingly complex organisms. Although generally rejected during his lifetime, his theory of inheritance remained popular into the 20th century. Lamarck also made his mark as a botanical and zoological systematist and a founder of invertebrate paleontology.

Lavoisier, Antoine Laurent 1743–1794
French chemist whose synthesis of existing chemical knowledge, sound methodology, and a quantitative approach revolutionized 18th-century chemistry. Lavoisier proved that matter is neither created nor destroyed during chemical changes and discovered that air comprises two gases, naming them oxygen and azote (nitrogen). In demonstrating the role oxygen plays in combustion, Lavoisier laid the foundations for modern chemistry. He coproduced a system of chemical nomenclature (1787), wrote the first list of elements, and helped establish the metric system. He was guillotined during the French Revolution.

Leaky, Mary 1913–1996
See p.419.

Leavitt, Henrietta Swan 1868–1921
Celebrated for her contributions to astronomy. Working at the Harvard College Observatory, Leavitt conducted groundbreaking work on Cepheid variable stars and stellar magnitudes. In 1912, she established a direct link between time span and brightness in Cepheid variables, pulsating stars that vary regularly in luminosity in periods ranging from days to months. Her discovery led to an invaluable method of determining the distances of stars and galaxies. Leavitt was also involved in determining the photographic magnitude of stars, developing the Harvard Standard of photographic measurements (1913).

Le Chatelier 1850–1926
See p.236.

Lee, Tsung-Dao 1926–
Chinese-born US physicist best known for his work in elementary particles, statistical mechanics, and astrophysics. He co-won the 1957 Nobel Prize for Physics for demonstrating violations of the principle of parity conservation, leading to important refinements in particle physics theory. He also created the Lee model (a solvable model of quantum field theory). Lee joined the Institute for Advanced Study, Princeton (1951–1953) and was later professor of theoretical physics there (1960–1963).

Lehmann, Inge 1888–1993
See p.352.

Lilienthal, Otto 1848–1896
See p.290.

Lind, James 1716–1794
See p.218.

Linnaeus, Carolus 1707–1778
See p.123.

Lippershey, Hans c.1570–c.1619
Dutch eyeglass maker credited with inventing the first practical telescope (1608). The States General of the Netherlands refused to award Lippershey a patent but paid him with the proviso that the telescope was modified into a binocular device. Others recognized the telescope's importance in astronomy, however, and a lunar crater and planet have since been named after him. Lippershey is also credited with inventing the compound microscope, independently of others during the same period.

Lister, Joseph 1828–1912
See p.229.

Locke, John 1632–1704
English philosopher whose work influenced the development of political philosophy and empiricism. His most famous works include his *Essay Concerning Human Understanding* (1690) and *Two Treatises on Government*, published in 1690 concerning natural law and natural rights. Locke's ideas about liberty and the social contract influenced the US Declaration of Independence (1776).

Lorentz, Hendrik Antoon 1853–1928
Dutch physicist known for his work on the relationship between electricity, magnetism, and light. Lorentz shared the 1902 Nobel Prize in Physics with Pieter Zeeman for his theory of electromagnetic radiation based on the Zeeman effect. In 1904, he introduced the concept of local time (different times in different locations), creating transformation equations that paved the way for Einstein's theory of relativity. He also discovered, almost simultaneously with L V Lorenz, the

relation between the refraction of light and the density of the translucent body (the Lorentz-Lorenz formula). From 1919, he led a study of projected movements of sea water caused by the reclamation of the Zuyderzee; his calculations contributed to the science of hydraulics.

Lorenz, Edward 1917–2008
American mathematician and meteorologist who first recognized the practical effect of chaos theory, influencing a wide range of basic sciences. In the early 1960s, Lorenz realized that small differences in a dynamic system such as the atmosphere could trigger huge, often unpredictable changes. This led him to formulate the "butterfly effect." In 1991, Lorenz won the Kyoto Prize for establishing the theoretical basis of weather and climate predictability, as well as the basis for computer-aided atmospheric physics and meteorology. He was Professor Emeritus at Massachusetts Institute of Technology from 1987 until his death.

Losada, Manuel 1929–
Spanish biochemist and molecular biologist known for his research on the photosynthetic assimilation of nitrogen. Concerned with the practical aspects of scientific research, Losada has focused on the development of biological and biochemical systems that can transform solar energy into chemical energy, and the use of microalgae to produce valuable compounds, such as protein-rich biomass. Recently Losada has been involved in plant developmental biology at the Institute of Plant Biochemistry and Photosynthesis at the University of Seville, where he is director.

Lovelace (Countess of), Ada 1815–1852
English mathematician and daughter of poet Lord Byron, she was described as the first computer programmer. An acolyte of inventor and mathematician Charles Babbage, Lovelace devised programs for his analytical engine (1843). She detailed how it could be used to calculate Bernoulli numbers, foreseeing the functions of a machine such as a general-purpose computer and anticipating future developments, such as computer-generated music. A software language, Ada, was named after her in 1979.

Lovelock, James 1919–
See pp.412–13.

Lyell, Charles 1797–1875
See p.190.

Malpighi, Marcello 1628–1694
Italian physiologist and biologist who, through his microscope studies of plant and animal tissue, founded the sciences of microscopic anatomy and histology. In 1661 he observed the anatomy of the frog lung. He named capillaries, scrutinized the silkworm, and made a major contribution to embryology. He discovered papillae (taste buds) and was the first to attempt a thorough study of the anatomical details of the brain. Fascinated by botany, he also discovered stomata, the pores of leaves. He was professor of the University of Bologna (1661–1691) and nominated personal physician to Pope Innocent XII (1691).

Malthus, Thomas Robert 1766–1834
English clergyman and philosopher best remembered for his theories on population growth. He argued that natural growth in population tends to outstrip food supplies. Unless population growth could be restricted, overpopulation would be checked only by war or famine. His theory, the Malthusian growth model, had a major impact on political, economic, and social thought. While berated by some, others, such as Darwin, were hugely influenced by him.

Marconi, Guglielmo 1874–1937
Italian physicist and inventor who proved radio communication was possible with the development of a successful wireless telegraph (1896). Marconi then sent the first wireless signal across the English Channel (1899) followed by the first wireless signal across the Atlantic (1902). This made headlines worldwide, paving the way for important developments in radio communications, broadcasting, and navigation services. Marconi won the 1909 Nobel Prize for Physics with German physicist Ferdinand Braun. He patented several new inventions between 1902 and 1912, including his magnetic detector, which became the standard wireless receiver for many years. He later worked on the development of shortwave wireless communication.

Margalef i López, Ramón 1919–2004
A Spanish founder of the modern science of ecology who made a

huge contribution to limnology (study of lakes) marine biology, oceanography, and theoretical ecology. He became Spain's first professor of ecology (1967), setting up the department of ecology at the University of Barcelona. His book *The Theory of Information in Ecology* (1957) attracted international acclaim, as did his *Perspectives in Ecological Theory* (1968). Among his most important work were the application of information theory to ecological studies and the creation of mathematical models for population studies.

Magulis, Lynn 1938–2011
See p.420.

Marsh, Othniel Charles 1831–1899
American paleontologist who became professor of paleontology at Yale University in 1866 and director of the Peabody Museum of Natural History. Marsh led his students on field trips to the American West and discovered many new dinosaur fossils. In 1871, his party discovered the first pterodactyl found in the United States. By the mid-1870s, Marsh had an exceptional collection of early mammals.

Maudslay, Henry 1771–1831
See p.172.

Maurey, Matthew 1888–1873
See p.213.

Maxim, Hiram Stevens 1840–1916
Anglo-American inventor who perfected the first fully automatic machine gun. The Maxim gun, in which the recoil of each shot loaded and fired the next round, was introduced in 1888. He also built a large steam-powered flying machine, which almost took off. During his career, Maxim took out 250 patents for various designs.

Maxwell, James Clerk 1831–1879
See p.167.

May, Robert McCredie 1936–2020
Australian theoretical physicist, zoologist, and ecologist. He trained as a theoretical physicist/applied mathematician and studied the way populations are structured. He pioneered work on chaos theory and ecological systems and devised methods of estimating the number of species on earth. He held a Professorship jointly at Oxford University and at Imperial College, London, was a President of the Royal Society (who awarded him their Copley medal in 2007), a member of the UK Government's Climate Change Committee, and was made a peer in 2001.

McClintock, Barbara 1902–1992
See p.349.

Mead, Margaret 1901–1978
American social anthropologist noted for her pioneering studies on childhood, adolescence, and sexual diversity in Pacific communities. Her most famous work, *Coming of Age in Samoa* (1929), suggested that "civilized" peoples could learn much from "primitive" peoples. Her studies were controversial at the time. She pioneered academic involvement in television by fronting documentaries and was the first elected president for the American Association for Advancement of Science (1974).

Meitner, Lise 1878–1968
See p.324.

Mendel, Gregor 1822–1884
See p.204.

Mendeleev, Dimitri 1834–1907
See p.233.

Mercator, Gerardus 1512–1594
Flemish geologist, cartographer, and instrument maker, Mercator produced his first globe in 1541. In 1569, he produced the first map of the world for navigational use, with longitude and latitude, using a map projection. He was the first to make a map of the British Isles and Flanders. The lunar crater Mercator is named after him.

Michell, John 1724–1793
English philosopher and amateur astronomer. In 1760, Michell

constructed a theory of earthquakes as wave motions in the interior of the Earth. As an amateur astronomer, he was a talented telescope builder. In 1761, he was elected Fellow of the Royal Society. From 1762 to 1764, Michell held the Woodwardian Chair of Geology at Cambridge, and in 1767 he was appointed rector of St. Michael's Church of Thornhill, near Leeds, UK. While a rector, he continued to do scientific work. He suggested the idea of black holes in 1784 and devised apparatus for measuring the density of the Earth.

Michelson, Albert Abraham 1852–1931
American physicist whose experiment with Edward Morley in 1887 helped determine the speed of light. Their failure to detect any influence of the earth's motion on the velocity of light was the starting point for Einstein's theory of relativity. Regarded as a key finding in scientific history, the discovery was made on an instrument Michelson invented and is used today to measure the wavelengths of spectra. He was the first American to win a Nobel Prize for Physics in 1907. A crater on the moon is named after him.

Mohs, Freidrich 1773–1839
German mineralogist who devised a scale to measure and specify the hardness of a mineral. In 1801 Mohs was given a job identifying minerals in a collection that belonged to a wealthy banker. The scale he devised to catalogue these minerals became known as the Mohs Hardness Scale. His approach was criticized by the mineralogical establishment of the time, but it ultimately brought him fame. In 1810 Mohs became a professor of mineralogy in Gratz and in 1826 professor of mineralogy at the University of Vienna.

Molina, Mario 1943–2020
See p.426.

Morgan, Thomas Hunt 1866–1945
American biologist who discovered the roles played by chromosomes in heredity when conducting breeding experiments using Drosophila fruit flies. In 1933, he became the first scientist to be awarded the Nobel Prize for genetics. He later established a division of research at the California Institute of Technology, which attracted the talents of some of the best geneticists of the day, including seven future Nobel Prize winners.

Morley, Edward 1838–1923
American chemist and physicist who collaborated with Albert Michelson in 1887 on an experiment to discover the speed of light, leading to Einstein's theory of relativity. Morley also worked on the oxygen composition of the atmosphere, and the velocity of light in a magnetic field. He was president of the American Association for the Advancement of Science (1895) and president of the American Chemical Society (1899). The lunar crater Morley was named after him.

Morse, Samuel Finlay Breese 1791–1872
American artist and electrical engineer, best known for inventing the Morse Code, first demonstrated in 1838. In this code, short and long signals, dots and dashes, represent letters and numbers. Morse also introduced the relay, an electric switch that strengthened the signal at intervals, making long distance communication possible.

Moseley, Harry 1887–1915
English physicist whose law formed the basis of modern atomic and nuclear physics. Christened Henry, Moseley studied physics at Oxford University before joining Ernest Rutherford at Manchester University. Moseley provided an experimental basis for equating nuclear charge with what he called atomic number, making it possible to predict missing elements in the periodic table from discontinuities in the spectral series. He moved back to Oxford to continue his work but enlisted in the Royal Engineers at the outbreak of World War I. He was killed during the Battle of Gallipoli.

Murchison, Roderick Impey 1792–1871
Scottish geologist who played a major role in establishing parts of the geographical time scale, notably the Silurian, Permian, and Devonian Periods. The development of this time scale, in the absence of any absolute dating of rocks, was one of the greatest achievements of 19th-century geologists. He was knighted in 1846 and made a baron in 1866. He spent his later years devoted to the Royal Geographical Society, which he had helped found. The crater Murchison on the moon is named after him.

Muybridge, Eadweard 1830–1904
English photographer, known for his use of multiple cameras to capture motion. He became a successful photographer in the US, taking memorable pictures of the western landscape. In 1878, he used 24 cameras to film a horse galloping, *The Horse in Motion*, and began the transformation from stills to motion picture. He later returned to England where he published two popular books.

Napier, John 1550–1716
See p.88.

Newcomen, Thomas 1663–1729
English engineer and builder of the first steam engine to be put to practical use. It was erected in 1712 to pump water out of a coal mine near Dudley Castle, Staffordshire. During the next three-quarters of a century, hundreds of Newcomen engines were installed in Britain, greatly increasing coal production. They were eventually superseded by James Watt's low-pressure condensing engines.

Newton, Sir Isaac 1642–1727
See pp.110–11.

Nightingale, Florence 1820–1910
English nurse who transformed hospital nursing in Britain and is regarded as the founder of modern nursing. During the Crimean War, despite official opposition, she organized a nursing service for the soldiers who named her "the Lady with the Lamp." She founded a training school for nurses at St. Thomas' Hospital in London in 1861, influencing many parts of the world. In 1907, she was the first woman to be awarded the Order of Merit. After her death, she was immortalized in many films, statues, plays, and banknotes.

Nobel, Alfred 1833–1896
Swedish inventor of dynamite and gelignite, of smokeless gunpowder and the detonator, and the establisher of the Nobel prizes. As a young man, Nobel worked in the research laboratory of his father's explosives factory near Stockholm. After his young brother died when liquid explosive nitro-glycerine blew up, Nobel tried to make the substance less sensitive. He succeeded in 1866 and called it dynamite. He later produced gelignite, which was even safer. Nobel also worked on synthetic rubber, leather, and silk. The synthetic element nobelium was named after him. He willed most of his large fortune to establishing the Nobel Prizes, annual awards for the year's most outstanding work for physics, chemistry, medicine, literature, and peace. First presented in 1901, they are awarded annually on December 10, the anniversary of his death.

Nüsslein-Volhard, Christiane 1942–
See p.215.

Ochoa, Severo 1905–1993
American biochemist best known for being the first to synthesize ribonucleic acid (RNA) outside the cell. He also discovered several important metabolic processes. The formation of RNA from appropriate nucleotides was later used for the synthesis of artificial RNA. Ochoa was jointly awarded the 1959 Nobel Prize for Physiology or Medicine with Arthur Kornberg, who synthesized deoxyribonucleic acid (DNA). Ochoa received the US National Medal of Science in 1979, and the asteroid 117435 Severochoa is named in his honor.

Ørsted, Hans Christian 1777–1851
Danish physicist and chemist who, in 1820, showed that when a magnetic compass is placed near a wire and electricity is passed through it, the compass needle moves. His findings resulted in intensive research throughout the scientific community in electrodynamics. The CGS unit of magnetic induction (oersted) is named after him, as was the first Danish satellite, which was launched in 1999.

Ohm, Georg Simon 1789–1854
German experimental physicist who established Ohm's law, which states that the electric current through a conductor is proportional to the potential difference across it. The current is equal to the voltage divided by the resistance. Ohm used thermocouples for this work, as they provided a steadier voltage than the batteries available at the time. The SI unit of electrical resistance, the ohm, is named after him. He was awarded the Copley Medal from the Royal Society in 1841 and became a Fellow of the Society in 1842.

Olbers, Heinrich Wilhelm 1758–1840
German astronomer who discovered the asteroids Pallas and Vesta in 1802 and 1807 and found a method of calculating the orbits of comets. He posed the question of why the night sky is so dark if stars are distributed infinitely throughout space, which became known as Olbers's paradox. He assumed that dust clouds block light from the stars. Later, astronomers realized that the total amount of energy and matter is too small to light up the night sky. Olbers became famous for his discoveries and in 1811 was elected to assist at the baptism of Napoleon II of France.

Oort, Jan Hendrik 1900–1992
Dutch astronomer recognized as one of the greatest of the 20th century. Oort demonstrated that the Milky Way rotates, discovered that each of the stars in our galaxy was traveling independently through space, and found mysterious "dark matter" that makes up about 25 percent of the universe. He is best known for his suggestion that the solar system is surrounded by a vast cloud of perhaps 100 billion comets, known as the Oort cloud. Oort received some of the highest awards in astronomy, including the 1966 Vetlesen Prize from Columbia University.

Oppenheimer, (Julius) Robert 1904–1967
American theoretical physicist who developed electron-positron theory, the Oppenheimer-Phillips process, quantum tunneling, black holes, and cosmic rays. Oppenheimer was appointed director of the Manhattan Project in 1941. His efforts earned him the Presidential Medal of Merit in 1946. He was chairman of the General Advisory Committee of the Atomic Energy Commission, or AEC, from 1947 to 1952, but opposed the development of even more powerful bombs. In 1953, at the height of US anticommunist feeling, Oppenheimer was accused of having communist sympathies, but a decade later, he was awarded the Fermi Award as a gesture of rehabilitation.

Ostwald, Wilhelm 1853–1932
German chemist who is regarded as the founder of modern physical chemistry. He investigated catalysts, the speed of chemical reactions, reversible reactions, and the way electrolytes split up and conduct electricity. Ostwald also established that indicators were weak acids whose ionized form differs in color from their undissociated or nonionized form. In 1887, he accepted an invitation as professor of physical chemistry at Leipzig University where he remained until his retirement in 1906. His work was recognized by the Nobel committee who awarded him the Prize in Chemistry in 1909.

Otto, Nikolaus August 1832–1891
German technician who pioneered the first practical four-stroke internal combustion engine. He was inspired by reports of the engine of the French engineer Etienne Lenoir, in which coal gas igniting inside a cylinder drove a piston. In 1872, Otto joined German engineer Eugene Langen to produce a stationary gas engine for industry. It was a commercial success. He employed the same four-stroke cycle in a new gas engine in 1876, after which it was known as the Otto cycle. The same four-stroke principle was later used by Karl Benz and Gottlieb Daimler to power the first motor cars.

Oughtred, William 1575–1660
English mathematician and Anglican minister renowned for his invention of an early form of the slide rule. He invented many new mathematical symbols, such as x for multiplication. His work influenced the chemist Robert Boyle, the architect Christopher Wren, and the mathematician-physicist Isaac Newton.

Owen, Richard 1804–1892
English biologist and comparative anatomist. Owen is best remembered for coining the word dinosauria, meaning "Terrible Reptiles," and for his opposition to Darwin's theory of evolution by natural selection. In 1856, Owen became the first Superintendent of the British Museum's natural history departments. He campaigned for a new museum dedicated to natural history. Land was bought in South Kensington, and in 1881 Owen's vision became reality. The famous building designed by Alfred Waterhouse is a lasting monument to Owen.

Palade, George Emil 1912–2008
Romanian biologist who pioneered modern cell biology, earning him the Nobel Prize in Physiology or Medicine in 1974. Palade's discoveries were later used in understanding diseases and in the protein production that is the basis of the biotechnology industry.

In 1973, he moved to Yale, where he became the chairman of the new department of cell biology. In 1990, he became the first dean for scientific affairs at the School of Medicine at the University of California, San Diego.

Paracelsus 1493–1541
See p.71.

Parsons, Sir Charles 1854–1931
See p.252.

Pascal, Blaise 1623–1662
French mathematician and physicist whose greatest contribution to mathematics was his work on the theory of probability, developed from a problem sent to him by a noted gambler. In connection with this work, he discovered the pattern of numbers known as Pascal's triangle. Pascal's other work included major discoveries in geometry and the invention of the first digital calculating machine. He established the basic principles of hydraulic pressure, and the SI unit of pressure, the pascal, is named after him. Deeply religious, he dedicated his later years to spiritual learning and wrote several philosophical essays.

Pasteur, Louis 1822–1895
See p.244.

Pastor, Julio Rey 1888–1962
Spanish mathematician who pioneered a modern approach to the history and teaching of mathematics. While making important contributions to projective geometry, he promoted the science of mathematics throughout Argentina and Spain; his textbooks and essays were widely publicized. A crater on the moon is named after him, as well as several academic institutions in Spain.

Pauli, Wolfgang 1900–1958
American theoretical physicist who in 1925 established the Pauli exclusion principle, which states that no two electrons in an atom can exist in exactly the same state, with the same quantum numbers. Pauli, and independently Arnold Sommerfeld, devised an atomic model that explained the electrical and thermal properties of metals. Pauli was the first to recognize the existence of neutrino and helped complete Dirac's quantum electrodynamics. He was awarded the Lorentz Medal in 1930.

Pauling, Linus Carl 1901–1994
American chemist whose best-known work was on the structure of protein molecules. He found that fibrous proteins are helical, sometimes with several helices twisted together into a "super-helix." He also did research on the chemical action of antibodies. In the late 1940s, Pauling was outspoken on the dangers of nuclear weapons. He was ostracized for his uncompromising stand but went on to win two Nobel Prizes: the 1954 award for Chemistry and the 1962 Nobel Peace Prize. He devoted himself to fighting disease and educating the public about health issues.

Pavlov, Ivan Petrovich 1849–1936
Russian physiologist whose experiments on dogs led to the discovery of the conditioned reflex. His work has had a widespread influence on psychology. He made his findings in 1904 while studying dogs. Realizing that they salivated in anticipation of food, not just at the sight of it, Pavlov experimented with other stimuli. He developed the process of learning called classical conditioning and worked out many of the basic laws underlying learning. He was awarded the Nobel Prize in Physiology or Medicine in 1904.

Payne-Gaposchkin, Cecilia 1900–1979
See p.329.

Peral, Issac 1851–1895
Spanish sailor and inventor associated with the Peral submarine built for the Spanish Navy. In 1882, he was appointed professor of mathematical physics at the Escuela de Ampliación de Estudios de la Armada where he began work on designing a submarine. *El Peral*, built in Cadìz, was completed in 1888 and was designed for military use. In 1889, she became the first submarine to successfully fire Whitehead torpedoes while submerged. Internal politics, however, kept the Spanish Navy from pursuing the project. Discouraged, Peral left the navy in 1890.

Perkin, William 1838–1907
See p.240.

Perlman, Radia 1951–
See p.378.

Petit, Alexis Therese 1791–1820
French physicist who, along with Pierre Dulong, discovered the Dulong-Petit Law, which states that for a solid element, the product of relative atomic mass and specific heat capacity is approximately constant. Petit's life was cut short by tuberculosis.

Piot, Peter 1949–
Belgian epidemiologist who was part of the medical team that discovered the Ebola virus in central Africa. His studies helped show how the disease spread and affected people. He went on to investigate the transmission of HIV in Africa, becoming one of the leading researchers in the field. During the course of his career, he has been president of the International AIDS Society, Executive Director of the Joint UN Programme on HIV/AIDS, and director of the London School of Tropical Medicine and Hygiene.

Planck, Max 1858–1947
See p.314.

Plimsoll, Samuel 1824–98
English politician and social reformer, best remembered for the Plimsoll line, which indicated how low or high a ship is resting in the water, serving as a warning against overloading. He directed his efforts against what he called "coffin ships," unseaworthy vessels, and attempted to push through laws that would secure better standards of safety on ships in the future.

Poincaré, Henri 1854–1912
See p.350.

Posadillo, Gregorio Marañón y 1887–1960
Spanish physician, writer, and philosopher. He specialized in endocrinology and became professor at the Complutense University in Madrid in 1931. He studied the emotional effects of adrenaline and was a pioneer in the development of cognitive theories of emotion. A prolific writer, he founded the Institute of Medical Pathology and was the first director of the Centre of Biological Research in Madrid.

Priestley, Joseph 1733–1804
See p.146.

Proust, Joseph-Louis 1754–1826
French chemist best known for Proust's law. In 1785, Proust held teaching posts in Spain, where he began studying different types of sugar and was the first to identify glucose. Proust published his law of constant composition, which evolved into the law of definite proportions (Proust's law). Most chemists followed the work of Claude Berthollet, who did not believe that substances always combine in constant and definite proportions as Proust did. A friendly rivalry ensued, and Proust was proved right.

Ptolemy (Claudius Ptolemaeus) 90–168 CE
Greek astronomer, author of the encyclopedic treatise the *Almagest* (c.150 CE), containing a summary of Greek astronomical thought. It is through him that the work of the mathematician Hipparchus survives. Ptolemy brought together the work of earlier Greek astronomers to produce the Ptolemaic system of the universe, in which the sun and planets orbit a centrally placed earth. This theory held for nearly 1,400 years before it was abandoned in favor of the sun-centered system of the universe put forward by Copernicus in 1543.

Purkinje, Jan 1787–1869
Czech physiologist who pioneered studies in vision, the function of the brain and heart, embryology, cells and tissue. He is best known for his discovery, in 1837, of Purkinje cells, large nerve cells with branching extensions found in the cerebellum. In 1823, he recognized that fingerprints can be used as a means of identification and in 1839 discovered Purkinje fibers, the fibrous tissue that conducts the pacemaker stimulus along the inside walls of the ventricles to all parts of the heart. Purkinje introduced the term plasma, and protoplasm, which is used to describe young animal embryos.

Pythagoras c.580–500 BCE
See pp.32–33.

Rainwater, (Leo) James 1917–1986
American physicist who shared the 1975 Nobel Prize in Physics for his work identifying the asymmetrical shapes of certain atomic nuclei. During World War II, Rainwater was involved with the Manhattan Project to build the atomic bomb. After the war, he developed his theory that not all atomic nuclei were spherical, later confirmed by Neils Bohr and Ben Mottleson. He received the Ernest Orlando Lawrence Award for Physics in 1963 and was made professor of physics in 1982 at Columbia University.

Raman, Chandrasekhar Venkata 1888–1970
Indian physicist. Raman pioneered work on optics and the scattering of light, later called the Raman effect, and won the Nobel Prize for Physics in 1930. In 1949, he established the Raman Research Institute in Bangalore, where he worked until his death. He was elected a Fellow of the Royal Society in 1924, knighted in 1929, and in 1957 was awarded the Lenin Peace Prize.

Ramsay, William 1852–1916
English chemist who discovered the inert gases, beginning with argon in 1894. He isolated helium in 1895, neon, krypton, and xenon from liquid air in 1898, and in 1910 discovered the rare gas radon. His work had a great influence on atomic theory and in recognition of this Ramsay was knighted in 1902 and awarded the Nobel Prize for Chemistry in 1904.

Rees, Martin 1942–
English cosmologist and astrophysicist who has made important contributions to the origin of cosmic microwave background radiation, galaxy clustering, and formation and was one of the first to theorize that vast black holes power quasars. He is Professor of Cosmology and Astrophysics at the University of Cambridge, was knighted in 1992, holds the honorary title of Astronomer Royal (1995), was appointed to the House of Lords in 2005, and made a member of the Order of Merit in 2007. He is a foreign associate of the National Academy of Sciences and the American Academy of Arts and Sciences and has received many national and international awards in recognition of his work.

Richter, Charles 1900–1985
See p.361.

Römer, Ole Christensen 1644–1710
Danish astronomer best known for discovering light's finite velocity: the speed of light. Römer invented various scientific instruments, including a micrometer, planetaria, and an alcohol thermometer that influenced Daniel Fahrenheit's researches. In his position as Royal Mathematician, he introduced the first national system for weights and measures in Denmark in 1683.

Röntgen, Wilhelm 1845–1923
See p.294.

Ross, Ronald 1857–1932
Indian-born English physician who discovered the life cycle of malarial parasites (*Plasmodium*) in mosquitoes. Ross was subsequently awarded the Nobel Prize for Physiology or Medicine in 1902. He spent much of his career initiating measures for the prevention of malaria. He was knighted in 1911 and in 1926 became Director of the Ross Institute and Hospital for Tropical Diseases in London, which was founded in his honor.

Rubin, Vera 1928–2016
See p.398.

Rumford, Benjamin Thompson, Count 1753–1814
See p.179.

Rutherford, Ernest 1871–1937
See p.286.

Russell, Henry Norris 1877–1957
American astronomer who helped establish modern theoretical astrophysics. He analyzed the physical conditions and chemical compositions of stellar atmospheres; his assertion that there is an abundance of hydrogen is now accepted as one of the basic facts of cosmology. He is best associated with the Hertzsprung-Russell diagram (1910), a graph in which the color or temperature of stars is plotted in relation to their brightness.

Russell became president of the American Astronomical Society, the American Association for the Advancement of Science, and the American Philosophical Society. In 1921, he was awarded the gold medal of the Royal Astronomical Society.

Sabin, Florence 1871–1953
American medical scientist who was an important figure in her field, and a pioneer for women. In her research work, Sabin made significant contributions to knowledge of the histology of the brain and the development of the lymphatic systems and to the understanding of the pathology and immunology of tuberculosis. She became active in public health matters and modernized Colorado's health system, known as the Sabin health laws. She was the first woman to become a full professor at John Hopkins Medical School. She was also the first female president of the American Association of Anatomists.

Salam, Abdus 1926–1996
See p.401.

Salas, Margarita 1938–2019
Spanish pioneer in molecular biology. After working under Severo Ochoa in America, she returned to Spain to initiate research into molecular biology, a new field at the time. Married to the scientist Eladio Viñuela, both were responsible for boosting Spanish research in molecular biology and biochemistry. She was a member of the Royal Academy of Exact, Physical and Natural Sciences of Spain and president of the Severo Ochoa Foundation. In 2007, she was named a member of the National Academy of Sciences in the United States, becoming the first Spanish woman to join that institution.

Salk, Jonas Edward 1914–1995
American medical researcher who found a vaccine for polio. Benefiting from the work of Harvard's John Enders, who had developed a way to grow polio in test tubes, Salk began a human trial of the vaccine in 1955. When it was made public, he became famous overnight, and polio was virtually eradicated. In 1963 Salk founded the Salk Institute for Biological Studies and in 1977 was awarded the US Presidential Medal of Freedom.

Sanger, Frederick 1918–2013
English biochemist whose research on the structure of insulin earned him the Nobel Prize for Chemistry in 1958, and which was a key influence on the work of Francis Crick in researching the nature of DNA. Sanger himself turned to DNA and by 1975 had developed the "dideoxy" method for sequencing DNA molecules, also known as "the Sanger method." He used this technique to develop the first fully sequenced DNA-based genome. This was important for the Human Genome Project and earned him a second Nobel Prize in Chemistry in 1980, which he shared with Walter Gilbert and Paul Berg.

Santacilia, Jorge Juan y 1713–1773
Spanish mariner, mathematician, and scientist. In 1734, he accepted a mission to measure the length of a degree of the longitudinal meridian at the Equator in South America and to determine the roundness of the Earth. He found that Earth is not exactly spherical but slightly flattened at the poles. His *Examen Marítimo* (1771), in which he applied Newtonian mechanics to naval engineering, is regarded as one of the greatest naval treatises of the 18th century.

Schiaparelli, Giovanni Virginio 1835–1910
Italian astronomer who in 1877 reported "canals" on Mars. Other astronomers thought the canals were artificial, though Schiaparelli had not claimed this. An asteroid and craters on Mars and the moon are named after him.

Schickard, Wilhelm 1592–1635
German astronomer who invented one of the first calculating machines. In 1631, he was appointed professor of astronomy at the University of Tübingen. There, his research also included mathematics and surveying. He was a renowned wood and copper engraver and he made significant advances in mapmaking. In 1623, long before Blaise Pascal and Gottfried Leibniz, Schickard invented a calculating machine that was used by Johannes Kepler. Known as the Speeding clock, it was used for calculating astronomical tables. Schickard died of bubonic plague; the lunar crater Schickard was named in his honor.

Schrödinger, Erwin 1887–1961
Austrian theoretical physicist who made significant contributions to quantum mechanics. He worked on wave mechanics, explaining the movement of an electron in an atom as a wave. The equation became known as Schrödinger's wave equation. After teaching at Berlin University, he left with the rise of Nazism and went to Oxford in 1933, where he learned that he had won the Nobel Prize with Paul Dirac. He emigrated to Ireland in 1940 to join the newly established Institute for Advanced Studies in Dublin. In 1956 Schrödinger returned to Vienna, where he later died of tuberculosis.

Schwann, Theodor 1810–1882
See p.215.

Sedgwick, Adam 1785–1873
English geologist, professor of geology at Cambridge University, who played a major role in establishing parts of the Geographical Time Scale, most notably the Cambrian and Devonian Periods. Sedgwick introduced the term Paleozoic for the era stretching from the Cambrian to the Permian Periods. He was an outspoken opponent of Darwin's theory of evolution, although the two remained friends. Sedgwick was awarded the Copley Medal in 1863, and the geological museum at Cambridge was renamed the Sedgwick Museum of Geology in his honor.

Semmelweis, Ignaz 1818–1865
See p.242.

Servet, Miguel 1511–1553
Spanish theologian, cartographer, physician, and humanist. In his role as physician, Servet came to differ from Galen in pulmonary circulation. Galen had proposed that aeration of the blood took place in the heart and assigned the lungs a minor function. Servet concluded that transformation of the blood occurred in the lungs. He was the first to publish a modern understanding of pulmonary respiration.

Shen Kuo 1031–1095
Chinese astronomer, mathematician, and scholar who discovered magnetic declination, transforming navigation. Shen mentions his findings in his famous work, *Mengxi bitan*. His planetary observation project, which involved measuring planets' positions thrice nightly for five years, was on a scale unequaled in Europe for 500 years.

Shizhen, Li 1518–1593
Chinese physician, regarded as one of the greatest in Chinese history. His courtesy name was Dongbi. He hailed from the Hubei Province of China and lived during the Ming Dynasty. In his work as a doctor, he became frustrated in the errors and omissions he found in textbooks and decided to compile a new comprehensive book. For this, he traveled all over China and experimented with many herbs and minerals. His *Bencao Gangmu* (Great Compendium of Herbs) was translated into several languages and introduced into Japan and later Europe.

Shockley, William Bradford 1910–1989
American who fathered the transistor and brought silicon to Silicon Valley. His coinvention of a "semiconductor sandwich" in 1947, with John Bardeen and Walter Brattain, was quickly adapted and radios of all shapes and sizes flooded the market. In 1956, they shared a Nobel Prize in Physics. Shockley set out to capitalize on his invention by developing a silicon transistor. He played a key role in the industrial development of the region at the base of the San Francisco Peninsula. In his later life, Shockley was a professor at Stanford and controversially became an outspoken advocate of eugenics.

Shoujing, Guo 1231–1316
Chinese astronomer, mathematician, and engineer of the Yuan Dynasty, best known for the Shoushi Calendar. This calendar had 365.2425 days in a year, which, remarkably, was only 26 seconds out from the time it takes Earth to go around the sun. Shoujing also invented and improved astronomical instruments, including a tower sundial, raising its height and dramatically increasing its accuracy. He is renowned for building the artificial Kunming Lake in Beijing as a reservoir. The Asteroid 2012 Guo Shou-Jing is named after him.

Simiao, Sun 581–682 CE
Chinese physician and alchemist, hugely influential in Chinese medicine. He wrote two important Chinese compilations of prescriptions and placed great emphasis on gynecology, pediatrics, and medical ethics, stipulating that all patients should be treated as equal. This doctrine was spread throughout China and acted as an early version of the Hippocratic oath.

Skinner, BF 1904–1990
American psychologist who discovered practical applications for behaviorism in the form of programmed learning and psychotherapy. Skinner experimented on rats and pigeons, inventing a small box with a lever inside, the Skinner box, to deliver stimuli. The animal learned to associate pressing the lever with receiving a food reward. He founded a school for experimental analysis of behavior and is regarded as one of the leading psychologists of the 20th century.

Smith, Adam 1723–1790
Scottish economist whose ideas formed the basis of the economic theories that became paramount in the 19th century, and have influenced economic thought ever since. The culmination of his life's work was the publication in 1776 of his book *An Inquiry Into the Nature and Causes of the Wealth of Nations*. He believed that wealth derived principally from labor; the more labor that was required to produce a commodity, the higher its value. His book also formed the basis of the policies of laissez faire and free trade, which became dominant in the 19th century.

Smith, William 1769–1839
English geologist who founded the science of stratigraphy. A surveyor and canal engineer who studied geology during his travels around Britain, Smith showed that rock strata could be correlated from locality to locality by examining their fossil contents. Smith's *Geographical Map of England and Wales, with Part of Scotland* (1815) was the first regional geographical map.

Snell, Willebrord 1580–1626
See p.116.

Snow, John 1813–1858
English obstetrician who worked to convince other doctors that cholera was caused by drinking contaminated water. He published an article in 1839 outlining his theory, but doctors remained unconvinced and were certain that cholera was spread by breathing "vapors" in the atmosphere. Snow believed that sewage was the culprit, and when a major epidemic broke out with a water pump at its source, he had the pump handle removed. The outbreak ceased, but it was years before Snow was taken seriously. Scientists now consider Snow to be the pioneer of public health research in the field of epidemiology.

Sorenson, Soren 1868–1939
See p.239.

Spallanzani, Lazzaro 1729–1799
See p.226.

Spemann, Hans 1869–1941
German embryologist. Spemann studied how embryological cells become specialized and differentiated in the process of forming a complete organism. He developed a method for operating on objects often less than two millimeters in diameter, founding the techniques of modern microsurgery. His greatest contribution, for which he won the 1935 Nobel Prize in Physiology or Medicine, was his discovery of the "organizer" effect. Spemann found that when an area containing an "organizer" is transplanted into an undifferentiated host embryo, the transplanted area can induce the host to develop into an entirely new embryo.

Spinoza, Baruch 1632–1677
Dutch-Jewish philosopher of rationalism, who advocated the intellectual love of God, whom he saw as the perfect being. Spinoza saw everything in the universe as a manifestation of this perfect being. As his view could not be identified with the Jewish God, he was expelled from his own community and was equally unpopular with Christians. Spinoza is best known for his *Ethics*, a monumental work that puts forward his vision, and is regarded as one of the greatest rationalists of 17th-century philosophy, having a great impact on writers and scientists in the 20th century.

Spitzer, Lyman 1914–1997
See p.397.

491

Sprengel, Christian Konrad 1750–1816
German botanist whose studies of sex in plants helped establish the study of pollination ecology. Sprengel's observations were driven by a belief that everything in nature had a purpose, including every part of a plant. He discovered the role played by insects and wind in the cross-pollination of plants, and also described dichogamy, in which stigma and antlers on the same flower ripen at different times to ensure cross-fertilization. He published his findings in *The Newly Revealed Mystery of Nature in the Structure and Fertilization of Flowers* (1793), but his work failed to gain credence until it was later discovered by Charles Darwin.

Steptoe, Patrick Christopher 1913–1988
English obstetrician and gynecologist whose technique for uniting an egg and a sperm in a Petri dish and implanting them into the mother's womb led to breakthroughs in fertility treatment. The first "test tube" baby born using in vitro fertilization was Louise Brown in 1978. For their efforts, Steptoe and colleague Robert Edwards were both named Commanders of the British Empire, and in 1987 Steptoe was made a Fellow of the Royal Society.

Stevens, Nettie 1861–1912
See p.308.

Stringfellow, John 1799–1883
English pioneer aviator under whose supervision the world's first powered flight took place in the UK in 1848. His model flying machine was powered by two contra-rotating propellers driven by one of Stringfellow's powerful steam engines. The first attempt took place indoors, but a lack of proper balance resulted in failure. The second attempt was more successful; the flying machine left a guide wire and flew straight and true for about 30ft (10m).

Sturgeon, William 1783–1850
English electrical engineer who in 1825 invented the electromagnet, which was essential in developing the telegraph, telephone, and electric motor. He left the army to become a teacher, and specialized in lecturing on electromagnetism. His findings were published in several scientific magazines. In 1840, he became the Superintendent of the Royal Victoria Gallery of Practical Science in Manchester. Starved of funds, the gallery later shut down.

Swammerdam, Jan 1637–1680
Dutch microscopist who made major discoveries in anatomy. His greatest contribution to biology was the understanding of insects and their development. Using dissections, he showed that the various stages of the growth of an insect, from egg to adult, are different forms of the same creature. He was the first to describe red blood cells, discovered valves in the lymphatic vessels, known as Swammerdam valves, described the ovarian follicles of mammals at the same time as physician Reinier de Graaf (1672), and devised new techniques for preserving and examining specimens.

Swan, Joseph 1828–1914
See p.250.

Talbot, (William) Henry Fox 1800–1877
English chemist and photographer who made ground-making improvements to developing, fixing, and printing. Fox Talbot discovered that an image could be produced after a very short exposure to light and chemically developed into a useful negative. With the negative image it was possible to repeat the process of printing. Fox Talbot patented his process in 1841, and the following year was rewarded with a medal from the Royal Society.

Tansley, Arthur 1871–1955
English botanist who pioneered the study of plant ecology and was responsible for developing the concept of the ecosystem. Tansley became the first president of the British Ecological Society, which he helped to establish, an organization that gained an international reputation for ecological science. He also helped set up and lead the Council for the Promotion of Field Studies as well as the Nature Conservancy Council in Britain.

Tesla, Nikola 1856–1943
See p.269.

Thales c.625–546 BCE
Greek philosopher, considered the first of the natural philosophers. He formed part of what became later known as the Miletus School with Anaximander and Anaximenes. He was the first to consider the nature of the elements and concluded that water was the fundamental principal of the universe. He accurately predicted an eclipse of the sun in 585 BCE and realized that the moon shone only by reflecting the sun's light. The idea of geometry as a systematic science involving rigorous proofs arose from him. Five theorems are attributed to Thales.

Thomson, Joseph John 1856–1940
See p.140.

Townes, Charles Hard 1915–2015
American physicist and educator best known for inventing the maser and laser. At Columbia University, he continued his research into microwave physics and detected the first complex molecules in interstellar space. He was elected to the National Academy of Sciences in 1956 and jointly awarded the Nobel Prize in Physics in 1964. In 1967, he was appointed professor at the University of California and was chairman of the advisory committee for the first human landing on the moon.

Trevithick, Richard 1771–1833
English engineer and pioneer of the successful high-pressure steam engine. This was the first type of steam engine to combine high power with low bulk and weight, making it more versatile than the Watt type that preceded it. Trevithick was the first engineer to design a boiler strong enough to withstand the new engine's high steam pressures. He made Britain's first self-propelled road vehicle in 1801 and the first railroad locomotive two years later. But Trevithick preferred to experiment with new ideas rather than develop his inventions and, as a result, he died penniless.

Turing, Alan 1912–1954
See pp.344–45.

van der Waals, Johannes Diderik 1837–1923
Dutch physicist who first proposed an equation of state for real gases to explain why ordinary gases do not obey the ideal gas laws, such as Boyle's law, at high pressures. He also put forward the Law of corresponding states, which served as a guide during experiments that ultimately led to the liquefaction of hydrogen and helium. For his achievements, he won the Nobel Prize in Physics in 1910.

van Helmont, Jan Baptista 1580–1644
Flemish physiologist, physician, and chemist who was the first to postulate on the existence of gases distinct from air. He coined the word "gas," possibly from the word "chaos." He believed that the prime elements of the universe were air and water, rejecting ancient ideas on the four elements. Van Helmont is also famous for his experiment on a willow tree that he planted and grew for five years to determine where plants gain their mass.

van Leeuwenhoek, Anton 1632–1723
See p.94.

Vaughan, Dorothy 1910–2008
See p.371.

Vega, Jurij 1754–1802
Slovenian mathematician and physicist, author of the first complete table of logarithms. His book *Treasury of all Logarithms* published in 1794, became an essential aid in computations and teaching. Vega also made a name for himself when he calculated the mathematical constant to an accuracy of 140 decimal places and is well known for his work on ballistics and astronomy. Vega's death was shrouded in mystery. His body was found in the Danube River; many suspect that he was murdered. A crater on the moon is named after him.

Verenadsky, Vladimir 1863–1945
See p.278.

Vesalius, Andreas 1514–64
See p.73.

Virchow, Rudolf Karl Ludwig 1821–1902
German physician and pathologist who popularized the medical rule that every cell originates from another cell. In *Cellular Pathology*, he explained his discovery that disease occurs at a cellular level. He conducted medical research into the cause of pulmonary thromboembolism and named leukemia. Also well known for his anthropological work, Virchow wrote a study in 1885 that challenged the belief common at the time of Aryan or Nordic racial superiority, showing that Europeans belonged to a mixture of races.

Volta, Alessandro 1745–1827
Italian physician, best known for his work in electricity. In 1775, Volta devised the electrophorus, a device that produced a static electric charge. In 1799, he developed the voltaic pile, a forerunner of the electric battery, which produced a steady stream of electricity. Napoleon made him a Count in 1810, and in honor of his work, the electrical unit known as the volt was named after him.

von Baeyer, Adolf 1835–1917
German organic chemist who synthesized the dye indigo and determined its structure in the 1880s. For this and his other contributions to the synthetic dyestuffs industry, he was awarded the 1905 Nobel Prize in Chemistry.

von Behring, Emil Adolph 1854–1917
German physician who developed serum therapy. He discovered that the body produces antitoxins and developed treatments for diseases such as diphtheria. Together with Japanese bacteriologist Shibasaburo Kitasato he applied the serum therapy technique to tetanus. Von Behring also introduced early vaccination techniques against diphtheria and tuberculosis. He won the first Nobel Prize in Physiology or Medicine in 1901.

von Braun, Wernher 1912–1977
German rocket scientist and champion of space exploration. Von Braun was the leader of the team that developed the V–2 ballistic missile for the Germans during World War II. In 1955, he became an American citizen and worked for NASA where he was director of the Marshall Space Flight Center and the chief architect of the Saturn V launch vehicle that propelled the Apollo spacecraft to the moon. He received the National Medal of Science in 1975.

von Fraunhofer, Joseph 1787–1826
German physicist best known for discovering the dark lines found in the sun's spectrum. In 1813, von Fraunhofer independently rediscovered William Hyde Wollaston's dark lines in the solar spectrum, which are now known as Fraunhofer lines. He described a great number of the 500 or so lines he could see using self-designed instruments. Fraunhofer lines were later used to reveal the chemical composition of the sun's atmosphere. Von Fraunhofer was honored with membership to the Bavarian Academy of Sciences and is considered the founder of the German optical industry. He was knighted in 1824.

von Guericke, Otto 1602–1686
German inventor and philosopher who was the mayor of Magdeburg, Germany. In 1650, Guericke invented an air pump to help him create a partial vacuum. In 1663, he demonstrated the power of a vacuum with his Magdeburg Hemispheres to Emperor Ferdinand III. During public demonstrations, teams of horses would famously attempt to pull the large hemispheres apart. In 1660, he developed a machine that produced static electricity; the first electric generator, which led to later experiments with electricity. The Otto von Guericke University of Magdeburg, Germany, is named after him.

von Helmholtz, Hermann 1821–1894
German physician and physicist who made significant contributions to the understanding of the nervous system, the functions of the ear and eye, and in physics the concept of the conservation of energy. In 1851, Helmholtz revolutionized ophthalmology with the invention of the ophthalmoscope. The following year, he successfully measured the speed of the nerve impulse. He moved to Berlin in 1871 to work as professor of physics and attracted many promising students, among them Heinrich Hertz, who developed the propagation of electromagnetic oscillations in space.

von Humboldt, Alexander 1769–1859
German naturalist and explorer credited with being a founder of modern geography and who helped popularize science through his book *Kosmos*. Humboldt's expeditions to South and Central America (1799–1804) and Russia (1829), his studies, and the data he gathered radically changed western science in the 19th century. His interest in the Earth's geomagnetic fields triggered one of the first examples of international scientific cooperation, the setting up of permanent observatories around the world. Humboldt made an invaluable contribution to comparative climatology and conducted pioneering work on the relationship between local geography and its flora and fauna. He also discovered and studied the Peru Current (originally called the Humboldt Current) off the west coast of South America.

von Laue, Max 1879–1960
German physicist who discovered that X-rays could be diffracted by crystals, which opened the door to X-ray crystallography and the study of the structure of crystals. This marked the advent of solid-state physics that played a vital role in the development of modern electronics. He also supported Albert Einstein's theory of relativity and conducted research on superconductivity, quantum theory, and optics. Von Laue became director of the University of Berlin's Institute for Theoretical Physics in 1919 and director of the Max Planck Institute, Berlin in 1951.

von Leibniz, Gottfried 1646–1716
German philosopher and mathematician. Leibniz published his discovery of differential and integral calculus (1684), which he made independently of Isaac Newton. He also invented the binary system, the foundation on which digital technology is based, and a calculating machine (1671). Leibniz made important contributions to metaphysics, logic, and physics and anticipated ideas that emerged later in geology, medicine, and psychology. Among his major works is *New Essays Concerning Human Understanding* (1704).

von Liebig, Justus 1803–1873
German chemist known for his work in organic, pharmacological, and agricultural chemistry. He pioneered the use of nitrogen-based fertilizers in agriculture, helped found Germany's new dye industry, and built a laboratory for chemical research that was copied worldwide. In later years, he found a way to process beef extract from carcasses, which was given the trade name OXO.

von Mayer, Julius Robert 1814–1878
German physician and physicist who discovered the concept of energy conservation but was unable to prove it. He was the first to determine the mechanical equivalent of heat, but it took scientists years to confirm his findings, notably James Joule in 1847, leading to the first law of thermodynamics. Mayer was awarded the Copley Medal in 1871 from the Royal Society and was granted the honor of adding "von" to his name.

von Siemens, Werner 1816–1892
German electrical engineer who played a significant role in the development of the telegraph industry. Von Siemens made the world's first pointer telegraph and electric dynamo. Together with Johann Georg Halske, he set up a company, Telegraphen-Bau-Anstalt von Siemens & Halske, in Berlin in 1847. Siemens AG is now one of the largest electro-technological firms in the world. The SI unit of electrical conductance, the siemens, is named after him.

Waksman, Selman Abraham 1888–1973
Ukrainian-born American biochemist who discovered several important antibiotics. In 1939 Waksman and his colleagues undertook to identify soil organisms, producing soluble substances that might be useful in the control of infectious diseases, now known as antibiotics. Within a decade, 10 antibiotics were isolated and characterized, including actinomycin, streptomycin, and neomycin. Streptomycin was the first effective treatment for tuberculosis. Eighteen antibiotics were discovered under Waksman's general supervision. He won widespread recognition for his work, receiving the Lasker Award in 1948, the Legion of Honor in 1950, and the Nobel Prize in Physiology or Medicine in 1952.

Wallace, Alfred Russell 1823–1913
See p.201.

Warburg, Otto Heinrich 1883–1970
German biochemist, physiologist, and medical doctor noted for his differentiation of a cancer cell from a normal, healthy cell. This was such a major finding that he was awarded the Nobel Prize for Physiology or Medicine in 1931. Warburg studied chemistry under Emil Fischer in Berlin and medicine under Ludolf von Krehl, before serving in the Prussian Horse Guards during

World War I. In 1931, he also discovered the nature and mode of action of the respiratory enzyme. Warburg's discovery opened up new ways in the fields of cellular metabolism and cellular respiration.

Watson, James Dewey 1928–
American geneticist, who, with Francis Crick, discovered the double helix structure of the nucleic acid DNA in 1953. He shared a Nobel Prize in 1962 with Crick and Maurice Wilkins. He wrote highly regarded textbooks and, like Crick, was widely honored. He received the Lasker Award, Copley Medal, and the Presidential Medal of Freedom (1961) and was made an Honorary Knight Commander in the Order of the British Empire. From 1988 to 1992, Watson took a leading role in the Human Genome Project.

Watson-Watt, Robert 1892–1973
English inventor of radar. His first interest was meteorology where he used short-wave radio to detect the location of thunderstorms. By combining this direction finding technique with the ranging capabilities of ionosondes, he devised a system that was capable of detecting airplanes. He called it Radio Detection and Ranging (RADAR). By the end of 1938, radar systems were in place along the south coast of Britain and were used during the Battle of Britain in 1940 to detect enemy aircraft. Watson-Watt was elected a Fellow of the Royal Society in 1941 and knighted in 1942 for his pivotal work in radar and its role in World War II.

Watt, James 1892–1973
See p.149.

Weber, Wilhelm Eduard 1804–1891
German physicist, who, together with mathematician Carl Friedrich Gauss, invented the first electromagnetic telegraph. Weber made sensitive magnetometers to measure magnetic fields, and instruments to measure direct and alternating currents. The SI unit of magnetic flux, the weber, is named after him. Weber's later years were spent at the University of Göttingen, Germany, where he worked in electrodynamics and the electrical structure of matter. He developed the electrodynamometer and defined the absolute unit of electrical resistance. He was elected a member of the Royal Swedish Academy of Sciences in 1855 and received the Copley Medal in 1859.

Wegener, Alfred 1880–1930
See p.356.

Weinberg, Steven 1933–
American physicist who is best known for his work in unified field theory. Four forces were believed to drive the laws of physics: gravity, electromagnetism, the strong force (which holds an atom's nucleus together), and the weak force (which breaks an atom apart). Weinberg theorized that the electromagnetic and the weak forces are the same at extremely high energy levels. This was a major step in finding a single equation to explain all the matter and forces in nature. Weinberg, along with Sheldon Glashow and Abdus Salam, was awarded the Nobel Prize in Physics, 1979. Weinberg currently holds the Josey Regental Chair in Science at the University of Texas, where he is a member of the physics and astronomy departments. He is considered by many to be the preeminent living theoretical physicist.

Weismann, August 1834–1914
See p.309.

Whipple, Fred Lawrence 1906–2004
American expert on the solar system who in 1949 proposed the "dirty snowball" theory of the origin of the comets. This visualizes comets as balls of frozen gas mixed with particles of dust. During World War II, he invented a device for cutting tinfoil into chaff, to confuse enemy radar. In recognition of this, he was awarded a Certificate of Merit in 1948. Whipple, who organized the tracking networks for the first US satellites, also discovered that sporadic meteors, like shower meteors, are debris from comets. In 1955, he became director of the Smithsonian Astrophysical Observatory (SAO), where he remained until 1973. He was honored with the Gold Medal from the Royal Astronomical Society in 1983.

Whitworth, Joseph 1803–1887
English engineer, the pioneer of commercial manufacture of precision machine tools. Before Whitworth, engineers made their own precision machine tools. By 1850, Whitworth's lathes, drilling, and planing machinery were world-renowned for their accuracy. Whitworth also developed a series of standard sizes for components and gauges to be used by engineers. His name is still associated with standard screw-thread sizes; British Standard Whitworth. He designed the Whitworth Rifle to replace the Enfield, and a method for making artillery from soft steel that was less likely to shatter under the high pressure of firing. He was a Fellow of the Royal Society and a generous benefactor to educational institutions.

Wilson, C.T.R. 1869–1959
Scottish physicist and meteorologist, Charles Thomson Rees Wilson is best known for his invention of the cloud chamber, an instrument used for decades in the study of subatomic particles. Wilson, who studied physics at Cambridge University, spent two weeks in 1894 measuring the weather on Ben Nevis, Scotland's highest mountain. As he said in his Nobel banquet speech, "Morning after morning I saw the sun rise above a sea of clouds and the shadow of the hill on the clouds below surrounded by gorgeous colored rings." This inspired him to attempt cloud formation in the laboratory. In 1911, he demonstrated that a charged particle passing through supersaturated air produced a visible trail that could be photographed and studied. In 1927, Wilson received the Nobel Prize for his work in particle physics.

Woese, Carl 1928–2012
American microbiologist and physicist whose identification of the archaea changed the way life is classified on earth. Before Woese most of biology had dealt with the larger creatures, which were organized into the three kingdoms of plants, animals, and fungi. By the 1970s, there were two additional microbial kingdoms: the prokaryotes (or bacteria) and the eukaryotes. Woese discovered a new kingdom found in the so-called 16s mitochondral RNA. He redrew the taxonomic tree, which has proved invaluable to conservationists and ecologists. In 2006, he was made a Foreign Member of the Royal Society.

Wöhler, Friedrich 1800–1882
See p.150.

Wright. Jane Cooke 1919–2013
See p.313.

Wright, Wilbur & Orville 1867–1912/1871–1948
American aviation pioneers who, inspired by the work of the German glider pioneer Otto Lilienthal, became the first men to achieve sustained and controlled flight in powered craft that were heavier than air. The Wright brothers, formerly bicycle manufacturers, designed a gas engine weighing about 170lb (77kg), making the aircraft sufficiently light and powerful. On December 17, 1903, they made four epoch-making flights in North Carolina. In 1908, they flew for an hour and carried a passenger for the first time. Their breakthrough was the invention of "three-axis control," which enabled the pilot to steer the aircraft effectively and to maintain its equilibrium.

Wu, Sau Lan 1941–
See p.383.

Yalow, Rosalyn Sussman 1921–2011
American medical physicist who collaborated with Soloman Berson to develop radioimmunoassay, a method that uses radioisotopes to measure substances in the blood that are present in the tiniest of quantities, such as hormones, vitamins, and enzymes. The technique revolutionized the treatment of conditions such as diabetes and infertility. In 1977, Yalow became only the second woman ever to win the Nobel Prize in Medicine.

Yingxing, Song 1587–1666
Chinese scientist and encyclopedist who lived during the late Ming Dynasty. He produced the *Tian Gong Kai Wu* (The Exploitation of the Works of Nature), considered one of China's technological classics, in 1637. It covered a wide range of subjects, from smelting, hydraulic devices and irrigation, silk, salt, sugar, ceramics, papermaking and ink, and transportation, with exquisitely detailed illustrations that helped explain Chinese production processes.

Youyou, Tu 1930–
See p.413.

Yukawa, Hideki 1907–1981
Japanese physicist who won the Nobel Prize in Physics in 1949 for his theory of elementary particles. His theory of how the nuclear force holds the nucleus together predicted the existence of the meson, a subatomic particle, that later became an important part of nuclear and high-energy physics. In 1955, Yukawa joined 10 other leading scientists in signing the Russell-Einstein Manifesto, calling for nuclear disarmament.

Zeiss, Carl 1816–1888
German optician famous for the company he founded, which bears his name. Zeiss became a noted lens-maker in the 1840s when his lenses were mostly used for microscopes. When cameras were invented, his company began manufacturing high-quality lenses for cameras, soon gaining an international reputation.

Zhang Heng 78–139 CE
See pp.54–55.

Zhenning, Yang 1922–
Chinese-American physicist who worked on statistical mechanics and symmetry principles, for which he received the Nobel Prize in Physics in 1957. Zhenning was an assistant to physicist Enrico Fermi at the University of Chicago. In 1965, he moved to New York's Stony Brook University, where he continued to work until 1999. He won the Rumford Prize in 1980 and the Albert Einstein Medal in 1995.

Zhongjing, Zhang 150–219 CE
Chinese herbal physician best known for his monumental work, *Shanghan Zabing Lun* (Treatise on Febrile Diseases Caused by Cold and Miscellaneous Diseases), considered one of the most influential works in the history of Chinese medicine. It contained new recipes and formulae, many of which are still used in traditional Chinese medicine.

Zhongzhi, Zu 429–500 CE
Chinese mathematician and astronomer who lived during the Song Dynasty of the Southern Dynasties. He produced the Daming Calendar in 465, calculated one year as 365.24281481 days, and derived two approximations of pi. He also achieved fame for his derivation and proof of the formula for the volume of a sphere.

Ziegler, Karl 1888–1966
German chemist who shared the 1963 Nobel Prize in Chemistry with the Italian Giulio Natta for developing organo-aluminum catalysts for the low-temperature polymerization of olefins, particularly in the production of ethylene. Ziegler conducted research into organo-metallic compounds, developing lead tetraethyl, the anti-knock additive used in gasoline.

Zu Chongzhi 429–500 CE
Chinese astronomer, mathematician, and engineer who created the Daming calendar and found several close approximations for pi. His calendar provided a more accurate number of lunations per year and, for the first time in China, took into account the precession of the equinoxes. It was eventually adopted in 510 CE. Zu's work with pi (he calculated its numerical value as between 3.1415926 and 3.1415927) remained unsurpassed until the 15th century. Zu Chongzhi also calculated the volume of a sphere.

Glossary

A

aberration A distortion of an image created by a lens or mirror, because the light rays do not meet at the focus.

absolute magnitude A measure of the actual luminosity of a star.

absolute scale A scale of temperature, also known as the Kelvin scale, that begins at absolute zero. Its unit of measurement is the kelvin.

absolute zero The lowest possible temperature: 0° K or -459.67° F (-273.15° C).

absorption 1. The taking up of one substance by another. 2. The letting in of radiation (the opposite of emission).

abyssal plain The essentially flat floor of the deep ocean, typically 3 miles (5 km) below the surface.

acceleration The rate of change of velocity.

acid A compound containing hydrogen that splits up in water to give hydrogen ions. Acidic solutions have a pH less than 7.

acid rain Rain that has become acidic due to water in the air reacting with certain chemicals (such as sulfur dioxide, nitrogen oxide, and carbon dioxide) emitted as a result of human activities, such as power production and vehicle emissions, or natural processes, such as volcanic emissions.

acoustics 1. The study of sound. 2. How sound travels around a room. For example, a concert hall needs to have good acoustics.

activation energy The amount of energy needed to start a chemical reaction.

active transport The use of energy to move molecules across a biological cell membrane against the concentration gradient, such as the chemical energy of ATP.

adaptation The way in which an organism gradually changes over many generations to become better suited to its particular environment.

additive Any substance added in small amounts, especially to food or drink, to improve it or inhibit its deterioration, e.g. to change color or taste or increase shelf life.

adhesion The force of attraction between the atoms or molecules of two different substances.

adhesive A sticky substance, such as paste or glue, used to join two surfaces together.

ADP Adenosine diphosphate. A compound that is formed when ATP releases energy.

advection fog A type of fog that forms where warm, moist air moves over a colder surface. It is often known as sea fog.

aerobic respiration A type of respiration that requires oxygen, unlike anaerobic respiration.

airfoil The special shape of an aircraft wing. It is more curved on the top than on the bottom, and produces lift as it moves through the air.

air resistance The force that resists the movement of an object through the air.

albedo How much an object, especially a planet or moon, reflects the light that hits it: its reflecting power.

alchemy A medieval science that tried, among other things, to find a way of changing metals, such as lead, into gold.

alga (plural: **algae**) An informal grouping of organisms, most of which photosynthesize, that do not have real roots or stems. Seaweeds are probably the best known examples.

aliphatic compound An organic compound made up of chains, rather than rings, of carbon atoms.

alkali A base that dissolves in water.

alkaline Describes a solution with a pH greater than 7.

allotropes Different forms of the same element. For example, diamond and graphite are allotropes of carbon.

alloy A mixture of two or more metals, or of a metal and a nonmetal.

alpha decay A type of radioactive decay in which atomic nuclei emit alpha particles. An alpha particle consists of two protons and two neutrons bound together, and is identical to a helium nucleus.

alternating current (AC) An electric current whose direction reverses at regular intervals.

alternator An electric generator that produces an alternating current.

alveolus (plural: **alveoli**) A tiny air sac in the lungs.

AM Amplitude modulation. The transmission of a signal by changing the amplitude of the carrier wave.

amalgam An alloy of mercury and another metal, such as tin.

amino acid Any of a group of organic chemicals in which the molecules each contain one or more amino groups (NH_2) and one or more carboxyl groups (COOH). Amino acids are the chemicals that make up proteins.

ammeter An instrument that measures electric current.

ammonification The conversion of nitrogen in rotting organic matter in the soil to ammonia by decomposers, such as bacteria and fungi.

amniocentesis A test during pregnancy in which a small amount of amniotic fluid is removed from inside the amniotic sac around the baby and examined for genetic, chromosome, and developmental abnormalities.

amp (ampère) The SI unit of electric current.

amplifier A device that changes and usually increases the amplitude of a signal, especially an electrical signal for audio equipment.

amplitude The size of a vibration, or the height of a wave, such as a sound wave.

anabolism A series of chemical reactions in living things, which builds up large molecules from small ones.

anaerobic respiration A type of respiration that does not require oxygen. It produces less energy than aerobic respiration, which does require oxygen.

analogue In computing, representing data in a directly measurable way, such as by a varying voltage or the position of a needle on a meter.

anaphase A stage in the natural division of living cells by mitosis. The paired chromosomes separate and begin moving to opposite ends of the cell.

anatomy The science that studies the structure of living things. Human anatomy is the study of the human body.

andesite Fine-grained brown or grayish volcanic rock.

anesthesia The medical inducement of pain relief by complete loss of sensation or consciousness, generally using drugs.

angiogram An X-ray of the blood vessels, usually taken to study blood flow.

angle of incidence The angle a light ray makes with the perpendicular to the surface it hits.

angle of reflection The angle a reflected light ray makes with the perpendicular to the reflecting surface.

anion A negatively charged ion.

anode A positive electrode.

anodizing The process of coating a metal object with a thin, protective layer of oxide, by electrolysis.

anthracite Hard coal that burns with hardly any flame or smoke.

antiretroviral drugs Drugs used for the treatment of retrovirus infections, principally HIV.

antibiotic A drug that is used to kill or inhibit the growth of bacteria that cause infections.

antibodies Special proteins produced in the body by white blood cells, to label foreign particles or antigens and stimulate the immune response.

anticyclone An area of high pressure air that often produces settled weather.

antigen Any substance that stimulates the body to produce antibodies and an immune response.

antimicrobial Capable of either killing microbes that cause infection or stopping them from multiplying.

antioxidant A compound added to foods and plastics, to prevent them from oxidizing and becoming stale or breaking down.

antiparticles A particle that is the same as a normal particle, except that it has an opposite electrical charge. If a particle and its antiparticle collide, they annihilate each other.

antiseptic An antimicrobial substance applied to the skin, to reduce the risk of infection.

apogee The point in the orbit of a moon or satellite at which it is farthest from the object it is orbiting.

apparent magnitude The brightness of a star as seen from Earth.

arteries Blood vessels that carry blood from the heart to other parts of the body.

arthropod An invertebrate, such as insect, spider, or centipede, that has an exoskeleton (external skeleton), a segmented body, and jointed appendages or feet.

artificial insemination The artificial insertion of sperm into the female reproductive tract to induce fertilization.

artificial selection The process by which humans change the genetic makeup of a species.

aseptic technique The performance of a medical or laboratory procedure under completely sterile conditions.

asexual reproduction Reproduction that involves only one parent.

asteroid A rocky body, also known as a minor planet or planetoid, that circles the Sun. Most asteroids are in the asteroid belt between Mars and Jupiter.

asthenosphere A soft layer of Earth's mantle.

astrolabe A historical astronomical instrument used by astronomers and sailors for locating the position of the Sun, Moon, planets, and stars.

astronaut A person who has been trained to be a crew member in a spacecraft.

astronomy The study of the stars, planets, and other bodies in space.

atmosphere The layer of gases that surrounds a planet.

atmospheric pressure The force per unit area exerted by the weight of the air above that area.

atoll A ring-shaped group of coral islands surrounding a lagoon, usually on top of a seamount.

atom The smallest part of an element that has the chemical properties of that element. An atom consists of a nucleus of protons and neutrons surrounded by orbiting electrons.

atomic nucleus The very dense region in the center of an atom, consisting of protons and neutrons.

atomic number The number of protons in the nucleus of an atom. All atoms of the same element have the same atomic number.

atomic weight Also known as the relative atomic mass, the ratio of the average mass of atoms of an element to $\frac{1}{12}$ of the mass of an atom of carbon-12.

ATP Adenosine triphosphate. A nucleotide molecule used by organisms for storing and transporting energy.

attractor A point, state, or set of states in a dynamic system toward which the system tends to move.

aurora borealis A display of lights in the night sky in Arctic regions, caused by the impact of electrically-charged particles from the Sun with Earth's atmosphere.

autoclave A strong, steam-heated container that is used for sterilization at high temperature and pressure.

autotrophic A plant that makes its own food by photosynthesis.

average See mean, median, and mode.

axis 1. An imaginary line around which an object rotates. 2. The line along which rock bends in a fold. 3. A reference line on a graph. 4. The second cervical vertebra of the spine.

B

background radiation 1. Low-intensity radiation emitted by radioactive substances in and around Earth. 2. Radiation (in the form of microwaves) detected in space that may have come from the Big Bang.

bacteriophage A parasitic virus that infects bacteria; also called a phage.

bacterium (plural: **bacteria**) A single-celled microscopic organism that does not have a membrane-enclosed nucleus or other organelles.

barchan A sand dune with a crest.

basalt A dark volcanic rock.

base 1. A compound that reacts with an acid to give water and a salt. 2. The number of different single-digit symbols used in a particular number system; e.g. binary has two digits, 0 and 1.

base pairs Two complementary chemical bases linked on either side of the double spiral of a DNA molecule. The order of base pairs spells out the DNA code.

base triplet See codon.

batholith A dome of igneous rock that solidifies in a huge underground mass.

battery A combination of two or more electrical cells connected together that produce electricity.

Beaufort scale A scale used to measure wind speed, ranging from 0 to 12 (calm to hurricane force).

bell curve The bell-shaped curve on a graph that represents the normal distribution of a set of data. The highest, middle point of the curve corresponds to the mean (average) and the curve slopes symmetrically downward on either side of this value.

beta decay A type of radioactive decay in which atomic nuclei emit beta particles (electrons or positrons).

Big Bang The theory that the Universe began with a massive explosion of matter. It is thought that everything in the Universe is still moving apart because of the explosion.

binary number system A method of representing numbers, typically for computing, in which only the digits 0 and 1 are used. Successive numbers are powers of 2.

binocular vision The ability of some animals to see objects in three dimensions, so that they are able to judge distance.

binomial system A system of giving an organism two names. The first name signifies the genus and the second name signifies the species.

biodegradable A substance that can decompose and become harmless naturally.

biogas A gas made when plant or animal waste rots down without the presence of air.

biogenic Produced by organisms.

biology The study of living things.

biomass 1. The total number of living organisms found in a given area. 2. Plant material, such as wood or peat, that is used as a source of energy.

biome A large ecosystem, such as a rain forest or desert.

biosphere The region of Earth and its atmosphere in which living things are found.

biotechnology The technology of manipulating living things on a microscopic level, often by modifying their genes.

bit In computing, a fundamental unit of information having just two possible values, as either of the binary digits 0 or 1.

bituminous Containing bitumen, a tarlike substance produced from petroleum.

black body A theoretical object that absorbs all the electromagnetic radiation that falls on it, and so appears black. At a particular temperature, a black body also radiates the maximum amount of energy possible at that temperature.

black dwarf The faded remains of a dead star. (*See also* white dwarf.)

black hole An extremely dense object in space. Its gravity is so strong that it pulls in anything around it, even light, so that it looks black.

black ice Thin, hard, transparent ice, especially on the surface of a road.

blastocyst A hollow ball of cells that is an early stage in the development of an embryo.

blood type Also known as blood group, any of several types into which blood can be classified based on the antigens on the surface of red blood cells.

boiling point The temperature at which a liquid becomes a gas.

bond The attraction between atoms, ions, or groups of atoms that holds them together in a molecule or solid.

Boolean algebra A symbolic system of algebra devised by George Boole that represents relationships between entities such as sets or objects. It can be used to solve logical problems and is fundamental to the way in which digital computers process data.

boson A class of subatomic particles that act as messengers carrying forces between other particles.

brine A strong solution of salt in water.

brown dwarf An object smaller than a star, but bigger than a planet, which never has quite enough mass to start hydrogen fusion in its core, so cannot shine.

Brownian motion The random movement of tiny particles

in a liquid or a gas, caused by molecules colliding with them.

bubble chamber A vessel filled with a superheated transparent liquid (usually liquid hydrogen), that was once used in physics to track the movement of electrically-charged particles.

buffer 1. A solution that is resistant to changes in pH. 2. An electric circuit used to join two other circuits.

butterfly effect A characterization of how, in chaos theory, small random incidents can have massive effects, such as a butterfly flapping its wings in South America leading to a storm in Europe.

byte A unit of information storage and transmission in computing. A byte consists of eight bits. A kilobyte is a thousand (10^3) bytes; a megabyte is a million (10^6) bytes; and a gigabyte is a billion (10^9) bytes.

C

CAD Computer-aided design.

calorie A unit of energy. The calorie used in food science is in fact a kilocalorie—1,000 calories.

calculus 1. The branch of mathematics that deals with infinitesimal changes. It comprises differential calculus, which is concerned with rates of change, and integral calculus, which can be used to calculate quantities such as areas and volumes.
2. Also called a stone, a hard mass that forms in the body, such as in the kidney, gallbladder, or urinary tract.

camouflage The color, markings, or body shape that help to hide an animal or plant in its surroundings.

capacitance The ability to store electric charge.

capacitor A device that is used to store electric charge temporarily.

capillaries Tiny blood vessels that carry blood to and from cells.

capillary action The movement of a liquid up or down due to attraction between its molecules and the molecules of the surrounding solid material.

carbohydrate Any of a group of organic compounds consisting of carbon, hydrogen, and oxygen that are a major dietary source of energy for animals. Carbohydrate-rich foods include sugar, potatoes, bread, and pasta.

carbon The chemical element with the symbol C and atomic number 6, which forms more compounds than any other element, including the important chemicals of life.

carbon cycle The continuous circulation of carbon, from the atmosphere (in carbon dioxide), through plants (trapped in carbohydrates by photosynthesis), then animals (that eat the plants), and finally back into the atmosphere (through respiration and decomposition).

carbon scrubbing A process for absorbing carbon dioxide from industrial waste, and spaceship and submarine life-support systems before it escapes into the air.

carburized The addition of carbon to the surface of iron and steel by heat treatment.

carnivore A meat eater.

cartilage Gristly connective tissue that makes up the soft parts of the vertebrate skeleton and is present in some joints. The skeletons of some fish, such as sharks and rays, are made entirely of cartilage.

cartography The science of map making.

cast A hollow in a rock, formed around a decomposed animal or plant. The cast is a mold in which minerals collect and solidify to form a fossil.

catabolism A series of chemical reactions in living things that break down large molecules into small ones, releasing energy in the process.

catalyst A chemical that speeds up a chemical reaction without being changed itself at the end of the reaction.

catalytic converter A device in a vehicle that uses a catalyst to change toxic exhaust gases into less harmful gases.

catenation The ability of a chemical element to form long chain polymer molecules.

cathode A negative electrode.

cathode ray tube A vacuum tube with a fluorescent screen

at one end and an electron gun at the other. The electron gun produces a stream of electrons (cathode rays), which, when they hit the screen, make it glow. Cathode ray tubes were used in old-fashioned televisions.

cation A positively charged ion.

cavitation 1. The formation of small holes (cavities) in something, such as the formation of bubbles in carbonated drinks and around moving marine propellers.
2. The natural enlarging of cracks in rocks by compressed air.

cecum A pouch in an animal's intestines which may house cellulose-digesting bacteria.

celestial body A natural object in space, such as a planet or star.

celestial pole One of two points in the celestial sphere about which stars appear to revolve when seen from Earth.

celestial sphere The imaginary sphere in which the stars seem to lie when seen from Earth.

cell The smallest unit of an organism that can exist on its own.

cell division The process by which one cell splits to produce two cells, called daughter cells.

cellulose The carbohydrate that forms the walls of plant cells.

Celsius The temperature scale named after Anders Celsius, in which water normally freezes at 0° and boils at 100°.

central nervous system The collection of nerves in the brain and spinal cord that acts to control animals' bodies.

center of mass The point at which the mass of an object or a group of objects can be assumed to be concentrated.

centrifugal force (or **centrifugal effect**) The force that appears to push outward on a body that is moving in a circle.

centrifuge A device used to separate substances of different densities, by spinning them at high speed.

centripetal force The force that pulls inward on a body moving in a circle.

Cepheid star A star that varies in brightness with great regularity, and acts as a distance marker in space.

ceramics Objects made of clay or porcelain fired in a kiln.

cerebellum A part of the brain at the back of the skull. Its primary role is to control movement and maintain balance.

cerebrum The largest part of the brain in humans, responsible for most conscious thought and activity. In humans, it is divided into two cerebral hemispheres and surrounds most of the rest of the brain.

cermet A material made from ceramic and metal. Cermets can withstand very high temperatures.

CERN *Conseil Européen pour Recherches Nucléaires*. The research center of the European Organization for Nuclear Research in Geneva.

CFC Chlorofluorocarbon. Any of a series of gases based on an alkane (such as methane or ethane) and containing chlorine and fluorine. Once widely used as refrigerants and propellants in aerosols, CFCs cause depletion of the ozone layer if allowed to escape into the atmosphere.

chain reaction A chemical or nuclear reaction in which the product of one step triggers the next step, which in turn triggers the next step, and so on. For example, in nuclear fission reactions, neutrons from the splitting of atomic nuclei go on to cause other nuclei to split.

chaos A state of disorder. A chaotic system appears to behave randomly and unpredictably, even though it follows deterministic laws. Chaos theory describes dynamic systems whose behavior is extremely sensitive to the initial conditions, such as the weather.

charged particle A subatomic particle or ion that has an electric charge.

chemical Any substance that can change when it is joined or mixed with another substance.

chemical bond *See* bond.

chemistry The science that deals with the structure, composition, and properties of substances and the changes they undergo as a result of reactions with other substances.

chemotherapy The treatment of cancer with drugs that kill cancer cells (cytotoxic drugs).

chlorophyll The green pigment found in many plants that absorbs light to provide the energy for photosynthesis.

chloroplasts Tiny bodies in some plant cells that contain chlorophyll.

chromatid One of the two identical strands of a chromosome after replication.

chromatography A method of separating a mixture by running it through a medium such as filter paper. Different parts of the mixture will move through the medium at different speeds and travel a greater distance.

chromosome A structure made of DNA and protein that is found in cells. A chromosome contains the genetic information (in the form of genes) of a cell.

chromosphere A layer of gases in the Sun's atmosphere that shines red.

cilium (plural: **cilia**) A tiny "hair" that projects from the surface of many small organisms. In humans, cilia line tracts inside the body such as the respiratory tract.

cipher A code in which letters in the original text are substituted with other letters, numbers, or symbols.

circuit A path around which an electric current can flow.

circulation The movement around something. Blood circulation is the continuous flow of blood around the body. Atmospheric circulation is the continuous movement of air in the atmosphere.

classification The process of arranging items in groups, or the system of grouping used for a particular collection of objects, such as living organisms or stars.

climate The prevailing weather conditions in an area over a long period of time.

clone Two or more identical organisms that share exactly the same genes.

cluster 1. A group of stars or galaxies.
2. A particle consisting of several atoms bound together.
3. A grouping of occurrences of a disease in a localized area.
4. A group of linked computers that work together as a single unit.

code A system in which some units, such as sequences of chemicals, letters, numbers, or symbols, are substitutes for, represent, or provide instructions for other things.

codon A sequence of three adjacent nucleotide bases that forms part of the genetic code. Most codons spell out the code for making a particular amino acid or for stopping or starting protein synthesis.

cohesion The force of attraction between two particles of the same substance.

coil An electric device made of a coil of electric wire, used for converting voltages, creating a magnetic field or inducing an electric current.

coke A fuel made by baking coal. It is mostly carbon and gives off far more heat than coal.

cold front The boundary between masses of cold and warm air, where the cold air is advancing and sharply undercutting the warm air.

collision boundary *See* convergent boundary.

colloid A mixture made up of tiny particles of one substance that are dispersed in another but do not dissolve.

colonization The occupation of an area by a population or species.

colony A large group of the same species of organism that live together.

coma 1. A cloud of gas and dust around the center of a comet.
2. A state of unconsciousness.

combustion A chemical reaction (burning) in which a substance combines with oxygen, producing heat energy.

comet A ball of frozen gas and dust that travels around the Sun. Some of the dust streams out from the comet to make a "tail."

commensalism Where two or more organisms live together and neither causes harm to the other.

community A group of people or animals that live in the same place.

commutator A rotating device that periodically reverses the direction of an electric current.

complexity theory The study of the behavior of extremely complicated dynamic systems, particularly how structure and order can arise in complex, apparently chaotic systems.

compound A substance containing atoms of two or more elements.

compressed air Air that is squeezed and kept under pressure greater than normal atmospheric pressure.

compression 1. Being pressed together or bunched up.

2. The increase in density of a fluid.

concave lens A lens that curves inward.

concentration The strength of a solution, or the amount of solute that is dissolved in a certain amount of solvent.

conception The coming together of a male sperm and a female ovum to create a zygote.

condensation The change of a gas or vapor into a liquid.

condenser 1. A cooling device for reducing a gas or vapor to a liquid.
2. Another name for a capacitor.

conduction The movement of heat or electricity through a substance.

conductor A substance through which heat or electric current flows easily.

cones Light-sensitive cells in the retina of the eye of humans and some other mammals that make it possible to see colors.

conjugation The transfer of genetic material between bacteria by direct cell-to-cell contact.

conjunctiva The mucus membrane that covers the front of the eye and lines the inside of the eyelids.

conservation 1. The preservation of any process or object.
2. The principle that a total quantity remains constant, whatever changes happen within a system.

constellation One of 88 regions of the night sky. Each constellation contains a group of stars joined by imaginary lines to represent an image.

continental drift The slow movement of the continents around the world over millions of years, carried on the tectonic plates that make up Earth's surface.

convection The transfer of heat through a fluid by currents within the fluid.

convergent boundary The line along which two tectonic plates that are moving toward each other meet. Also known as a collision boundary.

converging lens *See* convex lens.

convergent evolution The way in which different species evolve similar features because they are subjected to similar environmental conditions and selection pressures.

convex lens A lens that curves outward.

coordinate Each of a group of numbers used to specify the position of a point.

copper oxide A compound of copper and oxygen. There are two types: copper(I) oxide (Cu_2O, also called cuprous oxide) and copper(II) oxide (CuO, also called cupric oxide). Copper(I) oxide occurs naturally as the mineral cuprite and was the first known semiconductor.

coprolite Fossilized dung.

Coriolis effect The deflection of winds and ocean currents by the rotation of the earth.

corona The plasma atmosphere of the Sun or another star extending thousands of kilometers into space.

corrasion The wearing away of a surface by the action of rocks carried in ice or water.

corrosion Chemical attack of the surface of a metal.

cosine (cos) In a right-angled triangle, the cosine of an angle is the ratio of the side adjacent to the angle to the hypotenuse (the longest side).

cosmic rays High-energy, electrically-charged particles from space originating either in the Sun (lower energy), supernovae (medium energy), or beyond the galaxy (high energy).

cosmology The science of the origin and development of the Universe.

cotyledon A simple leaf that forms part of a developing plant. Also called a seed-leaf.

coulomb The SI unit of electrical charge: the quantity of electricity transported in 1 second by a current of 1 ampere.

covalent bond A chemical bond formed by atoms sharing one or more electrons.

CPU The central processing unit of a computer.

cracking The process of splitting larger molecules into smaller ones, often by heating with a catalyst.

cross-fertilization Fertilization of a plant with gametes from another plant of the same species.

crucible A container in which metals or other substances may be melted or heated to very high temperatures.

crust The thin rigid outer surface of Earth, made of rock.

cryptanalysis/cryptography The art of studying and deciphering coded messages.

crystal A solid whose constituent atoms, ions, or molecules are arranged in a regularly repeating pattern.

crystal lattice The repeating pattern of atoms or ions that forms a crystal.

crystallogram A pattern formed on a photographic plate by passing a beam of X-rays through a crystal.

cuneiform The wedge-shaped symbols of the writing system used in ancient Mesopotamia, Persia, and Ugarit.

curve A line on a graph that shows the varying relationship between two quantities.

cyclone Another name for a hurricane, especially for one over the Indian Ocean.

cylinder 1. A tube shape.
2. The chamber in an engine in which the piston slides up and down.

cystic fibrosis Hereditary disorder of the exocrine glands that causes excess mucus production, leading to clogging of the airways.

cytokinesis The division of the cytoplasm of a cell into two daughter cells at the end of meiosis or mitosis.

cytoplasm The contents of a cell, apart from the nucleus.

D

dark energy The little understood, repulsive force that accounts for about three quarters of the Universe's mass-energy and acts in the opposite direction to gravity, making the Universe expand.

dark matter Invisible matter detectable only by its gravitational effects on ordinary matter, believed to account for about a quarter of the Universe's mass.

decant To separate a mixture of a solid and a liquid by allowing the solid to settle and pouring off the liquid.

decay chain The series of transformations through different isotopes which a radioactive atom goes through as it decays.

decibel A unit used to measure the loudness of a sound.

decimal Relating to a number system based on the number ten, the power ten, or tenths.

decomposer A tiny organism, such as a bacterium, that breaks down dead matter.

decomposition 1. Organic decay.
2. Breaking larger molecules into smaller ones.

decryption The decoding of a coded message.

deep-sea trench A long depression in the ocean floor that forms where one tectonic plate is subducted beneath another.

defibrillator A device for restoring the rhythmic beating of the heart by the administration of a controlled electric shock.

denominator The number below the line of a fraction, the divisor.

density The mass per unit volume of a substance.

deposition The natural laying down of loose material, such as sediment, carried by moving water, wind, or ice.

depression An area of low air pressure that often brings bad weather.

dermis The layer of tissue in the skin below the outer epidermis.

desalination The removal of salt from seawater.

desertification The formation of a desert.

desiccate To dry out a substance by removing water from it.

desiccator A sealed container used to desiccate substances and keep them dry.

detector In electronics, the circuit in a radio receiver that separates out the sound signal from a radio wave.

detergent A substance which, when added to water, helps remove grease, oil, and dirt.

diagnosis The identification of an illness from its symptoms and signs.

dicotyledon A flowering plant that has two cotyledons (seed-leaves).

differentiation 1. In mathematics, the process of finding the derivative of a function, which is a measure of how a function changes.
2. In biology, the process by which cells or tissues change to become more specialized, as occurs during embryonic development, for example.

diffraction The bending of waves around obstacles or the spreading out of waves when they pass through a narrow aperture, such as when light waves pass through a diffraction grating.

diffusion The mixing of two or more different substances because of the random movement of the molecules.

digestion The breaking down of food into simpler molecules that can be utilized by the body.

digital Representing, storing, or manipulating sound, video, images, text, or data by discrete values rather than continuously varying ones. The most common way of doing this is by using binary encoding, in which a series of 1s and 0s is used to handle the data.

digital sound Sound recorded as a series of discrete pulses rather than as a continuous wave.

diode An electronic component that lets electricity flow in one direction only.

diploid cell A cell containing the full complement of chromosomes.

direct current (DC) An electric current that flows in one direction only.

discharge 1. The release or conversion of stored energy.
2. The removal of electrical charge from an object.
3. The emission of a fluid from a body.

disinfection Making something clean and free from infection, especially with the use of germ-killing chemicals.

displacement reaction A chemical reaction in which one kind of atom or ion in a molecule is replaced by another.

dissection The methodical cutting up of a body or plant to study its internal anatomy.

distillation The purification of a liquid by heating it until it vaporizes, then collecting the vapor and condensing it back into a liquid.

divergent boundary The line along which two tectonic plates are moving away from each other.

DNA (deoxyribose nucleic acid) The large, double-helix-shaped molecule that makes up the chromosomes that are found in almost all cells. It carries the genetic information in all known living organisms and in some viruses.

doldrums Area along the equator where the trade winds meet and form an area where there is very little wind.

domain In physics, a small, discrete region of magnetism within a magnet.

double helix A double spiral, particularly used in relation to the DNA molecule.

drag The force that slows down an object as it travels through a liquid or gas.

drone 1. A remote-controlled flightless aircraft.
2. A male bee that does no work but is able to fertilize the queen.

drought A long period with no rain.

ductile Ability of a substance to stretch out into a wire without losing its strength.

dwarf planet A small, planetlike object that is big enough to have become rounded by its own gravity, but is not big enough to have cleared surrounding space of objects.

dye A substance that colors a material.

dynamo A generator that produces direct current.

E

easterlies Prevailing winds that blow from the east.

echo A sound that is heard again because it bounces back off a solid object.

eclipse The shadow caused by a body blocking the light from another. See lunar eclipse and solar eclipse.

ecliptic The apparent path in the celestial sphere, through which the Sun passes during the course of a year.

ecology The study of the relationships between organisms and their environment.

ecosystem A distinct area in the biosphere, such as a lake or forest, that contains living things.

effort A force applied to move a load.

egg The female reproductive cell, or ovum.

elasticity The ability of a material to stretch and then return to its original shape.

electrode A solid electric conductor that collects or transmits electric current. Terminals in electrical components such as transistors and batteries are electrodes, as are the small disks attached to the body to monitor electrical activity of the heart in electrocardiography (ECG) and of the brain in electroencephalography (EEG).

electric charge The property of subatomic particles that makes them interact electromagnetically, either repelling or attracting each other.

electric circuit A complete loop of conducting material that carries an electric current and connects electrical elements, such as switches, resistors, and light bulbs.

electric current The flow of electrons or ions.

electrical resistance The tendency to restrict the flow of electricity.

electrolysis Chemical change in an electrolyte caused by an electric current flowing through it.

electrolyte A substance that conducts electricity when molten or in solution.

electromagnetic spectrum The complete range of electromagnetic radiation: gamma rays, X-rays, ultraviolet radiation, visible light, infrared radiation, microwaves, and radio waves.

electromagnetism The physics of the electromagnetic field created by the interaction of electricity and magnetism.

electromotive force (emf) The potential difference of a battery, cell, or generator. It pushes an electric current around a circuit.

electron A particle with a negative electric charge. Electrons orbit the nucleus of the atom in a "cloud." An electric current consists of a flow of electrons.

electron cloud A term used to describe the indeterminate position of an electron around the nucleus of an atom.

electron gun A device that produces a stream of electrons, called cathode rays. For use in a television, for example.

electron micrograph A magnified image of an object made by an electron microscope.

electron microscope A microscope that uses a beam of electrons to produce a magnified image of an object.

electron shell One of the layers in which electrons are arranged around the nucleus of an atom.

electrophoresis The separation of charged particles in a mixture.

electroplating Coating a metal surface electrolytically by passing an electric current through a liquid in which the metal is immersed.

electroscope An instrument that detects electric charge.

electrostatic field The field of force surrounding an electrically charged object.

electroweak theory The theory explaining the combined interaction of electromagnetism and the weak nuclear force on leptons, a group of subatomic particles that includes electrons.

element A substance that cannot be broken down into simpler substances.

ellipse An oval shape like a flattened circle, made by slicing across a cone at an angle.

embryo The first stage of development of newly conceived offspring. In humans, it covers the first eight weeks of pregnancy.

emission lines Bright lines in the spectrum of light emitted by a body.

emulsifier A substance that is used to make immiscible liquids blend.

emulsion Tiny particles of one liquid that are dispersed in another liquid.

encryption The process of coding a message.

endangered species A species of animal or plant that is at risk of becoming extinct, either because it is few in number or because of major threats to its habitat.

endocrinology The branch of medicine concerned with hormones and the endocrine glands.

endometrium The mucus membrane lining the uterus.

endoplasmic reticulum A system of membranes in a cell, on which chemical reactions take place.

endoscope An instrument for directly viewing inside the body. Endoscopes may be flexible or rigid, and comprise a light source, and a series of lenses or a miniature camera. Surgical instruments may be passed inside endoscopes to perform surgeries.

endoscopy A medical technique for seeing inside the body by inserting an endoscope. Surgery may also be performed during endoscopy.

endoskeleton The internal skeleton of a vertebrate.

endosperm The tissue in a seed that stores food.

endothermic reaction A reaction that results in the absorption of heat.

energy The capacity to do work.

entanglement In quantum physics, the linking of two particles as one, so that when the particles move apart a change in one instantly causes a change in the other.

environment The surroundings of an animal or plant.

enzyme A catalyst in living things that increases the speed of reaction in natural chemical processes.

epidermis The outer layer of the skin.

equation 1. A symbolic statement that the values of two mathematical expressions are equal to each other.
2. A symbolic representation of the changes that occur in a chemical reaction.

equator The imaginary circle around the middle of the earth, midway between the north and south poles.

equilibrium A state of physical or chemical balance.

equinox One of the two dates each year (about March 20/21 and September 22/23) on which the Sun is exactly above the equator and night and day are of equal length.

erosion The wearing away of Earth's surface due to the effects of weather, water, or ice.

erythrocytes Red blood cells.

escape velocity The minimum speed that a space rocket must reach to escape a planet's gravity.

estivation The deep sleep or immobility that some animals go into when the weather is very hot and dry.

ethology The study of animal behavior.

eukaryotic cell A cell with a nucleus.

Eurasia The combined land mass of Europe and Asia.

eutrophication Where an excess of nutrients, such as from fertilizers, seeps into water, causing the overgrowth of aquatic plants. This creates a shortage of oxygen in the water, killing animal life.

evaporation The changing of a liquid into a vapor as molecules escape from its surface.

evolution The change in organisms over time.

evolve To undergo evolution.

excretion The elimination of waste by organisms.

exosphere The outermost part of Earth's atmosphere, about 560 miles (900 km) above Earth's surface.

exoskeleton The hard outer skeleton of many invertebrates, such as insects.

exothermic A chemical reaction in which heat is released.

exponent The power, or superscript number to the right of a number, that indicates the number of times the number is to be multiplied by itself, as in $5^3 = 5 \times 5 \times 5$.

exponential growth Growth that accelerates with increasing speed: the larger something becomes, the faster its rate of growth.

expressed gene A gene that shows its effect in a living organism.

extinction The death of all the members of a species.

F

factorial The product of an integer and all the integers smaller than it; e.g. factorial 5 (5!) is $5 \times 4 \times 3 \times 2 \times 1 = 120$.

Fahrenheit Temperature scale named after Gabriel Fahrenheit, in which water at sea level normally freezes at 32° and boils at 212°.

family In biological classification, the taxonomic group into which an order is divided.

fault A break in Earth's crust.

faulting The sudden shifting of a block of rock due to the disturbance of Earth's crust.

fermentation The process by which yeast converts sugars into alcohol.

fermion One of the group of subatomic particles that are associated with matter, such as electrons, quarks, and protons, rather than those that carry force, such as bosons.

Ferrel cell The circulation of the atmosphere in mid-latitudes that brings westerly winds at surface level and returning easterly winds at high level.

fertilization The joining of male and female gametes.

fetus The unborn, developing offspring of a mammal. In humans, it covers pregnancy after the first eight weeks.

filigree Delicate ornamental tracery in gold, silver, or coppery wire.

filter A device that removes solid material from a liquid.

firmware The small, fixed programs that control electronic items such as remote controls and calculators, as well as computer keyboards and robots.

fluid A substance that can flow, such as a gas or liquid.

fluorescence Light that is given off by certain atoms when they are hit by ultraviolet radiation.

FM Frequency modulation. The transmission of a signal by changing the frequency of the carrier wave, such as a radio wave.

fold A bend in layers of rock.

folding The slow crumpling of layers of rock due to the lateral compression of Earth's crust.

food chain A series of organisms, each of which is consumed by the next.

food web The system of food chains in an ecosystem.

force Something that changes the movement or shape of an object.

force field The area in which a force can be detected.

forensic The application of scientific methods to the study of crime.

formula (plural: **formulae**) 1. A set of chemical symbols that represent the make up of a substance.
2. A set of mathematical symbols expressing a rule, principle, or method for determining an answer.

fossil Evidence of past life such as animal and plant remains, eggs, and footprints.

fossil fuel A fuel, such as coal or oil, that has been formed over millions of years from the remains of living things.

fractal A curve or geometrical pattern in which similar shapes recur at different scales.

fraction 1. A numerical quantity that is not a whole number. A fraction is expressed as one number divided by another, as in $1/2$, where the upper number is called the numerator and the lower one is called the denominator.
2. In chemistry, one of the portions into which a mixture is separated by a process such as distillation.

fracture zone The region on either side of divergent tectonic plate boundaries, in which rock is broken on a large scale by the movement of the plates.

Fraunhofer lines Dark lines in the Sun's spectrum caused by elements in the Sun's gases absorbing certain wavelengths of light.

freezing point The temperature at which a substance turns from liquid to solid.

frequency The number of waves that pass a point every second.

friction A force that resists or stops the movement of objects in contact with each other. The friction between an object and a fluid such as air or water is known as drag.

front The first part of an advancing mass of cold or warm air.

fundamental particle A basic subatomic particle, such as a quark, lepton, or boson, that is not thought to be made of smaller particles.

fungus (plural: **fungi**) Any of a group of eukaryotic organisms (those whose cells have nuclei) that reproduce by means of spores. The group includes molds, yeasts, smuts, and mushrooms.

fuse A safety device used in electrical circuits. It is a thin wire which melts if too much current passes through it.

G

Gaia hypothesis The concept that all the living things and physical components of Earth interact to form a complex self-regulating system, like a huge "organism."

galaxy A large group of stars, dust, and gas, all loosely held together by gravity. Our galaxy is called the Milky Way.

galvanize To coat iron with zinc to protect it from rust.

gamete A reproductive cell, such as a sperm or egg.

gametophyte The gamete (reproductive cell)-producing stage in the life cycle of plants such as ferns and mosses.

gamma rays A form of electromagnetic radiation with a very short wavelength.

ganglion (plural: **ganglia**) A group of nerve cells enclosed in a casing of connective tissue.

gas An airlike substance that expands to fill any space it enters.

gauge boson Any of the subatomic particles that carry the four fundamental forces of nature—electromagnetism, gravity, the weak nuclear force, and the strong nuclear force.

Geiger counter An instrument used to detect and measure radioactivity.

gene The basic unit of heredity in living things, typically a segment of DNA or RNA that provides the coded instructions for a particular protein.

gene code The sequence of nucleotide bases on DNA that codes for a particular gene.

gene map A plot of the sequence of genes along an entire strand of DNA.

generation A group of individuals born or living at the same time.

generator A device that converts mechanical energy into electrical energy.

genetic drift The process of change in the overall genetic composition of a population as a result of random events rather than by natural selection. Genetic drift causes a change in the relative prevalence of alleles (variations of a gene that controls the same trait) over time.

genetic engineering The process of artificially modifying the characteristics of an organism by manipulating its genetic material.

genetic fingerprinting The analysis of a DNA sample to identify who it belongs to.

genome The complete set of genes for an organism.

genotype The genetic makeup of an organism.

geochemistry The study of the chemistry and composition of the earth.

geology The science that deals with rocks, minerals, and the physical structure of the earth.

geomorphology The study of the physical features on Earth's surface.

geothermal Relating to the internal heat of Earth, or energy produced from it.

geothermal energy Energy harnessed from the heat in the interior of Earth.

germ 1. A microbe that causes disease.
2. The part of an organism from which a new organism can develop.

germination The early stages in the growth of a seed.

gibbous The phase of the Moon between a half and a fully illuminated disk.

glacier A slowly moving river of ice formed by the accumulation of hard-packed snow over many thousands of years.

gland An organ or group of cells that produces substances used by the body, such as hormones.

glaze A glassy substance fused on to the surface of pottery by heat to protect it.

global warming An increase in the average temperature of Earth's atmosphere. The recent increase in global temperatures is at least partly due to the greenhouse effect.

glucagon A hormone that stimulates the liver to turn stored glycogen into glucose when blood sugar levels are low; the opposite effect of insulin.

glucose A simple sugar that is the main carbohydrate source of energy in most living cells.

gluons Particles within protons and neutrons that hold quarks together.

glycogen A stored form of glucose in animal cells, made mainly in the liver and muscles.

granulation The appearance of dark spots or granules in the Sun's photosphere caused by convection currents.

gravitational force The force of gravitational attraction between matter.

gravitational lensing The "bending" of light from a distant bright object in space, such as a quasar, around a massive object en route, due to the massive object's gravity.

gravitational slingshot The extra acceleration a space probe acquires by swinging close to a planet and using the planet's gravity to alter its path and speed.

graviton A hypothetical particle responsible for gravity.

gravity The force of attraction between any two masses. It attracts all objects toward Earth, giving them weight.

great ocean conveyor The deep circulation of ocean waters around the world, driven by differences in density resulting from salt concentration and temperature differences. Also known as thermohaline circulation.

greenhouse effect The way in which certain gases in Earth's atmosphere, especially carbon dioxide, trap heat. The buildup of these gases leads to global warming.

greenhouse gases Gases that increase the atmosphere's capacity for preventing the escape of heat. They include carbon dioxide and methane.

grike An enlarged crack in limestone that is produced as the rock gradually dissolves in rainwater.

groynes Low walls or fences built along the seashore to prevent coastal erosion.

guttation The loss of water from the surface of a plant as liquid rather than vapor.

guyot An undersea mountain with a flat top.

gyre The complete circulation of ocean currents around an ocean basin.

gyroscope A fast-spinning wheel whose axis tends to stay pointing in the same direction once the wheel is spinning. A gyrocompass uses a gyroscope to point to true north. Gyrocompasses are more accurate than magnetic compasses and are widely used in navigation.

H

habitat The natural range of an animal or plant.

Hadley cell The circulation pattern of the atmosphere in the tropics, which brings north and southeasterly trade winds toward the equator at surface level and returns westerly winds at high level.

half-life 1. The time taken for radioactivity in a sample to drop to half of its original value.
2. The time taken for the concentration of a drug or other substance in an organism to naturally reduce to half its original level.

haploid cell A sex cell with half the number of chromosomes. See also diploid.

hard water Water that contains calcium and magnesium salts.

hardware The mechanism, wiring, and other physical components of a computer.

harmonics Waves whose frequencies are whole-number multiples of the basic frequency.

heart The hollow muscular organ in humans and other vertebrates that pumps blood around the body. Some invertebrates, such as mollusks and annelid worms, have a simpler organ that performs the same function.

hemisphere One half of a sphere. Earth is divided into the northern and southern hemispheres by the equator.

hemoglobin An iron-containing protein that is the oxygen-carrying component of red blood cells of vertebrates. It is also found in the tissues of some invertebrates.

herbivore An animal that only eats plants.

heredity The passing on of characteristics through the generations.

hertz (Hz) The SI unit of frequency. One hertz is one cycle per second.

hibernation The deep sleep or period of inactivity that some animals go into during the winter.

hierarchy An arrangement in ranks, according to factors such as complexity or power.

high frequency Short wavelength, rapid vibration.

HIV Human immunodeficiency virus, a retrovirus that causes the disease AIDS.

hologram A three-dimensional image created by the interference between two parts of a divided laser beam.

homeostasis The processes by which an animal keeps its internal environment (for example, temperature, blood pressure, and water balance) stable.

hominid Any member of the primate family Hominoidea, including humans.

hormones Chemical "messengers" that move around the bloodstream and control the functions of the body.

hot-spot volcano A volcano that forms away from tectonic plate boundaries above a hot convection current in Earth's mantle known as a mantle plume.

H-R diagram (Hertzsprung–Russell diagram) A graph that plots stars according to their absolute magnitude (real brightness), color, and temperature. Most stars fall along a line called the main sequence.

html An acronym for HyperText Markup Language, the main computer language used on the Internet.

http An acronym for HyperText Transfer Protocol, the call and response system used to link websites into the Internet.

Hubble's law The law stating that the distance of a galaxy is proportional to the speed at which it is moving away from us: the more distant the galaxy, the faster it is receding. This shows that the Universe is expanding.

Human Genome Project The worldwide science project completed in 2003 to map the entire sequence of genes in human DNA.

humidity The proportion of water vapor in the atmosphere.

humorism The ancient theory of the workings of the body according to the four humors or temperaments: sanguine, choleric, melancholic, and phlegmatic.

hurricane A huge, circular tropical storm in which there are wind speeds of 75 mph (120 km/h) or more.

hydraulic Describes a machine that operates by transferring pressure through a fluid.

hydraulic pressure The pressure created by a fluid pushed through a pipe.

hydrocarbon A chemical compound made up of hydrogen and carbon only.

hydroelectricity The generation of electricity by harnessing the energy in flowing water.

hydrogen The lightest and most abundant chemical element in the Universe. It has an atomic weight of just over 1 and makes up about 75% of the total mass of elements in the Universe.

hydrometer An instrument used to measure the density of a liquid.

hydroponics The process of growing plants without any soil in sand or gravel supplied with nutrients and water only.

hygiene Attention to cleanliness for the good of health.

hypha (plural: **hyphae**) One of many tiny threads that form the main body of a fungus.

hypotenuse The longest side of a right-angled triangle; the side opposite the right angle.

I

igneous rock Rock that is formed when molten magma cools and solidifies.

imaging A general term for any techniques that produces images. Modern imaging techniques often rely on computers to process raw imaging data from devices such as sensors and body scanners (for example, CT scanners).

immiscible Describes two liquids that do not blend together, such as water and oil.

immune system The body's natural defense mechanisms, which react to foreign material and organisms with effects such as inflammation and antibody production.

immunization Priming the body's immune system to fight against future infections by inoculation.

in vitro fertilization (IVF) An artificial method of conception in which egg cells are fertilized by sperm outside the uterus, in vitro (in glass).

indicator A substance that shows the pH of a solution by its color.

induction 1. The process by which an electrical conductor becomes electrically charged when near a charged object. 2. The process by which a magnetizable object becomes magnetized when in a magnetic field.
3. The process by which an electric current is produced in a circuit by varying the magnetic field linked with the circuit.

industrial plant The land, buildings, and machinery used to perform an industrial process.

inertia The tendency of an object to remain at rest or to keep moving in a straight line until a force acts on it.

infection An illness or medical problem caused by disease-causing organisms, such as bacteria and viruses.

infinity A quantity, distance, time, or number that is immeasurably large.

infrared radiation (IR) A type of electromagnetic radiation with a wavelength just longer than that of visible light but shorter than that of microwaves. It is commonly experienced as heat.

inheritance The range of natural characteristics and potential passed on by parents or ancestors.

inhibitor A substance that slows down a chemical reaction.

inoculation The deliberate introduction of disease-causing organisms into the body in a mild or harmless form to stimulate the production of antibodies that will provide future protection against the disease.

inorganic chemistry The branch of chemistry that deals with all chemicals except the large number of organic compounds (those that contain carbon–hydrogen bonds).

input Information fed into a computer.

insecticide Any chemical that kills insects.

insulator A material that reduces or stops the flow of heat, electricity, or sound.

insulin A hormone released by the pancreas that regulates blood sugar levels by controlling the concentration of glucose in the blood.

integrated circuit A tiny electric circuit made of components built into the surface of a silicon chip.

interference The disturbance of signals caused where two or more waves meet.

interglacial The period of warmer weather between two ice ages.

internal reflection The reflection of light when it passes from a dense to a less dense medium, such as from glass to air.

Internet The electronic information network linking computers around the world.

Internet service provider (ISP) A company that provides access to the Internet.

interphase The phase of the cell cycle during which a cell is not undergoing cell division. Cells spend most of their time in interphase.

interstellar space Space between the stars.

intestine The lower part of the digestive tract, from the stomach to the anus, in humans and other mammals.

invertebrate An animal with no backbone.

inverter A device used to convert direct current into alternating current.

invisible light Light made of waves too short or too long for the human eye to see, including infrared and ultraviolet.

ion An atom or group of atoms that has lost or gained one or more electrons to become electrically charged.

ionic compound A chemical compound, such as sodium chloride, in which ions are held together in a lattice by ionic bonds.

ionic bond A chemical bond formed when one or more electrons are passed from one atom to another, creating two ions of opposite charge that attract each other.

ionosphere The part of the atmosphere, 30–250 miles (50–400 km) above Earth's surface, that reflects radio waves.

iris The colored part of the eye that surrounds and controls the size of the pupil.

irradiation The use of radiation to preserve food.

irrational number Any number that cannot be expressed as a simple fraction. Written as a decimal, irrational numbers have infinitely many digits after the decimal point.

isobar A line on a weather map that connects points with the same atmospheric pressure.

isomer A chemical compound with the same formula as another compound, but a different physical structure.

isoseismal A line on a map that connects points at which earthquake shocks are of equal intensity.

isotope One of two or more atoms with the same atomic number but with different numbers of neutrons.

J

jet propulsion The pushing forward of an object by a stream of fluid or air.

jet stream Strong winds that circle Earth about 6 miles (10 km) above the surface.

joint 1. A structural line in sedimentary rocks.
2. An area of the body where bones meet.

joule The SI unit of work or energy, equal to the work done by a force of one newton moving one meter.

K

karyotype 1. The characterization of the chromosomes of a species or individual in terms of the number, size, and structure of the chromosomes.
2. A diagram or image of an individual's total complement of chromosomes arranged in their assigned numerical order.

Kelvin scale See absolute scale.

keratin The protein that makes up hair, horns, hoofs, nails, and feathers.

keyhole surgery Surgery performed through a very small incision, using special instruments and an endoscope.

kidney One of a pair of internal organs that filter waste products and excess water from the blood.

kinetic energy The energy an object has because of its movement.

kingdom The highest level in the classification of life forms.

Kyoto protocol An international agreement on climate change that sets industrialized countries binding targets for reducing greenhouse gas (GHG) emissions.

L

laccolith A mass of igneous rock that pushes the rock above into a dome shape.

lactose One of the sugars in milk.

Lamarckism The theory that evolution depends on the inheritance of characteristics acquired during an organism's life, named after Jean Baptiste Lamarck.

larva (plural: **larvae**) The second stage in the life of an insect, between the egg and the adult, such as a caterpillar.

laser (Light Amplification by the Stimulated Emission of Radiation). A device that emits an intense beam of light.

laser surgery Surgery performed with a laser beam, for example, reshaping the cornea to improve eyesight.

latent heat The heat needed to change a solid to a liquid or a liquid to a gas without a change of temperature.

latitude A measure of distance from the equator (the poles are at 90° latitude and the equator is at 0°). Lines of latitude are imaginary lines drawn around Earth, parallel with the equator.

LDR (Light-Dependent Resistor). A resistor whose resistance increases when the amount of light that hits it increases.

leaching The extraction of a soluble material from a mixture by passing a solvent through the mixture.

LED (Light-Emitting Diode). A diode that emits light when a current flows through it.

lens 1. Disk of glass ground to a precise shape so that it refracts light rays to a focus, to provide a sharp image or magnification.
2. The part of the eye that forms an image on the retina.

lepton A family of fundamental particles that are affected by electromagnetism, gravity, and the weak nuclear force, but not by the strong nuclear force, as quarks are.

leucocyte A general term for any white blood cell.

lever A rigid bar that pivots to move a load at one end when force is applied to the other.

lift The upward force produced by an aircraft's wings that keeps it airborne.

ligament A short, elastic band of fibers that connects two bones or cartilages at a joint.

light year The distance traveled by light in a year. It is equal to 5.9 million million miles (9.5 million million km).

lignin A polymer in the walls of the cells of trees and shrubs. It makes the plant woody.

lithosphere The layer of Earth that includes the crust and the upper mantle.

logarithm The power to which a base, such as 10, must be raised to produce a given number.

longitude A measure of distance around Earth, measured in degrees east or west of the prime meridian at Greenwich, London. Lines of longitude are imaginary lines drawn on Earth's surface between the poles.

longitudinal wave A wave in which the particles of the medium vibrate in the same direction as that in which the wave is traveling.

low frequency A radio frequency band with long waves that vibrate relatively slowly.

luminosity The amount of light given out by an object, such as a star.

lunar eclipse When the Moon moves into Earth's shadow so that it cannot be seen from Earth.

lymphocytes Any of a group of white blood cells that play a crucial role in immunity, by producing antibodies, for example.

lymphatic system A network of tubes and small organs that drains a fluid called lymph from the body's tissues into the bloodstream.

M

magma Liquid, molten rock in Earth's mantle and crust. It cools to form igneous rock.

magnetic field The area around a magnet in which its effects are felt.

magnetic poles 1. The two regions of a magnet where he magnetic effect is strongest.
2. The two variable points on Earth where Earth's magnetic field is strongest and toward which a compass needle points.

magnetism The invisible force of attraction or repulsion produced by a magnetic field.

magnetosphere The magnetic field around a star or planet.

magnification The degree to which a lens or other optical device makes something appear larger than it is.

main sequence A term for stars that fall within the main diagonal band on the Hertzsprung–Russell diagram. Such stars generate heat and light by converting hydrogen to helium by nuclear fusion in their cores. Most stars are of this type.

malleable A term for metal that can be hammered or pressed into shape.

mammal Warm-blooded vertebrate animal that gives birth to young that feed on their mother's milk.

manned maneuvering unit (MMU) A backpack used by astronauts to move around in space.

mantissa The part of a logarithm after the decimal point.

mantle The large part of Earth's interior between the core and the crust, made of warm, often semi-molten rock.

mass The amount of matter in an object.

mass wasting The downslope movement of rock and weathered material in processes such as landslides and rockfalls.

matter Anything that has mass and occupies space.

mean The middle or expected value of a set, typically found by adding all the values together then dividing by the number of values in the set.

meander A looplike bend in a river.

mechanical advantage A measure of the effectiveness of a machine: the ratio of the force produced by the machine (output) to the force applied to it (input).

median 1. Situated along the middle of the body, in the plane that divides the body into right and left halves.
2. In a set of values, the middle value when all the values are arranged in order of size.

meiosis The type of cell division that results in daughter cells that have half the number of chromosomes of the parent cell. Meiotic division produces egg and sperm cells.

melanin A brown pigment that is found in the skin, hair, and eyes.

melting point The temperature at which a solid becomes liquid.

membrane A thin skin.

meniscus The curved upper surface of a liquid in a thin tube.

menstrual cycle The recurring cycle of changes in women and other female primates that prepares the body for reproduction, including ovulation and menstruation.

Mercalli scale A scale that is used to measure an earthquake's intensity.

Mesolithic The middle part of the Stone Age, ending typically in the development of farming, between 11,000 and 7,000 years ago in Europe.

mesopause The part of the atmosphere about 50 miles (80 km) above Earth's surface. It is the upper limit of the mesosphere.

mesosphere 1. The layer of Earth's mantle underneath the asthenosphere.
2. The layer of the atmosphere above the stratosphere.

metabolism The sum total of chemical processes within a living organism.

metal Any of several elements that are usually shiny solids and good conductors of electricity and heat.

metallic bond A bond formed between metal atoms. The metal's electrons flow freely around the atoms.

metamorphic rock Rock that has been changed by great heat and pressure underground.

metamorphosis A change of form, such as from a caterpillar to a butterfly.

metaphase The stage of mitosis and meiosis, following prophase and preceding anaphase, during which the chromosomes are aligned across the middle of a cell.

meteor A tiny piece of dust or rock from space that burns up as it enters Earth's atmosphere, producing a streak of light.

meteoroid A small body of rock in space that will become a meteor if it enters Earth's atmosphere, and a meteorite if it reaches the ground.

meteorite A piece of rock or metal from space that enters Earth's atmosphere and reaches the ground without burning up.

meteorology The study of the weather.

microchip A small wafer of semiconducting material with an integrated circuit.

microclimate The climate of a small area, such as a valley.

micrograph A photograph taken using a microscope.

microorganism A tiny organism which can be seen only with the aid of a microscope.

microprocessor An integrated circuit that performs all the central processing tasks of a computer.

microscope An instrument that produces magnified images of very small objects.

microwave A type of electromagnetic radiation. Microwaves are very short radio waves.

mid-ocean ridge A ridge down the middle of the ocean floor, created by volcanic material erupting from the gap between diverging oceanic plates.

migration 1. The movement of animals to find food, a place to breed, or a more suitable climate.
2. The movement of one or more atoms, or of a chemical bond, from one position to another in a molecule .
3. The movement of ions between electrodes in electrolysis.
4. The change from one hardware or software system to another.

mimicry 1. Where a species of plant or animal evolves to look like another.
2. The copying of one individual's behavior by another.

mineral A naturally occurring inorganic solid.

mineralogy The study of minerals.

mirage An optical illusion produced by light bending through layers of air with different densities.

miscible Describes two or more liquids that can be blended together.

mitochondrion (plural: **mitochondria**) An organelle that produces energy for a cell.

mitosis Cell division in which the nucleus divides to produce two cells, each with the same number of chromosomes as the parent cell.

mixture A substance that contains two or more elements or compounds that are not combined chemically.

mode 1. Patterns of vibrations.
2. The value that occurs most frequently in a set of values.

modem A device for converting computer output and input into a form suitable for transmitting through telephone lines or optical fibers.

modulation The transmission of a signal by changing the characteristics of a radio wave (called the carrier wave).

mole The amount of a substance that contains the same number of atoms or molecules as there are in 0.4 oz (12 g) of carbon-12.

molecule The smallest unit of an element or compound, made up of at least two atoms.

momentum A quantity equal to an object's mass multiplied by its velocity.

monocotyledon A flowering plant with a single cotyledon (seed-leaf). See also dicotyledon.

monomer A molecule that can be bonded to other identical molecules to form a polymer.

monsoon A strong wind that changes direction according to the season, bringing torrential rain from the sea to areas such as India and Bangladesh.

moon A small body that orbits a planet.

moraine Rocks and debris that have been deposited by ice.

mordant dyes Dyes that need another chemical to be added to fix them to a fabric.

mouse In computing, a handheld device that is used to control a cursor on a computer monitor.

MRI (Magnetic Resonance Imaging). A noninvasive form of diagnostic medical imaging, in which a patient is placed in a strong magnetic field and the absorption and transmission of radio signals from the body is analyzed.

multidimensions The theoretical idea that there are many more than the standard three dimensions in space.

muscle A band of strong fibers in the body that is able to contract in response to nerve signals and thereby provide movement.

mutation A random change in the chromosomes of a cell.

mutualism Any close relationship between two species in which both benefit.

myelin A fatty material found around nerve fibers.

myofibril Stretchy threads found in muscle cells.

N

nacreous Relating to the mother-of-pearl made inside shellfish such as oysters.

nanometer (nm) One billionth of a meter (10^{-9}m).

nanotechnology Minute technology on the atomic or molecular scale, less than 100 nanometers.

natural logarithm (e) A logarithm in which the base is the irrational number e (about 2.71828).

natural selection The biological process by which inheritable characteristics that increase an individual's chances of survival and reproduction are passed on to the next generation.

nebula (plural: **nebulae**) A visible cloud of dust and gas in space.

nectar A sugary liquid found in the flowers of some plants, on which pollinators feed.

nematocyst A long, coiled thread that shoots out of a stinging cell, such as in a sea anemone.

nephron One of the million or so minute purification and filtration units in the kidney.

nerve A fiber that carries electrochemical "messages" (nerve impulses) from one part of the body to another.

neuron A nerve cell.

neutralize To make an acid or alkali into a neutral solution that is neither acidic nor alkaline.

neutron A particle in the nucleus of an atom that has no electrical charge.

neutron star A small, very dense star made mostly of neutrons, formed by the gravitational collapse of a giant star.

newton (N) An SI unit of force.

niche The position that a living thing occupies in an ecosystem.

nitrification The conversion of ammonia in the soil into nitrites or nitrates.

nitrogen A colorless, odorless, unreactive gas that makes up the majority of the atmosphere and plays an important part in the life of plants and the soil.

noble gases Gases such as helium, argon, and neon that have a complete complement of electrons in their outer shell and are very unreactive.

noctilucent cloud A high, luminous cloud of ice crystals seen at night, especially in summer at high latitudes.

nocturnal A term used to describe a living organism that is active at night and sleeps during the day.

nomenclature A body or system of names.

nuclear energy 1. Energy released by an atomic nucleus as a result of nuclear fission, nuclear fusion, or radioactive decay.
2. Electricity generated by a nuclear power station.

nuclear fission A nuclear reaction in which the nucleus of an atom splits into two smaller nuclei, releasing energy.

nuclear fusion A nuclear reaction in which the nuclei of light atoms, such as hydrogen, fuse to form a heavier nucleus, releasing energy.

nuclear reaction A change in the nucleus of an atom.

nucleolus A small, dense, round body inside the nucleus of a cell.

nucleus 1. The central part of an atom, made up of protons and neutrons.
2. A structure found in most plant and animal cells that contains the genetic material of the cell.

numerator The number above the line in a fraction, indicating the number that is divided.

nutrients Substances in food that are used by organisms for growth, maintenance, and reproduction.

nutrition The processes in which an organism takes in food and uses it for growth and maintenance.

O

obelisk A tapering, usually ancient, stone pillar.

observatory A building from which astronomers study space.

occlusion Where a cold front catches up with a warm front.

ohm (Ω) The SI unit of electrical resistance.

okta scale A scale for measuring cloud cover. One okta equals one-eighth cloud cover.

omnivore An animal that eats both plants and animals.

Oort cloud An immense spherical cloud surrounding the solar system and extending approximately three light years (about 30 trillion kilometers) from the Sun.

opaque Does not let light through.

optical fibers Thin glass fibers along which light travels. They are used in communications.

optics The branch of science that studies vision and the behavior of light, including how light is affected by lenses, mirrors, and other optical devices.

orbit The path of one body, such as a planet or satellite, around another body, such as a star or planet.

ore A naturally occurring rock from which metals can be extracted.

organ A group of tissues, usually grouped together in a discrete structure, that has a special function, such as the brain or the heart.

organelle Specialized structure that forms part of a plant or animal cell.

organic 1. A compound containing carbon.
2. Food production without the use of chemical fertilizers.

organic chemistry The chemistry of carbon compounds, mostly related to living things.

organism A living thing consisting of one or more cells.

oscillation A regular movement back and forth.

oscillator An instrument that produces an alternating current of known frequency.

oscilloscope An instrument that shows electrical signals on a screen.

osmosis The movement of water through a semi-permeable membrane from a weak solution to a more concentrated one.

ossify Turn to bone.

output Information from a computer.

ovulation The release of an ovum from the ovary about midway through the menstrual cycle.

ovum The egg cell.

oxidation When a substance gains oxygen or loses hydrogen, or an atom loses electrons in a chemical reaction.

oxide A compound formed between an element and oxygen.

oxidizing agent A substance that causes the oxidation of another substance.

oxygen A colorless, odorless, reactive gas that is the third most abundant element in the Universe, makes up 21% of the atmosphere, and is essential for most forms of life on Earth.

ozone An unstable, toxic form of oxygen with three atoms in its molecules. It is present in the upper atmosphere and absorbs ultraviolet radiation. At ground level it is a respiratory irritant.

P

P wave Primary wave. A fast-moving earthquake wave that alternately stretches and squeezes rocks as it moves.

pacemaker An electronic device implanted in the chest that delivers short electrical impulses via electrodes to regulate the heartbeat.

palynology The study of both living and fossil pollen and spores.

pancreas A gland behind the stomach that secretes digestive enzymes and hormones that regulate glucose levels.

pandemic cycle The now discredited idea that worldwide epidemics of diseases occur in cycles.

parabolic dish A specially shaped dish that collects and concentrates waves, such as radio waves.

parallax The apparent movement of objects against each other, such as the movement of nearby trees against background hills, as the observer moves.

parallel circuit A circuit in which there are at least two independent paths in a circuit to get back to the source.

parasite An organism that lives on and feeds off another organism, called the host, often until it destroys the host.

particle A tiny speck of matter.

particle accelerator A giant machine in which subatomic particles are accelerated around a tunnel by electromagnets, and smashed together at very high speeds.

particle physics The branch of physics that deals with subatomic particles.

parthenogenesis Reproduction without mating.

pasteurization The heating of food to destroy disease-carrying bacteria.

pathogen A microbe (microorganism) that causes disease.

payload The equipment carried into space by a spacecraft, such as a satellite.

penumbra A partial shadow, especially round the shadow of the Moon or Earth in an eclipse.

perigee The point in the orbit of a moon or satellite at which it is nearest the object it is orbiting.

periodic table A table of all the elements arranged in order of their atomic numbers.

pesticide A substance used to destroy insects and other pests of crops and domestic animals.

PET Positron Emission Tomography. A form of medical scan that creates an image by detecting positrons sent out by radioactive substances in the blood.

petrochemical Any chemical made from petroleum or natural gas.

petrology A science that deals with the study of rocks.

pH A measure of acidity or alkalinity of a solution.

phagocyte A type of body cell that engulfs and absorbs bacteria and other small particles.

pharmacology The branch of medicine concerned with drugs and their uses.

phases 1. The various visual aspects of a moon or planet, depending on how it is illuminated.
2. The three states in which matter occurs: solid, liquid, and gas or vapor.

phenotype The observable characteristics of an organism.

pheromones Chemical substances released by animals to communicate with one another by smell.

phloem Tissue that carries food in a plant.

photocell An electronic device that generates electricity when light falls on it, such as in a solar-powered calculator.

photochromicity The ability of an object, such as a lens, to darken or change color when exposed to light, and to return to its original color when the light is removed.

photoelectric effect The emission of electrons from the surfaces of some substances when light hits them.

photon The particle that makes up light and other electromagnetic radiation.

photosphere The visible surface of the Sun, which gives out almost all of its light.

photosynthesis The process by which plants make food from water and carbon dioxide, using energy from the Sun.

physics The science of the properties and nature of matter and the interactions of energy and matter.

physiology The study of how organisms work.

phytoplankton Tiny organisms that are part of plankton.

pi (π) The ratio of the circumference of a circle to its diameter, approximately 22 divided by 7, or about 3.14159.

piezoelectric effect The production of electricity by applying stress to certain crystals, such as quartz.

pigment A substance that gives color to a material, but unlike a dye, does not dissolve in it.

piston A sliding disk or short cylinder that is pushed up and down inside a cylinder to provide the power in an engine.

pitch The property of a sound that makes it high or low.

placebo An inactive substance given to a patient to compare its effects with a real drug.

plaintext Simple, unformatted text in a form that can be read by many different computer systems.

Planck constant (h) The ratio of the energy in one photon of electromagnetic radiation to its frequency. It is a fundamental constant in quantum physics.

planet A large body that orbits a star.

planetesimals Any small body circling the Sun from which planets are thought to have formed.

plankton Unicellular and juvenile organisms that live near the surface of the seas and inland waters.

plant Any organism that contains chlorophyll.

plaque A deposit on teeth where bacteria thrive.

plasma 1. The liquid part of the blood.
2. A hot, electrically charged gas, in which the electrons are free from their atoms.

plasma membrane The skin around a cell which regulates what molecules enter and what leave.

plasmid A normally circular strand of DNA in bacteria or protozoa.

platelet An irregular-shaped disc in the blood that releases chemicals to coagulate the blood.

plate tectonics The study of the plates that make up Earth's crust.

Pleiades The name of an open cluster of stars in the constellation of Taurus, one of the closest to Earth. Also known as M45 or the Seven Sisters.

polar cell The circulation of the atmosphere in polar regions in which cold descending air blows out from the poles at surface level, then returns at high level.

polar reversal The reversal of the direction of Earth's magnetic field.

polarized light Light in which the vibrations occur in only one plane.

pollination The deposition of pollen on a flower so that it can be fertilized and set seed.

pollution Dirtying or poisoning of the air, land, or water by substances such as waste chemicals from factories.

polygon A plane geometric figure with at least three straight sides and angles.

polyhedron A solid geometric figure with at least six flat faces.

polymer A substance made from long, chain-shaped molecules consisting of many identical monomer units.

population The number in a particular region, typically of inhabitants.

potential difference The difference in energy between two places in an electric field or circuit.

potential energy 1. Energy stored for use at a later time.
2. The stored energy that a body has because of its position or state.

positron Counterpart of an electron, with a positive charge.

power The rate of change of energy.

precipitate Tiny particles of solid in a liquid, made by a chemical reaction.

predator An animal that preys on others.

preservation The process of keeping something in its original state, or free from harm, erosion, or decay.

pressure Continual physical force pushing against a surface, or the force per unit area.

prey An animal that is hunted or eaten by another animal.

prime number Any positive that can be divided by itself and one to produce a whole number as the result.

prism 1. A solid geometric form with sides that form a parallelogram.
2. A prism-shaped block of glass, especially one with triangular sides, that is used to split white into the colors of the spectrum.

probability The likelihood of an event happening, normally expressed as a number between 0 and 1.

processor speed The rate at which the processor in a computer deals with data.

program A series of coded instructions to operate a computer.

prokaryotic cell A cell with no nucleus.

prominence A mass of glowing gas reaching out from the surface of the Sun.

prosthetic An artificial body part.

protein A substance found in foods such as meat, fish, cheese, and beans, which the body needs for growth and repair.

protein synthesis The process in which cells make proteins under instruction from DNA and RNA.

proton A particle in the nucleus of an atom that has a positive electric charge.

pulley A grooved wheel around which a rope or chain runs, used to lift heavy weights.

pulsar A dense star that emits regular pulses of radiation, usually radio waves.

Q

qualitative analysis Finding out what a substance is made of.

quantitative analysis Finding out how much of each ingredient is in a substance.

quantum chromodynamics (QCD) The quantum physics theory that deals with subatomic particles involved in the strong or "color" interaction in the nucleus.

quantum computing A computing system, as yet hypothetical, that makes use of quantum phenomena, such as quantum entanglement.

quantum electrodynamics (QED) The quantum physics theory that deals with the interactions between electrons, positrons, and photons.

quantum physics The branch of science that deals with subatomic particles and energy interactions in terms of minute discrete energy packets called quanta.

quantum theory The theory that light and other electromagnetic radiation is made up of a stream of photons, each carrying a certain amount of energy.

quarantine A time or place of isolation for people or animals that may carry an infectious disease.

quark One of a group of small particles that make up protons, neutrons, and other subatomic particles.

quasar The brilliant core of a young galaxy, probably a disk of hot gas around a massive black hole.

R

radar Radio detection and ranging. A way of detecting objects by sending out radio waves and collecting the "echoes."

radiation 1. An electromagnetic wave.
2. A stream of particles from a source of radioactivity. See also electromagnetic spectrum.

radio antenna An aerial or dish that picks up radio signals.

radioactive Emitting ionizing particles and radiation as atomic nuclei break up naturally.

radioactive dating A method of estimating the age of an object by measuring how much the radioactive isotopes in it have decayed.

radioactive decay The process in which unstable atomic nuclei emit ionizing particles and radiation as they break up.

radioactive tracers Substances that contain a radioactive atom to allow easier detection and measurement.

radioactivity The disintegration of the nuclei in an atom, causing radiation to be given off.

radiometric rock dating The process of finding an absolute age for rocks by detecting the stage of radioactive decay of particular isotopes in them.

radiosonde A package of instruments, including a miniature radio transmitter, carried into Earth's upper atmosphere by a weather balloon to gather meteorological information.

RAM Random access memory. Computer memory chips where information can be stored and retrieved.

rarefaction 1. Areas along a longitudinal wave, such as a sound wave, where the pressure and density of the molecules is decreased.
2. The lowering of the density of a gas.

ratio The proportional relation between two numbers.

reactants The substances that take part in a chemical reaction.

reaction 1. A force that is the same in magnitude, but opposite in direction to another force. Every force has a reaction.
2. Any change that alters the chemical properties of a substance or that forms a new substance.

reactivity The ability of a substance to take part in a chemical reaction.

real image An image formed where light rays focus. It can be seen on a screen. Compare to virtual image

recycling Using waste material again, thus saving resources and energy.

red giant A star near the end of its life, which has swollen in size and cooled.

red shift The stretching out of light (moving it toward the red end of the spectrum) from a galaxy that is moving away from Earth.

reducing agent A substance that causes the reduction of another substance.

reduction When a substance gains hydrogen or loses oxygen, or an atom gains electrons in a chemical reaction.

reflection The bouncing back of light, heat, or sound from a surface.

reflex An automatic reaction to something.

refraction The bending of light rays as they enter a different medium at an angle, such as from air to water.

refractive index The ratio of the speed of light in one medium to the speed of light in a second medium.

relative density The weight of a volume of a material divided by the weight of an equal volume of water.

relativity The description of space and time, energy and matter according to the theories of Einstein, which depend on the constancy of the speed of light.

repoussé The ancient art of decorating metal by hammering on the back of the piece.

reproduction The process of creating offspring.

resistance A measure of how much an electrical component opposes the flow of an electric current.

resistor A component in an electric circuit that opposes the flow of electricity.

resonance When the vibrations of an object become large because it is being made to vibrate at its "natural" frequency.

respiration The process in which oxygen is taken in by living things and used to break down food. Carbon dioxide and energy are produced.

resultant The overall force that results from two or more forces acting on an object.

retrovirus An RNA virus, such as HIV, the virus that causes AIDS. It replicates by inserting a DNA copy of its genes into the host cell.

reverberation Where an echo reaches a listener before the original sound has finished. It makes a sound seem to last longer.

rheostat A resistor whose resistance can be changed.

ria A long, narrow sea inlet caused by the flooding of a river valley.

ribosomes Tiny spherical bodies in the cytoplasm of cells, where proteins are made.

Richter scale The logarithmic scale for registering the magnitude of earthquake vibrations named after Charles F Richter.

rifting The opening up of a crack and the formation of a valley as a result of the separation of tectonic plates.

RNA Ribonucleic acid, a molecule related to DNA but that forms simpler, single, long strands of nucleotide bases. It plays a part in mediating between DNA and the rest of the cell.

robot A machine able to carry out a complex series of instructions according to a program.

ROM Read only memory. Computer memory in which information is stored permanently, so that it can be retrieved but not altered.

router An electronic networking device for forwarding data between computers and phone lines, often wirelessly, using radio waves.

S

S wave Secondary wave, an earthquake wave that travels only through solid rock as lateral or horizontal waves.

salt 1. A compound formed from the reaction of an acid with a base.
2. The common name for sodium chloride.

sap The liquid that flows through a plant, carrying food and water.

saprophyte An organism, such as a fungus or bacterium, that lives on dead or decaying matter.

satellite An object that orbits a planet. There are natural satellites, such as a moon, and artificial satellites, such as a craft used to reflect radio signals.

satellite dish A dish-shaped aerial that receives signals broadcast from satellites.

saturation The point at which no more solute can be dissolved in a solution.

scalar quantity A quantity that has only magnitude, such as mass and time.

SDSS Sloan Digital Sky Survey. A large scale astronomical project to map space, which covered more than a quarter of the sky and created three-dimensional maps containing more than 930,000 galaxies and over 120,000 quasars.

seamount An isolated undersea mountain on the deep ocean floor.

search engine A program for locating and retrieving data and documents from a computer network, especially the Internet.

secretion The release of specific substances from plant and animal cells.

sedimentary rock Rock formed when fragments of material settle on the floor of a sea or lake, for example. The layers are cemented together over time by a process called lithification.

sedimentation The geological process in which loose material is laid down by water, wind and moving ice.

seismic wave A wave that travels through the ground, such as from an earthquake or explosion.

seismology The study of earthquakes.

seismometer A device for recording and measuring earthquake waves.

selective breeding The process of choosing particular domestic animals for breeding to encourage the development of particular traits over the generations.

semiconductor A substance that has a resistance somewhere between that of a conductor and an insulator.

semipermeable membrane A membrane that lets small molecules through, but stops large molecules.

sequestration Short for carbon sequestration. Any process that uses the natural world to reduce the amount of carbon dioxide in the atmosphere, including fertilizing the ocean with iron to stimulate plankton growth.

sessile 1. Animals that cannot move around as adults, such as sea anemones.
2. Plants with no stalks, such as algae.

sex cell See gamete.

sextant A navigational instrument designed to measure the altitude of an object, such as the Sun above the horizon at noon.

sexual reproduction Reproduction that involves the combination of male and female gametes.

SI (Système International) unit A unit in the international system of measures based on the meter, kilogram, second, ampere, kelvin, candela, and mole.

sial The silica- and aluminium-rich upper layer of Earth's crust.

sickle-cell anemia A severe hereditary disease in which a mutated form of hemoglobin distorts red blood cells into a "sickle" shape.

silica A white or colorless compound of silicon that occurs naturally, such as quartz.

sima The silica- and magnesium-rich lower layer of Earth's crust.

sine (sin) The sine of an angle is the ratio of the side opposite to that angle in a right-angled triangle to the hypotenuse.

sinter 1. A hard siliceous or calcareous deposit around mineral springs.
2. A hard material created by the coagulation of coal powder when it is heated.

skeleton The frame of bone and cartilage in vertebrates that supports the body and protects its organs.

slag The waste matter separated from metals in smelting or from ores during processing.

slide rule A ruler with a central sliding bar designed to make quick calculations using logarithms.

smelting The process of extracting metal from ores, using heat to melt the pure metal out.

SNP Single nucleotide polymorphism. A small genetic change, or variation that can occur within a person's DNA sequence.

soft water Water that is free of dissolved calcium and magnesium salts.

software The programs used by a computer.

solar constant The amount of heat energy from the Sun received per unit area of Earth's surface.

solar eclipse An eclipse in which the Moon passes between Earth and the Sun so that the Sun, or part of it, cannot be seen from Earth.

solar flare A sudden burst of radiation from the Sun.

solar panel An object that collects energy from the Sun and uses it to heat water, for example, or to produce electricity.

solar system The Sun, the planets that orbit the Sun, their moons, and the other bodies in space whose movements are affected by the Sun's gravity.

solder An alloy (often made of tin and lead) used to join metal surfaces together.

solenoid A cylindrical core of wire that becomes a magnet when an electric current is passed through it.

solstice Each of the two times each year in midsummer and midwinter (around June 21 and December 21), when the Sun reaches its highest or lowest point in the sky at noon.

solubility The ability of a solute to be dissolved.

solute The substance that dissolves in a solvent to form a solution.

solution A liquid in which another substance is mixed evenly and invisibly.

solvent A substance that can dissolve other substances, or a liquid in which other substances are dissolved to form a solution.

sonar Sound navigation and ranging. A means of detecting objects and of navigating underwater by sending out sound waves.

somatic nuclear transfer A laboratory technique for creating a fertilized ovum using a "somatic" cell (ordinary body cell) to create a clone of the organism.

sonic boom A loud, explosive noise made by the shockwave from an object that is traveling faster than the speed of sound.

space age The era of space travel.

space probe An unmanned spacecraft sent from Earth to investigate the solar system.

space station A spacecraft, big enough for people to live and work on, which orbits Earth.

space-time The three dimensions of space combined with time in a single continuum.

species A group of organisms that look alike and can breed only with one another.

spectroscope An optical instrument that divides the light given off by an object into its spectrum.

spectroscopy The branch of science that is concerned with studying the radiation spectra emitted and absorbed by different substances.

SPECT Single photon emission computed tomography. A medical imaging technique that scans the patient with gamma rays to build up a complete 3D picture.

spectrum (plural: **spectra**) A particular distribution of wavelengths and frequencies, such as the electromagnetic spectrum.

specular reflection When light bounces off a surface at exactly the same angle as that at which it hits it.

speed The rate at which something is moving.

sperm One of the male reproductive cells released in semen during ejaculation, which must enter an ovum for fertilization to take place.

spinal cord A bundle of nerves running from the brain down through the spine.

standard model The principle theoretical framework for particle physics, combining the theories on how three of the four fundamental forces interact (electromagnetism and the strong and weak nuclear forces) with 12 basic particles (six quarks and six leptons—the electrons, muon and tauon, and three neutrinos).

star A celestial body that releases energy from the nuclear reactions in its core.

starch A polymer found in plants that is an important part of the human diet.

static electricity An electric charge held on an object, caused by the gain or loss of electrons.

stem cell Basic, undifferentiated cells that can develop into any other kind of cell.

sterilization The removal of bacteria from an object.

stoma (plural: **stomata**) A tiny opening in a plant's leaf or stem through which gases and water vapor pass.

stratigraphy The study of rock layers.

stratopause The boundary between the stratosphere and the mesosphere in Earth's atmosphere.

stratosphere The part of Earth's atmosphere between the troposphere and the mesosphere.

strong force The force that binds quarks, gluons, and other particles together in the nuclei of atoms to form protons, neutrons, and other particles.

strong interaction See strong force.

subatomic particle A particle smaller than an atom, such as a proton or a neutron.

subduction boundary A boundary between two tectonic plates where one plate is being subducted or pushed beneath another.

sublimation When a solid turns straight from a solid into a gas without becoming a liquid first.

submersible A craft that can travel underwater.

substance Any kind of matter.

succession The process of change from one ecosystem to another, such as from grassland to woodland.

sugars A group of soluble, sweet-tasting carbohydrates.

sunspot A cooler patch on the Sun's surface that appears darker than its surroundings.

super neutrino beam A very intense neutrino beam produced by a high power accelerator.

superconductor A substance that has no electrical resistance.

supernova (plural: **supernovae**) The explosion of a very large star at the end of its life.

supersonic Faster than the speed of sound.

surface tension An effect that makes a liquid seem as though it has an elastic "skin." It is caused by cohesion between the surface molecules.

suspension A mixture of tiny particles of solid matter or globules of liquid.

switch A device that turns an electric current on or off.

synapse A junction of two nerve cells, or between a nerve cell and a muscle fiber or a gland.

synthesis The combining of separate parts or different theories to make a whole.

synthetic An artificially created chemical.

synthesizer A musical instrument that creates sound electronically.

T

tangent 1. A straight line that just touches the outside of a circle or curve at a single point.
2. The tangent of an angle (tan) is the ratio of the side opposite that angle in a right-angled triangle to the side adjacent to it.

taxonomy The classification of living things.

tectonics The theory that Earth's surface is broken into a number of mobile slabs of rock called plates.

telemedicine A developing field of medicine in which problems are dealt with, or medical information transferred, over large distances using telecommunications technology.

telophase The final stage of cell division, in which two daughter cells are formed.

temperate Describes a climate that has mild summers and cool winters.

temperature A measure of how hot or cold something is.

terawatt (TW) A unit of power that is equal to one million watts or one thousand gigawatts.

terminal A connecting point on an electrical component.

theorem A concept that has been proved to be true or can be proved to be true.

thermal A current of rising hot air in the atmosphere.

thermistor A resistor whose resistance changes with temperature.

thermoplastic Plastics that can be continually reshaped whenever they are sufficiently heated.

thermoset Plastics that can only be shaped once by heating.

thermosphere The part of Earth's atmosphere between the mesosphere and the exosphere.

three dimensions (3D) Length, breadth, and depth.

timbre The quality of a musical sound.

tissue A group of similar cells that carry out the same function, such as muscle tissue, which can contract.

tissue typing The identification of antigens in tissue to minimize the possibility of rejection due to antigenic differences between a donor and recipient in organ transplantation.

titration A method of finding the concentration of a solution.

tomography Any technique for producing cross-sectional image ("slices") of parts of the body; for example, CT (computerized tomography) scanning.

torque A force that causes rotation. Known as "moment" by physicists.

totipotency The capability of a cell to develop into any kind of cell.

toxicology The study of toxic substances.

trace elements 1. Any element that occurs in very small concentrations.
2. Substances such as minerals, that are needed in only minute amounts by living things.

trade wind A wind that blows from the southeast or northeast all the year round in equatorial regions.

transcription The copying of sequences of genes from DNA to RNA.

transform boundary A boundary between tectonic plates where the plates are sliding horizontally past each other.

transformer A device that increases or decreases voltage.

transfusion The transfer of blood from a donor to a

recipient.

transistor A semiconductor device that acts as an electronic switch, amplifier, or rectifier.

translocation The movement a segment of a chromosome from one location to another, either on the same chromosome or to another chromosome.

translucent Allows some light through, but is not transparent.

transmission Movement through from one place to another.

transmutation 1. The evolutionary change of one species into another.
2. Conversion of one element into another via a nuclear reaction.

transparent Allowing nearly all light through; is "see-through."

transpiration The giving off of water vapor through the stomata of a plant's leaves or stem.

transplant The taking and implanting of tissue or organs from one part of the body to another, or from a donor to a recipient.

transverse wave A wave in which the particles of the medium vibrate at right angles to the direction in which the wave is traveling.

trigonometry The branch of mathematics concerned with triangles.

trophic level The level at which an animal is in a food chain.

tropical Describes a climate that is hot, with periods of heavy rainfall.

tropopause The boundary in Earth's atmosphere between the troposphere and the stratosphere.

troposphere The lowest layer of Earth's atmosphere, immediately above the ground.

tsunami Giant sea waves started by earthquakes or other major disturbances.

tunneling Passing through an apparent barrier, as in quantum tunneling, in which a particle appears to jump from one side of a barrier to another.

turbine A machine that is made to rotate in order to drive a generator.

typhoon The name that is given to a hurricane in the Pacific Ocean.

U

UHF Ultra-high-frequency radio waves.

ultrasound Sound with a frequency above that which the human ear can detect.

ultraviolet (UV) A type of electromagnetic radiation with a wavelength shorter than visible light.

umbra The dark central part of a shadow, where no light at all falls.

uncertainty principle The idea, formulated by Werner Heisenberg, that it is impossible to measure both the position and momentum of quantum objects, because the observation of one changes the other.

Universe All of space and everything it contains.

uplift The upward movement of a geological block, typically due to earthquakes or tectonic activity.

Upper Paleolithic The last period of the Old Stone Age, approximately 40,000 to 10,000 years ago.

upthrust A force pushing upward, opposite to the pull of gravity, such as buoyancy in water.

urine A yellowish waste fluid stored in the bladder of animals and discharged through the urethra.

V

vaccination See inoculation.

vacuole A small fluid-filled sac in the cytoplasm of a cell.

vacuum A space in which there is no matter.

vacuum tube A gas-tight glass tube from which all gas has been sucked out to create a vacuum.

valency The number of chemical bonds that an atom can make with another atom.

valve 1. A device, flap, or membrane for controlling the flow of gas or fluid through an opening.
2. One of the two sides of the shell of a bivalve mollusk.

3. A vacuum or gas-filled device for controlling the flow of an electric current.

variolation The production of smallpox pustules.

vector An organism that transmits a disease.

vector quantity A quantity that has both magnitude and direction, such as a force.

veins Vessels that carry blood from all parts of the body back to the heart.

velocity Speed in a particular direction.

ventricle One of the two large cavities of the heart.

Vernier scale A small movable, graduated scale added to a larger scale for increased accuracy in precision instruments such as barometers; named after Pierre Vernier.

vertebrate An animal with a spine.

vessel A duct carrying blood or another fluid through plants or animals.

VHF Very-high-frequency radio waves.

vibration A quick back-and-forth movement. For example, an earthquake makes Earth vibrate; sound makes the air vibrate.

virtual image An image formed where light rays appear to be focused, as in a reflection in a mirror.

virus A microscopic particle that invades cells to reproduce, and may cause disease.

viscosity A measure of how easily a substance flows.

viscous Thick, sticky, and slow-flowing.

visible light Light in the range of wavelengths that can be seen by the human eye.

vitalism The theory that life depends on some other force than purely chemical and physical.

vitamin An organic compound found in foods that is essential for good health.

volt The unit of potential difference.

voltaic cell A device that produces electricity by chemical change.

voltmeter or **voltammeter** A device used to measure potential difference.

volume 1. The amount of space something takes up.
2. The loudness of a sound.

vulcanization The hardening of rubber by heating it with sulfur.

W

waning Becoming smaller, as in the illuminated portion of the Moon visible from Earth.

warm front The boundary between masses of warm and cold air in which the warm air is advancing and gradually riding up over the cold air.

waterspout A column of water sucked up by a tornado moving over the sea.

watt (W) A unit of power (1 watt = 1 joule per second).

wave One of a series of repeated, often regular-moving variations in intensity or concentration, such as the compression of air created by sound as it moves.

wavelength The distance between successive wave crests.

waxing Growing, as in the illuminated portion of the Moon visible from Earth.

weak force The force in atomic nuclei involved in beta decay and affecting one of the four fundamental forces. Also known as the weak nuclear force or weak interaction.

weak interaction See weak force.

weathering The slow natural breakdown of rocks exposed to the weather.

weight The force with which a mass is pulled toward the center of Earth.

westerlies Prevailing winds blowing from the west.

whirlwind A column of air spinning rapidly in a funnel shape over land or water.

white dwarf The small, dense remains of a dead star.

WIMPs Weak Interacting Massive Particles. Hypothetical large particles that interact only via gravity, and may be prime candidates for dark matter.

WMAP Wilkinson Microwave Anisotropy Probe. A satellite space probe for measuring differences in the temperature of the Big Bang's remnant radiant heat across the full sky.

WMO World Meteorological Organization.

work The energy transferred when a force moves an object or changes its shape.

World Wide Web (www) A vast network on the Internet for gathering and exchanging data and documents through hypertext links.

X

X-ray A type of electromagnetic radiation with a wavelength shorter than ultraviolet radiation.

x-ray opaque dyes Dyes that block X-rays. They are used in medical imaging to show structures that do not show up clearly in conventional plain X-rays.

xylem The tissue that carries water through a plant.

Z

zeolite A natural or synthetic compound with an open structure that can act as a catalyst or a filter for individual molecules.

zeugen A ridge of hard rock formed by erosion.

Zhou The name of the dynasty of Chinese emperors who ruled from 11th-3rd centuries BCE, during which time philosophies such as Confucianism and Taoism were created, and Chinese writing originated.

zodiac The originally twelve (now thirteen) constellations that lie along the ecliptic, through which the Sun, Moon, and planets move.

zooplankton The tiny, often microscopic, animals in the sea that form part of plankton.

zygote A cell resulting from the fusion of two gametes, such as sperm and ovum at conception.

Index

Acknowledgments

Dorling Kindersley would like to thank the following for their kind assistance:

Smithsonian Enterprises

Kealy Gordon, Product Development Manager
Jill Corcoran, Director of Licensed Publishing
Brigid Ferraro, Vice President of Business Development and Licensing
Carol LeBlanc, President, SE

DK would also like to thank the following for their help in the preparation of this book: Jane Parker for the index; Juliet Jopson for additional consultancy on biology; Chris Bryan for additional consultancy on physics. Additional editorial help from Janet Mohun, Simon Tuite, Victoria Wiggins, Megan Hill, Miezan van Zyl, Nathan Joyce, David Summers, Peter Frances, Elizabeth Munsey, Cressida Malins, Ruth O'Rourke, Conor Kilgallon, Marcus Hardy, and Kajal Mistry. Additional design assistance from Maxine Pedliham, Francis Wong, Riccie Janus, Steve Woosnam-Savage, Dean Morris, and Anna Hall. Additional jacket help from Mark Cavanagh, and Claire Gell. Production help from Rob Dunn, and Rita Sinha. DK Cartography: Paul Eames. DK images: Romaine Werblow, Rose Horridge, and Emma Shepherd.

For this revised edition, **Dorling Kindersley** would like to thank Dr. Fiona Coward for her comments on human evolution; Daniel Crisp for help with the illustrations; Chhavi Nagpal for editorial assistance; Jaypal Singh Chauhan and Anita Yadav for DTP assistance; Juhi Sheth, Rakesh Kumar, and Priyanka Sharma for jackets assistance.

DATA CREDITS

435 Lunar eclipses 2021–27: Eclipse predictions by Fred Espenak, NASA/GSFC
436 Total solar eclipses 2021–30: Eclipse predictions by Fred Espenak, NASA/GSFC
243 National Library of Medicine: PMID: 32817606 / PMCID: PMC7439980 / DOI: 10.15585/mmwr.mm6933e2 r

PICTURE CREDITS

The publisher would like to thank the following for their kind permission to reproduce their photographs:

(Key: a-above; b-below/bottom; c-center; f-far; l-left; r-right; t-top)

Corner Tabs
16-63 Alamy Images: capt.digby (Ch 1). 68-123 Corbis: Lawrence Manning (Ch 2). 130-279 Getty Images: The Image Bank/Malcolm Piers (Ch 3). 286-373 Corbis: Science Faction/William James Warren (Ch 4). 378-419 Science Photo Library: Tony Craddock (Ch 5). 422-483 Corbis: Zefa/Josh Westrich (Reference). 484-512 Corbis: Science Faction/David Scharf (Endmatter).

1 NASA: JPL-Caltech. 2 Getty Images: Stefano Bianchetti / Corbis. 4-5 Science photo library: DAVID PARKER (tc). 6 Alamy Images: nagelestock.com (tr). Corbis: Bettmann (tl, bl). Getty Images: Comstock (cla). 7 Alamy Images: North Wind Picture Archives (tr). DK Images: Courtesy of the Natural History Museum, London/Colin Keates (tl). Getty Images: LOOK/Konrad Wothe (bl); Time Life Pictures/Mansell (tr). © The Metropolitan Museum of Art: Gift of J. Pierpont Morgan, 1917 (ca). 8 Corbis: Bettmann (bl); Hulton-Deutsch Collection (tl); Science Faction/Norbert Wu (br). Getty Images: National Geographic/James P. Blair (tr). Science Photo Library: Laguna Design (ca). 9 Chandra X-Ray Observatory: X-Ray: NASA/CXC/CfA/R.Tuellmann et al.; Optical: NASA/AURA/STScl (br). Corbis: EPA/Jose Jacome (tr); Getty Images: Space Frontiers / Stringer / Archive Photos (bl). Science Photo Library: CNRI (tl); Kevin Curtis (ca); NASA/R.B. Husar (cra). 10-11 Science Photo Library: CERN. 12 Getty Images: Comstock. 14 akg-images: James Morris (Astronomical Ceiling). Alamy Images: INTERFOTO Pressebildagentur (Herodotus Map); nagelestock.com (Callanish). The Bridgeman Art Library: Ashmolean Museum, University of Oxford (Abacus). Corbis: The Art Archive (Pythagoras); Christie's Images (Zhou Bronze); Dean Conger (Ziggurat); Reuters/Sergio Moraes (Eclipse); Sakamoto Photo Research Laboratory (Jomon Bowl); Werner Forman (Standard of Ur). DK Images: Courtesy of the

Science Museum, London/Dave King (Wheel). Getty Images: The Bridgeman Art Library (Faience). Wellcome Library, London: (Acupuncture). 14-15 Corbis: Edifice/Philippa Lewis (Background). 15 Alamy Images: North Wind Picture Archives (Ptolemy Map); The Print Collector (Anatomy of the Eye, Archimedes). Corbis: Stefano Bianchetti (Fibonacci). Getty Images: Altrendo Travel (Windmills); The Bridgeman Art Library/ Universitatsbibliothek, Gottingen (Gutenberg Bible). NASA: Marshall Space Flight Centre (Chinese Rocket). Photo Scala, Florence: Ministero Beni e Att. Culturali (Farnese Atlas). Science & Society Picture Library: (Chariot). TopFoto.co.uk: The British Library / HIP (Star Chart). 16 Corbis: Dean Conger (bc); Peter Johnson (tc). Getty Images: Jan Rendek (c). 16-17 Corbis: Ashley Cooper (c). 17 Alamy Images: INTERFOTO Pressebildagentur (crb). Corbis: Burstein Collection (bc); Sakamoto Photo Research Laboratory (tl). Getty Images: The Bridgeman Art Library (tr). 18 DK Images: Courtesy of The American Museum of Natural History/Lynton Gardiner (tl). 18-19 Corbis: Christie's Images (cb). 19 akg-images: Iraq Museum, Baghdad (tr). Alamy Images: Robin Weaver (br). DK Images: The British Museum, London/Nick Nicholls (tl). Getty Images: Gallo Images/Graeme Williams (c). 20 DK Images: Courtesy of the Science Museum, London/Dave King (tl, tc, tr). 21 Corbis: Werner Forman (b). FOTOE: (c). 22 Alamy Images: Mary Evans Picture Library (bc). Ancient Art & Architecture Collection: Ronald Sheridan (tr). Getty Images: Hulton Archive (cl). 23 Mary Evans Picture Library: AISA Media (tr). TopFoto.co.uk: The Granger Collection (l). www.heikenwaelder.at: (crb). 24 DK Images: CONACULTA-INAH-MEX. Authorized reproduction by the Instituto Nacional de Antropologia e Historia/Michel Zabe (cl); The Science Museum, London/Adrian Whicher (br). Wellcome Library, London: (tr, bc). 25 Corbis: Bettmann (tl); Historical Picture Archive (crb). DK Images: The Science Museum, London/ David Exton (bl). Wellcome Library, London: (cra). 26 akg-images: James Morris (cl). Alamy Images: nagelestock.com (tr). Corbis: Staffan Widstrand (crb). Science Photo Library: John Sanford (b). 27 Science Photo Library: Gordon Garradd. 28 Alamy Images: North Wind Picture Archives (tc). Photo Scala, Florence: Ministero Beni e Att. Culturali (bl). Science Photo Library: Royal Astronomical Society (tr). TopFoto.co.uk: The British Library / HIP (crb). 29 Alamy Images: Wolfgang Kaehler (br); Patrizia Wyss (tr). Dorling Kindersley: Whipple Museum of History of Science, Cambridge (c). 30 The Art Archive: (br). Corbis: The Art Archive (ca). Muséum des Sciences Naturelles, Brussels: (clb). 30-31 Alamy Images: The Print Collector (c). 32 Corbis: The Art Archive (r). Science Photo Library: Sheila Terry (bl). 33 akg-images: (cl). Science Photo Library: Royal Astronomical Society (bc). 34 Corbis: Free Agents Limited (cla). Museum of the History of Science, University of Oxford: Keiko Ikeuchi (tc). Image courtesy History of Science Collections, University of Oklahoma Libraries; copyright the Board of Regents of the University of Oklahoma: (bl). 35 Corbis: Bettmann (tr). *Euclid's Elements* by Adelard of Bath, c.1309-1316 (br). 36-37 Corbis: The Gallery Collection (b/Background). 37 Alamy Images: North Wind Picture Archives (tl). 38 Alamy Images: The London Art Archive (ca, bl). Corbis: Reuters/Sergio Moraes (tr). 38-39 Alamy Images: INTERFOTO Pressebildagentur (b). 39 Alamy Images: North Wind Picture Archives (tl). 40 Corbis: Derek Croucher (cl). DK Images: The British Museum, London/Peter Hayman (br); Courtesy of the Charlestown Shipwreck and Heritage Centre, Cornwall/Alex Wilson (cb/Scissors). Getty Images: Time Life Pictures/Mansell (tl). 41 Alamy Images: The Print Collector (t). 42 Getty Images: Johner Images (br). Rex Features: Dragon News (cla). Science & Society Picture Library: (bl). 43 Getty Images: The Bridgeman Art Library/Private Collection (tl, cra). 44-45 Alamy Images: The Print Collector. 46 Alamy Images: Martin Dalton (br). Corbis: Roman Soumar (cla). 47 Corbis: Edifice/Philippa Lewis (b). 48 DK Images: The British Museum, London/Peter Hayman (bl). 49 Getty Images: Riser/Jeffrey Coolidge (l). Science Photo Library: Professor Peter Goddard (br). 50 Corbis: Reuters/Daniel Aguilar (tr). Getty Images: AFP/Sanogo Issouf (clb); Altrendo Travel (b). 51 Corbis: Dean Conger (br); Eye Ubiquitous/James Davis (crb). DK Images: Courtesy of The Science Museum, London (ftl). Getty Images/iStock: Elenarts (cla). Science & Society Picture Library: (tc). 52 Corbis: Bettmann (cr); Gianni Dagli Orti (cla). Science Photo Library: Sheila Terry (b). 53 Alamy Images: Interfoto (tr). The Bridgeman Art Library: Derby Museum and Art Gallery, UK (bc). Corbis: Bettmann (c). 54 Alamy Images: Dennis Cox (bl).

Science Photo Library: (r). 55 China Tourism Photo Library: (br). Corbis: EPA/Thomas Rensinghoff (tl). DK Images: The Science Museum, London/John Lepine (tc, c). 56 Corbis: Christel Gerstenberg (bl). NASA: Marshall Space Flight Centre (c). 57 akg-images: Bibliothèque Nationale (b). Corbis: (tc). Press Association Images: NULL/ AP (cra). 58 Corbis: (br); Gianni Dagli Orti (clb). DK Images: © British Library Board. All Rights Reserved (Picture number 1022251.611) (cla); Courtesy of the London College of Printing (c). 59 Corbis: Bettmann (l). Dreamstime.com: Chernetskaya (br). Getty Images: The Bridgeman Art Library/Universitatsbibliothek, Gottingen (c). 60 Alamy Images: The Print Collector (t). Getty Images: Photographer's Choice/Sylvain Grandadam (bl). Image courtesy History of Science Collections, University of Oklahoma Libraries; copyright the Board of Regents of the University of Oklahoma: (cl). Science Photo Library: Sheila Terry (tr). 61 Alamy Images: The Print Collector (cl); Eric Robison (br). Corbis: Angelo Hornak (cr). Getty Images: Robert Harding World Imagery/Guy Edwardes (tl). 62 Alamy Images: The London Art Archive (br). The Art Archive: Bodleian Library, Oxford (*Pococke 375 folio 3v-4r*) (bl). 62-63 Wellcome Library, London: (t). 63 The Bridgeman Art Library: Giraudon/Musée Conde, Chantilly (cr). Corbis: Stefano Bianchetti (br). Getty Images: Time Life Pictures/Mansell (br). 66 Alamy Images: INTERFOTO Pressebildagentur (Paracelsus); The London Art Archive (Canon of Medicine). Corbis: Stefano Bianchetti (Copernicus World). DK Images: Courtesy of The Science Museum, London/Dave King (Microscope). Image courtesy History of Science Collections, University of Oklahoma Libraries; copyright the Board of Regents of the University of Oklahoma: (Anatomical Model). Science Photo Library: Jean-Loup Charmet (Sextant). 66-67 Wellcome Library, London: (Background). 67 The Bridgeman Art Library: The Science Museum, London (Napier Bones). Corbis: Bettmann (Gilbert, Newton, Galileo Telescope). DK Images: Courtesy of The Science Museum, London/Dave King (Newton Telescope). Library Of Congress, Washington, D.C.: (Spring). Image courtesy History of Science Collections, University of Oklahoma Libraries; copyright the Board of Regents of the University of Oklahoma: (Air Pump, Micrographia). Science & Society Picture Library: (Galileo Moon). Science Photo Library: Sheila Terry (Torricelli). 68 The Art Archive: Dagli Orti/Musée Ampre Poleymieux (bl). *Novum Organum* by Francis Bacon, 1650 (tc). 69 Alamy Images: INTERFOTO (t). Getty Images: Stock Montage (b). 70 Science & Society Picture Library: (cb). Wellcome Library, London: (t). 71 Alamy Images: INTERFOTO Pressebildagentur (br). Corbis: Bettmann (cr, bl). DK Images: The Science Museum, London/David Exton (cl, tc/Right). Wellcome Library, London: (tc/Left). 72 Alamy Images: Mary Evans Picture Library (bl). The Bridgeman Art Library: Archives Charmet/Bibliothèque des Arts Decoratifs, Paris (bc). DK Images: Courtesy of the Museum of Natural History of the University of Florence, Zoology section 'La Specola'/Liberto Perugi (r). 73 The Art Archive: Marc Charmet/Musée des Beaux Arts, Orléans (br). Corbis: Lawrence Manning (cra). Image courtesy History of Science Collections, University of Oklahoma Libraries; copyright the Board of Regents of the University of Oklahoma: (c). 74-75 Corbis: Stefano Bianchetti. 76 Science Photo Library: Jean-Loup Charmet (bl). TopFoto.co.uk: National Pictures (c). 76-77 Corbis: Stapleton Collection/Philip Spruyt (t). 77 Corbis: Bettmann (tr). TopFoto.co.uk: The Granger Collection (br). 78 Corbis: Hulton-Deutsch Collection (c). Science Photo Library: Eckhard Slawik (tr). 78-79 Corbis: Roger Ressmeyer (c). 79 Corbis: Roger Ressmeyer (c). 80 DK Images: Courtesy of The Science Museum, London/Clive Streeter (bl). 81 Alamy Images: Phil Degginger (bl). Corbis: Bettmann (tr). Dorling Kindersley: The Science Museum, London (c). 82 Alamy Images: The Print Collector (ca). Getty Images: The Bridgeman Art Library (r). Adam Hart-Davis: (bl). 83 Corbis: Stefano Bianchetti (tl). DK Images: Courtesy of The Science Museum, London/Dave King (cr). 84 Science & Society Picture Library: (cl). 85 Corbis: Roger Ressmeyer (crb). DK Images: Courtesy of The Science Museum, London/Dave King (l). Adam Hart-Davis: (tr). NASA: ESA, J. Hester and A. Loll (Arizona State University, ca). 86 Photo Scala, Florence: (l). 87 Alamy Images: Daniele Paglino / Alamy Stock Photo (bc). Getty Images: Topical Press Agency/Monty Fresco Jnr (tl). 88 The Bridgeman Art Library: Ashmolean Museum, University of Oxford (cl). DK Images: Courtesy of The Science Museum, London/Dave King (bc). Image courtesy History of Science Collections, University of Oklahoma

Libraries; copyright the Board of Regents of the University of Oklahoma: (bl). 89 The Bridgeman Art Library: The Science Museum, London (bl). Corbis: Kevin Schafer (tr). Getty Images: The Bridgeman Art Library/State Central Artilllery Museum, St Petersburg (br). 90 The Bridgeman Art Library: The Trustees of the Weston Park Foundation (bl). Corbis: Bettmann (cla, tc). Science Photo Library: (bc). 91 Science Photo Library: Steve Gschmeissner (crb). 92 Corbis: Historical Picture Archive (r). Image courtesy History of Science Collections, University of Oklahoma Libraries; copyright the Board of Regents of the University of Oklahoma: (clb, bc). 93 Alamy Images: Eric Nathan (crb). DK Images: Courtesy of The Science Museum, London/Dave King (tc). Getty Images: The Bridgeman Art Library/National Portrait Gallery, London (cra). Library Of Congress, Washington, D.C.: (bc). 94 DK Images: Courtesy of The Science Museum, London/Dave King (cra, bl). Image courtesy History of Science Collections, University of Oklahoma Libraries; copyright the Board of Regents of the University of Oklahoma: (br). Wellcome Library, London: (cla) 95 Getty Images: National Geographic/Paul Zahl (clb). Photolibrary: Phototake Science/Dr Gary D. Gaugler (cb). Corbis: Clouds Hill Imaging Ltd (bc). National Cancer Institute (tr); Omikron (br). 96 Corbis: Bettmann (bl). Science Photo Library: Sheila Terry (br). 96-97 Getty Images: Hulton Archive (t). 97 DK Images: Courtesy of The Science Museum, London/Dave King (br). 98 Getty Images: Hulton Archive (bc); Stock Montage (r). iStockphoto.com: Robin Brisco (clb). 99 Alamy Images: Mary Evans Picture Library (tl). Image courtesy History of Science Collections, University of Oklahoma Libraries; copyright the Board of Regents of the University of Oklahoma: (bc, br). 100-101 Corbis: Gianni Dagli Orti (c). 101 Corbis: Bettmann (bc). Getty Images: Time Life Pictures/Mansell (tr). 102 The Bridgeman Art Library: Private Collection (cl). 103 Getty Images: Stock Montage (cr). 104 Alamy Images: The Print Collector (tl). Getty Images: Popperfoto (ca). 105 NASA: JPL (cra). 107 NASA: JPL. 108-109 NASA: Hubble Heritage Team (STScl/AURA) , R.G. French (Wellesley College) , J. Cuzzi (NASA/Ames) , L. Dones (SwRI) , and J. Lissauer (NASA/Ames) (t). 109 DK Images: NASA (cl). NASA: JPL (tc). 110 Alamy Images: Lebrecht Music and Arts Photo Library (r); The Print Collector (bl). 111 Alamy Images: Neil McAllister (tl). Corbis: Bettmann (tl); Walter Geiersperger (bc). DK Images: Courtesy of The Science Museum, London/Dave King (cla). Getty Images: Hulton Archive (tc). Image courtesy History of Science Collections, University of Oklahoma Libraries; copyright the Board of Regents of the University of Oklahoma: (c). 112 Corbis: Frank Lukasseck (b). 113 Getty Images: AFP/DSK (br); Photographer's Choice/Joe McBride (b). 114 Corbis: amanaimages/Kawai Kazuo (cl). Image courtesy History of Science Collections, University of Oklahoma Libraries; copyright the Board of Regents of the University of Oklahoma: (cra). 115 Getty Images: LOOK/Konrad Wothe. 116 DK Images: The British Museum, London/Nick Nicholls. Science Photo Library: (bl). 117 Louise Thomas: (tr, cra). 118 Corbis: Demotix Live News/Demotix (cl). 119 Alamy Stock Photo: GL Archive (crb). Corbis: Bettmann (cla); Charles O'Rear (bc);). Science Photo Library: Detlev Van Ravenswaay (cb); Walter Pacholka, Astropics (tr). 120 The Art Archive: Dagi Orti/Pharaonic Village, Cairo (c). The Bridgeman Art Library: The Science Museum, London (c). Photo Scala, Florence: (bl). 121 Alamy Images: studiomode (cr). Corbis: Loop Images/Nadia Isakova (bc). Science Photo Library: National Physical Laboratory/Andrew Brookes (tr); Sheila Terry (c). 122 Corbis: Visuals Unlimited (Kingdom/Bacteria, Kingdom/Fungi). Getty Images: Stone/Guy Edwardes (Family/Dogs); Visuals Unlimited/Carolina Biological (Kingdom/Protoctists). Corbis: Peter Walker (cr). Courtesy of the Linnean Society of London (bl). DK Images: Graham High at Centaur Studios - Modelmaker/Dave King (bl/Dinosaur). 123 Science photo library: Eye of science (tl); David M. Phillips (tc); BIOPHOTO ASSOCIATES (tr). DK Images: Jane Burton (tc); Gary Ombler (cr); Dorling Kindersley (br). Corbis: CARLOS ORTEGA/epa (tr). 124–125 © The Metropolitan Museum of Art: Gift of J. Pierpont Morgan, 1917. 126 Alamy Images: Lebrecht Music and Arts Photo Library (Newton). Alamy Stock Photo: Granger Historical Picture Archive (crb). The Art Archive: (Newcomen). Corbis: Bettmann (Lind); Hulton-Deutsch Collection (Solar Eclipse). DK Images: Courtesy of The Science Museum, London/Clive Streeter (Leyden Jar). Getty Images: Derke/ O'Hara (Comet). Wellcome Library, London: (Hales). 126-127 Alamy

Images: North Wind Picture Archives (Background). 127 Alamy Stock Photo: Granger Historical Picture Archive (cla). Bridgeman Images: (bl). Corbis: Gianni Dagli Orti (Balloon). DK Images: Courtesy of The Science Museum, London/Dave King (Watt Engine). Getty Images: James Sharples (Priestley). Science Photo Library: John A. Ey III (Lightning). Wellcome Library, London: (Linnaea Borealis, Jenner). 128 Corbis: Bettmann (Gay-Lussac). DK Images: Courtesy of The Science Museum, London (Difference Engine); Courtesy of The Science Museum, London/Clive Streeter (Faraday Motor); The Science Museum, London (Dalton Atoms). Science & Society Picture Library: (Telegraph Machine). Science Photo Library: (Dalton Symbols). 128-129 Getty Images: The Bridgeman Art Library (Background). 129 Corbis: Louie Psihoyos (Archaeopteryx). DK Images: Courtesy of The Science Museum, London/Clive Streeter (Induction Motor); Courtesy of The Science Museum, London/Dave King (Edison Photograph); The Science Museum, London (Periodic Table). Photolibrary: Phototake Science/Dr Gary D. Gaugler (Anthrax Bacteria). 130-131 The Art Archive. 132 Alamy Images: North Wind Picture Archives (cra). Corbis: Bettmann (br). DK Images: Courtesy of The Science Museum, London/Clive Streeter (tl, clb). Science & Society Picture Library: (tr). 133 Corbis: Bettmann (bc). DK Images: Courtesy of The Science Museum, London/Dave King (b). Getty Images: The Bridgeman Art Library (tl). 134–135 Alamy Stock Photo: Granger Historical Picture Archive. 136 Dorling Kindersley: The Science Museum, London (ca). The Trustees of the British Museum: (tl). Getty Images: Thepalmer / Digitalvision Vectors (bl). 137 Alamy Stock Photo: Granger Historical Picture Archive (ca). Bridgeman Images: (c). Science Photo Library: David Parker (crb). 138 Corbis: Bettmann (cra); Michael Nicholson (cla). DK Images: The Science Museum, London (ca). Science Photo Library: (br); Sheila Terry (bc). 139 Corbis: Bettmann (crb). Science Photo Library: Charles D. Winters (tr). 140 The Bridgeman Art Library: The Royal Institution, London, UK (bl). Corbis: Terry W. Eggers (r). 140-141 Corbis: TongRo/ Beateworks (c). 141 DK Images: Rough Guides/Eddie Gerald (tc). Science Photo Library: Richard Folwell (cl); Pasieka (br); Charles D. Winters (bc). 142 Alamy Images: Peter Jordan (tl). Getty Images: Greg Pease (c). Science Photo Library: Alex Bartel (b). 142-143 Alamy Images: Martin Shields (bc). 143 Corbis: Moodboard (cl). Science Photo Library: Paul Rapson (tr). 144 Corbis: glcheng/RooM the Agency (c). 145 123RF.com: Vladimir Kondrachov 145cra (tr). Alamy Images: G.P. Bowater (cr). Still Pictures: UNEP/S. Compoint (t). The Art Archive: Eileen Tweedy (lc). Alamy Stock Photo: Suzanne Long (br). Corbis: Gilles Sabrié (br). 146 Getty Images: James Sharples (bc). TopFoto.co.uk: The Granger Collection (cl). 147 The Bridgeman Art Library: Roger-Viollet, Paris (tc). Photo Scala, Florence: HIP (tr). 148 Edinburgh University Library: (crb). Science Photo Library: Sheila Terry (l). 149 Corbis: Bettmann (tc). Edinburgh University Library: (cr, br). Mary Evans Picture Library: (bl). 150 Alamy Images: Trip (cla). Corbis: Michael Nicholson (bl). 151 Corbis: AgStock Images (cl). Dreamstime: Mesaluis (bc). 152 The Bridgeman Art Library: Archives Charmet/Bibliothèque des Arts Decoratifs, Paris (bl). 153 Corbis: Hulton-Deutsch Collection (c). Getty Images: ScienceFoto/G. Wanner (br). Science Photo Library: Philippe Psaila (tr). 154 Wellcome Library, London: (bl). 155 Wellcome Library, London: (br). 156-157 Wellcome Library, London. 158 Corbis: Bettmann (bl). DK Images: Courtesy of The Science Museum, London/Clive Streeter (cb). 159 Science Photo Library: John A. Ey III. 160 The Bridgeman Art Library: Collection of the New York Historical Society (br). Getty Images: The Bridgeman Art Library/Pennsylvania Academy of the Fine Arts, Philadelphia (r). 161 Corbis: Bettmann (tr, crb); Visions of America/Joseph Sohm (bl). DK Images: Courtesy of The Science Museum, London/Dave King (ca). Science Photo Library: Sheila Terry (ca). Alamy Images: Glasshouse Images (c).162-163 Corbis: Stefano Bianchetti. 163 DK Images: Courtesy of The Science Museum, London/Clive Streeter (bc). 165 DK Images: Courtesy of The Science Museum, London/Clive Streeter (tr). iStockphoto.com: MH (crb). 166 DK Images: Courtesy of The Science Museum, London/Dave King (b). 167 Corbis: Reuters/Mark Ralston (br); Baldwin H. Ward & Kathryn C. Ward (tr). DK Images: Courtesy of The Science Museum, London/Clive Streeter (tl). Science Photo Library: Alex Bartel (tr). 168 Corbis: Bettmann (bl); Car Culture (br). DK Images: Courtesy of The Science Museum, London/Clive Streeter (bl). 169 DK Images: Courtesy of The Science Museum, London/Clive Streeter (bl). 170

Corbis: Hulton-Deutsch Collection (r). Science Photo Library: Maria Platt-Evans (bl). 171 The Bridgeman Art Library: The Royal Institution, London (cl). Corbis: (tl). DK Images: Courtesy of The Science Museum, London/Clive Streeter (cra). Getty Images: The Bridgeman Art Library/The Royal Institution, London (bc). Science Photo Library: (crb). 172 DK Images: The British Museum, London/Alan Hills and Barbara Winter (bl); Courtesy of The Science Museum, London/Clive Streeter (r). Science & Society Picture Library: (bc). 173 Corbis: Roger Ressmeyer (cr). DK Images: Courtesy of the Fire Investigation Unit at Acton Police Station, London/Andy Crawford (tl). NASA: (tc); Katherine Stephenson, Standford University and Lockheed Martin Corporation (br). Science photo library: Caltech/MIT/LIGO Lab (br). 174 Alamy Stock Photo: Science History Images (clb). DK Images: Courtesy of The Science Museum, London (r). Getty Images: Hulton Archive / Stringer (cb). 175 Getty Images: STR / Jiji Press / AFP (cra). 176 Corbis: Zefa/Paul Freytag (b/ Chemical). Getty Images: Glowimages (b/Electrical); Iconica/ Nicholas Eveleigh (b/Elastic); Photographer's Choice/Joe McBride (b/Gravitational); Getty Images: Hulton Archive (tr). Alamy Images: Chronicle (tl); imageBROKER (cl). 176-177 Alamy Images: Fedor Selivanov (c). 177 Corbis: Gary Braasch (b/ Light); Pablo Corral (b/Sound); Reuters/Shaun Best (b/Kinetic); Francesco Acerbis (tr); Photographer's Choice/Eric Schnakenberg (b/Heat). SOHO/EIT (ESA & NASA): (b/Nuclear). 178 DK Images: Alamy/Brand X Pictures. 179 Corbis: Burstein Collection (cra); Zefa/Theo Allofs (tr). 180 Corbis: (ca). Science Photo Library: Sheila Terry (cb). Still Pictures: Henry Groskinsky (c). 181 Corbis: Underwood & Underwood (r). 183 Corbis: Bettmann (tc). 184 Corbis: EPA/Jose Jacome (bc). DK Images: Courtesy of the Natural History Museum, London/Colin Keates (fbr); Courtesy of the Natural History Museum, London/Harry Taylor (bc). Getty Images: National Geographic/Ralph Lee Hopkins (c); National Geographic/Jim Richardson (tr). Geological Sections of Strata in Thuringia by Johann Gottlob Lehmann, Germany, 1759 (cl). 185 Alamy Images: Scenics & Science (crb). Corbis: Science Faction/G. Brad Lewis (bl); Michael St. Maur Sheil (tr). Science Photo Library: Gregory Dimijian (c). 186 DK Images: Courtesy of the Natural History Museum, London/Colin Keates (b/Carboniferous). Getty Images: Hulton Archive (cra). 187 Alamy Stock Photo: The Natural History Museum (cra). Corbis: Reuters/Mike Segar (br). DK Images: Courtesy of the Natural History Museum, London/ Colin Keates (b/Triassic); Courtesy of the Natural History Museum, London/Harry Taylor (b/Quarternary). 188-189 Corbis: Louie Psihoyos. 190 Hulton-Deutsch Collection (tr). Getty Images: The Bridgeman Art Library (clb). 191 Corbis: Image Source (b). Getty Images: Stanford University (tc). 192 Science Photo Library: Pasieka (cra). 192-193 Corbis: Keren Su (b). 193 Alamy Images: geogphotos (tr). Corbis: Kazuyoshi Nomachi (tl); Zefa/Micha Pawlitzki (ftr). Science Photo Library: Duncan Shaw (tc). 194 Corbis: Karl-Heinz Haenel (c); Adam Jones (ca). 195 Corbis: Zefa/Theo Allofs (ca). Image courtesy History of Science Collections, University of Oklahoma Libraries; copyright the Board of Regents of the University of Oklahoma: (tr). Science Photo Library: Wayne Lawler (tl). 196 Alamy Images: K-Photos (br). Corbis: Bettmann (bl); Michael Freeman (cla). 197 Corbis: Rudy Sulgan (b). Getty Images: General Photographic Agency (tl); Science photo library: MATT MAZLOFF (cr). 198-199 The Natural History Museum, London. 200 Alamy Images: Arco Images GmbH (ca, br); David Hosking (bc); imagebroker (clb). Iggino Van Bael: (cl). Getty Images: The Image Bank/Pete Turner (br). 201 Alamy Images: imagebroker (clb). Iggino Van Bael: (cl). Corbis: Martin Harvey (bl). Getty Images: London Stereoscopic Company (cla). Photolibrary: Herbert Kehrer (bc/ Above). Science Photo Library: Christian Darkin (tr). Still Pictures: VISUM/Wolfgang Steche (bc/Below). 202 Corbis: The Gallery Collection (l). DK Images: Courtesy of the Natural History Museum, London/Dave King (cra). 203 akg-images: (ca). Corbis: Star Ledger/William Perlman (a). DK Images: Courtesy of the Natural History Museum, London/Down House/Dave King (tr); Courtesy of the Natural History Museum, London/Dave King (crb). Mary Evans Picture Library: (cl). Science Photo Library: (tc). 204 Getty Images: GAP Photos/FhF Greenmedia (tl); Time Life Pictures/Mansell (cra). The Natural History Museum, London: (r). 206 Corbis: Bettmann (bc). Science Photo Library: Royal Astronomical Society (bl). 207 Corbis: (br). NOAA: Central Library (tc). 208 Corbis: Alinari Archives (cla); Jim Reed Photography (cra). Getty Images: Photographer's Choice/Harald Sund (tr). Science Photo Library: (bl); Detlev van

Ravenswaay (br). 209 Corbis: Eye Ubiquitous/Bryan Pickering (cra); James Randklev (tr). NASA: Image courtesy the QuikSCAT team at NASA's Jet Propulsion Laboratory (clb). NOAA: National Weather Service (NWS) Collection (bc). 210 Corbis: Stefano Bianchetti (cl). NASA: Earth Sciences and Image Analysis Laboratory, Johnson Space Center (t). Image courtesy History of Science Collections, University of Oklahoma Libraries; copyright the Board of Regents of the University of Oklahoma: (bc). 211 Corbis: Niall Benvie (c); Daniel J. Cox (tc); Darrell Gulin (cb); Ultimate Chase/Mike Theiss (ca). Getty Images: The Image Bank/ Pete Turner (br). NASA: the TOMS Science Team & the Scientific Visualization Studio, GSFC (tr). 212 NOAA: Steve Nicklas, NOS, NGS (tr). 213 Medford Historical Society Collection (tl); Yogi, Inc. (tr). Science Photo Library: Lamont-Doherty Earth Observatory/W. Haxby (c). 214 Wellcome Library, London: (tr). 215 Alamy Stock Photo: Dpa Picture Alliance / Marijan Murat (br). Getty Images: Hulton Archive (tl). 216 Wellcome Library, London: (cla). 217 Getty Images: Bettmann (c). 218 Corbis: Bettmann (tc). DK Images: Courtesy of The Science Museum, London/Clive Streeter (bl). Wellcome Library, London: (br). 219 Alamy Images: PHOTOTAKE Inc. (cra). 220 Science Photo Library: Kent Wood (b). 221 Getty Images: Time Life Pictures/Mansell (tr). 223 The Bridgeman Art Library: Archives Charmet/Mairie de Sainte-Foy-La-Grande (br). Science Photo Library: Pasieka (bl). 225 Corbis: Hulton-Deutsch Collection (cra); Visuals Unlimited (tl, cla). DK Images: Courtesy of the Natural History Museum, London/Colin Keates (bl). 226 Alamy Images: Mary Evans Picture Library (cla). Essai de diotropique by Nicolaus Hartsoeker, 1694 (tr). 227 Corbis: Mediscan (tc). 228 Corbis: Visuals Unlimited (br). DK Images: Courtesy of The Science Museum, London/Dave King (bl). 228-229 Wellcome Library, London: (tc). 229 Corbis: Bernardo Bucci (br); Hulton-Deutsch Collection (tr). 230-231 DK Images: The Science Museum, London. 232 Lebrecht Music and Arts: Interfoto (tl). 233 Getty Images: Time & Life Pictures (cr). 234 Corbis: Hulton-Deutsch Collection (c); Michael Nicholson (cla). Science Photo Library: Martyn F. Chillmaid (cra). 235 Science Photo Library: Charles D. Winters (l). 236 Getty Images: Roger Viollet (bc). 237 Archiv der Max-Planck-Gesellschaft, Berlin-Dahlem: (bc). Science Photo Library: Cordelia Molloy (tc). 238 Corbis: Lawrence Manning (c/Distilled Water). Getty Images: David Woodfall (bl). 238-239 Science Photo Library: Bill Barksdale (bc). 239 Science Photo Library: (tr); Andrew Lambert Photography (tl); Garry Watson (c/Ammonia). 240 Science & Society Picture Library: Science Museum/ SSPL (bl). 241 Alamy Images: The Print Collector (br). Imperial War Museum: G P Lewis (tl). Alamy Stock Photo: Everett Collection Inc (cra). Science Photo Library: Charles D. Winters (crb). 242 Alamy Images: INTERFOTO Pressebildagentur (cl). Science Photo Library: Dr Kari Lounatmaa (c). Wellcome Library, London: (bl). 243 The Art Archive: Culver Pictures (tc). Science Photo Library: (l). 244 Corbis: Visuals Unlimited (cl, bl/Commas, bl/ Rods, bl/Spheres). Getty Images: Edward Gooch (cra); Visuals Unlimited/Dr Fred Hossler (bl/Cork-Screws). 245 Corbis: Australian Picture Library/Chris Boydell (c); Visuals Unlimited (cr/Enveloped, cr/Icosahedral, cr/Spiral-Helical). Getty Images: Stone/Hans Gelderblom (c/Complex). 246 Science Photo Library: Herve Conge, ISM (bl); Steve Gschmeissner (bc); Dr Gopal Murti (tl). 248 Alamy Images: Mary Evans Picture Library (bl). Science & Society Picture Library: (ca). 249 Getty Images: Three Lions/Evans (tl). Wellcome Library, London: (cr). 250 Corbis: Hulton-Deutsch Collection (bc); Roger Wood (t). DK Images: Courtesy of The Science Museum, London/Dave King (cr); Courtesy of The Science Museum, London/Clive Streeter (cl, r). 251 Corbis: Walter Hodges (c); Francis G. Mayer (tr). 252 DK Images: Courtesy of The Science Museum, London/Clive Streeter (cl). DK Images: Courtesy of The Science Museum, London/Clive Streeter (cl). Science Photo Library: Library of Congress/George Grantham Bain Collection (cra). 254 Getty Images: Hulton Collection (cr). 255 Corbis: Bettmann (bc). Getty Images: Car Culture (b); Riser/Jurgen Vogt (cra). 256 Science Photo Library: Science Source (bl). 257 Corbis: Arcaid/Richart Bryant (br). Courtesy of U.S. Navy: Photographer's Mate 3rd Class Jonathan Chandler (r). 258 Science photo library: EUROPEAN SPACE AGENCY (bl); EUROPEAN SPACE AGENCY/ESA/PACS/SPIRE/ Tracey Hill and Frederique Motte, Lab AIM Paris-Saclay, CEA/ Irfu-CNRS/INSU-University Paris Diderot (bc). 259 Corbis: Niall Benvie (cla/Left). Alamy Images: BSIP SA (c). Science photo library: NASA/DOE/FERMI LAT COLLABORATION (br). Photolibrary: Oxford Scientific/Owen Newman (cla/Right).

Science Photo Library: Dept. of Physics, Imperial College (cr); T-Service (bl). 260 Alamy Images: John Krstenansky (clb). Corbis: Leonard de Selva (cla). Getty Images: Stock Montage (bc). Science & Society Picture Library: (br). 261 Science & Society Picture Library: (bl). Dreamstime.com: Yalcinsonat (r). 262 Corbis: Hulton-Deutsch Collection (t, br). Science & Society Picture Library: (c). 263 Canon UK Ltd: (cla). DK Images: Canon UK Ltd (b/Canon); Judith Miller/Collectors Cameras (b/ Stereoscopic); Courtesy of The Science Museum, London/Dave King (b/Fox Talbot). Eugene Ilchenko: (b/Hasselblad). istockphoto.com: Mixmike (tr). Arne List: (b/Nikon F3). 264 DK Images: Courtesy of The Science Museum, London/Dave King (bl). 264-265 Getty Images: Hulton Archive (b). 265 Corbis: (cb); Bettmann (tc). DK Images: Courtesy of The Science Museum, London/Dave King (cr, br). 266 Science & Society Picture Library: (ca, bc). 267 Alamy Images: AF archive (bc). DK Images: Dave King (l). 268 DK Images: Judith Miller/Otford Antiques and Collectors Centre (a); Courtesy of The Science Museum, London/Clive Streeter (cla, bl). 269 Corbis: (tc). DK Images: Courtesy of The Science Museum, London/Dave King (c/ Left). Alamy Images: World History Archive (rc). 274 Getty Images: Hulton Archive/Sean Sexton (cla); Roger Viollet/Boyer (bl). Science Photo Library: Edward Kinsman (b). 275 Science Photo Library: Pasieka (cra). 276 FLPA: Sunset (tl). naturepl. com: Miles Barton (br). Science Photo Library: NASA (tr). 276-277 Getty Images: Robert Harding World Imagery/Steve & Ann Toon (b). 277 Alamy Images: INTERFOTO Pressebildagentur (br). Alamy Stock Photo: Avalon / Bruce Coleman Inc / Kay & Karl Ammann (b). 278 Alamy Images: RIA Novosti (c). Corbis: Reuters (cl). 280 Science Photo Library: Kevin Curtis. 282 Corbis: Bettmann (Stanley Steamer); Hulton-Deutsch Collection (Lilienthal). DK Images: Courtesy of The Science Museum, London/Clive Streeter (Vacuum Tube). Getty Images: AFP (Curie); Daniel Berehulak (Einstein Book). Science Photo Library: (Becquerel); Pasieka (Alzheimer Brain). Wellcome Library, London: (First X-Ray). 282-283 Science Photo Library: A. Barrington Brown (Background). 283 Corbis: Bettmann (Heisenberg). DK Images: Courtesy of the Imperial War Museum, London/Geoff Dann (Enigma Machine). Getty Images: Keystone (Einstein). Science Photo Library: Robert Gendler (Andromeda Galaxy); Royal Astronomical Society (Solar Eclipse); St Mary's Hospital Medical School (Penicillin). 284 Alamy Images: artpartner-images.com (Hologram). Corbis: (Manhattan Project); Bettmann (Pauling); Charles O'Rear (Richter). Getty Images: Popperfoto/Paul Popper (V-2). National Portrait Gallery, London: Peter Lofts Photography (Hodgkin). Science Photo Library: CNRI (Chromosomes). 284-285 Science Photo Library: Robert Gendler (Background). 285 Corbis: (Apollo 11). Getty Images: Three Lions/Evans (Vaccination); Time & Life Pictures/ Ralph Morse (Feynman). Science Photo Library: Peter Arnold Inc./Prof. E. Lorenz (Lorenz Attractor); Steve Gschmeissner (Cancer Cell); Dr Ken MacDonald (Ocean Floor); Pasieka (Insulin). 286 Getty Images: Topical Press Agency (bl). 287 Alamy Stock Photo: Archive PL (bc). Science Photo Library: David Parker (crb). dreamstime.com: Siarhei Dzenisevich (br). 288 Science Photo Library: (bc); Sheila Terry (ca). 289 Corbis: Bettmann (cra). 290 Alamy Images: Dennis Hallinan (cla). The Art Archive: Museo del Prado, Madrid/Laurie Platt Winfrey (clb). Corbis: Hulton-Deutsch Collection (bc). 290-291 Corbis: Underwood & Underwood (bc); EPA/Udo Weitz (br). 291 Corbis: GUILLAUME HORCAJUELO/epa (tr). DK Images: Courtesy of the Imperial War Museum, Duxford Andy Crawford (t/Concorde); Courtesy of The Shuttleworth Collection, Bedfordshire/Martin Cameron (t/Wright Flyer). 292 DK Images: Courtesy of The Science Museum, London/Clive Streeter (bl/Insert, bl, br). Science & Society Picture Library: (tc/Left, tc/Right). Corbis: Hulton-Deutsch Collection (bl). 292-293 Getty Images: Bletchley Park Trust (c). 293 Corbis: Bettmann (tl). Getty Images: Hulton Archive (c). 294 Corbis: Bettmann (bl). Science Photo Library: Jean-Loup Charmet (cb). Wellcome Library, London: (br). 295 akg-images: (tl). Getty Images: Nick Veasey (tr). 296 Alamy Images: Mary Evans Picture Library (bl). Getty Images: AFP (r). Wellcome Library, London: (bc). 297 Alamy Images: The Print Collector (t). The Bridgeman Art Library: Bibliotheque des Arts Decoratifs, Paris, France/ Archives Charmet (bc). Corbis: Ted Spiegel (cr). 298 Science Photo Library: (tr). 299 Corbis: Roger Ressmeyer (crb). DK Images: Courtesy of The Science Museum, London/Clive Streeter (cb). Library Of Congress, Washington, D.C.: (tc). 300-301 akg-images. 302 Corbis: Lester Lefkowitz

ACKNOWLEDGMENTS

(bc). 303 Science Photo Library: Royal Astronomical Society (bl). 304 Getty Images: AFP/Jean-Pierre Clatot (bc); Daniel Berehulak (clb); Topical Press Agency (r). 305 Alamy Stock Photo: Science History Images (bc). Corbis: Dave Bartruff (br); Hulton-Deutsch Collection (t). Getty Images: Keystone (cl). Science Photo Library: American Institute of Physics/Emilio Segre Visual Archives (cra). 306 Science Photo Library: M.I. Walker (br). 307 Corbis: Bettmann (cra); Clouds Hill Imaging Ltd. (tl). Science Photo Library: Steve Gschmeissner (crb). 308 Alamy Stock Photo: H.S. Photos (tr). Science Photo Library: Biophoto Associate (ca); CNRI (bl). 309 Photolibrary: Oxford Scientific (tl). Still Pictures: Ed Reschke (tc). 310-311 Science Photo Library: St Mary's Hospital Medical School. 312 Getty Images: Hulton Archive (tl). 313 Getty Images: Stone/Michael Rosenfeld (bl). iStockphoto.com: Richard Cano (br). Science Photo Library: Pasieka (cr). Shutterstock.com: Granger (tl). 314 Corbis: Bettmann (cra). 315 Image originally created by IBM Corporation: (br). Dr Tonomura: (t). 316 Corbis: Bettmann (tc). 317 Corbis: Macduff Everton (cr). Science Photo Library: ArSciMed (tc). 318-319 NASA: ESA, S. Beckwith (STScI) and the HUDF Team. 320 The Bridgeman Art Library: Bible Society, London, UK (a). 321 Science Photo Library: A. Barrington Brown. 322-323 Corbis. 324 Getty Images: Bettmann (cl). 324-325 Corbis: EPA (c). 325 Getty Images: Clement Mahoudeau / AFP (crb). SOHO/EIT (ESA & NASA): (cr). Alamy Images: US Navy Photo (tr). 326 Corbis: (clb, bl). Getty Images: Hulton Archive/Joe Munroe (r). 327 Getty Images: Time & Life Pictures/Ralph Morse (c). NASA: Kennedy Space Center (br). Science & Society Picture Library: (cra). © D-Wave Systems Inc. All rights reserved: (bl). 328 STScI/AURA/NASA: ESA (cra). 329 Alamy Stock Photo: Science History Images (clb). Science Photo Library: Emilio Segre Visual Archives / American Institute Of Physics (ESA & NASA): (t). 330 NASA: ESA/ CXC/ STScI/ J. Hester and A. Loll (Arizona State Univ.), R. Gehrz (Univ. Minn.) (br); HST/ASU/J. Hester and A. Loll (Arizona State Univ.), R. Gehrz (Univ. Minn.) (bl). Science Photo Library: Royal Greenwich Observatory (tr). 331 Science Photo Library: NASA / ESA / STScI / H.Richer,UBC (cl); Physics Today Collection / American Institute Of Physics (tc). 332 Alamy Images: Bob Gibbons (Acacia Tree). Corbis: Michael & Patricia Fogden (Tapir); Darrell Gulin (Aroid Vine); Steve Kaufman (Motmot); Frank Lane Picture Agency/Fritz Polking (Strangler Fig); Robert Pickett (Tarantula). DK Images: Courtesy of the Exmoor Zoo, Devon/Gary Ombler (Agouti). Getty Images: Digital Vision/Tom Brakefield (Jaguar); The Image Bank/Grant Fair (Butterfly); National Geographic/Nicole Duplaix (Coccoloba Tree); Visuals Unlimited/Steve Maslowski (Mouse). 333 Corbis: Daniel J. Cox (br); Raymond Gehman (bc). Getty Images: The Image Bank/A. & J. Visage (tr); Photographer's Choice/Sami Sarkis (tc); Alfred Eisenstaedt / The LIFE Picture Collection (cl). 334 Alamy Images: Dave Watts (tl); Photoshot License Ltd (bc). Corbis: Corbis (cl). 335 Corbis: Darrell Gulin (ftr); Mark A. Johnson (fcr); Wolfgang Kaehler (tr); Frans Lanting (ftl); Frank Lukasseck (fcl); David Samuel Robbins (cl); Michel Setboun (cr). Getty Images: Photographer's Choice/Darrell Gulin (tl). Alamy Images: Bill Coster (tr); Israel Hervás Bengochea (cl); Robert McGouey/Wildlife (br). DK Images: Mark Longworth (bl). Corbis: André Gilden/Nature in Stock (cl); Frank Krahmer (bc). 336 Alamy Images: Joe Tree (cb). Getty Images: Andy Sotiriou (clb). Science Photo Library: Humanities & Social Sciences Library / New York Public Library (cla). 337 akg-images: PictureContact (crb). Science Photo Library: Edward Kinsman (tl). TopFoto. co.uk: The Granger Collection (c). 338 Alamy Images: Florilegius (tl). Corbis: Bettmann (bl). Getty Images: Popperfoto/Paul Popper (bl). Science Photo Library: Ria Novosti (cl). Corbis: Bettmann (bl). Courtesy of Virgin Galactic: (cra). 338-339 Alamy Images: RGB Ventures / SuperStock (c). 339 Alamy Images: SpaceX (tl). 340 Alamy Stock Photo: Science History Images (bc). NASA: ESA/ The Hubble Heritage Team (STScI/ AURA, ca). Science Photo Library: Hale Observatories (br); Royal Astronomical Society (cl). 340-341 Science Photo

Library: Robert Gendler (c). 341 Science Photo Library: Mark Garlick (bc). 342 DK Images: Courtesy of the H. Keith Melton Collection (bc); Courtesy of the Royal Signals Museum, Blandford Camp, Dorset/Steve Gorton (bl). 343 Dreamstime.com: Yalcinsonat (crb). DK Images: Courtesy of the Imperial War Museum, London/Geoff Dann (l). Science & Society Picture Library: (tc). 344 Getty Images: Time & Life Pictures/Life Magazine (r). Science & Society Picture Library: (bc). TopFoto. co.uk: The Granger Collection, New York (cl). 345 Alamy Images: Alan Gallery (br). Photolibrary: Francis Firth Collection (tc). Rex Features: Andy Lauwers (ca). 346 Photolibrary: Imagestate/ Jewish Chronical (c). 346-347 Science Photo Library: A. Barrington Brown. 348 Alamy Images: Mary Evans Picture Library (bl). Science Photo Library: James King-Holmes (tr). 349 Getty Images: Bettmann (cr). 350 Corbis: Hulton-Deutsch Collection (cl). Getty Images: Science & Society Picture Library / SSPL (bc). Photolibrary: Phillip Hayson (r). 351 Wolfgang Beyer: (b). Science Photo Library: Peter Arnold Inc./Prof. E. Lorenz (crb); Gregory Sams (cra). 352 Corbis: Frans Lanting (b). Image courtesy History of Science Collections, University of Oklahoma Libraries; copyright the Board of Regents of the University of Oklahoma: (cla). Science Photo Library: (br). 354-355 Corbis: Roger Ressmeyer. 356 akg-images: (bl). Science Photo Library: Dr Ken MacDonald (br). 357 Corbis: Atlantide Phototravel (cl); Lloyd Cluff (fcl); EPA/Olivier Matthys (cr). NASA: JPL (fcr). 358 DK Images: Courtesy of the Natural History Museum, London/ Colin Keates (tr). 359 Getty Images: Stone/G. Brad Lewis (br). 360 Alamy Images: DOD Photo (bc). 361 Corbis: Charles O'Rear (bl). Science photo library: GSHAP (tl). 362 Horse-Hoeing Husbandry by Jethro Tull, 1752 (bl). 362-363 Corbis: Sylvain Saustier (tc). 363 Alamy Images: Bon Appetit (tr); Nigel Cattlin (br). Corbis: Michael Busselle (tc). Getty Images: Stock Montage / Archive Photos (bc); Visuals Unlimited/Inga Spence (clb). Photolibrary: Animals Animals/Patti Murray (fclb). Science Photo Library: Hank Morgan (cr). 364 Alamy Images: Wolfgang Kaehler. 365 Corbis: Bettmann (br). Getty Images: Jemal Countess / Stringer / Getty Images Entertainment (c). 366 Corbis: Jerry Cooke (cla). Dorling Kindersley: The Science Museum, London (tr). Getty Images: Taxi/Lester Lefkowitz (b). 367 © Intel Corporation: (tc). Science Photo Library: Gustoimages (c). 368 NASA: NASA (c). 369 Getty Images: DigitalGlobe (bl); Hulton Archive (fbl). Landsat 7 satellite image courtesy of NASA Landsat Project Science Office and USGS National Center for Earth Resources Observation Science: (bc). Alamy Images: Granger, NYC (tl); ScotStock (cr). NASA: NASA Goddard MODIS Rapid Response Team (cl). Corbis: ImageSat International/epa (bl); David Ebener/dpa (br). 370-371 Corbis. 371 Alamy Stock Photo: Alpha Historica. 372 Alamy Images: NASA Photo (bl); SPUTNIK (tc). Corbis: Hulton-Deutsch Collection (tc); NASA - digital version copyright/ Science Faction (tr). Getty Images: Popperfoto (ftc). 372-373 NASA: NASA Johnson Space Center Houston, Texas (c). 373 Getty Images: SVF2 (tc). NASA/ Handout/ CNP (tl). NASA Reuters (br). NASA: NASA (tr); NASA (br). Science photo library: SPUTNIK (cr). 374 Science Photo Library: Laguna Design. 376 Science photo library: ZEPHYR (bc). NASA: (Shuttle Launch). © Intel Corporation: (Intel 4004). DK Images: Courtesy of the Imperial War Museum, Duxford/Andy Crawford (Concorde). NASA (Mars). Getty Images: hemis.fr/Fred Derwal (Wind Farm). NASA: ESA/ CXC/ STScI/ J. Hester and A. Loll (Arizona State Univ.) , R. Gehrz (Univ. Minn.) (Supernova); Johns Hopkins University Applied Physics Laboratory/Carnegie Institution of Washington (Mercury). Science Photo Library: Mark Garlick (Superclusters); David Parker (Genetic Fingerprinting); Philippe Plailly (Base Pairs of Gene). 376-377 Corbis: Bettmann (Background). 377 Courtesy of Apple. Apple and the Apple logo are trademarks of Apple Computer Inc., registered in the US and other countries: (iPhone). Corbis: EPA/ Martial Trezzini (Hadron Collider); Karen Kasmauski (Dolly). Getty Images: AFP (Nanotubes); Stanford University (Zircon); Amy Sussman (Greene). NASA: (Chandra); JPL-Caltech/ K.

Gordon (University of Arizona) (Spitzer Telescope); Katherine Stephenson, Standford University and Lockheed Martin Corporation (Gravity Probe B). Science Photo Library: Fermilab (Top Quarks). 378 Alamy Stock Photo: Agefotostock (cr). Getty Images: Michael Smith (bc); Andreas Rentz (bl). 379 Philippe Bourcier - sysctl.org: (b). Getty Images: MARTIN BUREAU (cr). 380 akg-images: Sotheby's (cla). Alamy Images: age fotostock (cr). 380-381 Getty Images: Henrique Casinhas / SOPA Images / LightRocket. 381 Alamy Images: dpa picture alliance (tl). Corbis: Hank Morgan - Rainbow/Science Faction (rc); Corbis: Kim Kulish (bc). 382 Science Photo Library: Fermilab (tc). 383 © CERN Geneva: (cra). Getty Images: AFP/CERN (crb). Jeff Miller / University of Wisconsin-Madison: (bc). Science Photo Library: David Parker (ca). 384 Alamy Images: Adrian Muttitt (cr). National Portrait Gallery, London: Peter Lofts Photography (cl). Science & Society Picture Library: Science Museum (bc). 385 Corbis: Clements/ Hulton-Deutsch Collection (tc). Science & Society Picture Library: Science Museum (bc). TopFoto.co.uk: Topham / PA (bl). 386 Getty Images: Visuals Unlimited/Dr Dennis Kunkel (cla). www.glofish.com: (tr). Science Photo Library: © Estate of Francis Bello (bl). 387 Alamy Images: AGStockUSA, Inc. (crb). Science Photo Library: David Parker (br); J. C. Revy, ISM (tc/Above, tc/Below). 388-389 Science Photo Library: D. Phillips. 390 Science photo library: BIOPHOTO ASSOCIATES (tl). Corbis: Scott T. Smith (tc). Science photo library: W.A. RITCHIE/ ROSLIN INSTITUTE/EURELIOS (br). 391 Corbis: Karen Kasmauski (cb). Science photo library: ALEX HYDE (cl); MIKKEL JUUL JENSEN (tc). Alamy Images: Simone Montegrandi (c). 392 Corbis: Visuals Unlimited (b/Needle). Science Photo Library: Steve Gschmeissner (b/Human Hair); Laguna Design (b/Nanotube); David Mccarthy (b/DNA); Medical Rf. com (b/DNA, b/Red Blood Cells); Andrew Syred (b/Dust Mite). 393 Getty Images: AFP (tr). Science Photo Library: Peidong Yang, Lawrence Berkeley National Laboratory (bc). 394 Corbis: (br). NASA: Johns Hopkins University Applied Physics Laboratory/ Carnegie Institution of Washington (bl). Science Photo Library: JPL / NASA (bc). 395 NASA: New Horizons Spacecraft Data (tc); JPL-Caltech / MSSS (bl); JPL (bc); ESA, A. Simon (Goddard Space Flight Center), and M.H. Wong (University of California, Berkeley) (tc). 396 Corbis: (b). Caltech / Palomar Observatory: (cla). Science & Society Picture Library: Science Museum (cr). 397 Chandra X-Ray Observatory: NASA / UMass / Z.Li & Q.D.Wang (fcla). Getty Images: Al Fenn/ Time Life Pictures (br). NASA: (tl); Bill Schoening, Vanessa Harvey/ REU program/ NOAO/ AURA/ NSF (c); Infrared Processing and Analysis Center, Caltech/ JPL (cr); JPL-Caltech/ K. Gordon (University of Arizona) (bl); X-ray: CXC/CfA/R. Tuellmann et al.; Optical: AURA/STScI (tc); JPL-Caltech (cla). 398-399 ESO: M. Kornmesser / Nick Risinger (skysurvey.org). 398 Alamy Stock Photo: Yogi Black (clb). NRAO: ALMA (ESO / NAOJ / NRAO), M. MacGregor; NASA / ESA Hubble, P. Kalas; B. Saxton (NRAO / AUI / NSF) (cla). Visible light image: the NASA / ESA Hubble Space Telescope (cla). 399 Getty Images: Javier Soriano / AFP (br). NASA: Ames / JPL-Caltech / T Pyle (tc). Science Photo Library: Dr Seth Shostak (bc). 400 Princeton University: (tr). Science Photo Library: Emilio Segre Visual Archives / American Institute Of Physics (tr); Fermilab (bl). 400-401 Science Photo Library: Volker Springel / Max Planck Institute For Astrophysics (tc). 401 Science Photo Library: Celestial Image Co. (bc); W. Couch & R. Ellis (c). 402-403 Alamy Stock Photo: Science History Images. 403 Getty Images: Joe McNally / Hulton Archive. 405 Corbis: Bettmann (tr); EPA/Martial Trezzini (bc). 406 Getty Images: Amy Sussman (br). 407 Bathsheba Sculpture LLC: (bl). Houghton Mifflin Company: (br). 408 Corbis: Lester Lefkowitz (r). Getty Images: Roger Viollet/Harlingue (clb). Photolibrary: Phototake Science/Scott Camazine (tc). Science Photo Library: Newcastle Hospitals NHS Trust/Simon Fraser (bc). 409 Corbis: Mediscan (tl). Photolibrary: Nordic Photos/Chad Ehlers (crb). Science Photo Library: GE Medical Systems (bl); ISM (tr); Zephyr (bl). 410 Getty Images: Express/Stan Meagher (bc). Rex Features: Olivier Pirard (tl). Science Photo Library: Jean-Loup

Charmet (clb); Steve Gschmeissner (br). 411 Alamy Images: Medical-on-Line (ca/Artificial Lens). Corbis: Charles O'Rear (cb/ Pacemaker, br); Corbis: John Gress (cr). Photolibrary: AGE fotostock/Frank Siteman (c). Science Photo Library: (ca/Heart Valve); CC, ISM (tc); Dr P. Marazzi (cb/Joint Replacement). 412 Alamy Images: Reuters / Philippe Wojazer (cb). Corbis: Historical Picture Archive (cr). Science photo library: A. DOWSETT, PUBLIC HEALTH ENGLAND (bl). 413 Alamy Stock Photo: Xinhua / Wang Chengyun (bc) Getty Images: Joe Raedle / Getty Images News (tc). Science photo library: GUNILLA ELAM (cr). DR GARY GAUGLER/SCIENCE PHOTO LIBRARY MEDICINE (c). 414 Science Photo Library: Biozentrum, University of Basel (bl); Philippe Plailly (br). 415 Corbis: Louis Quail (cr); Sygma/ Ferdaus Shamim (bc). Science Photo Library: ICRF/James King-Holmes (tl). 416 Dreamstime.com: Katarzyna Bialasiewicz (tr). 417 Alamy Stock Photo: Dpa Picture Alliance / Kay Nietfeld (tr). Science Photo Library: AMI Images (cb). 418 Alamy Stock Photo: Puwadol Jaturawutthichai (bl); UPI / Shannon McPherron / MPI EVA Leipzig (cl); Universal Images Group North America LLC / DeAgostini / G. Cigolini (bc). Dorling Kindersley: Natural History Museum, London (br). U.S. National Library of Medicine, History of Medicine Division: National Center for Biotechnology Information (tr). 419 Dorling Kindersley: Natural History Museum, London (br). Getty Images: DEA / A. Dagli Orti / De Agostini (bc). Getty Images / iStock: Noctiluxx / E+ (tr). Shutterstock.com: Clifford Ling / ANL (c). 420 DK Images: DK Images (tl). NASA: NASA (tr). Science photo library: DETLEV VAN RAVENSWAAY (bc). 420-421 Alamy Images: USFS Photo (cr). 422 Science Photo Library: George Steinmetz (br). Press Association Images: AP Photo/Gregorio Borgia (cr). 423 Corbis: Barry Lewis (cra). Science photo library: Jesse Allen, NASA Earth Observatory/MODIS Land Group (tc). Still Pictures: Biosphoto/ Ittel Jean-Frédéric & Association Moraine (bl, br). 424 Corbis: Benjamin Rondel (c). Getty Images: hemis.fr/Fred Derwal (c). 424-425 Corbis: Paul A. Souders (cr). 425 © Pelamis Wave Power Ltd: (tc). Science Photo Library: Rosenfeld Images Ltd (cla). 426 Corbis: George Steinmetz (clb). NASA: Provided by the SeaWiFS project/Goddard Space Flight Center, and ORBIMAGE (br). Alamy Images: Wiskerke (tc). Shutterstock.com: Lenny Ignelzi / AP (cr). 427 Corbis: David Madison (tl). 428 Science Photo Library: NASA/R.B. Husar.

All other images © Dorling Kindersley
For further information see: www.dkimages.com